Horizons

TWELVE STEPS TO THE UNIVERSE

Beginning with the sidewalk scene
directly above, we zoom outward in
twelve steps. At each step we enlarge
our field of view to see an area 100
times larger in diameter. Only three
steps take us from a park bench to our
entire planet, and three more steps
reveal our solar system of planets
circling the sun. Only four more steps
enlarge our field of view to show our
Milky Way galaxy, and two steps
beyond that, we see a region of the
universe filled with millions of galaxies
similar to our own. See Chapter 1.

The Horsehead Nebula, a dark dust cloud in Orion.

Horizons

FOURTH EDITION

Exploring the Universe

Michael A. Seeds

Joseph R. Grundy Observatory
Franklin and Marshall College

Wadsworth Publishing Company
Belmont, California
A Division of Wadsworth, Inc.

For Janet and Kathryn

Astronomy Editor: Julie Butler
Development Editors: John Bergez, Jeannine Drew
Editorial Assistant: Leslie With
Production Editor: Gary Mcdonald
Designer: Ann Butler
Print Buyer: Barbara Britton
Art Editor: Kelly Murphy
Permissions Editor: Peggy Meehan
Copy Editor: Tom Briggs
Photo Researcher: Laren Crawford
Technical Illustrator: Illustrious
Cover Designer: Steve Markovich
Cover Photos: (background) NASA/Goddard Space
Flight Center; (foreground) NASA
Compositor and Color Separator: Black Dot Graphics
Printer: Arcata/Hawkins

Frontispiece: Copyright the Anglo-Australian Telescope Board. Photography by David Malin.

Contents Photos: p. vii, © Royal Observatory Edinburgh; p. xi, Astronomical Society of the Pacific; pp. xiii, xv, xx, National Optical Astronomy Observatories; pp. xvi, xxi, xxii, NASA; p. xvii © Association of Universities for Research in Astronomy, Inc., Kitt Peak National Observatory; p. xviii, courtesy Stephen M. Larson; p. xix, Anglo-Australian Telescope Board; p. xxiii, courtesy J. Trümper, Max-Planck Institute.

Unit Openers: Units 1, 2, and 3, Astronomical Society of the Pacific; Unit 4, NASA; Unit 5, Pat O'Hara Photography.

Printed in the United States of America

1 2 3 4 5 6 7 8 9 10—97 96 95 94 93

Library of Congress Cataloging-in-Publication Data
Seeds, Michael A.
  Horizons, exploring the universe / Michael A.
Seeds.—4th ed.
    p. cm.
  Includes bibliographical references and index.
  ISBN 0-534-18936-9 (alk. paper)
  1. Astronomy.   I. Title.
QB45.S44   1993
520—dc20   92-29503

Preface

Few of my friends ever greet me with a cheerful, "What's new?" They know an astronomer's news can keep them occupied for hours. We've located arachnoids on Venus, ice on Mercury, arcs around Neptune, jets in Orion, craters on asteroids, neutron stars in binaries, quasars in clusters, and dark matter everywhere, but we still can't find those pesky neutrinos. Astronomy is a dynamic science where new discoveries regularly alter our perception of reality. That is why we love it and why our students find it exciting. In this new edition of *Horizons: Exploring the Universe*, I have tried to present astronomy as a unified system of understanding that relates our students' personal existence to the basic processes in the universe, but I have also tried to express some of the excitement we feel when someone says, "What's new?"

MINIMUM TANGLE

The body of astronomical knowledge resembles a marionette with tangled strings. Each fact, observation, or principle is connected by inferences, assumptions, and theories to other facts, observations, and principles. The teacher's task is to untangle this marionette and make clear the logic of the connections. Just as a marionette has inherent in its design an untangled state, the body of astronomical knowledge has a state that makes the logic of the inferences, assumptions, and theories clearest. In this book, I have attempted to present astronomy in that state of minimum tangle.

By considering stars, galaxies, cosmology, and the solar system in that order, we simplify many of the evidential arguments and intuitive concepts so important to understanding astronomy. For example, by discussing star formation before the solar system, students recognize the solar nebula for what it is, a natural by-product of star formation. More important, by studying stars, galaxies, and cosmology first, students return to the solar system with a cosmic perspective on nature that places them and their planet into the universe, rather than constructing a universe around them.

Facts can be meaningful only when they are synthesized into a consistent description of nature. Thus, this book views the universe as a small set of natural processes that are responsible for a wide variety of phenomena and that explain a diverse assortment of objects. For example, galaxies, star clusters, individual stars, and planets are all expressions of the same process—gravitational contraction. Only the scale is different. This emphasis on processes presents astronomy not as a collection of unrelated facts but as a unified body of knowledge.

In addition, this text presents astronomy as a case study in science. It distinguishes between observation and theory, between evidence and conclusion. We want our students to understand how science creates, tests, improves, or discards models of natural phenomena and thus more closely approximates a complete understanding of natural processes.

CHANGES IN THE FOURTH EDITION

Of course this new edition contains the latest discoveries and images in astronomy. But I have also taken this opportunity to make less obvious changes in the book that will make it a more useful tool for both instructor and student. Many instructors have written with suggestions and questions, and careful reviews by instructors have helped identify problems and find solutions. In addition, for this edition, we obtained detailed reviews by three recognized experts in the fields of stellar evolution, galaxies and cosmology, and solar system astronomy. Those technical reviews led not only to minor changes in wording but also to major revisions of entire chapters. Finally, the editorial department at Wadsworth has provided careful guidance on the use of language. The major changes in the book were thus drawn from many sources. These changes include the following:

- To guide the student through the book, I have divided the book into five units. Unit 1, "Exploring the Sky," is carefully rewritten to lead the student from a personal view of the sky as the ceiling of the world, to an astronomer's view of the sky as the depths of the universe.
- Mathematical boxes have been rewritten to clearly distinguish among the presentation of the method, examples, and solutions. All math boxes now contain at least one solved example.
- Through detailed comments from instructors, reviewers, and editors, I have made many changes in the text to help students follow arguments

and understand diagrams. Many of these changes are minor, but their cumulative effect will be substantial.

- Guided by a technical review, I have rewritten Chapter 14, "Galaxies with Active Nuclei," not only to include the latest discoveries, but also to reflect a contemporary consensus on the nature of active galactic nuclei. The discussions of star formation, stellar evolution, neutron stars, the origin of the solar system, and the physics of planetary surfaces and atmospheres also benefited from these technical reviews.
- This new edition has provided an opportunity to further develop graphs and diagrams. With the help of the Wadsworth editorial and art staff, I have improved existing diagrams and added new diagrams, including the filmstrip figures that have been so successful in illustrating temporal processes. We have added color to many diagrams not only to make them more attractive, but also to illustrate important concepts where color has pedagogical value.
- While Chapter 13, "Galaxies," now contains new sections on dark matter and the distribution of galaxies through space, discussion of these concepts is spread throughout the chapters of Unit 3, "The Universe of Galaxies."
- Based on the advice of the technical reviewer and a number of instructors, I have revised Chapter 15, "Cosmology," to emphasize the true nature of the red shift, including a new filmstrip feature, and the geometry of the expansion. A new section on the origin of structure concludes the chapter.
- A newly revised and expanded section on Earth's atmosphere discusses the greenhouse effect and the ozone layer as environmental issues on Earth, so they can later be understood when we meet them on Venus, Mars, and Titan.
- The news from the Magellan probe has stimulated a complete reanalysis of the surface geology of Venus in Chapter 17. In a real sense, the images sent back by Magellan have made Venus into a newly discovered planet.
- Chapter 18, "Worlds of the Outer Solar System," concludes with a rewritten discussion of the origin of Pluto and the existence of small ice worlds in the outer solar system.

CONTINUING FEATURES

Many useful features have been retained and updated. "Perspectives," which appear at the end of several chapters, introduce new and interesting ideas that allow students to review and apply the principles covered in the chapter. These Perspectives might discuss the development of a theory, the synthesis of hypotheses from data, the testing of theories by observation, or the meaning of statistical evidence.

Study aids for each chapter include a chapter summary, a list of new terms, review questions, discussion questions, problems, and recommended readings. The first time a new term appears in the text it is set in **boldface**. These terms are defined in the glossary. The review questions are nonquantitative and could lead to essay answers that can be found in the text. Discussion questions may go slightly beyond the text and ask the student to consider the implications of the material in the chapter. These discussion questions may be useful to stimulate class discussion. The problems are quantitative or involve mathematical reasoning. (Answers to even-numbered problems appear at the end of the book, and to odd-numbered problems in the Instructor's Manual.) The recommended reading, which is intended for the student, ranges from *National Geographic* to *Science*. Instructors may wish to guide students in selecting appropriate reading material.

ANCILLARIES

- An *Instructor's Manual* contains course outlines, suggestions for planetarium programs, main concepts, outlines, test questions for each chapter, as well as sources of supplies, films, books, and charts.
- *Testing software* for both the IBM-PC and MacIntosh environments is available.
- *Wadsworth Astronomy Transparencies* consist of one-, two-, and four-color acetates showing illustrations from Wadsworth astronomy texts.

- *Voyages Through Space and Time* by Jon Wooley, Eastern Michigan University, contains projects for use with VOYAGER, the Interactive Desktop Planetarium, and is for sale to students. Cross-references to *Voyages* appear at the ends of pertinent chapters in the text, where its use would lend further understanding to topics covered in the chapters.
- *Introductory Astronomy Exercises* by Dale Ferguson, Baldwin-Wallace College, contains a wide variety of laboratory exercises and is for sale to students.
- *The Laser Disc Universe* is a HyperCard stack that ties *Horizons* to the widely used laser discs *Astronomy, The Sun,* and *Voyager Gallery* from Optical Data.
- *Color slides* are available in two different sets: Portrait of the Solar System and A Color Tour of the Milky Way.

Available through an adoption incentive program:

- *Software* for classroom demonstrations or laboratory instruction is available in two formats: *The Sky* for IBM-PCs and compatibles, and VOYAGER, the Interactive Desktop Planetarium for MacIntosh.
- The *videotape series* "The Astronomers" is available as individual tapes or as part of a series.

For more information about any of the above, contact your Wadsworth sales representative.

ACKNOWLEDGMENTS

My thanks to the many students and teachers who have responded so enthusiastically to *Horizons: Exploring the Universe*. Their comments and suggestions have been very helpful in completing this new edition. I would especially like to thank the numerous reviewers whose careful analysis and thoughtful suggestions have been invaluable in refining *Horizons: Exploring the Universe* as a teaching tool.

The people listed in the illustration credits were very kind in providing photographs and diagrams. Special recognition goes to the following, who were always ready to help locate unusual images from their institutions: Frank Bash, McDonald Observatory; Michael Belton, National Optical Astronomy Observatories; Patricia Bridges, U.S. Geological Survey; Jeff Butler, Lancaster County Planning Commission; Linda Carroll, U.S. Geological Survey; Steven Charlton, National Optical Astronomy Observatory; Don and Cheryl Davis; Bob Freid, Stamps With Teeth; Helen S. Horstman, Lowell Observatory; Charles Keller, Los Alamos National Laboratory; William Livingston, National Solar Observatory; David Malin, Anglo-Australian Observatory; Wallace Ravven, California Association for Research in Astronomy; Patricia A. Ross, National Space Science Data Center; Stephen Saunders, Jet Propulsion Laboratory; Janet Sandland, Royal Observatory, Edinburgh; Rudolph E. Schild, Center for Astrophysics; Patricia Shand, Lick Observatory; Martin Slade; Jet Propulsion Laboratory; Patricia Smiley, National Radio Astronomy Observatory; Brad Smith, National Space Science and Data Center; Jeff L. Stoner, National Optical Astronomy Observatory; Harvey Tananbaum, Center for Astrophysics; Donald E. Wilbur, Pennsylvania Department of Transportation; Denise R. Whitehead, National Optical Astronomy Observatory.

My appreciation also goes to the following institutions for their assistance in providing figures: The Anglo-Australian Observatory, Ames Research Center, *The Astrophysical Journal*, Bell Laboratories, Brookhaven National Laboratories, California Association for Research in Astronomy, Celestron International, High Altitude Observatory, Jet Propulsion Laboratories, Johnson Space Center, Lick Observatory, Lowell Observatory, Lunar and Planetary Laboratory, Martin Marietta Aero-Space Corporation, Mount Wilson and Las Campanas Observatories, National Aeronautics and Space Administration, National Optical Astronomy Observatories, National Radio Astronomy Observatory, Palomar Observatory, Pennsylvania Department of Transportation, Royal Observatory, Edinburgh, U.S. Geological Survey, and Yerkes Observatory. Many Voyager, Pioneer, and Mariner photographs were provided by the National Space Science Data Center.

It has been a pleasure to work with and learn from the Wadsworth people. Special thanks go to the editorial and production staff including Anne Scanlan-Rohrer, John Bergez, Jeannine Drew, Leslie With, Ann Butler, Gary Mcdonald, Kelly Murphy, Barbara Britton, and Peggy Meehan.

Finally I would like to thank my wife, Janet, and my daughter, Kathryn. They have done their best to tolerate the long hours, late nights, and disrupted weekends.

Mike Seeds
Lancaster, Pennsylvania

Manuscript Reviewers for This and Previous Editions

Timothy Beers, Michigan State University
David Buckley, East Stroudsburg University
Thomas Bullock, West Valley College
Mark J. Comella, Duquesne University
Neil F. Comins, University of Maine
John J. Cowan, University of Oklahoma
Dale Cruikshank, NASA Ames Research Center
David Curott, University of North Alabama
Charles A. Eckroth, Saint Cloud State University
Felix Elles, Western Washington University
Eric Feigelson, Penn State University
Paul Feldker, St. Louis Community College
Martin L. Goodson, Delta College
Thomas Hughes, College of Alameda
Thomas L. Johnson, Ferris State College
Stephen Lattanzio, Orange Coast College
J. Gordon Likely, University of Minnesota
James LoPresto, Edinboro University
Valdar Oinas, Queensborough Community College
Edwin A. Olson, Whitworth College
Melvyn Oremland, Pace University
B. E. Powell, West Georgia College
Clifton W. Price, Millersville University
Lawrence Ramsey, Pennsylvania State University
John B. Shaefer, Geneva College
Michael Stewart, San Antonio College
David Theison, University of Maryland
Theodore D. Violett, Western State College of Colorado
Fred Walter, SUNY @ Stony Brook
Daniel Weedman, Pennsylvania State University
Louis Winkler, Pennsylvania State University
Kenneth Yoss, University of Illinois

To the Reader

You will approach retirement around the year 2040, and your children about 2065. Your grandchildren will not be retiring until almost 2100. You and your family will live through a century of exploration unlike any in the history of this planet. You will see explorers return to the moon, and your children could be the first colonists in lunar cities. Your grandchildren may reach Mars, mine the asteroid belt, explore the icy satellites of Jupiter and Saturn, or leave the solar system bound for the stars. A century ago the airplane had not been invented. Whatever humanity is like a century in the future, we can guess that it will be deeply involved in the exploration of the solar system. Astronomy, the study of the universe beyond the clouds, helps us understand what we will find when we leave earth.

Living in the next century might be enough justification for taking an astronomy course, but there are other reasons. The coming years will see tremendous advances in science and technology, advances that will confuse anyone not familiar with how science progresses from data to hypothesis to theory to natural law. Should your state permit nuclear waste disposal sites? Should you support construction of orbiting solar power stations? Should you give your children massive doses of vitamin C to combat colds? To resolve such technical issues, you need to apply some of the methods of science. Thus, as you study astronomy in the pages that follow, look at it as an example of scientific reasoning. Distinguish between data and theory, and notice how hypotheses are tested over and over.

Yet another reason for taking an astronomy course is to satisfy your natural curiosity. Having heard about black holes, the expanding universe, or the rings of Saturn, you may want to know more about them. Satisfying your own curiosity is the most noble reason for studying anything.

Curiosity might lead you to consider astronomy as a career, but you should know that the field is very small and jobs are hard to find. You might, however, consider astronomy as a hobby—an activity for personal satisfaction and enrichment. The magazines listed here will keep you up to date with the rapid advances in the field and give you some ideas for further projects, such as telescope building and astronomical photography:

Astronomy, 21027 Crossroads Circle, P.O. Box 1612, Waukesha, WI 53187

The Griffith Observer, 2800 East Observatory Road, Los Angeles, CA 90027

Mercury, Astronomical Society of the Pacific, 390 Ashton Ave., San Francisco, CA 94112

The Planetary Report, The Planetary Society, 65 N. Catalina Ave., Pasadena, CA 91106

Sky and Telescope, Sky Publishing Corporation, P.O. Box 9111, Belmont, MA 02178-9111

All these reasons for taking astronomy are reasonable, but the most important reason is astronomy's cultural value. The one reason you should study astronomy, the reason your school goes to the expense of teaching the course for you, is that astronomy tells you about your place in nature. It shows you our tiny planet spinning in space amid a vast cosmos of stars and galaxies. It takes you from the first moment of creation to the end of the universe. You will see our planet form, life develop, and our sun die. This knowledge has no monetary value, but it is priceless if you are to appreciate your existence as a human being.

Astronomy will change you. It will not just expand your horizons, it will do away with them. You will see humanity as part of a complex and beautiful universe. If by the end of this course you do not think of yourself and society differently, if you don't feel excited, challenged, and a bit frightened, then you haven't been paying attention.

M.A.S.

ABOUT THE AUTHOR

Mike Seeds is Professor of Astronomy at Franklin & Marshall College, as well as director of the college's Joseph R. Grundy Observatory. An active researcher on photometry of short-period variable stars, he includes among his other research interests archeoastronomy and telescope automation. He is the Principal Astronomer in charge of the Phoenix 10, the first robotic telescope. In 1989 he received the Christian R. and Mary F. Lindback Award for Distinguished Teaching. In addition to writing textbooks, Seeds frequently contributes to journals and creates educational computer programs for students in his own courses. He has also published educational software for toddlers! Seeds is the author of *Foundations of Astronomy, Third Edition,* for Wadsworth, as well as *Astronomy: Selected Readings* (Benjamin/Cummings, 1980), and (with Joseph R. Holzinger) *Laboratory Exercises in Astronomy* (Macmillan, 1976).

Brief Contents

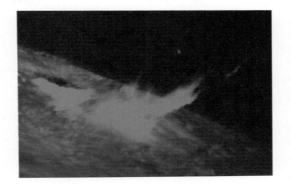

Contents

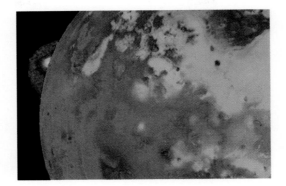

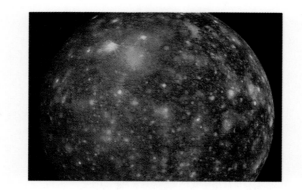

UNIT   2

THE STARS   109

6 • Atoms and Starlight   111

7 • The Sun   133

8 • The Properties of Stars   150

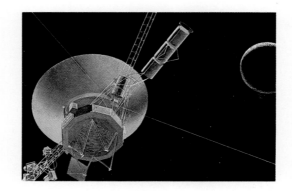

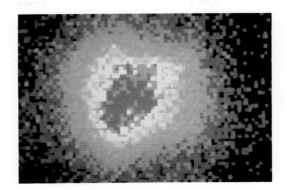

OBSERVATIONAL ACTIVITIES

The

Sky

CHAPTER

1

The Scale of the Cosmos

The longest journey begins with a single step.

Confucius

We are going on a voyage out to the end of the universe. Marco Polo journeyed east, Columbus west, but our voyage will take us away from our home on Earth, out past the moon, sun, and planets, past the stars we see in the sky and past billions more that we cannot see without the aid of the largest telescopes. We will journey through great whirlpools of stars to the most distant galaxies visible from Earth—and then we will continue on, carried only by experience and imagination—looking for the structure of the universe itself.

Besides journeying through space, we will also travel in time. We will explore the past: see the sun and planets form, search for the formation of the first stars and the origin of the universe. We will also explore forward in time to watch the sun die and the earth wither.

Though we may find an end to the universe, a time when it will cease to exist, we will not discover an edge. Our universe may extend in all directions without limit. Such vastness dwarfs our human dimensions, but not our intelligence or imagination.

Astronomy is more than the study of stars and planets. It is the study of the universe in which we exist. We live on a small planet circling a small sun drifting through the universe, but astronomy can take us out of ourselves and thus help us understand what we are. Our study of astronomy introduces us to sizes, distances, and timespans far beyond our common experience. The comparisons in this chapter are designed to help us grasp their meaning.

In our voyage through space, we can take comfort from a single important idea. No matter how big the universe is, it obeys a small set of simple rules. That is, the universe is rational and can be understood through thoughtful, careful study.

(Photo by author)

How big is a star? The answer—roughly 1 million miles in diameter—is meaningless. Such a large number tells us nothing. How can we humans, only 5 to 6 or so feet tall, hope to understand the vastness of the universe? The secret lies in the single word *scale*.

To illustrate the scale of astronomical bodies, to fit ourselves into the universe, we will journey from a campus scene to the limits of the cosmos in 12 steps. Each step will widen our view by a factor of 100. That is, each successive picture in this chapter will show a region of the universe that is 100 times wider than the preceding picture.

This scene shows a region about 52 feet in diameter. It is occupied by a student, a sidewalk, some lawn, two trees, and a few bushes—all objects whose size we can understand. Only 12 steps separate this scene from the universe as a whole.

Field of view enlarged 100 times from previous image. (Pennsylvania Department of Transportation, Bureau of Design)

WE NOW INCREASE our field of view by a factor of 100, to an area 1 mile across. The area of the preceding photograph is shown by a small red square (arrow). Individual people, trees, and sidewalks vanish, but now we can see a college campus and the surrounding streets and houses.

Although we have begun our adventure using feet and miles, we will switch to the metric system of units because it makes arithmetic simpler (see Appendix C). One mile is 5280 feet, and each foot is 12 inches, so 1 mile is $5280 \times 12 = 63,360$ inches. In the metric system, this calculation is easier. One kilometer (km) is 1000 meters (m), and each meter is 100 centimeters (cm). Thus 1 km is $1000 \times 100 = 100,000$ cm.

One mile equals 1.609 km, so our photograph is about 1.6 km in diameter. (See Appendix C for conversions.) Only 11 more steps separate us from the largest dimensions in the universe.

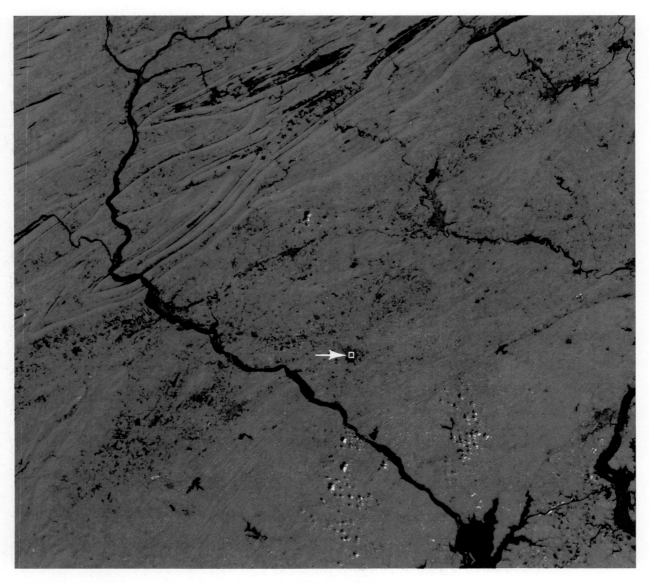

Field of view enlarged 100 times from previous image. (NASA infrared photograph)

OUR FIELD OF VIEW now spans 160 km (100 miles). At this scale, we see the natural features of the earth's surface. The Allegheny Mountains cross the photograph in the upper left, and the Susquehanna River flows southeast into Chesapeake Bay. In this infrared image, the foliage is red, the water black, and a few small puffs of cloud are white.

These features remind us that we live on an evolving planet. Forces in the earth's crust have pushed the mountain ranges up into parallel folds, like a rug wrinkled on a polished floor. The clouds remind us that the earth's atmosphere is rich in water, which falls as rain and erodes the mountains, washing material down the rivers and into the sea. The mountains and valleys that we know are only temporary features; they are constantly changing.

As we explore the universe we will see that it too is changing. We will see stars evolving and dying, and we will discuss the possible end of the universe.

Field of view enlarged 100 times from previous image. (NASA)

THE NEXT STEP in our journey shows our entire planet. It is 12,756 km (8000 miles) in diameter and rotates on its axis once a day. This image shows most of the daylight side of the planet, with the sunset line at the extreme right. The rotation of the earth carries us eastward across the daylight side.

A thin layer of water makes up the oceans, and the atmosphere is only a few miles deep. On the scale of this photograph, the depth of the atmosphere on which our lives depend is less than the thickness of a single piece of thread.

The water and air on our planet have made life possible, but we know of no other planet where such conditions exist. Only eight other planets orbit the sun in our **solar system,** and none has liquid water on its surface. In addition, we can see no other planets orbiting other stars. Such planets probably exist, but they are too distant to be visible. So far as we know, we are the only life in the universe.

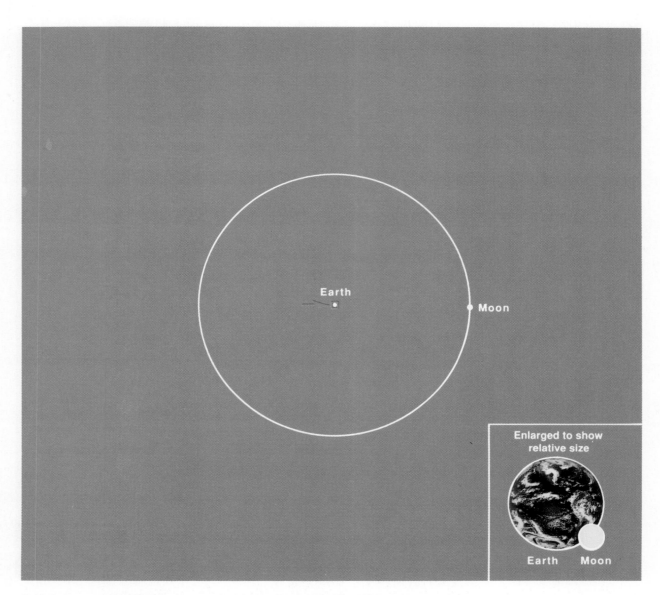

Field of view enlarged 100 times from previous image. (Earth image by NASA)

AGAIN WE ENLARGE our field of view by a factor of 100, and we see a region of the universe 1,600,000 km wide. Earth is the small white dot in the center, and the moon, only one-fourth the diameter of Earth, is an even smaller dot along its orbit 380,000 km from Earth.

Distances in space are so enormous that astronomy is known as the science of big numbers. Yet we will use numbers much larger than these to discuss the depths of the universe. Rather than write these numbers out completely, we will use scientific notation—a simple way to write big numbers without writing a great many zeros. For example, in scientific notation we would write 380,000 as 3.8×10^5. The 5 tells us to move the decimal point five places to the right. Notice that we could also write this as 38×10^4 or 0.38×10^6.

We can also use scientific notation to write very small numbers. If you are not familiar with scientific notation, consult Appendix C. The universe is too big to discuss without using it.

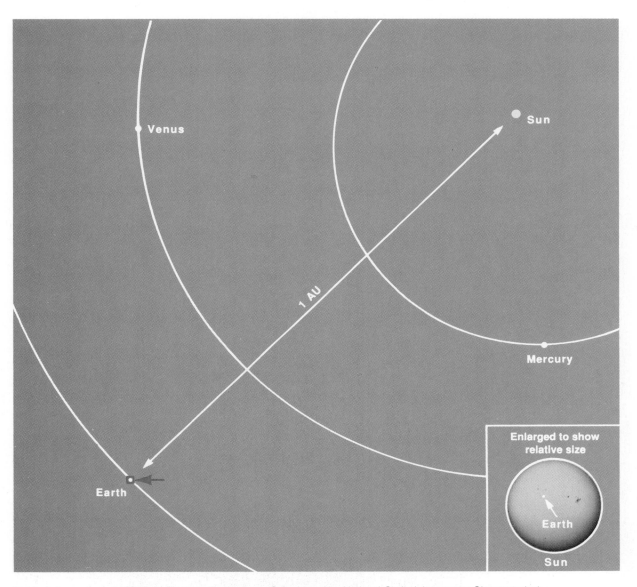

Field of view enlarged 100 times from previous image. (Solar image by National Optical Astronomy Observatories)

WHEN WE ONCE AGAIN ENLARGE our field of view by a factor of 100, Earth and the moon disappear into the small box at lower left. But now we see the sun and two other planets. The sun is 109 times larger in diameter than Earth (inset), but it, too, is nothing more than a dot on this diagram.

This figure spans 1.6×10^8 km. One way to deal with such large distances is to define new units. Astronomers use the average distance from Earth to the sun (about 1.5×10^8 km) as a unit of distance called the **astronomical unit (AU).** We can say that the average distance from Venus to the sun is about 0.7 AU, while the average distance from Mercury to the sun is about 0.39 AU.

We refer to *average* distances because the orbits of the planets are not perfect circles but ellipses. This is particularly apparent for Mercury: Its orbit carries it as close to the sun as 0.307 AU and as far away as 0.467 AU. Earth's orbit is more circular, and its distance from the sun varies by only 1.7 percent.

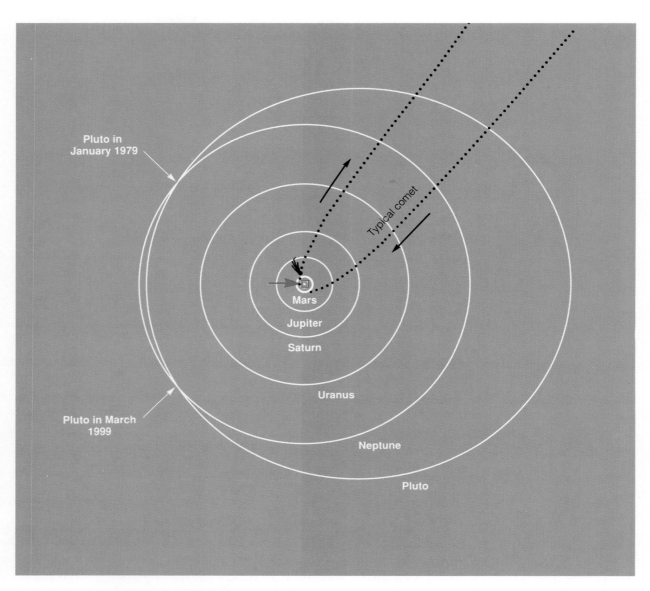

Field of view enlarged 100 times from previous image.

WHEN WE BEGAN our journey, our field of view was only 52 feet (about 16 m) in width. We now see the entire solar system. Our field of view is 1 trillion (10^{12}) times wider than in our first view.

The details of the preceding figure are lost in the tiny square at the center here. Mars, the next outward planet, lies only 1.5 AU from the sun. In contrast, the outer planets—Jupiter, Saturn, Uranus, Neptune, and Pluto—are so far from the sun that they are easy to place in this diagram. These are cold worlds far from the sun's warmth. Light from the sun takes over 4 hours to reach Neptune, which is slightly over 30 AU distant. By comparison, sunlight reaches Earth in only 8 minutes.

Notice that Pluto's orbit is so elliptical that it can come closer to the sun than Neptune can. In fact, Neptune is now farther from the sun than Pluto is and will remain the most distant planet in our solar system for the rest of this century.

Field of view enlarged 100 times from previous image.

WHEN WE AGAIN enlarge our field of view by a factor of 100, our solar system vanishes. The sun is visible as a point of light, but all the planets and their orbits are now crowded into the small square at the center.

Nor are any stars visible except for the sun. The sun is a fairly typical star, and it seems to be located in a fairly typical neighborhood in the universe. Although there are many billions of stars like the sun, none is close enough to be visible in this area, which is 11,000 AU in diameter. The stars are typically separated by distances about 10 times as large as this diagram.

It is difficult to imagine the isolation of the stars. If the sun were represented by a golf ball in New York City, then the nearest star would be another golf ball in Chicago. Except for the widely scattered stars and a few atoms of gas drifting between the stars, the universe is nearly empty.

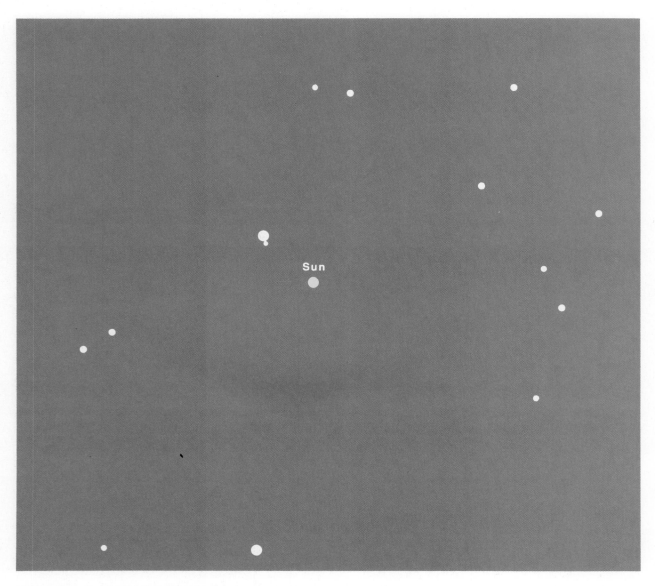

Field of view enlarged 100 times from previous image.

OUR FIELD OF VIEW has now expanded to a diameter of a bit over 1 million AU. The sun is located at the center, and we see a few of the nearest stars.

These stars are so far away it is not reasonable to quote the distances in astronomical units. We must define a new unit of distance, the light-year. One **light-year (ly)** is the distance that light travels in 1 year, roughly 10^{13} km or 63,000 AU. The star nearest to the sun is Proxima Centauri at a distance of 4.2 ly. Light from Proxima Centauri takes 4.2 years to reach Earth. The diameter of our field of view is now 17 ly.

In this diagram, the diameters of the dots represent the brightness of the stars and not their actual diameters. This is the custom in astronomical diagrams, and it is also how star images are recorded on photographs. Bright stars make larger spots on the photographic plate than faint stars, even though they may not be larger stars.

Field of view enlarged 100 times from previous image. This box ■ represents relative size of previous frame. (National Optical Astronomy Observatories)

AS WE EXPAND our field of view by another factor of 100, the sun and its neighboring stars vanish into the background of thousands of stars. The field of view is now 1700 ly in diameter.

Of course, no one has ever journeyed thousands of light-years from the sun to photograph the solar neighborhood, so we use a typical photo of the sky. Here the diameters of star images are related to brightness and not the true diameter of the stars. The sun is a relatively faint star, so we could not locate it on such a photo.

What we do *not* see is critically important. We do not see the thin gas that fills the spaces between the stars. Although that gas is thinner than the best vacuum on Earth, it is such clouds of gas that give birth to stars. Our sun formed from such a cloud about 5 billion years ago. We will see more star formation when we expand our field of view by another factor of 100.

Field of view enlarged 100 times from previous image. (Anglo-American Telescope Board)

IF WE EXPAND our field of view by another factor of 100, we see our own **Milky Way galaxy.** No one can travel far enough away to photograph our galaxy, so we use a photo of a similar galaxy with an arrow pointing to a representative location for the sun.

The sun and stars of the previous figure would seem lost among the 100 billion stars of the galaxy. Most stars are smaller and fainter than our sun; some are larger and more luminous. Why stars differ is one of the mysteries of the universe we will explore.

Our galaxy has graceful **spiral arms** where stars are born in great clouds of gas and dust. Our sun is presently passing through such a spiral arm.

Our galaxy is roughly 100,000 ly in diameter. Until about 70 years ago astronomers thought it was the entire universe, an island universe of stars in an otherwise empty vastness. Now we know that our galaxy is only one of billions of galaxies scattered throughout the universe.

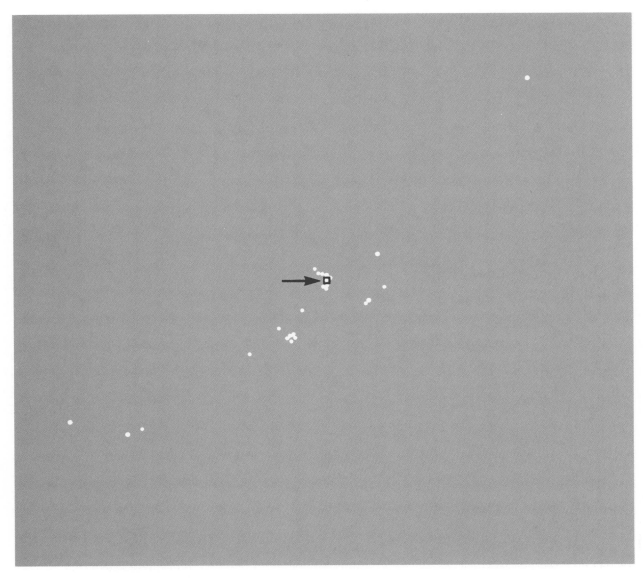

Field of view enlarged 100 times from previous image.

As we expand our field of view by another factor of 100, our galaxy becomes a luminous speck surrounded by other specks. This diagram includes a region 17 million ly in diameter. Each dot represents a galaxy.

Our galaxy (arrow) is one of a small cluster of galaxies called the **Local Group.** This cluster contains roughly two dozen galaxies scattered through a region about 6 million ly in diameter.

Among the galaxies we see here, a few are as large as our own galaxy, but most are smaller. A few have the beautiful spiral arms we see in our galaxy, but most do not. Among more distant galaxies we see a few that are twisted into peculiar shapes or wracked by violent eruptions. No one is sure what makes one galaxy different from another. One theory holds that the centers of some galaxies contain supermassive black holes, which are capable of swallowing stars whole. Whatever the truth, the evolution of galaxies must occasionally be marked by events of titanic violence.

Field of view enlarged 100 times from previous image. This box ■ represents relative size of previous frame. (Detail from galaxy map from M. Seldner, B. L. Siebers, E. J. Groth, and P. J. E. Peebles, *Astronomical Journal* 82 [1977])

IF WE EXPAND our field of view one final time, we see that our Local Group of galaxies is part of a larger supercluster, a cluster of clusters. Other galaxies are not scattered at random throughout the universe but lie in clusters within larger superclusters.

Representing the universe at this scale is a diagram in which each tiny dot represents the location of a single galaxy. We see superclusters linked to form long filaments outlining voids that seem nearly empty of galaxies. These appear to be the largest structures in the universe. Were we to expand our field of view once more, we would see a sea of filaments and voids. In puzzling over the origin of these structures, we are at the frontier of human knowledge.

Our problem in studying astronomy is to keep a proper sense of scale. Remember that each of the billions of galaxies contains billions of stars. Most of those stars probably have families of planets like our solar system, and on some of those billions of planets liquid water oceans and a protective atmosphere may have spawned life. It is possible that some other planets in the universe are inhabited by intelligent creatures who share our curiosity and our sense of wonder at the scale of the cosmos.

The vast scale of the universe might seem incomprehensible, but the true message of astronomy is reassuring. Throughout our study of astronomy, as we look at stars, planets, galaxies, and the universe itself, time after time we will see that, however vast the universe is, it is ruled by a small set of natural laws. These natural laws hold tremendous power to simplify; if we understand gravity, for instance, we can understand the orbital motion of everything from the earth's moon to the largest of galaxies. In the chapters that follow, we will search out the natural laws that rule the universe. The true message of astronomy is that the universe follows rules and that we can understand those rules.

SUMMARY

Our goal in this chapter has been to preview the scale of astronomical objects. To do so, we journeyed outward from a familiar campus scene by expanding our field of view by factors of 100. Only 12 such steps took us to the largest structures in the universe.

The numbers in astronomy are so large it is not convenient to express them in the usual way. Instead, we use the metric system to simplify our calculations and scientific notation to express big numbers more easily. The metric system and scientific notation are discussed in Appendix C.

We live on the rotating planet Earth, which orbits a rather average star we call the sun. We defined a unit of distance, the astronomical unit, to be the average distance from Earth to the sun. Of the eight other planets in our solar system, Mercury is closest to the sun, and Neptune is currently the most distant at about 30 AU.

The sun, like most stars, is very far from its neighboring stars, and this leads us to define another unit of distance, the light-year. A light-year is the distance light travels in 1 year. The nearest star to the sun is Proxima Centauri at a distance of 4.2 ly.

As we enlarged our field of view, we discovered that the sun is only one of 100 billion stars in our galaxy and that our galaxy is only one of billions of galaxies in the universe. Galaxies appear to be grouped together in clusters, superclusters, and filaments, the largest structures known.

As we explored we noted that the universe is evolving. The earth's surface is evolving, and so are stars. Stars form from gases in space, grow old, and eventually die. We do not yet understand how galaxies form or evolve.

Among the billions of stars in each of the billions of galaxies, many probably have planets, but even the stars nearest to the sun are too distant for us to see any planets they might have. We suppose that some of these planets are like the earth, and we wonder if a few are inhabited by intelligent beings like ourselves.

NEW TERMS

solar system

scientific notation

astronomical unit (AU)

light-year (ly)

Milky Way galaxy

spiral arm

Local Group

QUESTIONS

1. Why are astronomical units and light-years more convenient for measuring astronomical distances than miles or kilometers?

2. In what ways is our universe evolving?

3. Why do all stars, except for the sun, look like points of light as seen from Earth?

4. Why are we unable to see planets beyond the nine in our solar system?

5. Which is the outermost planet in our solar system? Why does this change?

6. In photographs, some stars look larger than others. What does this tell us about the stars?

7. How long does it take light to cross the diameter of our galaxy? of the Local Group of galaxies?

8. What are the largest known structures in the universe?

9. How many planets inhabited by intelligent life do you think the universe contains? Explain your answer.

PROBLEMS

1. How many inches are there in 100 yards? How many centimeters are there in 100 m?

2. If 1 mile equals 1.609 km, and the moon is 2160 miles in diameter, what is its diameter in kilometers?

3. The earth rotates once a day and has a radius of 6378 km. With what speed is the equator moving eastward in km/sec? in mph?

4. If sunlight takes 8 minutes to reach the earth, how long does moonlight take?

5. If the earth were transported to the center of the sun, would the moon's orbit lie inside or outside the surface of the sun?

6. How many suns would it take, laid edge to edge, to reach the nearest star?

7. How many kilometers are there in a light-minute? (**HINT:** The speed of light is 3×10^5 km/sec.)

8. How many galaxies like our own, laid edge to edge, would it take to reach the nearest large galaxy (which is 2×10^6 ly away)?

RECOMMENDED READING

CALDER, NIGEL. *Timescale*. New York: Viking Press, 1982.

DICKINSON, TERRENCE. *The Universe and Beyond*. Ontario: Camden House, 1986.

MORRISON, PHILIP, and PHYLIS MORRISON. *Powers of Ten*. New York: W. H. Freeman, 1982.

SAGAN, CARL. *Cosmos*. New York: Random House, 1980.

WEISKOPF, VICTOR. *Knowledge and Wonder: The Natural World As Man Knows It*. Cambridge, Mass.: MIT Press, 1979.

FILM/VIDEO

Powers of Ten. Santa Monica, Calif.: Pyramid Film and Video, 1989.

CHAPTER
2

The Sky

The sky is the rest of the universe as seen from our planet. When we look at the stars, we look up through a layer of air only a few miles thick. Above that, space is nearly empty, with the stars scattered light-years apart. In the previous chapter, we took a quick journey through the universe to survey its scale. Now we are ready to begin our detailed study of astronomy, our search for the natural laws that govern the universe. Our first step, in this chapter, is to try to understand why the sky looks the way it does.

As you read this chapter, keep in mind that we live on a planet. The stars are scattered into the void all around us, some very distant and some closer. Our planet rotates on its axis once a day, so from our viewpoint the sky appears to rotate around us each day. Not only does the sun rise in the east and set in the west, but so also do the stars. That daily mo-

The Southern Cross I saw every night abeam. The sun every morning came up astern; every evening it went down ahead. I wished for no other compass to guide me, for these were true.

Captain Joshua Slocum
Sailing Alone Around the World

tion is a reflection of the rotation of our planet.

2-1 THE STARS

On a dark night far from city lights, we can see a few thousand stars in the sky. Like ancient astronomers, we will try to organize what we see by naming groups of stars and individual stars, and by specifying the brightness of individual stars.

Constellations All around the world, ancient cultures gave names to groups of stars—constellations—to honor gods, heroes, and animals (Figure 2-1). Of course, each culture created its own unique set of constellations. The constellations we are familiar with in Western culture originated in Mesopotamia over 5000 years ago, with other constellations added by Babylonian, Egyp-

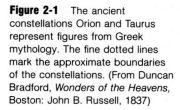

Figure 2-1 The ancient constellations Orion and Taurus represent figures from Greek mythology. The fine dotted lines mark the approximate boundaries of the constellations. (From Duncan Bradford, *Wonders of the Heavens,* Boston: John B. Russell, 1837)

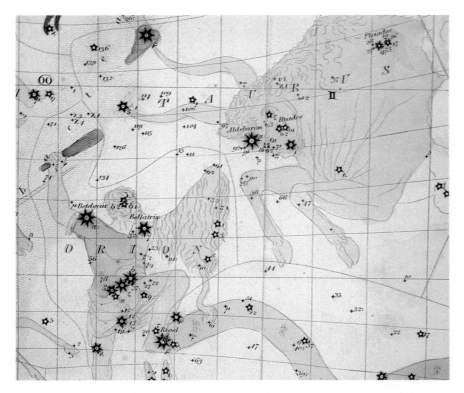

tian, and Greek astronomers during the classical age. Of these ancient constellations, 48 are still in use today.

To the ancients, a constellation was a loose grouping of stars that symbolized a certain figure, and constellation boundaries were not precisely defined (Figure 2-2a). Many of the fainter stars were not included in any constellation, and regions of the southern sky not visible to the ancient astronomers of the northern latitudes were not divided into constellations. In recent centuries, astronomers added 40 modern constellations to fill the gaps, and in 1928 the International Astronomical Union established 88 official constellations with clearly defined boundaries in the sky (Figure 2-2b). Thus a constellation now represents not a group of stars, but an area of the sky, and every spot on the sky belongs to one and only one constellation.

In addition to the 88 official constellations, the sky contains a number of less formally defined groupings called **asterisms.** The Big Dipper, for example, is a well-known asterism, but it is only part of the larger constellation Ursa Major (the

Great Bear). Another asterism is the Great Square of Pegasus (Figure 2-2), which includes three stars from Pegasus and one from Andromeda. (Box 2-1 and the star charts at the end of this book will introduce you to the brighter constellations.)

Although we identify constellations and asterisms by name, we must keep in mind that they are made up of stars that are usually not physically associated with one another. Some may be many times farther away than others and moving through space in different directions. The only thing they have in common is that they lie in approximately the same direction from Earth.

The Names of the Stars In addition to naming groups of stars, ancient astronomers named the brighter stars, and modern astronomers still use many of those names. Whereas the names of the constellations are in Latin, the common language of science in Renaissance Europe, most star names derive from ancient Arabic, though much altered by the passing centuries. The name of Betelgeuse, the bright red star in Orion, for example, comes from the Arabic *yad al-jawza,* meaning "hand of

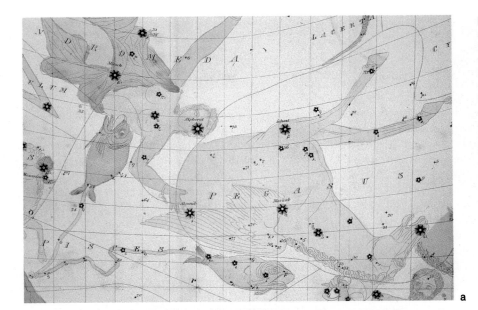

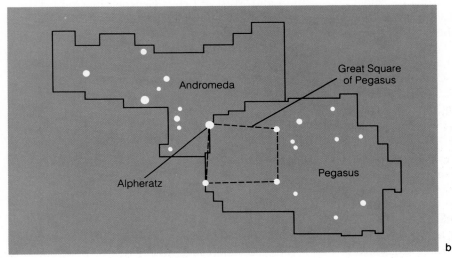

Figure 2-2 To the ancients, the star Alpheratz was one of the stars in the great square of Pegasus and was also the eye of Andromeda. (a) In this engraving from 1837, curved dotted lines mark the constellation boundaries and show that Alpheratz is part of Andromeda. (From Duncan Bradford, *Wonders of the Heavens*, Boston: John B. Russell, 1837) (b) Modern constellation boundaries are precisely defined by international agreement.

Jawza' (Orion)." Aldebaran, the bright red eye of Taurus the bull, comes from the Arabic *al-dabar an,* meaning "the follower (of the Pleiades)," and Fomalhaut comes from *fam al-hut,* meaning "mouth of the fish."

Naming individual stars is not very helpful because we can see thousands of them, and these names do not help us locate the star in the sky. Another way to identify stars is to assign Greek letters (see Appendix D) to the bright stars in a constellation in approximate order of brightness.

Thus the brightest star is usually designated α (alpha), the second brightest β (beta), and so on. For many constellations, the letters follow the order of brightness, but some constellations, by tradition, mistake, or the personal preferences of early chartmakers, are exceptions (Figure 2-3).

To identify a star by its Greek letter designation, we give the Greek letter followed by the genitive (possessive) form of the constellation name, such as α Canis Majoris. This both identifies the star and the constellation and gives us a clue to the relative

A Guide to the Constellations

Constellations are difficult to learn because the constellations above the horizon change with the seasons and with the time of night. To simplify the process, use the descriptions in this box in conjunction with the monthly star charts at the end of this book to find one of the key constellations. We assume you are observing in the evening, a few hours after sunset.

In the summer, soon after sunset, look for a very bright star nearly overhead. This star is Vega in the key constellation Lyra (the Lyre). Consult the appropriate star chart from the back of the book to find Hercules, Corona Borealis (the Northern Crown), and Bootes (the Bear Driver) to the west of Lyra. East is Cygnus (the Swan—also known as the Northern Cross), and southeast is Aquila (the Eagle).

Early in the autumn, Lyra is nearly overhead as darkness falls, but by mid-October it is in the western sky and Pegasus (the Winged Horse) becomes the key constellation high in the east.

Once you've found Pegasus, look for Andromeda.

As winter comes, Pegasus moves into the western sky in the evenings. Starting in December, look for Orion (the Hunter) in the southeast sky. When you have found Orion, you can find the surrounding constellations, including Canis Major (the Big Dog), which contains the brightest star in the sky, Sirius.

As winter passes and spring approaches, Orion moves into the southwestern quadrant of the evening sky. Beginning in late March, look for the sickle shape of the key constellation Leo (the Lion). West of Leo is Cancer (the Crab), a faint constellation that is hard to find. East of Leo, look for the kite shape of Bootes (the Bear Driver). By late spring and early summer, Leo is in the western sky in the evening. Beginning in June, look again for the summer key constellation, Lyra, in the east.

---

brightness of the star. Compare this with the ancient name for this star, Sirius, which tells us nothing about location or brightness.

This method of identifying a star's brightness is only approximate. In order to discuss the sky with precision, we must have an accurate way of referring to the brightness of stars, and for that we must consult one of the first great astronomers.

The Brightness of Stars Hipparchus, a Greek astronomer who lived about 2100 years ago (Figure 2-4), divided the stars into six classes. The brightest were first-class stars, and those slightly fainter were second-class stars. Continuing down to the faintest stars he could see, the sixth-class stars, he recorded his classifications in a great star catalogue that became a basic reference in ancient astrono-

my. His method, slightly modified, is still in use today.

In spite of its value, Hipparchus' method may seem a bit confusing. First, when early astronomers translated the catalogue into Latin, they used the word *magnitudo*, meaning "size." In English, this became magnitude, even though it refers to the brightness of the stars and not to their size. Thus the **magnitude scale** is the astronomer's brightness scale.

The second source of confusion is that the fainter the star, the larger the magnitude number. For example, 6th-magnitude stars are fainter than 1st-magnitude stars. This may seem backwards at first, but think of it as Hipparchus did. The brightest stars are first-class stars, and the fainter stars are second- and third-class, and so on.

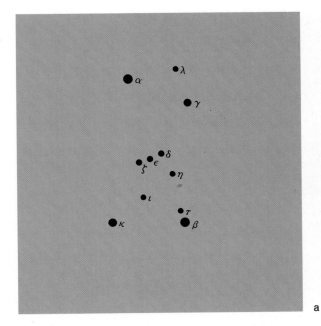

a

Figure 2-3 (a) The brighter stars in each constellation are assigned Greek letters in approximate order of brightness. In Orion, β is brighter than α and κ is brighter than η. (b) A long-exposure photograph reveals the many faint stars that lie within Orion's constellation boundaries. These are members of the constellation, but they do not have Greek letter designations. Note the differences in the colors of the stars. The crosses around the bright stars are caused by an optical effect in the telescope. (Photo courtesy William Hartmann)

b

Figure 2-4 Hipparchus (2nd century B.C.) was the first great observational astronomer. Among other things, he constructed a catalogue listing 1080 of the brightest stars by position and dividing them into six brightness classes now known as magnitudes. He is honored here on a Greek stamp that also shows one of his observing instruments.

Modern astronomers have made a major improvement in Hipparchus' magnitude system by measuring stellar brightness with sensitive instruments. For example, instead of merely saying that θ (theta) Leonis is a 3rd-magnitude star, they can say specifically that its magnitude is 3.34.

If we measure the brightness of all of the stars in Hipparchus' first brightness class, we discover that some are brighter than 1.0. For example, Vega (α Lyrae) is so bright its magnitude, 0.04, is almost zero. A few are so bright the magnitude scale must extend into negative numbers (Figure 2-5). On this scale, Sirius, the brightest star in the sky, has a magnitude of -1.47.

These are known as **apparent visual magnitudes.** These magnitudes refer to how bright the stars look and do not compensate for their distance from Earth. A star that is a million times more

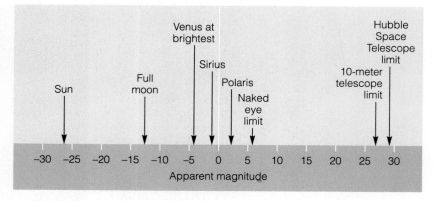

Figure 2-5 The scale of apparent visual magnitudes extends into negative numbers to represent the brighter objects, and to positive numbers larger than 6 to represent objects fainter than the human eye can see.

luminous than the sun might appear very faint if it were very far away, and a star that is much fainter than the sun might look bright if it were nearby. In Chapter 8, we will develop a magnitude scale that takes distance into account and tells us how bright the stars really are. Apparent visual magnitude only tells us how bright they *appear*. (Box 2-2 discusses magnitudes in more detail.)

2-2 THE CELESTIAL SPHERE

A Model of the Sky Ancient astronomers thought of the sky as a great, hollow, crystalline sphere surrounding Earth. The stars, they imagined, were attached to the inside of the sphere like thumbtacks stuck in the ceiling. The sphere rotated once a day, carrying the sun, moon, planets, and stars from east to west.

We know now that the sky is not a great, hollow, crystalline sphere. The stars are scattered through space at different distances, and it isn't the sky that rotates once a day, but Earth that turns on its axis. Although we know that the crystalline sphere is not real, it is convenient as a model of the sky. As long as we keep the true nature of the sky in mind, we can use the model to analyze the appearance and motions of the sky.

We will call this model of the sky the **celestial sphere,** an imaginary hollow sphere of very large radius surrounding the earth and to which the stars seem to be attached (Figure 2-6). We must use a vary large radius for our celestial sphere so that no part of Earth is significantly closer to a given star than any other part. Then it does not matter where on Earth we go; the sky always looks like a great sphere centered on our position.

If we watch the sky for a few hours, we can see movement. As the rotation of Earth carries us eastward, the sun moves across the sky and sets in the west. As it gets dark, we can see the stars, and in a few hours it becomes obvious that the rotation of Earth is making the sky rotate westward. As some constellations set in the west, others rise in the east.

Angles on the Sky One way we might use the celestial sphere is to describe the location of objects. Just as we might tell a friend that we own a cabin 40 km north of Flagstaff, we might want to tell our friend to look for the moon a specific distance north of a certain star. We can't measure distances in the sky in kilometers, however. Rather, we need to measure angles on the celestial sphere.

Astronomers often use angles to describe distance across the sky. They might say, for instance, that the angular distance between the moon and the bright star Vega is 8°, meaning that if we point one arm at the moon and the other arm at Vega, the angle between our arms is 8°. We know the star is hundreds of light-years away, and we know that the moon is much closer, so the true distance between them is immense. But if we imagine them painted

BOX 2-2
Magnitudes

Brightness is subjective. How bright a star looks depends on such things as the physiology of the eye and the psychology of perception. For purposes of measurement, we should use the more precise term *intensity*—a measure of the light energy from the star that hits 1 square centimeter (cm²) in 1 second. If two stars have intensities I_A and I_B, then the ratio of their intensities is I_A/I_B.

Modern astronomers have defined the magnitude scale so that two stars that differ by 5 magnitudes have an intensity ratio of exactly 100. Thus two stars that differ by 1 magnitude must have an intensity ratio of $100^{1/5}$, or 2.512 —that is, the light of one star must be about 2.5 times more intense. Two stars that differ by 2 magnitudes will have an intensity ratio of 2.512×2.512, or about 6.3, and so on (Table 2-1).

Example A: Suppose star C is 3rd magnitude and star D is 9th magnitude. What is the intensity ratio? *Solution:* The magnitude difference is 6 magnitudes, and the table shows the intensity ratio is 250. Therefore light from star C is 250 times more intense than light from star D.

A table is convenient, but for more precision we can express the relationship as a simple formula. The intensity ratio I_A/I_B is equal to 2.512 raised to the power of the magnitude difference $m_B - m_A$:

$$\frac{I_A}{I_B} = (2.512)^{(m_B - m_A)}$$

Example B: If the magnitude difference is 6.32 magnitudes, what is the intensity ratio? *Solution:* The intensity ratio must be $2.512^{6.32}$. A pocket calculator tells us the answer: 337.

When we know the intensity ratio and want to find the magnitude difference, it is convenient to solve the formula for the magnitude difference:

$$m_A - m_B = 2.5 \log(I_B/I_A)$$

TABLE 2-1	
Magnitude and Intensity	
Magnitude Difference	**Intensity Ratio**
0	1
1	2.5
2	6.3
3	16
4	40
5	100
6	250
7	630
8	1600
9	4000
10	10,000
.	.
.	.
.	.
15	1,000,000
20	100,000,000
25	10,000,000,000
.	.
.	.
.	.

Example C: Suppose that light from Sirius is 24.2 times more intense than light from Polaris. What is the magnitude difference? *Solution:* The magnitude difference is 2.5 log(24.2). Our pocket calculator tells us the logarithm of 24.2 is 1.38, so the magnitude difference is $2.5 \times 1.38 = 3.4$ magnitudes.

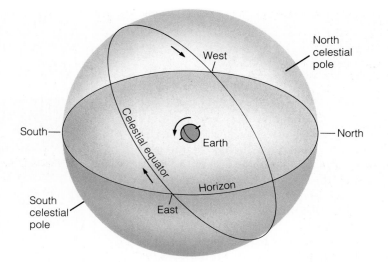

Figure 2-6 The celestial sphere models the appearance of the sky. The poles mark the pivots, and the equator divides the sky in half. Those objects below our horizon are invisible. The earth rotates eastward and makes objects in the sky appear to rise along the eastern horizon and set along the western horizon.

on the celestial sphere, we can think of their separation as an "angle *on* the sky." Thus we can discuss the angular distance between two objects even when we don't know their true distance from each other.

We measure angles in degrees, minutes of arc, and seconds of arc. There are 360° in a circle and 90° in a right angle. Each degree is divided into 60 **minutes of arc** (sometimes abbreviated 60′). If you view a 25¢ piece face-on from the length of a football field, it has an angular diameter of about 1 minute of arc. Each minute of arc is divided into 60 **seconds of arc** (abbreviated 60″). If you view the ball in the tip of a ballpoint pen from the length of a football field, it has an angular diameter of about 1 second of arc.

We can establish some angles on the sky that will be helpful in estimating other angles. The sun and moon are each about 0.5° in diameter. The pointer stars of the Big Dipper are about 5° apart, and the bowl of the Big Dipper is about 30° from the north celestial pole.

While it is sometimes convenient to locate things with respect to a bright star, we need to establish some reference points and lines on the sky. They will provide the basis for a precise way of locating objects.

Reference Marks on the Sky Just as we use the earth's poles and equator as reference marks on

the earth, we can use corresponding reference marks on the sky. The celestial poles and celestial equator are defined by the earth's rotation.

If we watch the night sky for a few hours, we can see the stars moving westward (Figure 2-7). In the northern sky, they appear to revolve around a point called the **north celestial pole,** the point on the sky directly above the earth's North Pole. The **south celestial pole** is the corresponding point directly above the earth's South Pole. If the earth's axis could be extended out to the celestial sphere, it would touch at the celestial poles.

Another important reference mark on the celestial sphere is the **celestial equator,** an imaginary line around the sky directly above Earth's equator (see Figure 2-6). The celestial equator divides the sky into two equal hemispheres and is everywhere 90° from the celestial poles.

The orientation of the celestial poles and equator with respect to our horizon depends on our latitude. To be precise, the angular distance from the horizon up to the north celestial pole equals the latitude of the observer. For example, if we stood in the ice and snow of the earth's North Pole, our latitude would be 90°N, and the north celestial pole would be directly overhead (Figure 2-8). If we walked southward, our latitude would decrease and the north celestial pole would sink closer to the northern horizon. When we finally stood on the earth's equator, our latitude would be 0°, and

Figure 2-7 (a) A time exposure of an hour or so shows stars as streaks due to the rotation of the earth. (National Optical Astronomy Observatory) (b) From the middle latitudes of the United States, about 40°N, we find the stars of the northern constellations circling the north celestial pole. (c) In the eastern sky, stars rise at an angle to the horizon. (d) In the south, the stars circle the south celestial pole, which is invisible below the southern horizon. Compare with Figure 2-6.

the north celestial pole would lie on our northern horizon. This relationship, by the way, makes it simple for navigators to find their latitude by measuring the angle between the northern horizon and the north celestial pole.

The star Polaris happens to lie very near the north celestial pole and thus hardly moves as the sky rotates. At any time of night, in any season of the year, Polaris stands above the northern horizon and is consequently known as the North Star. (See star charts at end of book.)

From a latitude typical of the United States, constellations around Polaris never disappear below the horizon and are thus known as **north circumpolar constellations** (Figure 2-9). At higher latitudes, more constellations are north circumpolar, and at lower latitudes, there are fewer north circumpolar constellations. There are also **south**

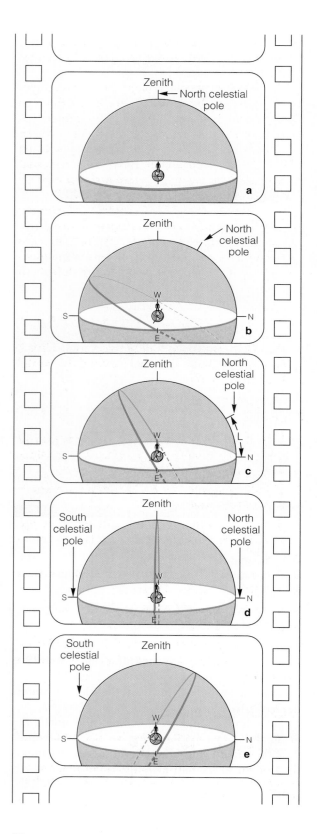

Figure 2-8 At the earth's North Pole (a), we would see the north celestial pole directly overhead, and the celestial equator would circle the horizon. As we journeyed southward (b), the angle between the north celestial pole and the northern horizon (L) would always equal our latitude (c). At the earth's equator (d), we would see the celestial poles on our horizon and the celestial equator would pass through our zenith. From southern latitudes (e), we would see the south celestial pole above our horizon.

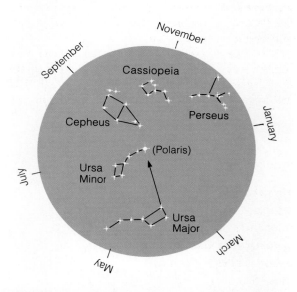

Figure 2-9 The north circumpolar constellations as seen from a latitude typical of the United States. To use the chart, face north soon after sunset and hold the chart in front of you with the current month at the top.

circumpolar constellations around the south celestial pole that never rise above the southern horizon as seen by an observer at a middle northern latitude.

The reference marks we have established—the celestial poles and the celestial equator—are the basis for a celestial coordinate system that works much like the system of latitude and longitude on Earth (see Appendix B). That system makes it possible to be precise in locating objects in the sky.

We now have a model of the sky—the celestial sphere—complete with reference marks and angular measurements to guide us. We can locate constellations, name the stars, and express their brightness numerically. We are now ready to discuss the principal motions of the earth, moon, and sun—the subject of the next chapter.

SUMMARY

Astronomers divide the sky into 88 areas called constellations. Although the constellations originated in Greek and Middle Eastern mythology, the names are Latin. Even the modern constellations, added to fill in the spaces between the ancient figures, have Latin names. The names of stars usually come from ancient Arabic, though modern astronomers often refer to a star by constellation and Greek letters assigned according to brightness within each constellation.

The magnitude system is the astronomer's brightness scale. First-magnitude stars are brighter than 2nd-magnitude stars, which are brighter than 3rd-magnitude stars, and so on. The magnitude we see when we look at a star in the sky is its apparent visual magnitude, which does not take into account its distance from us.

The celestial sphere is a model of the sky, carrying the celestial objects around the earth. Because Earth rotates eastward, the celestial sphere appears to rotate westward on its axis. The northern and southern celestial poles are the pivots on which the sky appears to rotate. The celestial equator, an imaginary line around the sky above Earth's equator, divides the sky in half.

Astronomers often refer to angles "on" the sky as if the stars, sun, moon, and planets were equivalent to spots painted on a plaster ceiling. Then the angle on the sky between two objects is independent of the true distance between the objects in light-years.

NEW TERMS

constellation

asterism

magnitude scale

apparent visual magnitude

celestial sphere

minute of arc

second of arc

north and south celestial poles

celestial equator

north and south circumpolar constellations

REVIEW QUESTIONS

1. Why have astronomers added modern constellations to the sky?

2. What information does a star's Greek letter designation often contain?

3. Give two reasons why the magnitude scale might seem confusing.

4. How do we define the locations of the celestial poles and the celestial equator?

DISCUSSION QUESTIONS

1. Describe the difference between the ancient and the modern conception of the celestial sphere.

2. How have you thought of the sky? as a ceiling? as a dome overhead? as a sphere around the earth? as a limitless void?

PROBLEMS

1. If light from one star is 40 times more intense than light from another star, by what do they differ in magnitudes?

2. If two stars differ by 8.6 magnitudes, what is their intensity ratio?

3. If star A is 4th magnitude and star B is 6th magnitude, which is brighter? What is their intensity ratio?

4. By what factor is sunlight more intense than moonlight? (HINT: See Figure 2-5.)

5. Sketch the celestial sphere and label the poles, equator, and horizon.

6. Which constellations are north circumpolar for an observer at the earth's equator? at the earth's North Pole?

RECOMMENDED READING

ALLEN, RICHARD HINCKLEY. *Star Names: Their Lore and Meaning.* New York: Dover, 1963.

HOLZINGER, J. R., and M. A. SEEDS. *Laboratory Exercises in Astronomy.* Ex. 5, 17, 23, and Appendix A. New York: Macmillan, 1976.

KUNITZSCH, PAUL. "How We Got Our 'Arabic' Star Names." *Sky and Telescope* 65 (January 1983), p. 20.

KUNITZSCH, PAUL, and TIM SCOTT. *Short Guide to Modern Star Names and Their Derivations.* Otto Harrassowitz Verlag, 1986.

LOVI, GEORGE, and WIL TIRION. *Men, Monsters and the Modern Universe.* Richmond, Va.: Willman-Bell, 1989.

Mag 6 Star Charts. Barrington, N.J.: Edmund Scientific Co., 1982.

MENZEL, D. H. *A Field Guide to the Stars and Planets.* Boston: Houghton Mifflin, 1964.

NORTON, A. P. *A Star Atlas.* Cambridge, Mass.: Sky, 1964.

Seasonal Star Charts. Northbrook, Ill.: Hubbard Press, 1972.

WHITNEY, CHARLES A. *Whitney's Star Finder.* New York: Knopf, 1980.

FURTHER EXPLORATION

VOYAGER, the Interactive Desktop Planetarium, can be used for further exploration of the topics in this chapter. This chapter corresponds with the following projects and chapters in *Voyages Through Space and Time:* Project 2, The Horizon System; Project 3, Constellations and Planets; Project 4, The Equatorial System; Chapter 2, Locating Objects on the Earth and in the Sky; Chapter 3, Telling Time by the Stars.

3

Cycles of the Sky

The Aztecs sacrificed humans and offered their beating hearts to the brilliant star that appeared sometimes in the evening sky and sometimes in the morning sky—the object we know as the planet Venus. The Aztecs saw in the cycle of Venus a metaphor for life, death, and rebirth, and they associated Venus with their principal god, Quetzalcoatl (kwét-zal-ko-áh-tl), who had died, journeyed through the underworld, and returned to life. The Aztecs linked the cycles of their lives on Earth directly to the cycles they saw in the sky.

Even today our lives are linked to sky cycles. The rotation of the earth on its axis makes the sun appear to rise and set, and our bodies follow that cycle of day and night in sleeping and waking. The moon revolves around the earth in its orbit passing through a cycle of phases, and we divide our calendar into months of roughly one lunar cycle. The orbital mo-

Even a Man who is pure in heart and says his prayers by night may become a wolf when the wolfbane blooms and the moon shines full and bright.

Proverb from old Wolfman movies

tion of the earth around the sun causes a year-long cycle of seasons by which we regulate our agricultural, political, and social lives.

The cycles in the sky are the cycles of the sun and moon. In this chapter, we will study those motions and see our earth as a rotating planet circling its sun. While we may be studying the sky, we will be learning about our world.

3-1 THE CYCLE OF THE SUN

Perhaps the most obvious cycles in the sky are those that involve the motion of the sun. Of course, those motions are produced by the rotation and revolution of the earth.

Astronomers distinguish between **rotation,** the turning of a body about an axis, and **revolution,** the motion of a body about a point located outside the body. Thus we say that Earth rotates on its axis once a day and revolves around the sun once a year.

Earth's rotation produces the daily cycle of sunrise and sunset. The annual revolution of Earth around the sun makes the sun appear to move around the sky in a year-long cycle that causes the seasons. To understand that motion, we must imagine that we can make the sun fainter.

The Ecliptic　When the sun is above the horizon, its brilliance makes Earth's atmosphere glow with scattered light. Because blue light scatters off of air molecules slightly better than red light does, we see the sky filled with blue light "bouncing about" in all directions. No matter where we look in the daytime sky, we see this blue light, and we cannot see the fainter stars. If we could make the sun a million times fainter, it would be about as bright as the full moon, and we would be able to see stars even when the sun was above the horizon.

If we noted the position of the sun among the stars day by day, we would quickly notice that it was moving eastward. In early January, for instance, we would see the sun against the background of the stars in Sagittarius (Figure 3-1). Earth is moving, however, so each day we would see the sun slightly eastward of its previous posi-

tion. By February it would have moved into the constellation Capricornus, and by March it would be in Aquarius.

Of course, it is not quite correct to say that the sun is "in Aquarius." The sun is only 93 million miles away, and the stars of Aquarius are at least a million times farther away. But in March of each year, we would see the sun against the background of the stars in Aquarius, and thus we could say, "The sun is in Aquarius."

If we continued watching the sun against the background of stars throughout the year, we could plot its path on a star chart. After one full year, we would see the sun begin to retrace this line as it continued its annual cycle of motion around the sky. This line, the apparent path of the sun around the sky, is called the **ecliptic.** Another way to define the ecliptic is to say it is the projection of the earth's orbit on the sky. If the celestial sphere were a great screen illuminated by the sun at the center, then the shadow cast by Earth's orbit would be the ecliptic (Figure 3-1). Yet a third way to define the ecliptic is to refer to it as the plane of Earth's orbit. These three definitions of the ecliptic are equivalent, and it is worth considering them all because the ecliptic is one of the most important reference lines on the sky. We will use it, for instance, to discuss the seasons.

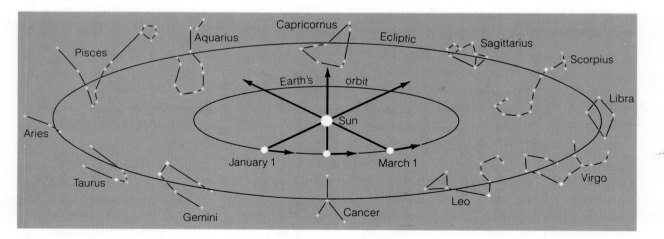

Figure 3-1　As Earth moves around its orbit, we see the sun in front of different constellations. The sun appears to move around the ecliptic, the projection of Earth's orbit onto the sky.

The eastward motion of the sun along the ecliptic is a reflection of Earth's motion around its orbit. Earth completely circles the sun in 365.24 days, and consequently the sun circles the sky following the ecliptic in the same number of days. This means the sun, traveling 360° around the ecliptic in 365.24 days, travels about 1° eastward each day. The sun is about 0.5° in diameter, so it travels twice its own diameter per day.

The sun's motion along the ecliptic is complicated by the fact that Earth does not rotate perpendicular to the plane of its orbit. Its axis of rotation is tipped 23.5° from perpendicular. Earth, like a spinning top, holds its axis fixed in space as it moves around the sun (Figure 3-2). We saw in the previous chapter that Earth's axis always points toward a spot on the sky very near Polaris, the North Star. Although the direction of Earth's axis does drift slowly (see Box 3-1), we will not notice any change in our lifetimes.

Because Earth is tipped 23.5° in its orbit, the ecliptic is tipped 23.5° from the celestial equator. Recall that the celestial equator is the projection of Earth's equator, and that the ecliptic is the projection of Earth's orbit. Since Earth tipped 23.5°, its equator is tipped 23.5° from the plane of its orbit. When we project this on the sky, we find that the ecliptic and celestial equator meet at an angle of 23.5° (Figure 3-4). (Find the ecliptic on the star charts at the end of this book.)

The ecliptic and celestial equator cross at two places on the sky called equinoxes (Figure 3-4). The **vernal equinox** is the place where the sun crosses the celestial equator moving northward, and the **autumnal equinox** is the place where it crosses moving southward. The sun crosses the vernal equinox on or about March 21, and it crosses the autumnal equinox on or about September 22. The exact dates of the equinoxes can vary by a day or two because of leap year and other factors.

We can identify two other reference marks on the ecliptic by noting where the sun is farthest from the celestial equator. About June 22, the sun is farthest north at the point called the **summer solstice**. The **winter solstice** is the point where the sun is farthest south, about December 22.

Note that the equinoxes and solstices are points on the sky, but the same words refer to the times when the sun crosses those points. You might hear

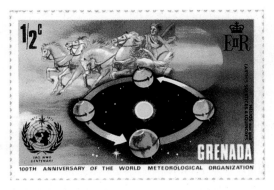

Figure 3-2 Earth's axis of rotation is inclined 23.5° from the perpendicular to its orbit. We have seasons on Earth because the axis remains fixed as Earth orbits the sun. This stamp from Grenada shows the earth at four locations in its orbit: summer solstice (right), autumnal equinox (background), winter solstice (left), vernal equinox (foreground).

someone say, "This year the vernal equinox occurs at 3:02 A.M. on March 21." Whether we think of them as places or times, the equinoxes and solstices are important because they mark the beginning of each of the seasons.

The Seasons The seasonal temperature depends on the amount of heat we receive from the sun. To hold the temperature constant, there must be a balance between the amount of heat we gain and the amount we radiate to space. If we receive more heat than we lose, we get warmer; if we lose more than we gain, we get cooler.

The motion of the sun around the ecliptic tips the heat balance one way in summer and the opposite way in winter. Because the ecliptic is inclined with respect to the celestial equator, the sun spends half the year in the northern celestial hemisphere and half the year in the southern celestial hemisphere (Figure 3-4). When the sun is in the northern celestial hemisphere, the northern half of Earth receives more direct sunlight—and therefore more heat—than the southern half. This makes North America, Europe, and Asia warmer.

The seasons are reversed in Earth's southern half (see Figure 3-2). While the sun is in the northern celestial hemisphere warming North America, South America becomes cooler. Southern Chile

Precession

If we could watch the sky for a few hundred years, we would discover that the north celestial pole is moving slowly with respect to Polaris. The celestial poles and the celestial equator, our supposedly fixed reference marks, are moving very slowly because of the slow change in the direction of Earth's axis of rotation. This slow toplike motion is called **precession.** Earth's axis sweeps around in a cone, taking almost 26,000 years for each sweep (Figure 3-3).

Because Earth's precession has such a long period, it has little effect over a few hundred years. During our lifetimes, the north celestial pole will draw slightly closer to Polaris, pass its nearest point about A.D. 2100, and then begin to move away. Only by making careful observations could we detect this motion.

Precession is caused by the gravitational pull of the sun and moon. Because Earth is not a perfect sphere—it has a slight bulge around its equator—the sun and moon pull on it, trying to make it spin upright in its orbit. This forces Earth's axis to precess. The same thing happens to a child's toy top. Gravity tries to make it fall over, and the spinning top precesses.

One result of precession is the change of the pole star. As the north celestial pole moves around the sky, it sometimes comes close to one star or another. It just happens to be near Polaris now. Ancient Egyptian records show that 5000 years ago the north celestial pole was near the star Thuban (α Draconis). In about 13,000 years, the pole will have moved away from Polaris and will be near the bright star Vega (α Lyrae).

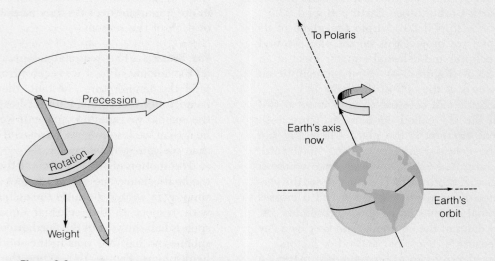

Figure 3-3   (a) The weight of a spinning gyroscope tends to make it fall over, and as a result it precesses in a conical motion about a vertical line. (b) The sun's gravity tends to twist Earth's axis upright in its orbit, and as a result it precesses. In 13,000 years, Earth's axis of rotation will point toward Vega.

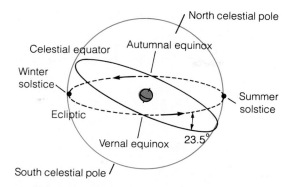

Figure 3-4 The ecliptic (dashed line), the sun's apparent path through the sky, crosses the celestial equator at the equinoxes. The solstices mark the most northerly and most southerly points.

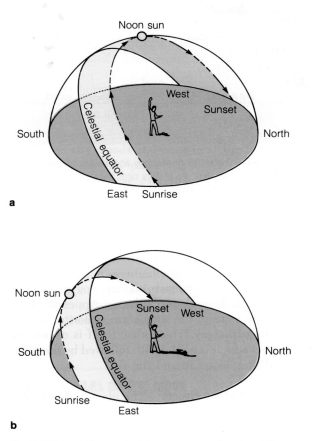

Figure 3-5 (a) The path of the sun across the sky at the summer solstice. (b) The sun's path at the winter solstice.

has warm weather on New Year's Day and cold in July.

To see how the sun can give us more heat in summer, think about the path the sun takes across the sky between sunrise and sunset. Figure 3-5 shows these paths when the sun is at the summer solstice and at the winter solstice as seen by a person living at latitude 40°, a good average latitude for most of the United States. Notice that at the summer solstice, the sun rises in the northeast, moves high across the sky, and sets in the northwest (Figure 3-5a). But at the winter solstice, the sun rises in the southeast, moves low across the sky, and sets in the southwest (Figure 3-5b). Two features of these paths tip the heat balance.

First, the summer sun is above the horizon for more hours of each day than the winter sun. Summer days are long and winter days are short. Since the sun is above our horizon longer in summer, we receive more energy each day.

Second, the sun stands high in the sky at noon on a summer day. It shines almost straight down, as shown by our small shadows. On a winter day, however, the noon sun is low in the southern sky. Each square meter of the ground gains little heat from the winter sun because the sunlight strikes the ground at an angle and spreads out, as shown by our longer shadows (Figure 3-6). These two effects work together to tip the heat balance and produce the seasons.

We mark the beginning of the seasons by the position of the sun. Spring begins at the moment the sun crosses the celestial equator going north (the vernal equinox). Summer begins at the moment the sun reaches its most northerly point (the summer solstice), and autumn begins when the sun crosses the celestial equator going south (the autumnal equinox). We mark the official beginning of winter when the sun reaches its most southerly position (the winter solstice).

Of course, the weather does not turn warm the instant spring begins. The ground, air, and oceans are still cool from winter, and they take a while to warm up. Likewise, in the autumn, the ground, air, and oceans slowly release the heat stored through the summer. Because of this thermal lag, the average daily temperatures lag behind the solstices by about 1 month. Although the sun crosses the sum-

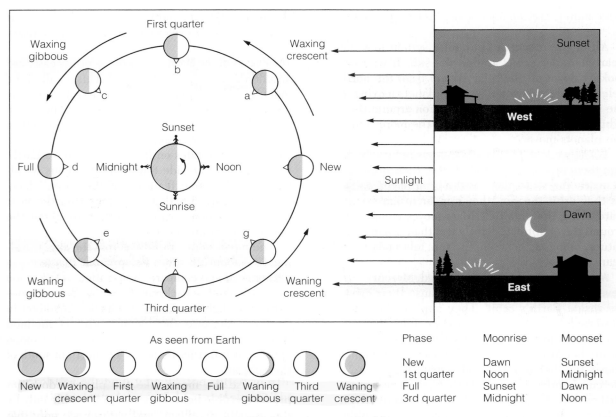

As seen from Earth

	Phase	Moonrise	Moonset
	New	Dawn	Sunset
	1st quarter	Noon	Midnight
	Full	Sunset	Dawn
	3rd quarter	Midnight	Noon

Figure 3-8 The phases of the moon are produced by the varying amounts of the illuminated surface we can see. Earth is shown as seen from above the North Pole. Lower images show the moon as it appears from Earth. Letters refer to photos in Figure 3-9. Note how the moon rotates to keep the same side facing Earth. Drawings at right show how the waxing crescent moon appears in the evening sky and the waning crescent moon appears in the dawn sky.

(shrinks) through gibbous phase to *third quarter,* then through crescent to new moon. To distinguish between the gibbous and crescent phases of the first and second half of the cycle, we refer to *gibbous waning* and *crescent waning* when the moon is shrinking, and to *gibbous waxing* and *crescent waxing* when it is growing.

The cycle of lunar phases takes 29.53 days, the synodic period of the moon, or about 4 weeks. Thus new moon, first quarter, full moon, and third quarter occur at nearly 1-week intervals. In general, an object's **synodic period** is its orbital period with respect to the sun. (*Synodic* comes from the Greek words meaning "together" and "path.") To

see why the moon's synodic period is longer than its sidereal period, imagine we begin observing at new moon—that is, when the moon is near the sun in the sky. After 27.322 days, the moon's sidereal period, it has circled the sky and returned to the same place among the stars where it was last new, but the sun has moved about 27° eastward along the ecliptic. The moon needs slightly more than 2 days to catch up with the sun and reach new moon again. Thus the moon's synodic period is longer than its sidereal period.

To summarize, let us follow the moon through one cycle of phases. At new moon, the moon is nearly in line with the sun and sets in the west with

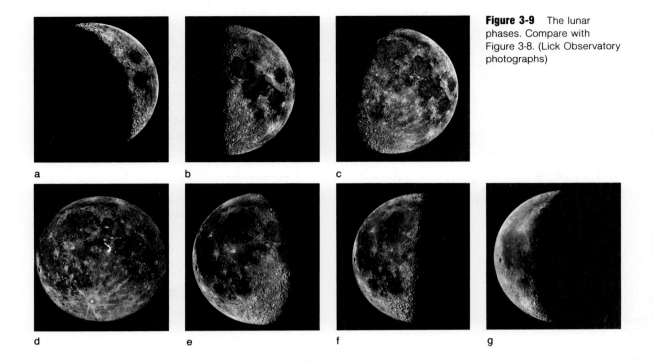

a b c

d e f g

the sun. Thus we see no moon at new moon. A few days after new moon, we see the waxing crescent above the western horizon soon after sunset. On such evenings we might be able to see the dark part of the moon in addition to the brighter crescent. This has been called "the old moon in the new moon's arms," and is caused by earthshine, sunlight reflected from Earth and illuminating the night side of the moon.

Each evening the crescent moon is fatter and higher above the horizon, until, about 1 week after new moon, it reaches first quarter and stands high in the southern sky at sunset. The first-quarter moon does not set until about midnight (see Figure 3-8). In the days following first quarter, the moon waxes fatter, becoming gibbous waxing. Each evening we find it farther east among the stars, and it sets later and later. About 2 weeks after new moon, the moon reaches full, rising in the east as the sun sets in the west. The full moon is visible all night, setting in the west at sunrise.

The waning phases of the moon may be less familiar because the moon is not visible in the early evening sky. As it wanes through gibbous, it rises later and later; by the time it reaches third quarter, it does not rise until midnight. The waning crescent does not rise until even later, and if we wish to see the thin waning crescent just before new moon, we must get up before sunrise and look for the moon above the eastern horizon.

Almost everyone is familiar with the changing phases of the moon, but those who live near the seashore are probably familiar with another phenomenon related to the lunar cycle—the periodic advance and retreat of the ocean tides.

Tides The moon's gravity has dramatic effects on the earth. The side of Earth facing the moon is about 4000 miles closer to the moon than the center of Earth is, and the moon's gravity pulls on the near side of Earth more strongly than on Earth's center. Though we think of our planet as solid, it is not perfectly rigid, so the moon's gravity draws the rocky surface of the near side up into a bulge a few centimeters high.

Seawater responds to the force of the moon's gravity by flowing into a bulge of water on the side of Earth facing the moon. There is also a bulge on

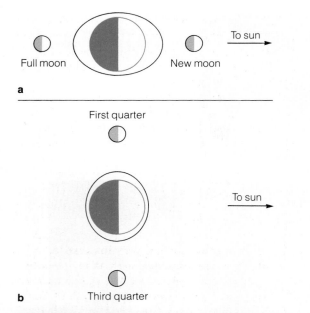

Full moon   New moon

a

First quarter

To sun

To sun

b   Third quarter

Figure 3-10 (a) When the moon and sun pull in the same direction, their tidal forces add and the tidal bulges are larger. Thus spring tides occur at new moon and full moon. (b) When the moon and sun pull at right angles, their tidal forces do not add and the tidal bulges are smaller. Such neap tides occur at first- and third-quarter moon.

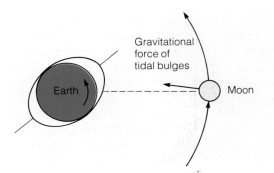

Gravitational force of tidal bulges

Earth   Moon

Figure 3-11 The rotation of Earth drags the tidal bulges ahead of the Earth–moon line (exaggerated here). The gravitational attraction of these masses of water pulls the moon forward in its orbit, forcing its orbit to grow in size.

the side away from the moon, which develops because the moon pulls more strongly on Earth's center than on the far side. Thus the moon pulls Earth away from the oceans, which flow into a bulge on the far side.

We can see dramatic evidence of this effect if we watch the ocean shore for a few hours. Though Earth rotates on its axis, the tidal bulges remain fixed with respect to the moon. As the turning Earth carries us into a tidal bulge, the ocean water deepens and the tide crawls up the beach. Later, when Earth carries us out of the bulge, the water becomes shallower and the tide falls. Because there are two bulges on opposite sides of Earth, the tides should rise and fall twice a day on an ideal coast.

In reality, the tide cycle at any given location can be quite complex because of the latitude of the site, shape of the shore, winds, and so on. Tides in the Bay of Fundy, for example, occur twice a day and can exceed 40 feet, while the northern coast of

the Gulf of Mexico has only one tidal cycle, of roughly 1 foot, each day.

The sun, too, produces tidal bulges on Earth. At new moon and at full moon, the moon and sun produce tidal bulges that add together (Figure 3-10a) and produce extreme tidal changes; high tide is very high, and low tide is very low. Such tides are called **spring tides,** even though they occur at every new and full moon, and not just in the spring. **Neap tides** occur at first- and third-quarter moon, when the moon and sun pull at right angles to each other (Figure 3-10b). Then the tides caused by the sun reduce the tides caused by the moon, and the rise and fall of the ocean is less extreme than usual.

Tidal forces can have surprising effects. The friction of the ocean waters with the seabeds slows the rotation of the earth and makes the day grow by 0.0016 seconds per century. Fossils of marine animals confirm that only 400 million years ago Earth's day was 22 hours long. In addition, Earth's gravitational field exerts tidal forces on the moon, and, although there are no bodies of water on the moon, friction within the flexing rock has slowed the moon's rotation to the point that it now keeps the same face toward Earth (see Figure 3-8).

Tidal forces can also affect orbital motion. Friction with the ocean beds drags the tidal bulges eastward out of a direct Earth–moon line (Figure 3-11). These tidal bulges contain a large amount of mass, and their gravitational field pulls the moon forward in its orbit. As a result, the moon's orbit is growing larger, and it is receding from Earth at about 3 cm per year, an effect that astronomers can measure by bouncing laser beams off reflectors left on the lunar surface by the Apollo astronauts.

These and other tidal effects are important in many areas of astronomy. In later chapters, we will see how tidal forces can pull gas away from stars, rip galaxies apart, and melt the interiors of satellites orbiting near massive planets. For now, however, we must consider other aspects of the lunar cycle. The stately progression of the lunar phases and the ebb and flow of the ocean tides are commonplace, but occasionally something peculiar happens: The moon darkens and turns copper-red in a lunar eclipse.

Lunar Eclipses   A lunar eclipse occurs at full moon when the moon moves through the shadow of Earth. Since the moon shines only by reflected sunlight, we see the moon gradually darken as it enters the shadow.

Earth's shadow consists of two parts. The **umbra** is the region of total shadow. If we were in the umbra of Earth's shadow, we would see no portion of the sun. If we moved into the **penumbra**, however, we would be in partial shadow and would see part of the sun peeking around the edge of Earth. Thus in the penumbra the sunlight is dimmed but not extinguished.

If the orbit of the moon carries it through the umbra, we see a total lunar eclipse (Figures 3-12 and 3-13). As we watch the moon in the sky, it first moves into the penumbra and dims slightly; the deeper it moves into the penumbra, the more it dims. In about an hour, the moon reaches the umbra, and we see the umbral shadow darken part of the moon. It takes about an hour for the moon to enter the umbra completely and become totally eclipsed. Totality, the period of total eclipse, may last as long as 1 hour 45 minutes, though the timing of the eclipse depends on where the moon crosses the shadow.

When the moon is totally eclipsed, it does not disappear completely. While it receives no direct sunlight, it does receive some sunlight refracted (bent) through Earth's atmosphere. If we were on the moon during totality, we would not see any part of the sun because it would be entirely hidden behind Earth. However, we would be able to see Earth's atmosphere illuminated from behind by the sun. The red glow from this "sunset" illuminates the moon during totality and makes it glow coppery red (Figure 3-14).

If the moon does not move completely into the umbra, we see a partial lunar eclipse (see Figure 3-12). The part of the moon that remains outside the umbra receives some direct sunlight, and the glare usually prevents our seeing the faint coppery glow of the part of the moon in the umbra.

A penumbral lunar eclipse occurs when the moon passes through the penumbra but misses the umbra entirely. Since the penumbra is a region of partial shadow, the moon is only partially dimmed. A penumbral eclipse is not very impressive.

While there are usually no more than one or two lunar eclipses each year, it is not difficult to see one. We need only be on the dark side of Earth when the moon passes through Earth's shadow. That is, the eclipse must occur between sunset and sunrise at our location. Table 3-1 will allow you to determine which upcoming total and partial eclipses will be visible from your location.

Solar Eclipses   We who live on planet Earth can see a phenomenon that is not visible from most planets. It happens that the sun is 400 times larger than the moon and, on the average, 390 times farther away, so the sun and moon have nearly equal angular diameters—0.5°. (See Box 3-2.) Thus the moon is just the right size to cover the bright disk of the sun and cause a **solar eclipse.** If the moon covers the entire disk of the sun, we see a total eclipse. If it covers only part of the sun, we see a partial eclipse.

Whether we see a total or partial eclipse depends on whether we are in the umbra or the penumbra of the moon's shadow (Figure 3-16). The umbra of the moon's shadow barely reaches Earth and casts a small circular shadow never larger than 270 km (168 miles) in diameter. If we are standing in that umbral spot, we are in total shadow, unable to see any part of the sun's surface, and the eclipse is total. But if we are located outside the umbra, in the penumbra, we see part of the sun peeking around the edge of the moon and the eclipse is partial. Of course, if we are outside the penumbra, we see no eclipse at all.

Figure 3-12 During a total lunar eclipse, the moon's orbit carries it through the penumbra and completely into the umbra. During a partial lunar eclipse, the moon does not completely enter the umbra. Compare the cross section of Earth's shadow with Figure 3-13.

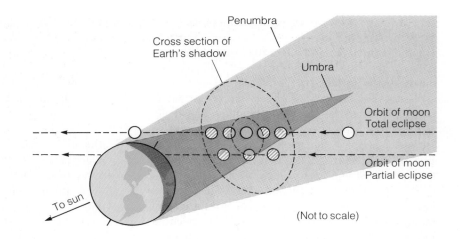

Penumbra

Cross section of Earth's shadow

Umbra

Orbit of moon
Total eclipse

Orbit of moon
Partial eclipse

To sun

(Not to scale)

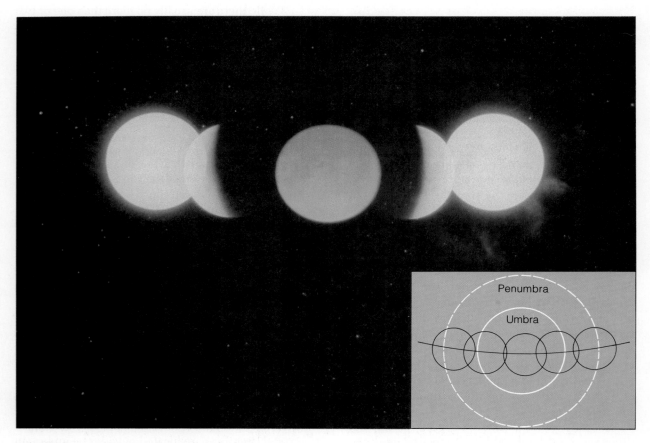

Figure 3-13 This multiple-exposure photo of a lunar eclipse spans 5 hours and shows the moon passing through Earth's umbra and penumbra (inset) from the right. The totally eclipsed moon (center) was 10,000 times dimmer than the full moon, so the exposure was lengthened to record the fainter image. Photographic effects make the moon's orbit appear curved here. Compare with Figure 3-12. (© 1982 by Dr. Jack B. Marling)

Figure 3-14 During a total lunar eclipse, the moon turns coppery red as sunlight, refracted by Earth's atmosphere, illuminates the moon in a red, sunset glow. An astronaut on the moon during such an eclipse would see this red light coming from a circular sunset completely encircling Earth. (Celestron International)

Because of the orbital motion of the moon and the rotation of Earth, the moon's shadow sweeps rapidly across Earth in a long, narrow path of totality. If we want to see a total solar eclipse, we must be in the path of totality. When the umbra of the moon's shadow sweeps over us, we see one of the most dramatic sights in the sky, the totally eclipsed sun.

The eclipse begins as the moon slowly crosses in front of the sun. It takes about an hour for the moon to cover the solar disk, but as the last sliver of sun disappears, dark falls in a few seconds. Automatic street lights come on, drivers of cars turn on their headlights, and birds go to roost. The sky becomes so dark we can even see the brighter stars.

The darkness lasts only a few minutes because the umbra is never more than 270 km (168 miles) in diameter and sweeps across Earth's surface at over 1600 km/hr (1000 mph). The sun cannot remain totally eclipsed for more than 7.5 minutes, and the average period of totality lasts only 2 or 3 minutes.

When the moon covers the bright surface of the sun, called the **photosphere,** we can see the bright gases of the **chromosphere** just above the photosphere, and the **corona,** the sun's faint outer atmosphere (Figure 3-17). The corona is a low-density, hot gas that glows with a pale white color. Streamers caused by the solar magnetic field streak the corona. The chromosphere is often marked by eruptions on the solar surface called **prominences** and is bright pink. The corona, chromosphere, and prominences are visible only when the brilliant photosphere is covered. As soon as part of the photosphere reappears, the fainter corona, chromosphere, and prominences vanish in the glare, and

TABLE 3-1
Total and Partial Eclipses of the Moon 1993–2010

Date	Time* of Mideclipse (GMT)	Length of Totality (Min)	Length of Eclipse (Hr: Min)
1993 June 4	13:02	96	3:38
1993 Nov. 29	6:26	46	3:30
1994 May 25	3:32	Partial	1:44
1995 Apr. 15	12:19	Partial	1:12
1996 Apr. 4	0:11	86	3:36
1996 Sept. 27	2:55	70	3:22
1997 Mar. 24	4:41	Partial	3:22
1997 Sept. 16	18:47	62	3:16
1999 July 28	11:34	Partial	2:22
2000 Jan. 21	4:45	76	3:22
2000 July 16	13:57	106	3:56
2001 Jan. 9	20:22	60	3:16
2001 July 5	14:57	Partial	2:38
2003 May 16	3:41	52	3:14
2003 Nov. 9	1:20	22	3:30
2004 May 4	20:32	76	3:22
2004 Oct. 28	3:05	80	3:38
2005 Oct. 17	12:04	Partial	0:56
2006 Sept. 7	18:52	Partial	1:30
2007 Mar. 3	23:22	74	3:40
2007 Aug. 28	10:38	90	3:32
2008 Feb. 21	3:27	50	3:24
2008 Aug. 16	21:11	Partial	3:08
2009 Dec. 31	19:24	Partial	1:00
2010 June 26	11:40	Partial	2:42
2010 Dec. 21	8:18	72	3:28

*Times are Greenwich Mean Time. Subtract 5 hours for Eastern Standard Time, 6 hours for Central Standard Time, 7 hours for Mountain Standard Time, and 8 hours for Pacific Standard Time. From your time zone, lunar eclipses that occur between sunset and sunrise will be visible, and those at midnight will be best placed.

totality is over. The moon moves on in its orbit, and in an hour the sun is completely visible again.

Just as totality begins or ends, a small part of the photosphere can peek out from behind the moon through a valley at the edge of the lunar disk. Although it is intensely bright, such a small part of the photosphere does not completely drown out the fainter corona, which forms a silvery ring of light with the brilliant spot of photosphere gleaming like a diamond (Figure 3-18). This **diamond ring effect** is one of the most spectacular of astronomical sights, but it is not visible during every

The Small-Angle Formula

The angular diameter of an object is related to its linear diameter and its distance by the small-angle formula (Figure 3-15). *Linear diameter* is the distance between an object's opposite sides. The linear diameter of the moon, for instance, is 3476 km. The *angular diameter* of an object is the angle formed by two lines extending from opposite sides of the object and meeting at our eye. Clearly, the farther away an object is, the smaller its angular diameter.

In the small-angle formula, we always express angular diameter in seconds of arc, and we always use the same units for distance and linear diameter:

$$\frac{\text{angular diameter}}{206{,}265^*} = \frac{\text{linear diameter}}{\text{distance}}$$

We can use this formula to find any of these three quantities if we know the other two.

The number 206,265 is the number of seconds of arc in a radian. When we divide by 206,265, we convert the angle from seconds of arc into radians.

Example A: Suppose we see an automobile 1 km away and note that its angular diameter is about 13 minutes of arc. What is its linear diameter? *Solution:* Remember that 1 km is 1000 m and that 1 minute of arc equals 60 seconds of arc. Thus 13 minutes of arc equals 780 seconds of arc:

$$\frac{780}{206{,}265} = \frac{\text{linear diameter}}{1000}$$

So the linear diameter of the car is 780,000/206,265, which equals 3.8 m; it must be a limousine.

Example B: The moon has a linear diameter of 3476 km and is about 384,000 km away. What is its angular diameter? *Solution:* We can leave linear diameter and distance in kilometers, and find the angular diameter in seconds of arc:

$$\frac{\text{angular diameter}}{206{,}265} = \frac{3476}{384{,}000}$$

The angular diameter is 1870 seconds, or 31 minutes, of arc—about 0.5°.

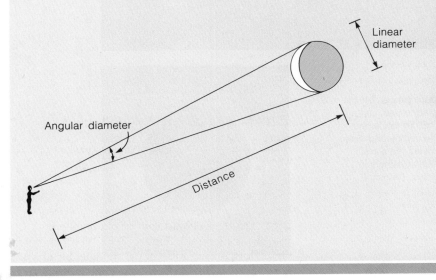

Figure 3-15 The small-angle formula relates angular diameter, linear diameter, and distance. Angular diameter is the angle formed by lines extending from our eye to opposite sides of the object—in this figure, the moon. Linear diameter and distance are typically measured in kilometers or meters.

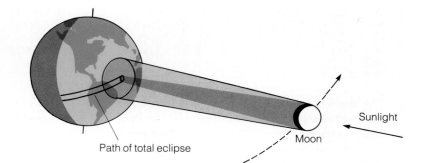

Figure 3-16 Observers in the path of totality see a total solar eclipse when the umbral shadow sweeps over them. Those in the penumbra see a partial eclipse.

Sunlight

Moon

Path of total eclipse

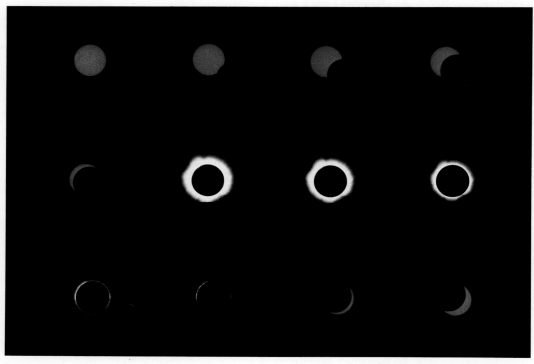

a

b

Figure 3-17 (a) A sequence of photographs (upper left to lower right) shows the stages of a total solar eclipse. Longer exposures were needed during totality to record the fainter corona. (Grundy Observatory) (b) The white corona and pink prominences are visible when the moon covers the sun's brilliant photosphere during totality. Note the streamers in the corona caused by the sun's magnetic field. (William P. Sterne, Jr.)

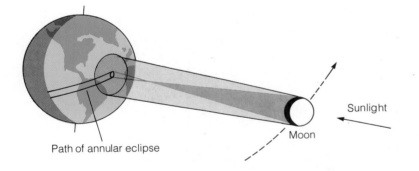

Figure 3-18 The diamond ring effect can sometimes occur momentarily at the beginning or end of totality if a small segment of the photosphere peeks out through a valley at the edge of the lunar disk. (National Optical Astronomy Observatory)

Figure 3-19 If the moon is near the farther part of its orbit, the umbral shadow does not reach Earth, resulting in an annular eclipse.

Sunlight

Moon

Path of annular eclipse

solar eclipse. Its occurrence depends on the exact orientation and motion of the moon.

Because the moon's orbit is slightly elliptical, its distance from Earth varies. **At perigee,** the closest point to Earth, the moon is 5.5 percent closer than average, and at **apogee,** the farthest point from Earth, it is 5.5 percent farther away. If the moon crosses between the sun and Earth while it is in the farther part of its orbit, its umbra does not reach all the way to Earth (Figure 3-19). If the umbra does not touch Earth, there can be no path of totality and no total eclipse. Under these circumstances, we see an **annular eclipse.**

During an annular eclipse, the moon looks slightly smaller because it is slightly farther away. It is too small to completely cover the sun, and we see the sun's bright photosphere around the edge of the moon in a brilliant ring, or *annulus* (Figure 3-20). Annular eclipses are less impressive than total solar eclipses because the bright annulus of the photosphere blinds us to the fainter corona, chromosphere, and prominences. An annular eclipse will be visible from the central United States on May 10, 1994 (Figure 3-21). A list of future total and annular eclipses is given in Table 3-2.

Figure 3-20 The annular eclipse of May 30, 1984. A bright ring of photosphere remains visible around the moon, and the corona and prominences are too faint to be visible. (Laurence Marschall)

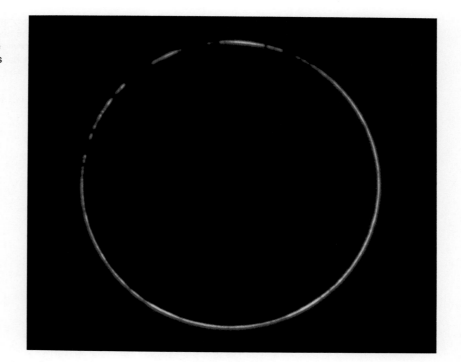

Figure 3-21 The annular eclipse of May 10, 1994. The moon's shadow will sweep across the central United States in the late morning, producing an annular eclipse as seen from within the central path. The eclipse will be partial as seen from the rest of the United States, Canada, and Mexico.

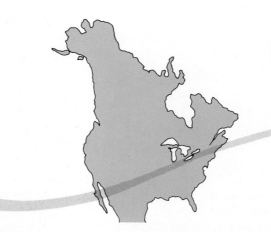

Precise prediction of eclipses requires detailed calculations, but we can make general predictions if we understand the cycles in which eclipses occur. Many ancient peoples recognized those cycles and were able to predict eclipses. (See Box 3-3.)

The cycles of eclipses, tides, lunar phases, and the seasons are immediate experiences for the inhabitants of Earth, but a less obvious cycle may be the most important of all. As described in the "Per-

spective" section that concludes this chapter, small periodic changes in the orbit of the earth may affect Earth's climate and cause the ice ages.

In this chapter, we have studied the cycles of the sun and moon, and in passing have seen our earth spinning on its axis and whirling around the sun. Knowledge of such movements is only a few centuries old. In the next chapter, we will tell the story of how the people of Earth discovered that they live on a moving planet.

Total and Annular Eclipses
of the Sun, 1993*–2010

Date	Total/ Annular (T/A)	Time of Mideclipse** (GMT)	Maximum Length of Total or Annular Phase (Min:Sec)	Area of Visibility
1994 May 10	A	17^h	6:14	Pacific, N. America, Atlantic
1994 Nov. 3	T	14^h	4:24	S. America, Atlantic
1995 Apr. 29	A	18^h	6:38	Pacific, S. America
1995 Oct. 24	T	5^h	2:10	Asia, Borneo, Pacific
1997 Mar. 9	T	1^h	2:50	Siberia
1998 Feb. 26	T	17^h	4:08	Pacific, N. of S. America, Atlantic
1998 Aug. 22	A	2^h	3:14	Sumatra, Borneo, Pacific
1999 Feb. 16	A	7^h	1:19	Indian Ocean, Australia
1999 Aug. 11	T	11^h	2:23	Atlantic, Europe, S.E. and S. Asia
2001 June 21	T	12^h	4:56	Atlantic, S. Africa, Madagascar
2001 Dec. 14	A	21^h	3:54	Pacific, Central America
2002 June 10	A	24^h	1:13	Pacific
2002 Dec. 4	T	8^h	2:04	S. Africa, Indian Ocean, Australia
2003 May 31	A	4^h	3:37	Iceland, Arctic
2003 Nov. 23	T	23^h	1:57	Antarctica
2005 Apr. 8	AT	21^h	0:42	Pacific, N. of S. America
2005 Oct. 3	A	11^h	4:32	Atlantic, Spain, Africa
2006 Mar. 29	T	10^h	4:07	Atlantic, Africa, Turkey
2006 Sept. 22	A	12^h	7:09	N.E. of S. America, Atlantic
2008 Feb. 7	A	4^h	2:14	S. Pacific, Antarctica
2008 Aug. 1	T	10^h	2:28	Canada, Arctic, Siberia
2009 Jan. 26	A	8^h	7:56	S. Atlantic, Indian Ocean
2009 July 22	T	3^h	6:40	Asia, Pacific
2010 Jan. 15	A	7^h	11:10	Africa, Indian Ocean
2010 July 11	T	20^h	5:20	Pacific, S. America

*There are no total or annular solar eclipses in 1993.
**Times are Greenwich Mean Time. Subtract 5 hours for Eastern Standard Time, 6 hours for Central Standard Time, 7 hours for Mountain Standard Time, and 8 hours for Pacific Standard Time.
^{h}hours.

Eclipse Seasons

We see a solar eclipse when the moon passes between Earth and the sun, that is, when the lunar phase is new moon. We see a lunar eclipse at full moon. However, we don't see eclipses at every new moon and every full moon. Why not?

Figure 3-22 is a scale drawing of the umbral shadows of Earth and the moon. Notice that they are extremely long and narrow. Earth, moon, and sun must line up almost exactly or the shadows miss their mark and there is no eclipse.

To be eclipsed, the moon must enter Earth's shadow. However, because its orbit is tipped (Figure 3-23a), the moon often misses the shadow, passing north or south of it, and no lunar eclipse occurs. Also, in order to produce a solar eclipse, the moon's shadow must sweep over Earth. The inclination of the moon's orbit, however, means that it often reaches new moon with its shadow passing north or south of Earth, and there is no solar eclipse.

For an eclipse to occur, the moon must reach full or new moon at the same time it passes through the plane of Earth's orbit; otherwise, the shadows miss (Figure 3-23b). The points

where it passes through the plane of Earth's orbit are called the **nodes** of the moon's orbit, and the line connecting these is called the line of nodes. Twice a year this line of nodes points toward the sun, and for a few weeks eclipses are possible at new moon and full moon (Figure 3-23c). These intervals when eclipses are possible are called eclipse seasons, and they occur about six months apart.

If the moon's orbit were fixed in space, the eclipse seasons would always occur at the same time each year. The moon's orbit precesses slowly, however, because of the gravitational pull of the sun on the moon, and the precession slowly changes the direction of the line of nodes. The line turns westward, making one complete rotation in 18.61 years. As a result, the eclipse seasons occur about 3 weeks earlier each year. The motion of the line of nodes, combined with the periodicity of the lunar phases, means that every 6585.3 days the eclipse seasons start over and the same pattern of eclipses repeats (6585.3 days equals 18 years 11.3 days or 18 years 10.3 days depending on the number of leap days during the interval). Because this cycle, termed the **Saros cycle**, con-

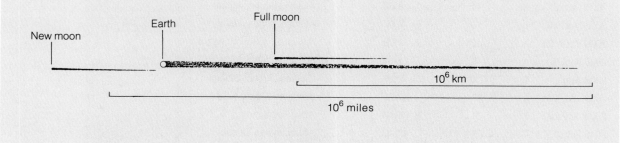

New moon   Earth   Full moon

10^6 km

10^6 miles

Figure 3-22 This scale drawing of the umbral shadows of Earth and the moon shows how easy it is for the shadows to miss their mark at full moon and new moon. Diameters of Earth and the moon are exaggerated by a factor of 2 for clarity.

tains about one-third of a day more than an integer number of days, an eclipse visible in North America will recur after 18 years 11.3 days, about one-third of the way around the world, in this case in the eastern Pacific. Many ancient peoples recognized the Saros cycle from their records of previous eclipses and were able to predict when eclipses would occur, even though they did not understand what the sun and moon were or what alignments produced eclipses.

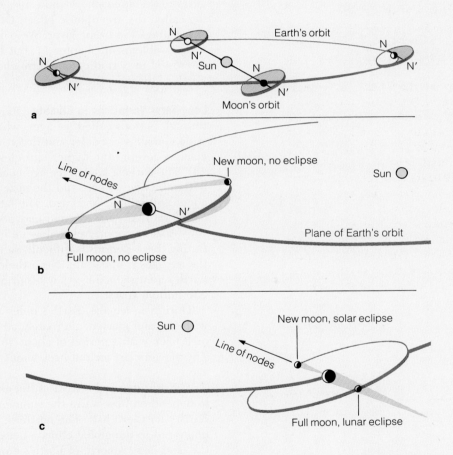

Figure 3-23 (a) The moon's orbit is tipped about 5° to Earth's orbit. The nodes N and N' are the points where the moon passes through the plane of Earth's orbit. (b) If the line of nodes does not point at the sun, the shadows miss and there are no eclipses at new moon and full moon. (c) At those parts of Earth's orbit where the line of nodes points toward the sun, eclipses are possible at new moon and full moon.

Climate and Ice Age

Earth's Climate   Weather is what happens today, but climate is the average of what happens over tens of years. Occasional hot summers, floods, droughts, and other variations in weather are probably only random, short-term changes. Changes in the climate are slow.

Some changes are caused by humans. We add gases to the atmosphere and cause a slow warming called the greenhouse effect. Other waste gases attack the ozone layer. We will discuss these problems in Chapter 17, but here we are interested in periodic, long-term changes in Earth's climate—the ice ages.

Long-Term Variations in Climate   Earth has gone through periods called ice ages, when the worldwide climate was cooler and dryer, and thick glaciers covered the higher latitudes (Figure 3-24). The earliest known ice age occurred roughly 570 million years ago, and the next occurred about 280 million years ago. The latest began only 3 million years ago and may not have ended yet—that is, we may even now be in an ice age. Dating these periods is difficult, so the timing of the ice ages is uncertain. Nevertheless, some Earth scientists believe they occur about every 250 million years.

During an ice age, Earth's poles are covered with ice, and glaciers alternately advance and melt back with a period of about 40,000 years. The glaciers last melted back about 10,000 years ago.

The advance and retreat of the glaciers seem to depend on how warm the summers are in Earth's northern half. The northern half is more important to the global climate because most of the land mass is north of Earth's equator. If the summers are not warm enough, the previous winter's snow and ice fail to melt completely and accumulate year after year, building into advancing glaciers. However, if the summers are warm enough to melt the previous winter's snow and ice, the ice sheets do not grow.

In 1920, Yugoslavian meteorologist Milutin Milankovitch proposed that small changes in Earth's rotation and revolution affect the heat balance and modify climate. Not until the 1960s, when studies of ocean sediment showed that past climatic variations fit the predictions of the **Milankovitch hypothesis,** was the theory widely considered as an explanation of long-term changes in Earth's climate.

The Changing Shape of Earth's Orbit   Earth has an orbit that is slightly elliptical and passes perihelion about January 3 each year, but the slight increase in heat that we get by being 1.7 percent closer to the sun is wiped out by the seasonal temperature variation from the inclination of Earth's axis. January is a cold month for Earth's northern half, even though Earth is slightly closer to the sun.

By studying the motion of Earth and the other planets, astronomers have discovered that the shape of Earth's orbit changes with a period of about 93,000 years. When the orbit gets more elliptical, the variation in the distance from Earth to sun is more than 1.7 percent, which may be enough to make our winters milder and summers cooler. If the summers in the northern half of Earth are cooler, ice and snow accumulate and glaciers advance. When the orbit becomes more circular, summers are warmer and ice sheets melt back.

This sounds like a good theory, but glacial periods occur every 40,000 years, not every 93,000 years. There must be more to the ice ages than just variation in the shape of Earth's orbit.

Precession   The precession of Earth's axis is slowly changing the seasons. Earth is now tipped toward the sun in June, producing summer in the northern half. But in 13,000 years, precession will have tilted Earth the other way, away from the sun in June, producing winter in the northern half of the planet. We will of course adjust

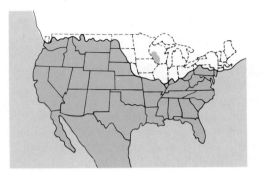

Figure 3-24   Only 20,000 years ago much of the northern United States was covered by glacier ice. The solid line indicates the farthest advance of the ice.

our calendar to move the months with the seasons, keeping June a summer month. The important point is not the month but the place in Earth's orbit where winter occurs.

It happens that winter in Earth's northern half now occurs when Earth is near perihelion. Because it is 1.7 percent closer to the sun, our winters are very slightly less severe than they would be if Earth's orbit were circular.

In 13,000 years, however, precession will have moved winter to the other side of Earth's orbit, we will be 1.7 percent farther from the sun, and winters will be slightly colder. Summer will occur near perihelion, we will be 1.7 percent closer to the sun, and summers will be warmer. If summer in Earth's northern half becomes warmer, the ice should melt faster, and the glaciers should recede. Thus we might expect Earth's climate to change with the same period as precession, about 26,000 years.

It is not that simple, however. The effects of precession combine with the effects of the changing shape of Earth's orbit. If the orbit becomes almost circular, then it doesn't matter when perihelion occurs because Earth won't be significantly closer to the sun. But if the orbit becomes more elliptical, then the time of perihelion is

Perspective (continued)

important. The fact that the two variations work together with different periods complicates the problem. A third variation in Earth's motion adds another complication.

The Inclination of Earth's Axis Not only can astronomers study the changing shape of Earth's orbit and its axial precession, they can also study its changing inclination. Earth has not always been tipped at an angle of 23.5°: The inclination varies with a period of about 41,000 years from 22° to 28°. Since the ecliptic is the projection of Earth's orbit onto the sky, a change in Earth's axial tilt changes the inclination of the ecliptic to the celestial equator (see Figure 3-4). If this inclination becomes smaller, the sun does not travel as far north in the sky during summer, producing cooler summers and favoring the accumulation of ice sheets.

It is tempting to point to the 40,000-year glacial cycle and identify it with the 41,000-year variation in Earth's inclination, but we would probably be wrong to do so. We must remember that there are at least three factors working to change the climate, each with a different period. When the three processes work together to produce cool summers, ice sheets may accumulate, producing a glacial period. But the pattern of the advance and retreat of the glaciers is very complex because of the three effects at work.

The Milankovitch theory has not been widely studied until recently, when teams studying microscopic fossils in ocean floor sediments found evidence of periodic climate changes that seem to fit the theory. Also, the Viking spacecraft that visited Mars in the late 1970s returned data suggesting that the Martian climate varies with small, periodic changes in the planet's motions (see Chapter 17). Most Earth scientists now agree that variations in Earth's rotation and revolution do affect the climate, but they do not agree on the importance of the effect.

SUMMARY

Because Earth orbits the sun, the sun appears to move eastward along the ecliptic through the constellations. Because the ecliptic is tipped 23.5° to the celestial equator, the sun spends half the year in the northern celestial hemisphere and half the year in the southern celestial hemisphere, producing the seasons. The seasons are reversed south of Earth's equator.

Because we see the moon by reflected sunlight, its shape appears to change as it orbits Earth. The lunar phases wax from new moon to first quarter to full moon, and wane from full moon to third quarter to new moon. A complete cycle of lunar phases takes 29.53 days.

The moon's gravitational field exerts tidal forces on Earth that pull the ocean waters up into two bulges, one on the side of Earth facing the moon and the other on the side away from the moon. As the rotating Earth carries the continents through these bulges of deeper water, the tides ebb and flow. Friction with the seabeds slows Earth's rotation, and the gravitational force the bulges exert on the moon force its orbit to grow larger.

When the moon passes through Earth's shadow, sunlight is cut off and the moon darkens in a lunar eclipse. If the moon only grazes the shadow, the eclipse is partial, or penumbral, and not total.

If the moon passes directly between the sun and Earth, it produces a total solar eclipse. During such an eclipse, the bright photosphere of the sun is covered, and the fainter corona, chromosphere, and prominences become visible. An observer outside the path of totality sees a partial eclipse. If the moon is in the farther part of its orbit, it does not cover the photosphere completely, resulting in an annular eclipse.

The motion of Earth changes in ways that can affect the climate. Changes in orbital shape, in precession, and in axial tilt can alter the planet's heat balance and may be responsible for the ice ages and glacial periods.

NEW TERMS

rotation

revolution

ecliptic

precession

vernal equinox

autumnal equinox

summer solstice

winter solstice

perihelion

aphelion

morning and evening
    stars

sidereal period

synodic period

spring and neap tides

lunar eclipse

umbra

penumbra

solar eclipse

photosphere

chromosphere

corona

prominence

diamond ring effect

perigee

apogee

annular eclipse

node

Saros cycle

Milankovitch hypothesis

REVIEW QUESTIONS

1. If Earth did not rotate, could we still define the ecliptic? Why or why not?

2. What would our seasons be like if the earth were tipped 35° instead of 23.5°? What would they be like if Earth's axis were perpendicular to its orbit?

3. Why isn't the summer solstice the hottest day of the year?

4. Why don't the planets move exactly along the ecliptic?

5. How does the moon slow Earth's rotation, and how does Earth slow the moon's revolution?

6. Do the phases of the moon look the same from every place on Earth, or is the moon full at different times in different locations?

7. Why isn't there an eclipse at every new moon and every full moon?

8. Why have more people seen a total lunar eclipse than have seen a total solar eclipse?

9. Why should the eccentricity of Earth's orbit make winter in the northern hemisphere different from winter in the southern hemisphere?

10. How could small changes in the inclination of Earth's axis affect world climate?

DISCUSSION QUESTIONS

1. Could planets in other planetary systems have ecliptics? Could they have seasons?

2. If people on Earth saw a total solar eclipse, what would astronauts on the moon see?

PROBLEMS

1. Draw a diagram like Figure 3-5 and show the path of the sun across the sky at the vernal equinox.

2. If Earth is about 5 billion (5×10^9) years old, how many precessional cycles have occurred?

3. Identify the phases of the moon if on March 21 the moon were located at (a) the vernal equinox, (b) the autumnal equinox, (c) the summer solstice, (d) the winter solstice.

4. Identify the phases of the moon if at sunset the moon were (a) near the eastern horizon, (b) high in the south, (c) in the southeast, (d) in the southwest.

5. About how many days must elapse between first-quarter moon and third-quarter moon?

6. Draw a diagram showing Earth, the moon, and shadows during (a) a total solar eclipse, (b) a total lunar eclipse, (c) a partial lunar eclipse, (d) an annular eclipse.

7. Phobos, one of the moons of Mars, is 20 km in diameter and orbits 5982 km above the surface of the planet. What is the angular diameter of Phobos as seen from Mars? (HINT: See Box 3-2.)

8. A total eclipse of the sun was visible from Canada on July 10, 1972. When will this eclipse occur again? From what part of the earth will it be total?

9. When will the eclipse described in Problem 8 next be total as seen from Canada?

OBSERVATIONAL ACTIVITY:
THE ORBIT OF THE MOON

Each clear night, locate the moon and plot its course against the background of stars in the appropriate star chart at the end of this book. Label the location of the moon on nights of first-quarter, full, or third-quarter moon.

Over a period of a month or two, your observations will mark out the orbit of the moon in the sky.

Sketch a smooth line through your observations to show that orbit. Although the moon's orbit gradually precesses with a period of 19 years, it will not move appreciably during a semester.

You will notice that the moon reaches first-quarter, full, and third-quarter phase farther east each month. This is because the sun is moving eastward along the ecliptic, and the phases of the moon are related to the location of the sun.

RECOMMENDED READING

BOK, BART J., and LAWRENCE JEROME. *Objections to Astrology.* Buffalo, N.Y.: Prometheus Books, 1975.

BRACHER, KATHERINE. "Getting to the 1860 Solar Eclipse." *Sky and Telescope* 61 (February 1981), p. 120.

EVANS, D. L., and H. J. FREELAND. "Variations in the Earth's Orbit: Pacemaker of the Ice Ages?" *Science* 198 (November 4, 1977), p. 528.

GRIBBIN, J. "Why Does Earth's Climate Change?" *Astronomy* 6 (February 1978), p. 18.

HAWKINS, G. S. *Stonehenge Decoded.* Garden City, N.Y.: Doubleday, 1965.

HAYS, D. D., J. IMBRIE, and J. J. SHACKLETON. "Variations in the Earth's Orbit: Pacemaker of the Ice Ages." *Science* 194 (December 10, 1976), p. 1121.

IMBRIE, J., and K. P. IMBRIE. *Solving the Mystery.* Short Hills, N.J.: Enslow, 1979.

KELLY, IVAN. "The Scientific Case Against Astrology." *Mercury* 9 (November/December 1980), p. 135.

MARSCHALL, LAURENCE A. "Shadow Bands—Solar Eclipse Phantoms." *Sky and Telescope* 67 (February 1984), p. 116.

MEEUS, JEAN. "Solar Eclipse Diary: 1985–1995." *Sky and Telescope* 68 (October 1984), p. 296.

MENZEL, D. H. *A Field Guide to the Stars and Planets.* Boston: Houghton Mifflin, 1964.

SHIPMAN, H. L. "Megaliths and the Moon: Eclipse Prediction in the Stonehenge Era." *The Griffith Observer* 38 (February 1974), p. 7.

STEPHENSON, F. RICHARD. "Historical Eclipses." *Scientific American* 247 (October 1982), p. 170.

WHITEMAN, MARK. "Eclipse Predictions on Your Computer." *Astronomy* 14 (November 1986), p. 67.

See also current issues of *Sky and Telescope* and *Astronomy* for information about coming eclipses.

FURTHER EXPLORATION

VOYAGER, the Interactive Desktop Planetarium, can be used for further exploration of the topics in this chapter. This chapter corresponds with the following projects in *Voyages Through Space and Time:* Project 3, Constellations and Planets; Project 5, The Inner Planets; Project 6, The Outer Planets; Project 7, Precession; Project 8, The Earth's Seasons; Project 9, Types of Time; Project 10, The Equation of Time; Project 11, Sidereal Time; Project 12, Lunar Phases; Project 13, Conjunction and Occultation; Project 14, Lunar Eclipses; Project 15, Solar Eclipses.

CHAPTER
4

The Origin of Modern Astronomy

The story of modern astronomy begins with the death of the great Polish astronomer Nicolaus Copernicus in May 1543 and the almost simultaneous publication of his model of the universe. That model revolutionized not only astronomy but all science, and it inspired a new consideration of our place in nature. We will trace this story over the 99 years from 1543 and Copernicus' death to 1642, the year that also saw the death of another great astronomer, Galileo, and the birth of one of the greatest scientists in history, Isaac Newton.

Two themes run through our story: the place of the earth in the cosmos and the nature of planetary motion. The debate over the place of the earth involved theological questions and eventually brought Galileo Galilei before the Inquisition. The nature of planetary motion involved the most basic physics of motion and eventually led Isaac Newton to understand gravitation and motion. Thus the Copernican model of the universe led to the birth of modern science.

The passions of astronomy are no less profound because they are not noisy.

John Steinbeck
The Short Reign of Pippin IV

To understand why the Copernican model was so important, we must backtrack to ancient Greece and meet the two great authorities of ancient astronomy, Aristotle and Ptolemy. It was the Greeks who first struggled to understand the place of the earth and the nature of planetary motion.

4-1 PRE-COPERNICAN ASTRONOMY

Aristotle (Figure 4-1), a Greek philosopher who lived from 384 to 322 B.C., taught and wrote on philosophy, history, politics, ethics, poetry, drama, and other subjects. Because of his sensitivity and insight, he became the great authority of antiquity.

Figure 4-1 Aristotle (384–322 B.C.), honored on this Greek stamp, wrote on such a wide variety of subjects and with such deep insight that he became the great authority on all matters of learning. His opinions on the nature of Earth and the sky were widely accepted for almost two millennia.

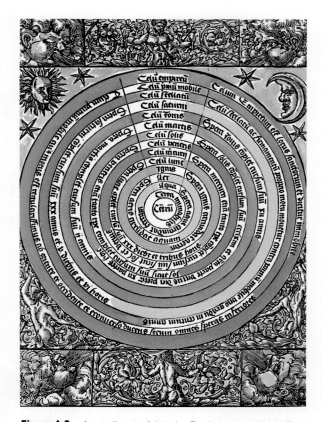

Figure 4-2 According to Aristotle, Earth is motionless (*Terra immobilis*) at the center of the universe. Earth is surrounded by spheres of water, air, and fire (*ignus*), above which lie spheres carrying the celestial bodies beginning with the moon (*lune*) in the lowest celestial sphere. This woodcut is from C. Cornipolitanus' book *Chronographia* of 1537. (From the Granger Collection, New York)

Much of what Aristotle wrote about scientific subjects such as physics, astronomy, biology, and so on, was incorrect. But we must remember that Aristotle was not a scientist. He was attempting to understand the universe by constructing a system of formal philosophy. Modern scientists do not expect physical objects such as the sun and moon to have intrinsic properties such as purity and spiritual perfection, but the methods of modern science, depending so heavily on unbiased observation and supporting evidence, were invented 2000 years after Aristotle. To the extent that his model of the universe explained Earth and the heavens as part of his intellectual system of philosophy, his theory was a success and was accepted for almost 2000 years.

Aristotle believed the universe consisted of Earth, corrupt and changeable, and the heavens, perfect and immutable. He knew Earth was spherical, but he concluded, for philosophical reasons, that it sat immobile at the center of the universe (Figure 4-2). Subsequent astronomers, following Aristotle's philosophy, argued that the earth could not move because they saw no parallax in the positions of the stars.

Parallax is the change in the apparent position of an object due to a change in the location of the observer. It is actually an old friend, though you may not have recognized its name, for you use parallax to judge the distance to things. To see how this works, close your right eye and use your thumb, held at arm's length, to cover some distant object, a building perhaps. Now look with your right eye. Your thumb seems to move to the left, uncovering the building (Figure 4-3). This apparent shift is parallax, and your brain uses it to estimate distances to objects around you.

For centuries astronomers reasoned that if Earth moved around the sun, we would view the stars from different places in Earth's orbit at different

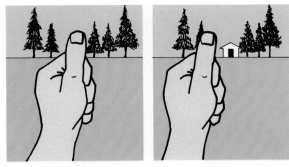

Seen by left eye   Seen by right eye

Figure 4-3 To demonstrate parallax, close one eye and cover a distant object with your thumb held at arm's length. Look with the other eye and your thumb appears to have shifted position.

times of the year. Then parallax would distort the shapes of the constellations; they would look largest when Earth was nearest that part of the sky. Because they did not see this—that is, because they saw no parallax—they concluded that Earth did not move. Actually, they saw no parallax because the stars are much farther away than they supposed, and the parallax is much too small to be visible to the naked eye.

In his astronomy, Aristotle adopted a belief originated by Pythagoras, a Greek mathematician and philosopher who had lived about 150 years earlier. Pythagoras taught that the seven heavenly bodies—five planets, the sun, and moon—were carried by separate crystalline spheres whose various rotations gave rise to the motions of the sun, moon, and planets. Above these spheres was the sphere of fixed stars. Aristotle thought Earth was immobile, and consequently he believed all of the spheres swirled westward around Earth each day. Thus most Greek philosophers viewed the sky as a perfect, heavenly machine not many times larger than Earth itself.

One reason for the survival of Aristotle's views was the work of the mathematician Claudius Ptolemaeus, usually referred to as Ptolemy (Figure 4-4). His nationality and birth date are unknown, but he lived and worked in the Greek settlement at Alexandria in about A.D. 140. There he studied mathematics and astronomy and developed a mathemati-

Figure 4-4 The muse of astronomy guides Ptolemy (c. A.D. 140) in his study of the heavens. (Courtesy Owen Gingerich and the Houghton Library)

cal model of the universe based on the teachings of Aristotle.

The Ptolemaic model, like Aristotle's, was a **geocentric** (Earth-centered) **universe.** In addition, it incorporated the Greek belief that the heavenly bodies moved perfectly. Since the only perfect motion is uniform motion and the only perfect curve is a circle, Ptolemy assumed the planets moved with **uniform circular motion,** a concept first proposed by Plato (c. 427–347 B.C.). But simple, circular paths centered on Earth do not account for the motions of the planets in the sky. The planets sometimes move faster and sometimes slower, and occasionally they appear to slow to a stop and move backward (westward) for a time in **retrograde motion** (Figure 4-5).

To describe the complicated planetary motions and yet preserve a geocentric model with uniform circular motion, Ptolemy adopted a system of wheels within wheels. The planet moved in a small

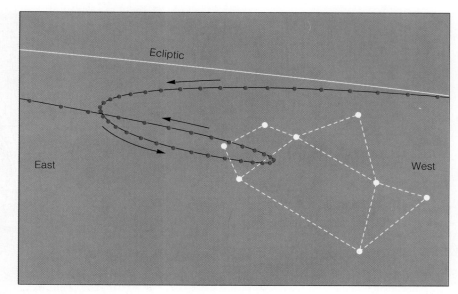

Figure 4-5 The word *planet* comes from the Greek for "wanderer," referring to the motion of the planets against the background of fixed stars. Here the motion of Mars along the ecliptic near the teapot shape of Sagittarius is shown at 4-day intervals. Though the planet usually moves eastward, it sometimes slows to a stop and moves westward in retrograde motion.

Ecliptic

East

West

circle called an **epicycle,** and the center of the epicycle moved along a larger circle around Earth called a **deferent.** By adjusting the size of the circles and the rate of their motion, Ptolemy could account for most planetary movement. But as a final adjustment, he placed Earth off center in the deferent circle and specified that the center of the epicycle would appear to move at constant speed only if viewed from a point called the **equant,** located on the other side of the deferent's center (Figure 4-6). Thus, by using a few dozen circles of various sizes rotating at various rates, Ptolemy's system reproduced the motions of the planets (Figure 4-7).

About A.D. 140, Ptolemy included this work in a book that he called *Mathematical Syntaxis,* but that is now known as *Almagest.* With Ptolemy's death, the classical astronomy of Greece ended forever. The breath of civilization passed to invading Arabs, who dominated most of the known world for nearly 1000 years. Arabian astronomers translated, studied, and preserved many classical manuscripts, and in Arabic, Ptolemy's book became *Al Magisti* (Greatest). Beginning in the 1200s, European Christians drove the Arabs out of Spain and recovered their classical heritage through Arabic translations. In Latin, Ptolemy's book became *Almagestum* and thus our modern *Almagest.* For 1000 years, Arab astronomers studied and preserved

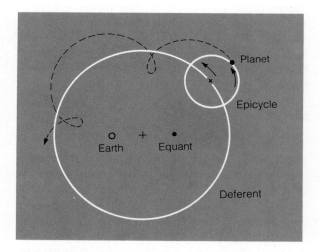

Figure 4-6 Ptolemy accounted for a planet's motion by placing it on a small circle (epicycle) that moved along a larger circle (deferent). Viewed from the equant, the center of the epicycle would have moved at constant speed.

Ptolemy's work, but they made no significant improvements in his theory.

At first the Ptolemaic system predicted the positions of the planets with fair accuracy, but as centuries passed, errors accumulated and Arabian and later European astronomers had to update the system, computing new constants and adjusting the epicycles. In the middle of the 13th century, a

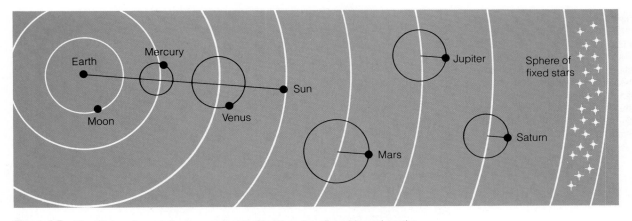

Figure 4-7 The Ptolemaic system was geocentric (Earth-centered) and based on the uniform circular motion of epicycles. Notice that the centers of the epicycles of Mercury and Venus always lie on the Earth–sun line.

team of astronomers supported by King Alfonso X of Castile worked for 10 years revising the Ptolemaic system and publishing the result as the *Alfonsine Tables*. It was the last great adjustment of the Ptolemaic system.

4-2 COPERNICUS

Nicolaus Copernicus was born in 1473 (Figure 4-8) in what is now Poland. In 1543, the year of his death, the Ptolemaic system, in spite of many revisions, was still a poor predictor of planetary positions. Yet because of the authority of Aristotle, it was the officially accepted theory of the universe.

The Catholic church had adopted the astronomy of Aristotle as part of church dogma. According to the Aristotelian universe, the most perfect region was in the heavens, and the most imperfect at the center of the earth. This matched perfectly the Christian geometry of heaven and hell, so anyone who criticized the Ptolemaic system risked a charge of heresy.

Throughout his life, Copernicus had been associated with the church. His uncle, by whom he was raised and educated, was an important bishop in Poland, and after studying canon law and medicine

in some of the major universities in Europe, Copernicus became a canon of the church at the age of 24. He served as secretary and personal physician to his powerful uncle for 15 years. When his uncle died, Copernicus went to live in quarters adjoining the cathedral in Frauenburg. Because of this long association with the church and his fear of persecution, he hesitated to publish his revolutionary ideas in astronomy.

In fact, his ideas were already being discussed long before the publication of his book. His interest in astronomy had begun during his college days, and he apparently doubted the Ptolemaic system even then. About 1507, at the age of 34, he wrote a short pamphlet that discussed the motion of the sky and outlined his hypothesis that the sun, not Earth, was the center of the universe and that Earth rotated on its axis and revolved around the sun. To avoid criticism and possible charges of heresy, he distributed his pamphlet in handwritten form to friends in the scientific community. By 1515 it was well known, and by 1530 church officials were asking about his work.

Copernicus worked on his book *De Revolutionibus Orbium Coelestium* over a period of many years and was still refining the calculations at the time of his final illness. Not only did he not want to publish an incomplete work, he was hesitant to publish his work until he was safely beyond the

Figure 4-8 Nicolaus Copernicus (1473–1543) proposed that the sun and not Earth was the center of the universe. These stamps were issued in 1973 to commemorate the 500th anniversary of his birth. Note the sun-centered model on the U.S. stamp and the pages from his book *De Revolutionibus* on the East German (DDR) stamp.

reach of his critics. This was a time of rebellion in the church—Martin Luther was speaking harshly about fundamental church teachings, and others, both scholars and scoundrels, were questioning the authority of the church. It is understandable that the Church did not welcome more criticism, even on matters as abstract as astronomy. Remember, also, that one possible penalty for heresy was to be burnt at the stake. Thus Copernicus did not approve the publication of his book until he realized he was dying. Although he did see printed proof pages, he never saw the book in its final bound form.

De Revolutionibus did not prove that the Ptolemaic model was wrong. It placed the sun at the center of the universe and thus reduced Earth to a mere planet orbiting the sun with the others. Copernicus quoted a number of criticisms of the Ptolemaic model while defending his own, but his arguments were not conclusive. His model did not predict the positions of the planets more accurately than Ptolemy's. Although Copernicus had rejected geocentrism, he had kept uniform circular motion.

Thus his model had to include small epicycles for each planet. The 16th-century *Prutenic Tables* (1551), a collection of tables that predicted planetary positions according to the Copernican system, were not significantly more accurate than the 13th-century *Alfonsine Tables.* Both could be in error by as much as 2°, which is four times the angular diameter of the moon.

The Copernican *model* was inaccurate. It included deferents and small epicycles, and thus did not precisely describe the motions of the planets. But the Copernican *hypothesis*, that the universe was **heliocentric** (sun-centered), was correct. (See Box 4-1.) Why that hypothesis gradually won acceptance in spite of the inaccuracy of the epicycles and deferents is a question historians still debate. There are probably a number of reasons, including the revolutionary temper of the times, but the most important factor may be the elegance of the idea. Placing the sun at the center of the universe produced a symmetry among the motions of the planets that was pleasing to the eye as well as to the intellect (Figure 4-9). No longer did Venus and

Hypothesis, Model, Theory, and Law

Even scientists misuse the words *hypothesis, theory, model,* and *law.* In this book, we will try to distinguish between them because they are key elements in the pursuit of scientific understanding.

A **hypothesis** is a single assertion or conjecture that must be tested. It could be true or false. "All Texans love chile," for example, is a hypothesis. Copernicus asserted that the universe was heliocentric. Thus his assertion was a hypothesis subject to testing.

A **model** is a description of a natural phenomenon and cannot be right or wrong. Thus we can use a model, such as the celestial sphere, even though it is not physically correct. The Copernican hypothesis was the central idea around which Copernicus attempted to build a model of the universe. To the extent that it described the motion of the planets, the model was successful.

A **theory** is a system of rules and principles that can be applied to a wide variety of circumstances. A theory may have begun as a hypothesis, but it was tested, expanded, and generalized. We would be correct to refer to the *theory* of gravity, but many scholars would argue that the Copernican hypothesis was not a theory. It lacked a precise description of orbital motion and mutual gravitation, features added later by Kepler and Newton.

A **natural law** is a theory that almost everyone accepts as true. Such a theory has been tested over and over, and applied to many situations. Thus natural laws are the most fundamental principles of scientific knowledge.

Scientists work by developing a hypothesis and testing it, often by building a model based on the hypothesis. If the model is a successful description of nature, the hypothesis is more likely to be true. As a hypothesis is tested repeatedly, expanded, and applied to many circumstances, we begin to refer to it as a theory. If the theory is so intensively tested that it is almost universally accepted as true, we refer to it as a natural law.

Mercury revolve around empty points located between Earth and the sun. Now they, like the rest of the planets, moved in orbits around the sun.

In addition, the Copernican model explained the retrograde (westward) motion of the planets in a straightforward way. Earth moves faster along its orbit than the planets that lie farther from the sun. Consequently, Earth periodically overtakes and passes these planets. Imagine that you are in a race car, driving rapidly along the inside lane of a circular race track. As you pass slower cars driving in the outer lanes, they fall behind, and if you did not know you were moving, it would seem that the cars in the outer lanes occasionally slowed to a stop and then backed up for a short interval. The same thing happens as Earth passes a planet such as Mars. Although Mars moves steadily along its orbit, as seen from Earth, it appears to slow to a stop and move westward (retrograde) as Earth passes it (Figure 4-10). Because the planetary orbits do not lie in precisely the same plane, a planet does not resume its eastward motion in precisely the same path it followed earlier. Consequently, it describes a loop whose shape depends on the angle between the orbital planes. This simple explanation of retrograde motion did not prove that Copernicus was correct, but it was a much simpler explanation than that provided by the Ptolemaic theory.

Although astronomers throughout Europe read and admired *De Revolutionibus,* they did not usually accept the Copernican hypothesis. The mathematics was elegant, and the astronomical observations and calculations were of tremendous value, but few astronomers believed that the sun actually

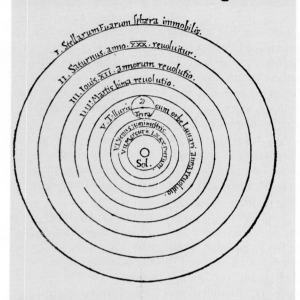

NICOLAI COPERNICI

net,in quo terram cum orbe lunari tanquam epicyclo contineri diximus.Quinto loco Venus nono menfe reducitur.   Sextum deniqʒ locum Mercurius tenet,octaginta dierum fpacio circū currens.In medio uero omnium refidet Sol.   Quis enim in hoc

pulcherrimo templo lampadem hanc in alio uel meliori loco po neret,quàm unde totum fimul pofsit illuminare?Siquidem non inepte quidam lucernam mundi,alij mentem,alij rectorem uo∘ cant.Trimegiftus uifibilem Deum,Sophoclis Electra intuentē omnia.Ita profecto tanquam in folio regali Sol refidens circum agentem gubernat Aftrorum familiam.Tellus quoque minime fraudatur lunari minifterio,fed ut Ariftoteles de animalibus ait, maximam Luna cum terra cognatione habet.Cōcipit interea à Sole terra,& impregnatur annuo partu.   Inuenimus igitur fub hac

Figure 4-9   The Copernican universe as reproduced in *De Revolutionibus*. Earth and all of the known planets moved in separate orbits centered on the sun, surrounded by an outer, immobile sphere of fixed stars. (Yerkes Observatory photograph)

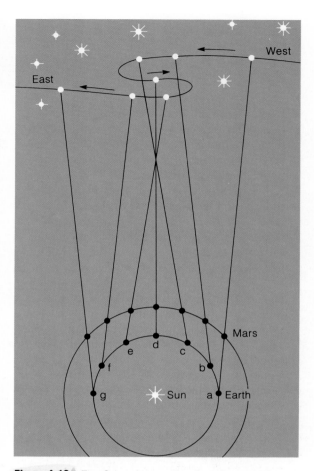

Figure 4-10   The Copernican explanation of retrograde motion. As Earth overtakes Mars (a–c), it appears to slow its eastward motion. As Earth passes Mars (d), it appears to move westward (retrograde). As Earth draws ahead of Mars (e–g), it resumes its eastward motion against the background stars. Compare with Figure 4-5. The positions of Earth and Mars are shown at equal intervals of 1 month.

was the center of the universe and that the earth moved. *De Revolutionibus* did not resolve the problems of the place of the earth and the nature of planetary motion.

4-3 TYCHO BRAHE

The great observational astronomer of our 99-year story was a Danish nobleman, Tycho Brahe (Figure 4-11), born December 14, 1546, only 3 years after the publication of *De Revolutionibus*. In histories of astronomy, the great astronomers are often referred to by their last name, but Tycho Brahe is usually called Tycho. Were he alive today, he would no doubt object to such familiarity from his obvious inferiors—he was well known for his vanity and lordly manners.

Tycho's college days were eventful. He was officially studying law with the expectation that he would enter Danish politics, but he made it clear to

Figure 4-11 Tycho Brahe (1546–1601), a Danish nobleman, established an observatory at Hveen and measured planetary positions with high accuracy. His artificial nose is suggested in this engraving.

his family that his real interests were in astronomy and mathematics. It was also during his college days that he became involved in a duel over some supposed insult and received a wound that disfigured his nose. For the rest of his life, he wore false noses made of gold and silver, and stuck on with wax (see Figure 4-11). The disfigurement probably did little to improve his disposition.

Tycho's first astronomical observations were made while he was a student. In 1563, Jupiter and Saturn passed very near each other in the sky, nearly merging into a single point on the night of August 24. Tycho found that the *Alphonsine Tables* were a full month in error and that the *Prutenic Tables* were in error by a number of days. These discrepancies dismayed Tycho and sparked his interest in the motions of the planets.

Then in 1572 a "new star" (now called Tycho's supernova) appeared in the sky, shining more brightly than Venus. Tycho carefully measured its position and concluded that it displayed no parallax. To understand the significance of this observation, we must note that the rotation of Earth carries us eastward and that the positions of celestial objects near Earth, such as the moon, change slightly through the night because the observer is in different positions. Tycho believed Earth was stationary and that the heavens were rotating westward, but that makes no difference. Objects moving

with the heavens near a stationary Earth should still have measurable parallax. That he detected no change in the position of the new star through the night proved it was farther away than the moon. Therefore, he concluded, it was a new star in the heavens. Since Aristotle and Ptolemy held that the heavens were perfect and unchanging, the new star led Tycho to question the Ptolemaic system. He summarized his results in a small book, *De Stella Nova* (The New Star), published in 1573.

The book attracted the attention of astronomers throughout Europe, and soon Tycho was summoned to the court of the Danish King Frederik II and offered funds to build an observatory on the island of Hveen just off the Danish coast. Tycho also received a steady source of income as landlord of a coastal district from which he collected rents. (He was not a popular landlord.) On Hveen, Tycho constructed a luxurious home with four towers especially equipped for astronomy, and populated it with servants, assistants, and a dwarf to act as jester. Soon Hveen was an international center of astronomical study (Figure 4-12).

Tycho's great contribution to the birth of modern astronomy was not theoretical. Because he could measure no parallax for the stars, he concluded that Earth had to be stationary, thus rejecting the Copernican hypothesis. However, he also rejected the Ptolemaic model because of its inaccurate predictions. Instead, he devised a complex model in which Earth was the immobile center of the universe around which the sun and moon moved. The remaining planets orbited the sun. This Tychonic model was popular for a short time.

The true value of Tycho's work was observational. Because he was able to devise new and better instruments, he was able to make highly accurate observations of the positions of the stars, sun, moon, and planets. Tycho had no telescopes—they were not invented until the next century—so his observations were made by the naked eye peering along sights on his large instruments (Figure 4-13). In spite of these limitations, he measured the positions of 777 stars to better than 4 minutes of arc and measured the positions of the sun, moon, and planets almost daily for the 20 years he stayed on Hveen.

Unhappily for Tycho, King Frederik II died in 1588, and his young son took the throne. Suddenly Tycho's temper, vanity, and noble presumptions

Figure 4-12 Tycho's palatial observatory and grounds at Hveen. (Courtesy Yerkes Observatory)

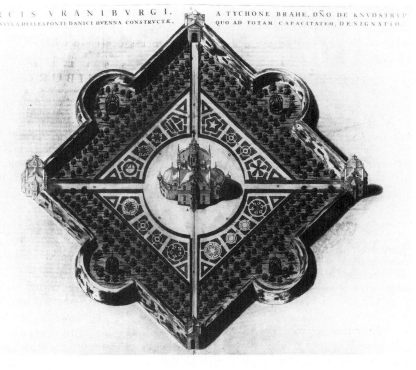

threw him out of favor. In 1596, taking most of his instruments and books of observations, he went to Prague, the capital of Bohemia, and became imperial mathematician to the Holy Roman Emperor Rudolph II. His goal was to revise the *Alphonsine Tables* and publish the revision as a monument to his new patron. It would be called the *Rudolphine Tables*.

Tycho did not intend to base the *Rudolphine Tables* on the Ptolemaic system, but rather on his own Tychonic system, proving once and for all the validity of his hypothesis. To assist him, he hired a few mathematicians and astronomers, including one Johannes Kepler. Then in November 1601, Tycho overate at a nobleman's home and collapsed.

Figure 4-13 Much of Tycho's success was due to his skill in designing large, accurate instruments such as this one known as the mural quadrant. In this engraving, the figure of Tycho at the center, his dog, and the scene in the background are a mural painted on the wall within the arc of the quadrant. The observer at the extreme right (Tycho) peers through a sight out the loophole in the wall at the upper left and thus measures the object's altitude above the horizon. (The Granger Collection, New York)

Before he died, 9 days later, he asked Rudolph II to make Kepler imperial mathematician. Thus the newcomer, Kepler, became Tycho's replacement (though at one-sixth Tycho's salary).

4-4 JOHANNES KEPLER

No one could have been more different from Tycho Brahe than Kepler (Figure 4-14). He was born on December 27, 1571, to a poor family in a region now included in southwest Germany. His father was unreliable and shiftless, principally employed as a mercenary soldier fighting for whomever paid enough. He finally failed to return from a military expedition, either because he was killed or because he found circumstances more to his liking elsewhere. Kepler's mother was apparently an unpleasant and unpopular woman. She was accused of witchcraft in later years, and Kepler had to defend her in a trial that dragged on for 3 years. She was finally acquitted but died the following year.

Kepler was the oldest of six children, and his childhood was no doubt unhappy. The family was not only poor and often lacking a father, but it was also Protestant in a predominantly Catholic region. In addition, Kepler was never healthy, even as a child, so it is surprising that he did well in the pauper's school he attended, eventually winning a scholarship to the university at Tübingen, where he studied to become a Lutheran pastor.

During his last year of study, Kepler accepted a job in Graz teaching mathematics and astronomy, a job he resented because he knew little about the subjects. Evidently he was not a good teacher, either—he had few students his first year, and none at all his second. His superiors put him to work teaching a few introductory courses and preparing an annual almanac that contained astronomical, astrological, and weather predictions. Through good luck, in 1595 some of his weather predictions were fulfilled, and he gained a reputation as an astrologer and seer. Even in later life he earned money from his almanacs.

While still a college student, Kepler had become a believer in the Copernican hypothesis, and at Graz he used his extensive spare time to study

Figure 4-14 Johannes Kepler (1571–1630) derived three laws of planetary motion from Tycho Brahe's observations of the positions of the planets. This Romanian stamp commemorates the 400th anniversary of Kepler's birth. Ironically, it contains an error—the orientation of the moon.

astronomy. By 1596, the same year Tycho left Hveen, Kepler was ready to solve the mystery of the universe. That year he published a book called *The Forerunner of Dissertations on the Universe, Containing the Mystery of the Universe.* The book, like nearly all scientific works of that age, was written in Latin, and is now known as *Mysterium Cosmographicum.*

By modern standards, the book contained almost nothing of value. It began with a long appreciation of Copernicanism and then went on to speculate on the reasons for the spacing of the planetary orbits. Kepler felt he had found the underlying architecture of the universe in the five regular solids*—the cube, tetrahedron, dodecahedron, icosahedron, and octahedron. Because these five solids were the only regular solids, he supposed they were the "spacers" between the planetary orbits (Figure 4-15). Kepler advanced astrological, numerological, and even musical arguments for his theory.

The second half of the book is no better than the first, but it has one virtue—as Kepler tried to fit the five solids to the planetary orbits, he demonstrated that he was a talented mathematician and that he was well versed in astronomy. He sent copies to Tycho and to Galileo in Rome, and both

A regular solid is a three-dimensional body each of whose faces is the same. A cube is a regular solid each of whose faces is a square.

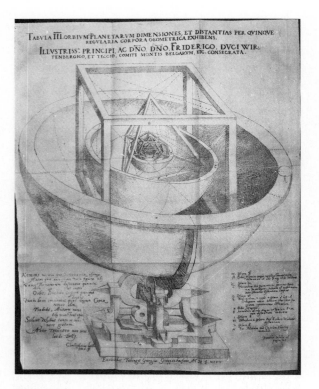

Figure 4-15 Kepler believed the five regular solids were the spacers between the spheres containing the planetary orbits. His book *Mysterium Cosmographicum* contained this fold-out illustration of the spheres and spacers. (Courtesy Owen Gingerich and the Houghton Library)

recognized his talent in spite of the mystical content of the book.

Life was unsettled for Kepler because of the persecution of Protestants in the region, so when Tycho invited him to Prague in 1600, he came readily, eager to work with the famous astronomer. Tycho's sudden death in 1601 left Kepler in a position to use the observations from Hveen to analyze the motions of the planets and complete the *Rudolphine Tables*. Tycho's family, recognizing that Kepler was a Copernican and guessing that he would not follow the Tychonic system in completing the *Rudolphine Tables*, sued to recover the instruments and books of observations. The legal wrangle went on for years. Tycho's family did get back the instruments Tycho had brought to Prague, but Kepler had the books and he kept them.

Whether Kepler had any legal right to Tycho's

records is debatable, but he put them to good use. He began by studying the motion of Mars, trying to deduce from the observations how the planet moved. By 1606, he had solved the mystery, this time correctly. The orbit of Mars (and all planets) was an ellipse with the sun at one focus. Thus he abandoned the 2000-year-old belief in the circular motion of the planets. But the mystery was even more complex. The planets did not move at constant speed along their orbits—they moved faster when close to the sun and slower when farther away. Thus Kepler abandoned both uniform motion and circular motion and finally solved the problem of planetary motion.

Kepler published his results in 1609 in a book called *Astronomia Nova* (New Astronomy). Like Copernicus' book, *Astronomia Nova* was written in Latin for other scientists and was highly mathematical. In some ways the book was surprisingly advanced. For instance, Kepler discussed the force that holds the planets in their orbits and came within a paragraph of stating the principle of mutual gravitation.

In spite of the abdication of Rudolph II in 1611, Kepler continued his astronomical studies. He wrote about a supernova that had appeared in 1604 (now known as Kepler's supernova) and about comets, and he authored a textbook about Copernican astronomy. In 1619, he published *Harmonice Mundi* (The Harmony of the World) in which he returned to the cosmic mysteries of *Mysterium Cosmographicum*. The only thing of note in *Harmonice Mundi* is his discovery that the radii of the planetary orbits are related to the planet's orbital period. That and his two previous discoveries are now recognized as Kepler's three laws of planetary motion (Table 4-1).

Kepler's first law states that the orbits of the planets around the sun are ellipses with the sun at one focus. An **ellipse** is defined as a figure drawn around two points called the *foci* in such a way that the distance from one focus to any point on the ellipse and then back to the other focus equals a constant. This makes it very easy to draw ellipses with two thumbtacks and a loop of string. Press the thumbtacks into a board, loop the string about them, and place a pencil in the loop. If you keep the string taut as you move the pencil, it traces out an ellipse (Figure 4-16a). The closer together the

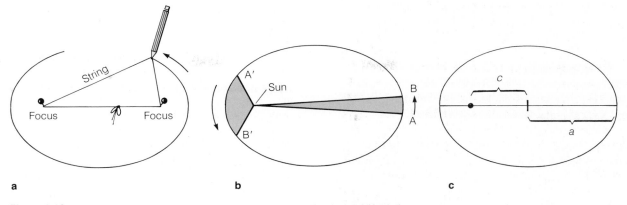

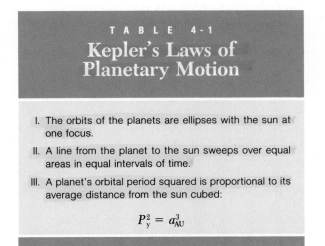

Figure 4-16 (a) Drawing an ellipse with two tacks and a loop of string. (b) Kepler's second law: "A line from a planet to the sun sweeps over equal areas in equal intervals of time." (c) Kepler's third law: The average distance from a planet to the sun equals *a*, the semimajor axis of its orbit. The eccentricity equals *c/a*. A circle is an ellipse of eccentricity 0.

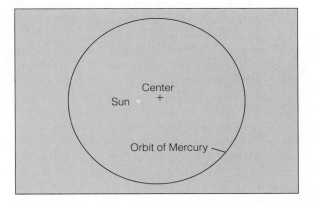

Figure 4-17 The orbits of the planets are nearly circular. Of the planets known to Kepler, Mercury has the most elliptical orbit. Pluto's orbit is only slightly more elliptical.

thumbtacks, the more circular the ellipse. Though Kepler was able to determine the elliptical shape of the planetary orbits, they are nearly circular. Of the planets known to Kepler, Mercury has the most elliptical orbit, but even it deviates only slightly from a circle (Figure 4-17).

Kepler's second law states that a line from the planet to the sun sweeps over equal areas in equal intervals of time (Figure 4-16b). This means that when the planet is closer to the sun and the line connecting it to the sun is shorter, the planet must move more rapidly if the line is to sweep over the same area. Thus the planet in Figure 4-16b would move from point A′ to point B′ in one month, sweeping out the area shown. But when the planet was farther from the sun, one month's motion would carry it from A to B.

Kepler's third law states that a planet's orbital period squared is proportional to its average distance from the sun cubed. Because of the way the planet moves along its orbit, its average distance from the sun is equal to half of the long diameter of the elliptical orbit, the semimajor axis *a* (Figure 4-16c). If we measure this quantity in astronomical units (AU) and the orbital period in years (y), we can summarize the third law as:

$$P_y^2 = a_{AU}^3$$

For example, Jupiter's average distance from the sun is roughly 5.2 AU. The semimajor axis cubed would be about 140.6, so the period must be the square root of 140.6, about 11.8 years.

It is important to notice that Kepler's three laws are empirical. That is, they describe a phenomenon without explaining why it occurs. Kepler derived them from Tycho's extensive observations, not from any fundamental assumption or theory. In fact, Kepler never knew what held the planets in their orbits or why they continued to move around the sun. His books are a fascinating blend of careful observation, mathematical analysis, and mystical theory.

In spite of Kepler's recurrent involvement with astrology and numerology, he continued to work on the *Rudolphine Tables.* At last, in 1627, they were ready, and he financed their printing himself, dedicating them to the memory of Tycho Brahe. In fact, Tycho's name appears in larger type on the title page than Kepler's own. This is especially surprising when we recall that the tables were based on the heliocentric model of Copernicus and the elliptical orbits of Kepler, and not on the Tychonic system. The reason for Kepler's evident deference was Tycho's family, still powerful and still intent on protecting the memory of Tycho. They even demanded a share of the profits and the right to censor the book before publication, though they changed nothing but a few words on the title page.

The *Rudolphine Tables* was Kepler's masterpiece, the precise model of planetary motion that Copernicus had sought but failed to find. The accuracy of the *Rudolphine Tables* was strong evidence that both Kepler's model for planetary motion and the Copernican hypothesis for the place of the earth were correct. Copernicus would have been pleased.

Kepler died on November 15, 1630. During his life he had been a Copernican and had written a number of books proclaiming his belief, but he had never been seriously persecuted because he lived in northern Europe, beyond the reach of the Inquisition. Others were not so lucky. The last astronomer in our story was a Copernican, too, but he had the misfortune to be born in southern Europe, where the Inquisition held sway.

4-5 GALILEO GALILEI

Copernicus, Tycho, Kepler, and Galileo were all mathematicians and astronomers whose lives are forever linked to the birth of modern astronomy. But Galileo did something the others did not—he used a telescope. Galileo and one of his telescopes appear on modern Italian banknotes (Figure 4-18). He was the first astronomer to observe the heavens with a telescope and consider the implications of what he saw. His observations satisfied him that the universe could not be geocentric. At a time when it was heresy to do so, Galileo became the great defender of the Copernican theory.

Galileo was born in Pisa in 1564 and thus was a contemporary of Tycho and Kepler. At the university, he studied medicine, although his true interests were mathematics, mechanics, and astronomy. We can be sure he was talented in mathematics because, even though family finances forced him to leave school without a degree, he obtained a position as a professor of mathematics at the University of Pisa only 4 years later.

In Pisa, his teaching assignment included astronomy, and he must have taught the Ptolemaic system. He may even then have been a Copernican, but he would have been foolish to try to teach Copernicus' hypotheses. Italy was no place to introduce unorthodox ideas that might challenge church teachings. Perhaps while he taught astronomy, he became convinced that the universe was Copernican and not Ptolemaic.

In 1609, Galileo obtained some lenses and built a telescope (Figure 4-19), following descriptions of similar instruments built by Dutch lens makers. He did not invent the telescope, nor was he the first to turn one toward the sky, but he was the first to study the telescopic appearance of the heavenly bodies and consider their implications for the models of the universe then popular. In that sense, he was the first astronomer to use a telescope for science, and some historians thus refer to him as the father of modern science.

In 1610, he published a book called *Sidereus Nuncius* (The Sidereal Messenger), in which he

Figure 4-18 Galileo Galilei (1564–1642), remembered as the great defender of Copernicanism, also made important discoveries in the physics of motion. He is honored here on an Italian 2000-lira note. The reverse side shows one of his telescopes at lower right and a modern observatory above it.

Figure 4-19 Galileo's telescopic discoveries generated intense interest and controversy. Some critics refused to look through a telescope lest it deceive them. (Courtesy Yerkes Observatory)

described what he saw through his telescope and showed how his observations supported the Copernican theory.

One of his important discoveries was that Venus, like the moon, goes through phases. In the Ptolemaic model, Venus revolves around an epicycle located between Earth and the sun. Thus an observer on Earth would never see the planet fully illuminated by the sun. That is, it would always be seen as a crescent. But Galileo saw Venus go through a complete set of phases, which proved that it did indeed revolve around the sun (Figure 4-20).

In addition, he discovered four satellites orbiting the planet Jupiter. Some critics of Copernicus had said Earth could not move because the moon would be left behind. But Jupiter moved yet kept its satellites, so Galileo's discovery proved that Earth too could move and keep its moon. Also, the accepted teaching was that everything revolved around the earth. Galileo's telescope revealed that some objects orbited Jupiter. Thus there could be other centers of motion.

Finally, the telescope challenged Aristotle's contention that the heavens were perfect and unchanging. Galileo observed spots on the sun, raising the suspicion that the sun was less than perfect. By noting the movement of the spots, he concluded that the sun rotated on its axis, just as Copernicus said Earth did. When he observed the moon, he found not a polished, perfect sphere but a craggy, mountainous terrain surrounding flat areas later mistaken for seas.

Sidereus Nuncius was a popular book, and it made Galileo famous. He soon left his teaching position and went to Florence, where he became personal philosopher and mathematician to the Grand Duke of Tuscany. When Galileo visited Rome in 1611 he gave well-attended lectures and had long and friendly discussions with the powerful Cardinal Barberini. But he also made enemies. Personally, Galileo was outspoken, forceful, and sometimes tactless, and he enjoyed debate. Most of all he enjoyed being right. Thus in the years following publication of *Sidereus Nuncius*, in lectures, debates, and letters, he offended important people who questioned his telescopic discoveries.

By 1616, Galileo was the center of a storm of controversy. Some said he was mistaken (or worse), and others refused even to look through a telescope. Pope Paul V decided to end the disruption and so, when Galileo visited Rome in 1616, Cardinal Bellarmine interviewed him and ordered him to cease his astronomical work. Books relevant to Copernicanism, including *De Revolutionibus*, were banned.

In some respects, the years that followed gave Galileo hope that he could resume his work. His interviews with the pope and Cardinal Bellarmine in 1616 had been stern but not disrespectful, and in a final interview before he left Rome, the pope had shown open friendship. Then in 1621 Pope Paul V died, as did his successor, Pope Gregory XV, in 1623. The next pope was Galileo's friend Cardinal Barberini, who took the name Urban VIII. Galileo visited the new pope, and though the

Figure 4-20 (a) In the Ptolemaic universe, Venus moved around an epicycle centered on the Earth–sun line (see Figure 4-7) and always appeared as a crescent. (b) Galileo's telescope showed that Venus goes through a full set of phases, proving that it must orbit the sun, as in the Copernican universe.

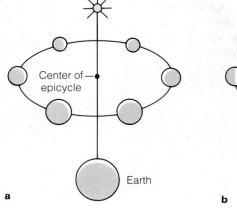

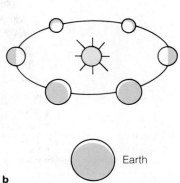

order of 1616 was not revoked, Urban VIII remained friendly.

Thus in 1624 Galileo began to write a book defending the Copernican theory, a project that was finally completed in the last days of 1629. After some delay, the book was approved by the local censor in Florence and by the head censor of the Vatican in Rome. It was printed in 1632.

The book was called *Dialogo Dei Due Massimi Sistemi* (Dialogue Concerning the Two Chief World Systems) (Figure 4-21). It was written as a conversation between three friends who debate the Copernican and Ptolemaic hypotheses. Salviati, a swift-tongued defender of Copernicus, dominates the book. Sagredo is a reasonably intelligent but uninformed believer in the Ptolemaic universe. Simplicio, the third character in the book, is the dismal defender of Ptolemy. In fact, Simplicio does not seem very bright.

Galileo seriously misjudged his safety in writing the book. For one thing, it was not an unbiased account of the two systems, although he later claimed it was. More importantly, either intentionally or unintentionally, he exposed the pope's authority to possible ridicule. Pope Urban VIII was fond of arguing that, as God was omnipotent, He could construct the universe in any form while making it appear to us to have a different form. Galileo placed this argument in the mouth of Simplicio, and the enemies of Galileo showed the passage to the pope as an example of Galileo's disrespect. The pope thereupon ordered Galileo to Rome to meet the Inquisition.

Upon his arrival, Galileo was interrogated by the Inquisition four times. He was not tortured, but he was threatened with torture. At first Galileo tried to maintain his pride and defend his ideas, but he was soon cowed by the Inquisition. He must have thought often of Giordano Bruno, tried, condemned, and burnt at the stake in Rome in 1600. Bruno had been an outspoken critic of the church in many respects, but one of his offenses was Copernicanism. The problem faced by the Inquisition, however, was that the book had been approved by the censors, so in the end they had to return to the orders given Galileo in 1616. The official record of the interview between Galileo and Cardinal Bellarmine included the statement that Galileo was "not to hold, teach, or defend in any way" the principles of Copernicus.* Galileo's assertion that his dialogue was an unbiased discussion and not a defense of any one hypothesis was worthless even if it had been true.

On June 22, 1633, at the age of 70, Galileo knelt before the Inquisition and read a recantation admitting his errors. Tradition has it that as he rose he whispered, *"E pur si muove"* (still it moves), referring to Earth, but it is unlikely that anyone with any sense would risk such defiance.

Galileo was sentenced to life imprisonment. Perhaps through the intervention of the pope, he was held in confinement at his villa where he could

Because the document is unsigned, some scholars suspect it was forged and that Galileo never received those instructions in 1616.

Figure 4-21 Aristotle, Ptolemy, and Copernicus discuss astronomy in this frontispiece from Galileo's book *Dialogue Concerning the Two Chief World Systems.* (Courtesy Owen Gingerich and the Houghton Library)

meet his family, though other visitors were forbidden. During these years, he studied mechanics and physics, and even wrote a book on the subject that was published in Holland in 1638. During his last few years, he was allowed a few visitors, and two young scientists came to stay with him. As he was blind by then, he no doubt enjoyed their discussions. At last, on January 8, 1642, after 10 years' imprisonment, 99 years after the death of Copernicus, Galileo died.

The church's condemnation of Galileo has been a subject of heated debate for 350 years. Many have argued that it shows that religion is incompatible with science and is incapable of incorporating new discoveries into religious teachings. In 1979, however, Pope John Paul II ordered a reexamination of the case against Galileo. In 1990, the pope spoke at Pisa, birthplace of Galileo, "whose scientific work, imprudently opposed at the beginning, is now recognized by all as an essential step in the methodology of research and, in general, on the path toward the knowledge of the world of nature." Galileo has been forgiven, and the quest for the place of the earth is finally complete.

4-6 ISAAC NEWTON AND MODERN ASTRONOMY

The New Astronomy We date the origin of modern astronomy from the 99 years between the deaths of Copernicus and Galileo because it was an age of transition. That period marked the transition between the Ptolemaic universe and the Copernican, but it also marked a transition in the nature of astronomy in particular and science in general. Before the events of our story, scientific principles were drawn not from observation but from philosophical judgments of what the universe should be like. Thus Aristotle believed the heavenly bodies were perfect because he felt they should be. In such an atmosphere, scientific discoveries and observations had to be bent to fit expectations.

The discoveries of Kepler and Galileo were accepted in the 1600s because the world was in transition. Astronomy was not the only thing changing during this period. The Renaissance is commonly taken to be the period between 1300 and 1600,

and thus the 99 years of this history lie at the culmination of the reawakening of learning in all fields (Figure 4-22). The world was open to new ideas and new observations. Martin Luther remade Christianity, and other philosophers and scholars reformed their areas of human knowledge. Had Copernicus not published his ideas, someone else would have suggested that the universe was heliocentric. History was ready to shed the Ptolemaic model.

In addition, this period marks the beginning of the modern scientific method. Beginning with Copernicus, Tycho, Kepler, and Galileo, scientists depended more and more on empirical evidence—that is, on observation and measurement. This trend, too, is linked to the Renaissance and its advances in metalworking and lens making. Before the time in which our story began, no astronomer had looked through a telescope because one could not be made. By 1642, not only telescopes but also other sensitive measuring instruments made science into something new and precise. Also, the growing number of scientific societies increased the exchange of observations and hypotheses among scientists and stimulated more and better work. The most important advance, however, was the application of mathematics to scientific questions. Kepler's work demonstrated the power of mathematical analysis, and as the quality of these numerical tools improved, the progress of science accelerated. Thus our story is not just the story of the birth of modern astronomy, but that of modern science as well.

Isaac Newton Galileo died in January 1642. Some 11 months later, on Christmas day 1642,* a baby was born in the English village of Woolsthorpe. His name was Isaac Newton (Figure 4-23), and his life represented the first flower of the seeds planted by the four astronomers of our story.

Newton was a quiet child from a farming family, but his work at school was so impressive that his uncle financed his education at Trinity College, where he studied mathematics and physics. In

*Because England had not yet reformed its calendar, December 25, 1642, in England was January 4, 1643, in Europe. It is only a small deception to use the English date and thus include Newton's birth in our 99-year history.

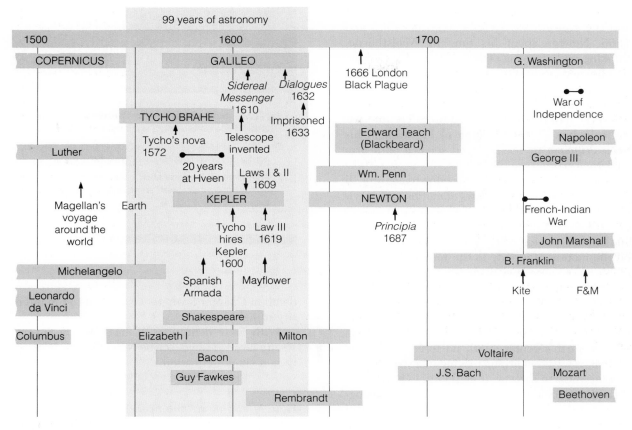

99 years of astronomy

1500   1600   1700

COPERNICUS   GALILEO   G. Washington

Sidereal   *Dialogues*
Messenger   1632
1610   Imprisoned   War of
1633   Independence

TYCHO BRAHE

Tycho's nova   Telescope   Edward Teach   Napoleon
1572   invented   (Blackbeard)

Luther   George III

20 years   Laws I & II   Wm. Penn
at Hveen   1609

Earth   KEPLER   NEWTON   French-Indian
War

Magellan's   Tycho   Law III   *Principia*   John Marshall
voyage   hires   1619   1687
around the   Kepler   B. Franklin
world   1600

Michelangelo   Spanish   Mayflower   Kite   F&M
Armada

Leonardo
da Vinci   Shakespeare

Columbus   Elizabeth I   Milton

Bacon   Voltaire

Guy Fawkes   J.S. Bach   Mozart

Rembrandt   Beethoven

1666 London
Black Plague

Figure 4-22 The 99 years between the death of Copernicus in 1543 and the birth of Newton in 1642 marked the transition from the ancient astronomy of Ptolemy and Aristotle to the revolutionary theory of Copernicus. This period saw the birth of modern scientific astronomy.

Figure 4-23 Isaac Newton (1642–1727), working from the discoveries of Galileo and Kepler, derived three laws of motion and the principle of mutual gravitation. He and some of his discoveries are honored on this English 1-pound note.

1665, the Black Plague swept through England and the colleges were closed. During 1665 and 1666, Newton spent his time in Woolsthorpe, thinking and studying. It was during these years that he made most of his discoveries in optics, mechanics, and mathematics. Among other things, he studied optics, developed three laws of motion, divined the nature of gravity, and invented differential calculus. The publication of his work in his book *Principia* in 1687 placed science on a firm analytical base.

It is beyond the scope of this book to analyze all of Newton's work, but his laws of motion and gravity had an important impact on the future of astronomy. From his study of the work of Galileo, Kepler, and others, Newton extracted three laws that related the motion of a body to the forces acting on it (Table 4-2). These laws made it possible to predict exactly how a body would move if the forces were known.

In thinking about the motion of the moon, Newton realized that some force had to pull the moon toward Earth's center. With no force to alter its motion, it would continue moving in a straight line and leave Earth forever. It could circle Earth only if Earth attracted it. Newton's insight was to recognize that the force that held the moon in its orbit was the same as the force that made apples fall from trees. The force of gravity was universal—that is, all objects attract all other objects with a force that depends only on their masses and the distance between their centers.

The **mass** of an object is a measure of the amount of matter in the object, usually expressed in kilograms. Mass is not the same as weight. An object's *weight* is the force that Earth's gravity exerts on the object. Thus an object in space far from Earth might have no weight, but it would contain the same amount of matter and would thus have the same mass that it would have on Earth.

Newton also realized that the distance between the masses was important. He recognized that the force of gravity had to decrease as the square of the distance between the objects increased. Specifically, if the distance from, say, Earth to the moon were doubled, the gravitational force between them would decrease by a factor of 2^2, or 4. If the distance were tripled, the force would decrease by a factor of 3^2, or 9. This relationship is known as the **inverse square law**. (We will discuss it in more

TABLE 4-2
Newton's Three Laws of Motion

I. A body continues at rest or in uniform motion in a straight line unless acted upon by some force.

II. A body's change of motion is proportional to the force acting on it and is in the direction of the force.

III. When one body exerts a force on a second body, the second body must exert an equal and opposite force back on the first body.

detail in Chapter 8 where we apply it to the intensity of light.)

With these definitions of mass and the inverse square law, we can describe Newton's law of gravity in a simple equation:

$$F = -G\frac{Mm}{r^2}$$

Here M and m are the masses of two objects, r is the distance between their centers, G is the gravitational constant, and F is the force of gravity acting between the two objects.

Orbital Motion In developing the three laws of motion and the theory of gravity, Newton resolved the problem Kepler had toyed with—why the planets move along their orbits. In fact, Newton's work explained orbital motion and made it possible to analyze the motions of the heavens with great precision.

To illustrate the principle of orbital motion, imagine that we place a large cannon at the top of a mountain, point the cannon horizontally, and fire it (Figure 4-24). The cannonball falls to earth some distance from the foot of the mountain. The more gunpowder we use, the faster the ball travels, and the farther from the foot of the mountain it falls. If we use enough powder, the ball travels so fast it never strikes the ground. Our planet's gravity pulls it toward Earth's center, but Earth's surface curves away from it at the same rate at which it falls. Thus we say it is in orbit. If the cannonball is high above

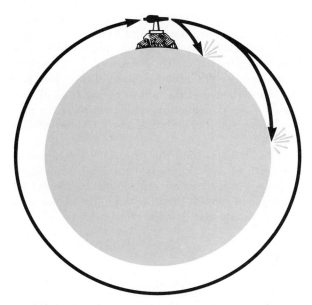

Figure 4-24 A cannon on a high mountain could put its projectile into orbit if it could achieve a high enough velocity.

the atmosphere, where there is no friction, it will fall around Earth forever.

Simple physics says that an object in motion tends to stay in motion in a straight line unless acted upon by some force. Thus the cannonball in this example travels in a curve around Earth only because Earth's gravity acts to pull it away from its straight-line motion.

Newton's explanation of orbital motion is only one example of the power of his conception of nature. His laws of motion and gravity were *general* laws that described the motions of all bodies under the action of external forces. In addition, the laws were *productive* because they made possible specific calculations that could be tested by observation. For example, Newton's laws of motion could be used to derive Kepler's third law from the law of gravitation.

Newton's discoveries remade astronomy into an analytical science in which astronomers could measure the positions and motions of celestial bodies, calculate the gravitational forces acting on them, and predict their future motion.

Were we to trace the history of astronomy after Newton, we would find scientists predicting the motion of comets, the gravitational interaction of the planets, the motions of double stars, and so on.

Astronomers built on the discoveries of Newton, just as he had built on the discoveries of Copernicus, Tycho, Kepler, and Galileo. It is the nature of science to build on the discoveries of the past, and Newton was thinking of that when he wrote, "If I have seen farther than other men, it is because I stood upon the shoulders of giants."

When we gaze at the night sky, we too stand on the shoulders of giants. We understand what we see because we understand the work of Copernicus, Kepler, and Newton. But modern astronomers gaze at the sky with powerful, modern tools, and those tools are the subject of the next chapter.

SUMMARY

Classical astronomy was based on the writings of the Greek philosopher Aristotle. He taught that Earth was the immobile center of the universe and that the stars and planets were carried around Earth by great crystalline spheres. This model of the universe was given mathematical form about A.D. 140 in the *Almagest*, the great work of Ptolemy. Ptolemy preserved the classical belief in uniform circular motion by using epicycles, deferents, and equants.

In contrast to the geocentric universe of classical astronomy, the universe devised by Copernicus was heliocentric, or sun-centered. One advantage of a heliocentric universe is that retrograde motion, the occasional westward motion of the planets, is easily explained. Copernicus did not publish his book *De Revolutionibus* until 1543, the year he died. The teachings of Aristotle had become official church doctrine, and as a critic, Copernicus could have been charged with heresy.

The Danish astronomer Tycho Brahe did not accept the Ptolemaic or the Copernican model but rather developed his own, in which the sun and moon circled Earth, and the planets circled the sun. Although his hypothesis was not correct, Tycho made precise observations of planetary positions that later led to a true understanding of planetary motion.

Johannes Kepler, Tycho Brahe's assistant, inherited the Danish astronomer's records in 1601 and used his observations to uncover three laws of planetary motion. Kepler discovered that the planets follow ellipses with the sun at one focus, that they move faster when near the sun, and that a planet's period squared is proportional to its orbital radius cubed.

Galileo Galilei was a great defender of the Copernican hypothesis. Galileo was the first person to use a telescope to observe the heavens and to recognize the significance of what he saw. His discoveries of the phases of Venus, the satellites of Jupiter, the mountains of the moon, and other phenomena helped undermine the Ptolemaic universe. In 1633, Galileo was finally condemned before the Inquisition for refusing to halt his defense of Copernicanism.

Born in 1642, the same year that Galileo died, Isaac Newton used the work of Kepler and Galileo to discover three laws of motion and the theory of universal gravitation. These laws made it possible to understand the orbital motions of the planets as the consequence of the sun's gravity. In addition, Newton's work made it possible to analyze the motion of any celestial body and predict its path in the future.

The 99 years from the death of Copernicus to the birth of Newton marked the birth of modern science. From that time on, science depended on evidence to support theories and relied on the analytic methods first demonstrated by Newton.

NEW TERMS

parallax	hypothesis
geocentric universe	model
uniform circular motion	theory
retrograde motion	natural law
epicycle	ellipse
deferent	mass
equant	inverse square law
heliocentric universe	

REVIEW QUESTIONS

1. Why did early astronomers conclude that Earth did not move?

2. Draw and label a diagram showing an epicycle, deferent, and equant.

3. In what ways were the systems of Ptolemy and Copernicus similar?

4. Why did the Copernican hypothesis win gradual acceptance?

5. When Tycho observed the new star of 1572, he could detect no parallax. What did he conclude from this?

6. How did the *Alphonsine Tables*, *Prutenic Tables*, and *Rudolphine Tables* differ?

7. What are Kepler's three laws of planetary motion?

8. Review Galileo's telescopic discoveries and explain why they supported the Copernican universe.

9. Galileo was condemned by the Inquisition, but Kepler, also a Copernican, was not. Why not?

10. How did Newton's discoveries change astronomy?

DISCUSSION QUESTIONS

1. Why did Copernicus delay publication of his book? Did his model accurately predict the positions of the planets? Could he have felt some loyalty to the church?

2. Why might Tycho have hesitated to hire Kepler? Why did Tycho appoint Kepler, a Copernican, his scientific heir?

PROBLEMS

1. If you lived on Mars, which planets would describe retrograde loops? Which would never be visible as crescent phases?

2. Galileo's telescope showed him that Venus has a large angular diameter (61 seconds of arc) when it is a crescent and a small angular diameter (10 seconds of arc) when it is nearly full. Use the small-angle formula to find the ratio of its maximum distance to its minimum distance. Is this ratio compatible with the Ptolemaic universe shown in Figure 4-7?

3. Galileo's telescopes were not of high quality by modern standards. He was able to see the rings of Saturn, but he never reported seeing features on Mars. Use the small-angle formula to find the angular diameter of Mars when it is closest to

Earth. How does that compare with the maximum diameter of Saturn's rings?

4. If a planet had an average distance from the sun of 10 AU, what would its orbital period be?

5. If a space probe were sent into an orbit around the sun that brought it as close as 0.5 AU and as far away as 5.5 AU, what would its orbital period be?

6. Pluto orbits the sun with a period of 247.7 years. What is its average distance from the sun?

7. Comet Halley has an orbital period of about 75 years. What is its average distance from the sun?

RECOMMENDED READING

ARMITAGE, A. *Copernicus: The Founder of Modern Astronomy*. New York: A. S. Barnes, 1962.

BERRY, A. *A Short History of Astronomy*. New York: Dover, 1961.

BOORSTIN, DANIEL J. *The Discoverers*. New York: Vintage Books, 1983.

BROAD, WILLIAM J. "Priority War: Discord in Pursuit of Glory." *Science* 211 (January 30, 1981), p. 465.

———. "A Bibliophile's Quest for Copernicus." *Science* 218 (November 12, 1982), p. 661.

BRONOWSKI, J. *The Ascent of Man*. Chapters 6 and 7. Boston: Little, Brown, 1973.

BULLOUGH, VERN L. (Ed.) *The Scientific Revolution*. New York: Krieger, 1970.

CHRISTIANSON, J. "The Celestial Palace of Tycho Brahe." *Scientific American* 204 (February 1961), p. 118.

COHEN, I. B. *The Newtonian Revolution*. London: Cambridge University Press, 1980.

———. "Newton's Discovery of Gravity." *Scientific American* 244 (March 1981), p. 167.

DRAKE, S. "Newton's Apple and Galileo's Dialogue." *Scientific American* 243 (August 1980), p. 150.

DURHAM, FRANK, and ROBERT D. PURRINGTON. *Frame of the Universe*. New York: Columbia University Press, 1983.

FERMI, L., and G. BERNARDINI. *Galileo and the Scientific Revolution*. Greenwich, Conn.: Fawcett, 1965.

FERRIS, TIMOTHY. *Coming of Age in the Milky Way*. New York: Morrow, 1988.

GINGERICH, O. "Johannes Kepler and the Rudolphine Tables." *Sky and Telescope* 42 (December 1971), p. 328.

———. "Copernicus and Tycho." *Scientific American* 229 (December 1973), p. 86.

———. "Tycho Brahe and the Great Comet of 1577." *Sky and Telescope* 54 (December 1977), p. 452.

———. "The Galileo Affair." *Scientific American* 247 (August 1982), p. 133.

———. "Ptolemy and the Maverick Motion of Mercury." *Sky and Telescope* 66 (July 1983), p. 11.

———. "Laboratory Exercises in Astronomy—The Orbit of Mars." *Sky and Telescope* 66 (October 1983), p. 300.

HALL, A. RUPERT. *Philosophers at War*. London: Cambridge University Press, 1980.

KOESTLER, A. *The Watershed: A Biography of Johannes Kepler*. Garden City, N.Y.: Doubleday, 1960.

KOYRE, A. *From the Closed World to the Infinite Universe*. New York: Harper & Row, 1958.

KUHN, T. S. *The Copernican Revolution*. Cambridge, Mass.: Harvard University Press, 1957.

McPEAK, WILLIAM J. "Tycho Brahe Lights Up the Universe." *Astronomy* 18 (December 1990), p. 28.

MOORE, P. *Watchers of the Sky*. New York: Putnam, 1973.

RAVETZ, J. "The Origins of the Copernican Revolution." *Scientific American* 215 (October 1966), p. 88.

SHEA, W. R. *Galileo's Intellectual Revolution*. New York: Macmillan, 1972.

THOREM, VICTOR E. *The Lord of Uraniborg*. New York: Cambridge University Press, 1990.

WALLACE, WILLIAM A. (Ed.) *Reinterpreting Galileo*. Washington, D.C.: Catholic University Press, 1986.

WESTFALL, R. S. "Newton and the Fudge Factor." *Science* 179 (February 23, 1973), p. 751. See also *Science* 180 (June 15, 1973), p. 1118.

WILSON, C. "How Did Kepler Discover His First Two Laws?" *Scientific American* 226 (March 1972), p. 93.

FURTHER EXPLORATION

VOYAGER, the Interactive Desktop Planetarium, can be used for further exploration of the topics in this chapter. This chapter corresponds with the following projects and chapters in *Voyages Through Space and Time:* Project 5, The Inner Planets; Project 6, The Outer Planets; Chapter 1, The Motions of Celestial Objects.

Astronomical Tools

He burned his house down for the fire insurance and spent the proceeds on a telescope. . . .

Robert Frost, "The Star-Splitter"

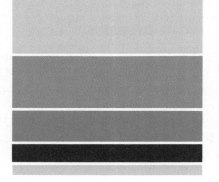

Starlight is going to waste. Every night, light from the stars falls on trees, oceans, roofs, and empty parking lots, and it is all wasted. To an astronomer, nothing is so precious as starlight. It is our only link to the sky, and the astronomer's quest is to gather as much starlight as possible and extract from it the secrets of the stars.

The telescope is the emblematic tool of the astronomer because its purpose is to gather and concentrate light for analysis. Nearly all of the interesting objects in the sky are faint sources of light, so modern astronomers are driven to build the largest possible telescopes to gather the maximum amount of light. Thus our discussion of astronomical tools is concentrated on large telescopes and the specialized tools used to analyze light.

If we wish to gather visible light, a normal telescope will do, but, as we will see in this chapter, visible light is only one kind of radiation. We can extract information from other forms of radiation by using specialized telescopes. Radio telescopes, for example, give us an entirely different view of the sky. Some of these specialized telescopes can be used from Earth's surface, but some must go into orbit above Earth's atmosphere. Telescopes that observe X rays, for instance, must be placed in orbit.

In addition to telescopes to gather light, astronomers need instruments to analyze light. Such instruments measure brightness, record images, or form spectra.

Astronomers no longer study the sky by mapping constellations, charting the phases of the moon, or following the retrograde motion of the planets. Modern astronomers use the most sophisticated telescopes and instruments to analyze starlight. Thus we begin this chapter with a discussion of the nature of light.

5-1 RADIATION: INFORMATION FROM THE SKY

Although we see light every day, it is not obvious what it is or how it works. To understand the nature of light, we must borrow a few concepts from physics.

Light as a Wave and a Particle If you have admired the colors in a soap bubble, you have seen light behave as a wave. But when that light enters the light meter on a camera, it behaves as a particle. How it behaves depends on how we observe it—it is both wave and particle.

We experience waves whenever we hear sound. Sound waves are a mechanical disturbance that travels through the air from source to ear. Sound requires a medium, so on the moon, where there is no air, there can be no sound. In contrast, light is made up of electric and magnetic waves that can travel through empty space. Unlike sound, light does not require a medium and thus can travel through a perfect vacuum.

Because light is made up of both electric and magnetic fields, we refer to it as **electromagnetic radiation.** As we will see later, visible light is only one form of electromagnetic radiation. These oscillating electric and magnetic fields travel through space at about 300,000 km/sec (186,000 mi/sec). This is commonly referred to as the speed of light c, but it is in fact the speed of all such radiation.

Electromagnetic radiation is a wave phenomenon—that is, it is associated with a periodically repeating disturbance, a wave. We are familiar with waves in water: If we disturb a pool of water, waves spread across the surface. Imagine that we use a meter stick to measure the distance between the successive peaks of a wave. This distance is the **wavelength,** usually represented by the Greek letter lambda (λ) (Figure 5-1).

While light does behave as a wave, it also behaves as a particle. We refer to a "particle of light" as a **photon,** and we can think of a photon as a bundle of waves.

The amount of energy a photon carries depends inversely on its wavelength. That is, shorter wavelength photons carry more energy, and longer wavelength photons carry less. We can express this relationship in a simple formula:

$$E = \frac{hc}{\lambda}$$

Here h is Planck's constant (6.6262×10^{-20} joule sec) and c is the speed of light (3×10^8 m/sec). The important point in this formula is not the value of the constants but the inverse relationship between the energy E and the wavelength λ. As λ gets smaller, E must get larger. Thus a photon of visible light carries a very small amount of energy, but a photon with a very short wavelength can carry much more energy.

The Electromagnetic Spectrum A spectrum is an array of electromagnetic radiation in order of wavelength. We are most familiar with the spectrum of visible light—which we can see in rainbows—but the visible spectrum is merely a small segment of the much larger electromagnetic spectrum (Figure 5-2).

The average wavelength of visible light is about 0.0005 mm. We could put 50 light waves end to end across the thickness of a sheet of household plastic wrap. It is too awkward to measure such short distances in millimeters, so we will measure the wavelength of light in nanometers (nm). One nanometer is 10^{-9} m. The wavelength of visible light ranges from 400 nm to 700 nm. (See Box 5-1.)

Just as we sense the wavelength of sound as pitch, we sense the wavelength of light as color.

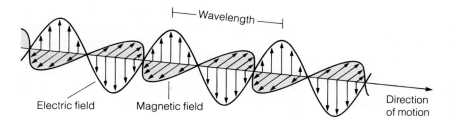

Electric field   Magnetic field   Direction of motion

Figure 5-1 Electric and magnetic fields travel together through space as electromagnetic radiation. The wavelength, commonly represented by the Greek letter lambda (λ), is the distance between successive peaks in the wave.

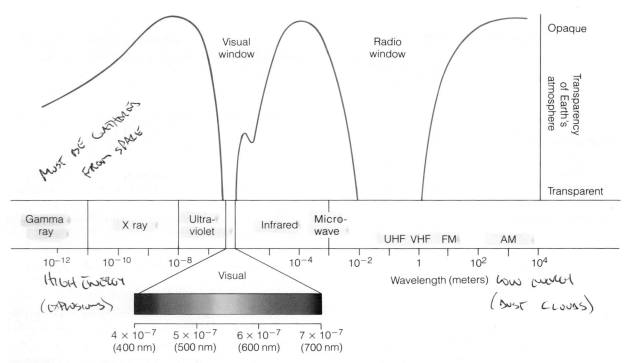

Figure 5-2 The electromagnetic spectrum includes all wavelengths of electromagnetic radiation. Earth's atmosphere is relatively opaque at most wavelengths. Visual and radio windows allow light and some radio waves to reach Earth's surface.

Light near the short-wavelength end of the visible spectrum (400 nm) looks violet to our eyes, and light near the long-wavelength end (700 nm) looks red (see Figure 5-2).

Beyond the red end of the visible spectrum lies infrared radiation, where wavelengths range from 700 nm to about 1 mm. Our eyes are not sensitive to this radiation, but our skin senses it as heat. A "heat lamp" is nothing more than a bulb that gives off large amounts of infrared radiation.

Beyond the infrared part of the electromagnetic spectrum lie radio waves. The radio radiation used for AM radio transmissions has wavelengths of a few kilometers, while FM, television, military, governmental, and ham radio transmissions have wavelengths that range down to a few meters. Microwave transmissions, used for radar and long-distance telephone communications, for instance, have wavelengths from a few centimeters down to about 1 mm.

The distinction between the wavelength ranges is not sharp. Long-wavelength infrared radiation

and the shortest microwave radio waves are the same. Similarly, there is no clear division between the short-wavelength infrared and the long-wavelength part of the visible spectrum. It is all electromagnetic radiation.

At wavelengths shorter than violet, we find ultraviolet radiation, with wavelengths ranging from 400 nm down to about 10 nm. At even shorter wavelengths lie X rays and gamma rays. Again, the boundaries between these wavelength ranges are not clearly defined.

X rays and gamma rays can be dangerous, and even ultraviolet photons have enough energy to do us harm. Small doses produce a suntan; larger doses can cause sunburn, and extreme doses might produce skin cancers. Contrast this to the lower-energy infrared photons. Individually, they have too little energy to affect skin pigment, a fact that explains why you can't get a tan from a heat lamp. Only by concentrating many low-energy photons in a small area, as in a microwave oven, can we transfer significant amounts of energy.

Measuring the Wavelength of Light

The wavelength of light is so short, we must use special units of measure when discussing it. Here we will use nanometers, to be consistent with the International System. One nanometer is 10^{-9} m.

Another unit that astronomers use commonly, and a unit that you will see in other references on astronomy, is the **angstrom (Å)** (named after the Swedish astronomer Anders Jonas Ångström). One angstrom is 10^{-10} m, and visible light has wavelengths between 4000 Å and 7000 Å. One nanometer contains 10 angstroms.

You may also find radio astronomers describing wavelengths in centimeters or millimeters, and infrared astronomers often refer to wavelengths in micrometers (or microns). One micrometer (μm) is 10^{-6} m.

We are interested in electromagnetic radiation because it brings us clues to the nature of stars, planets, and other celestial objects. Only a small part of this radiation, however, in the form of visible light and some radio waves, can get through Earth's atmosphere and reach the surface of Earth; other wavelengths are absorbed. The highest parts of the atmosphere absorb X rays, gamma rays, and some radio waves, and a layer of ozone (O_3) at an altitude of about 30 km absorbs ultraviolet radiation. In addition, water vapor in the lower atmosphere absorbs infrared radiation. The two wavelength regions in which our atmosphere is transparent are called **atmospheric windows.** Obviously, if we wish to study the sky from Earth's surface, we must look out through one of these windows.

Having described the nature of electromagnetic radiation and the electromagnetic spectrum, we can now study the tools astronomers use to analyze radiation.

5-2 OPTICAL TELESCOPES

Conventional astronomical telescopes are designed to gather light in the visible part of the spectrum. These optical telescopes depend on either lenses or mirrors to focus the light.

Refracting Telescopes In a **refracting telescope,** the main light-gathering element is a lens that refracts (bends) the light. Because of the shape of the lens, light striking the edge bends more than does light striking the central area, and the light that enters the lens from some distant object comes together to form a small, inverted image (Figure 5-3). The **focal length** of a lens is the distance from the lens to the point where it focuses parallel rays of light (Figure 5-4).

To build a refracting telescope, we need a lens of relatively long focal length to form an image of the object we wish to view. This lens is often called the **objective lens** because it is closest to the object. To view the image, we add a lens of short focal length called an **eyepiece** to enlarge the image and make it easy to see (Figure 5-5). Thus the eyepiece acts as a magnifier. By changing eyepieces we can easily change the magnification of the telescope.

Refracting telescopes suffer from a serious optical defect that limits what we can see through them. When light is refracted through glass, shorter wavelengths bend more than longer wavelengths, and blue light comes to a focus closer to the lens than does red light (Figure 5-6a). If we focus the eyepiece on the blue image, the red light is out of focus, producing a red blur around the image. If we focus on the red image, the blue light blurs. This color separation is called **chromatic aberration.**

Figure 5-3 A lens forms an image by refracting (bending) light. A curved mirror can form an image by reflecting light. In both cases, the image of distant objects is inverted.

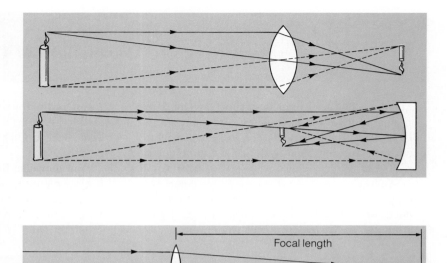

Figure 5-4 The focal length of a lens is the distance from the lens to the point where parallel rays of light come to a focus. The lens above has a longer focal length than the lens below.

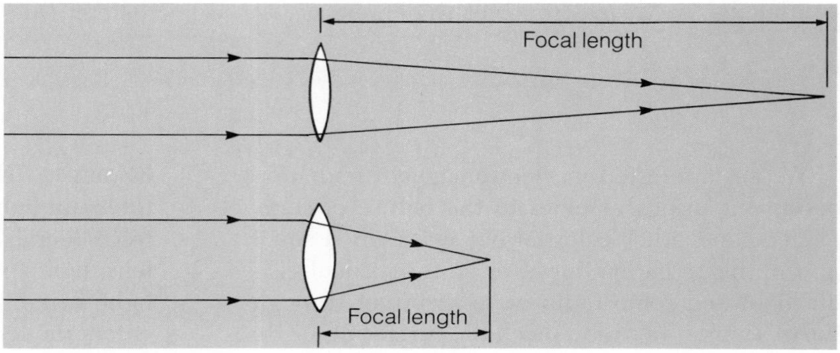

Focal length

Focal length

A telescope designer can partially correct for this effect by replacing the single objective lens with one made of two lenses ground from different kinds of glass. Such lenses, called **achromatic lenses,** can be designed to bring any two colors to the same focus (Figure 5-6b). Since our eyes are most sensitive to red and yellow light, we might bring these two colors to the same focus, but blue and violet would still be out of focus, producing a hazy blue fringe around bright objects.

Refracting telescopes were popular through the 19th century, but they are no longer economical. Achromatic lenses are expensive because they contain four optical surfaces instead of two. Also, refractors cannot be made larger than about 1 m in diameter because such large pieces of optically pure glass are hard to make, and because such large lenses, supported only at their edges, sag under their own weight. The largest refractor in the world is the 1-m (40-inch) telescope at Yerkes Observatory in Wisconsin. Its lens weights half a

ton. A larger refractor will probably never be built, not only because of problems inherent in refractors, but also because of the advantages of reflecting telescopes.

Reflecting Telescopes A **reflecting telescope** uses a concave mirror, the **objective mirror,** to focus starlight into an image. Objective mirrors are usually made of glass or quartz covered with a thin layer of aluminum to act as a reflective surface. Unlike household mirrors, which have the reflective layer on the back, astronomical mirrors have the layer on the front surface so the light does not enter the glass. This means that reflectors have no chromatic aberration. The light reflects off of the aluminum surface on the front surface of the mirror, and all wavelengths come to the same focus.

The objective mirror forms an image at the location called the **prime focus** at the upper end of the telescope tube. As it is often inconvenient to

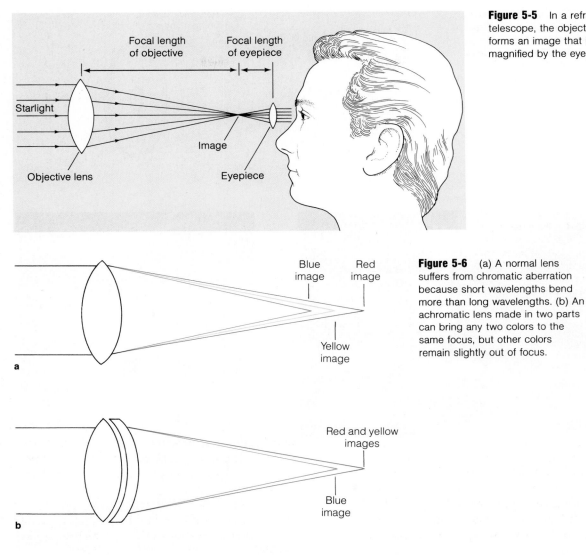

Figure 5-5 In a refracting telescope, the objective lens forms an image that is magnified by the eyepiece.

Focal length of objective

Focal length of eyepiece

Starlight

Objective lens

Image

Eyepiece

Figure 5-6 (a) A normal lens suffers from chromatic aberration because short wavelengths bend more than long wavelengths. (b) An achromatic lens made in two parts can bring any two colors to the same focus, but other colors remain slightly out of focus.

Blue image

Red image

Yellow image

a

Red and yellow images

Blue image

b

view the image there, a smaller, **secondary mirror** reflects the light to a more accessible location. In one popular arrangement, the secondary mirror reflects the light back down the telescope tube through a hole in the center of the objective (Figure 5-7). This is called a **Cassegrain telescope** (Figure 5-8).

Although many modern telescopes are Cassegrains, other focal arrangements are possible (Figure 5-9). To use the largest telescopes, the observer can ride in a prime-focus cage, an arrangement

that is good for observing faint objects. Some telescopes bring light to a **Newtonian focus** (named after Isaac Newton, the inventor of this arrangement), but the Newtonian focus of a large telescope is inconveniently far above the floor. Some telescopes can direct the light to a **coudé focus** in a separate room where a large spectrograph can be mounted. (*Coudé* comes from the French word for "elbow" and refers to the bent path of the light.) A **Schmidt camera** is a telescope specifically designed for recording wide-field, low-distortion

Figure 5-7 In a Cassegrain reflecting telescope, the objective mirror forms an image that is magnified by the eyepiece. For convenient viewing, a secondary mirror reflects the light through a hole in the objective.

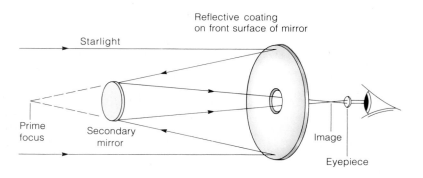

Reflective coating on front surface of mirror

Starlight

Prime focus

Secondary mirror

Image

Eyepiece

a

Figure 5-8 (a) Astronomer C. R. Lynds adjusts an instrument mounted on the Cassegrain observing cage beneath the objective mirror of the 4-m Mayall telescope at Kitt Peak National Observatory. (b) The 4-m (158-inch) Mayall telescope. During observations, astronomers run the telescope and instruments mounted on it remotely from a nearby control room. (©Association of Universities for Research in Astronomy, Inc., Kitt Peak National Observatory)

b

photographs (Figure 5-10). Many large telescopes can switch the light path from the prime focus to the Cassegrain, and sometimes to the coudé focus, but a Schmidt camera cannot be changed to another arrangement. The **Schmidt-Cassegrain telescope** combines the Schmidt's thin correcting lens with the Cassegrain's perforated mirror and is very popular with amateur astronomers.

Whatever their focal arrangement, nearly all modern telescopes use mirrors, not only because mirrors lack chromatic aberration, but also because they are more economical. A mirror has only one optical surface. Since light does not enter the mirror, the glass need not be perfectly clear. Also, a mirror can be supported over its entire back surface to reduce sagging. Nevertheless, the weight of these large mirrors is still a problem.

New-Generation Telescopes The largest conventional telescope in the world is the 6-m (236-inch) reflector on Mount Pastukhov in the former Soviet Union (Figure 5-11a), and the second largest is the

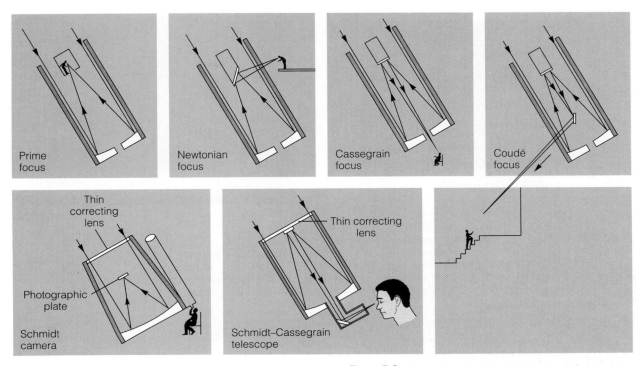

Prime focus

Newtonian focus

Cassegrain focus

Coudé focus

Thin correcting lens

Photographic plate

Schmidt camera

Thin correcting lens

Schmidt–Cassegrain telescope

Figure 5-9 Many large telescopes can be used at a number of focal positions. Schmidt cameras, however, are used only for photography. The Schmidt-Cassegrain arrangement is commonly found in small telescopes.

5-m (200-inch) Hale Telescope on Mount Palomar. Both use a single, thick mirror. The 5-m mirror alone weighs 14.5 tons, and its mounting weighs 500 tons. Such massive mirrors are very expensive to grind to shape and very difficult to support. New technologies, however, are paving the way for a new generation of telescopes.

One such technology involves the grinding of large mirrors, which can take years and be very expensive. However, at the University of Arizona, a new oven has been developed that will reduce casting costs. The oven turns like a merry-go-round, forcing the molten glass outward in the

Figure 5-10 A Schmidt camera is used solely for photography. The observer places a photographic plate inside the camera through an access hatch about midway between top and bottom, and then aims and guides the camera using the finder and guide telescopes located to the right in this photo. (National Optical Astronomy Observatories)

a

b

Figure 5-11 (a) The world's largest conventional telescope is the 6-m reflector in the former Soviet Union, shown on a Soviet stamp. It contains a mirror made from a single thick casting. (b) The six 1.8-m mirrors of the Multiple Mirror Telescope move on a single mounting and are the equivalent of a 4.5-m telescope. Both telescopes have alt-azimuth mountings. (Courtesy of the Whipple Observatory, a joint facility of the University of Arizona and the Smithsonian Institution)

mold to form a concave upper surface. Once cooled, the spun-cast mirror can be ground quickly to final shape. The oven is able to cast mirrors up to 8 m in diameter.

Another way to reduce costs is to make the mirror in segments—small segments are less expensive and weigh less. The Multiple Mirror Telescope (MMT) uses six round mirrors, each 1.8 m (72 inches) in diameter, on one mounting (Figure 5-11b). The mirrors combine their light to produce the equivalent of a 4.5-m (176-inch) telescope. The Keck Telescope in Hawaii uses 36 hexagonal mirror segments held in alignment by a computer to form a mirror 10 m (400 inches) in diameter (Figure 5-12a). The telescope has been so successful that a second telescope is being built to work in tandem with the first. A special-purpose Spectroscopic Survey Telescope (SST) being built by Penn State and University of Texas astronomers will have an 8-m mirror made up of 85 computer-controlled segments in a novel mounting (Figure 5-12b).

Thinner telescope mirrors also weigh less, but such "floppy mirrors" cannot hold their shape on their own. However, computerized support systems can control the shape of the mirror in what astronomers now call **active optics.** The New Technology Telescope (NTT) at the European Southern Observatory in Chile, for instance, contains a

3.58-m (141-inch) mirror that is only 24 cm (10 inches) thick. Such thin, lightweight mirrors not only cost less to build but also require less massive mountings.

Spin casting and active optics are making a new generation of giant telescopes possible. The six mirrors of the MMT will be replaced by a single, spun-cast, 6.5-m mirror in 1994. A number of teams from universities and observatories are planning to build 8-m telescopes, including the European Southern Observatory's Very Large Telescope (VLT), which will actually consist of four 8.2-m (323-inch) telescopes located on a mountaintop in Chile.

Not all new-generation telescopes are large. Telescopes as small as 25 cm (10 inches) have become powerful research instruments thanks to computer control. Six Automatic Photometric Telescopes (APTs) now observe from the APT Service Observatory atop Mount Hopkins near Tucson, Arizona. Every clear night of the year, desktop computers at the observatory guide the telescopes to measure the brightness of selected stars. During the day, the computers report their observations and receive further instructions from astronomers over phone lines. Such APTs can be quite small (Figure 5-12c) and inexpensive, so a number of others are under construction.

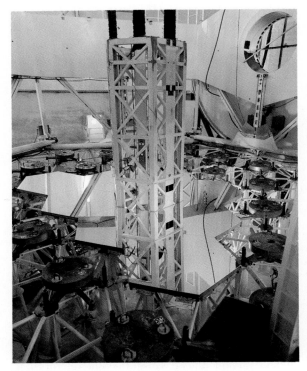

a

b

c

Figure 5-12 New-generation telescopes. (a) The Keck Telescope in Hawaii contains an array of mirror segments with a total diameter of 10 m. Here 9 segments have been installed in the telescope. (California Association for Research in Astronomy) (b) An artist's conception of the Spectroscopic Survey Telescope shows how it will use an 8-m array of mirrors to focus light on the end of an optical fiber, which will conduct the light to a remote spectrograph. The telescope will observe stars for about an hour as they pass overhead. (University of Texas at Austin, McDonald Observatory) (c) Standing only as tall as a typical astronomer, this 0.75-m Automatic Photometric Telescope can observe variable stars every clear night for months without human attention. Such robotic telescopes are changing the way astronomers work. (Russell Genet)

Telescope Mountings   Large or small, all astronomical telescopes need mountings to hold them steady, aid in locating objects, and keep them pointed properly. Because an astronomical telescope magnifies the image, a small vibration can cause blurring. Even the vibrations of someone walking across the observatory floor can have serious consequences. Thus mountings tend to be massive and expensive. Nearly all telescope mountings have indicators that permit the operator to aim the telescope at preselected objects by using the astronomical coordinate system (see Appendix B).

Although mountings must be firm, they must also move easily and smoothly to follow the stars. The eastward rotation of Earth makes objects in the sky appear to move westward. If we made no allowance for this motion, a star would move out of the field of view in a minute or less. To keep a star in view, a telescope mounting must contain a motor and gears, called a **sidereal drive,** which move the telescope smoothly westward to track the star. Most telescope mountings have one axis parallel to Earth's axis to facilitate this westward motion parallel to Earth's equator. Such mountings are called **equatorial mountings** (Figure 5-13).

Improvements in computer technology are making telescope mountings and observatories much less expensive. Instead of building a large, awkward equatorial mounting, astronomers can mount a telescope like a cannon. That is, the mounting can move the telescope in altitude (perpendicular to the horizon) and in azimuth (parallel to the horizon). Such **alt-azimuth** mountings are possible because a computer can convert the astronomical coordinates of a star into instructions to aim the telescope and can then move the telescope to track the star.

Such computer-controlled alt-azimuth mountings are stronger and smaller, and do not require buildings as large as those needed for equatorial mountings. In fact, the MMT and the Keck Telescope make the telescope mounting part of the building: As the telescope moves, the entire building rotates.

The Powers of a Telescope A telescope can aid our eyes in only three ways. These are called light-gathering power, resolving power, and magnifying power.

Most interesting celestial objects are faint sources of light, so we need a telescope that can gather large amounts of light to produce a bright image. **Light-gathering power** refers to the ability of a telescope to collect light. Catching light in a telescope is like catching rain in a bucket—the bigger the bucket, the more rain it catches (Figure 5-14). This is why astronomers use large telescopes and why they refer to telescopes by their diameters.

The second power, **resolving power,** refers to the ability of the telescope to reveal fine detail.

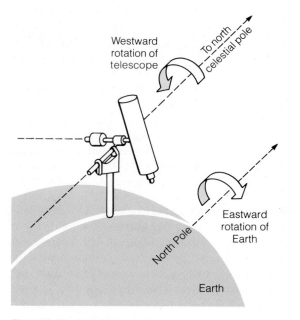

Figure 5-13 Westward motion around the polar axis of an equatorial mounting counters the eastward rotation of Earth and keeps the telescope pointed at a given star.

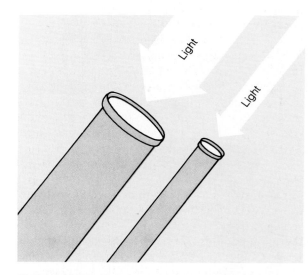

Figure 5-14 Gathering light is like catching rain in a bucket. A large-diameter telescope gathers more light and has a brighter image than a smaller telescope of the same focal length.

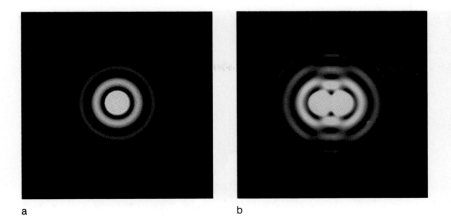

a　　　　　　　　　　　　　b

Whenever light is focused into an image, a blurred fringe surrounds the image (Figure 5-15). Because this **diffraction fringe** surrounds every point of light in the image, we can never see any detail smaller than the fringe. There is nothing we can do to eliminate diffraction fringes; they are produced by the wave nature of light. If we use a large-diameter telescope, however, the fringes are smaller and we can see smaller details. Thus the larger the telescope, the better its resolving power.

Two other factors—optical quality and atmospheric conditions—limit the detail we can see. A telescope must contain high-quality optics to achieve its full potential resolving power. Even a large telescope shows us little detail if its optics are marred with imperfections. Also, when we look through a telescope, we look through miles of turbulent air in Earth's atmosphere, which makes the image dance and blur, a condition called **seeing.** A related phenomenon is the twinkling of a star. The twinkles are caused by turbulence in Earth's atmosphere, and a star near the horizon, where we look through more air, will twinkle more than a star overhead.

On a night when the atmosphere is unsteady, the stars twinkle, the images are blurred, and the seeing is bad. Even under good seeing conditions, the detail visible through a large telescope is limited, not by its diffraction fringes, but by the air through which the observer must look. A telescope performs better on a high mountaintop, where the air is thin and steady, but even there Earth's atmosphere limits the detail the telescope can reveal.

The third and least important power of a tele-scope is **magnifying power,** the ability to make the image bigger. Since the amount of detail we can see is limited by the seeing conditions and the resolving power, very high magnification does not necessarily show us more detail. Also, we can change the magnification by changing the eye-piece, but we cannot alter the telescope's light-gathering or resolving power.

If you visit a department store to shop for a telescope, you will probably find telescopes described according to magnification. One may be labeled an ''80-power telescope'' and another a ''40-power telescope.'' However, the magnifying power really tells us little about the telescopes. Astronomers identify telescopes by diameter because that determines both light-gathering power and resolving power.

Because astronomical telescopes are so sensitive, we often find observatories located on mountaintops far from cities. The air above high mountaintops is thin, so it is more transparent, the sky is darker, and the stars are brighter. Also, the winds blow smoothly over some mountains, so there is little turbulence to ruin the seeing. In addition, mountaintops located far from cities suffer less from **light pollution,** the brightening of the night sky by light scattered from artificial outdoor lighting. Many of our cities are so brightly lit by street lights and advertising that the night sky glows with scattered light (Figure 5-16), and faint stars and galaxies are invisible. Even some of our largest observatories are threatened by growing light pollution from nearby population centers, and astronomers work actively with city planners to reduce outdoor lighting. As a consequence, astronomers

Figure 5-16 This satellite view of the continental United States at night shows the light pollution produced by outdoor lighting. The glow from such lighting makes faint stars and galaxies invisible and forces astronomers to place their telescopes as far from population centers as possible. Even so, some observatories are threatened by spreading light pollution. (U.S. Air Force photograph)

Figure 5-17 Kitt Peak National Observatory. The top of the mountain is crowded with telescope domes to take advantage of the thin, dry, stable air 2100 m (6900 ft) above sea level. The 4-m telescope dome is in the foreground. (National Optical Astronomy Observatories)

The Powers of the Telescope

Light-gathering power is proportional to the area of the telescope objective. A lens or mirror with a large area gathers a large amount of light. Because the area of a circular lens or mirror of diameter D is $\pi(D/2)^2$, we can compare the areas of two telescopes, and therefore their relative light-gathering powers, by comparing the square of their diameters. That is, the ratio of the light-gathering powers of two telescopes A and B is equal to the ratio of their diameters squared:

$$\frac{LGP_A}{LGP_B} = \left(\frac{D_A}{D_B}\right)^2$$

Example A: Suppose we compare a 4-cm telescope with a 24-cm telescope. How much more light will the large telescope gather? *Solution:*

$$\frac{LGP_{24}}{LGP_4} = \left(\frac{24}{4}\right)^2 = 6^2 = 36 \text{ times more light}$$

Example B: Our eye acts like a telescope with a diameter of about 0.8 cm, the diameter of the pupil. How much more light can we gather if we use a 24-cm telescope? *Solution:*

$$\frac{LGP_{24}}{LGP_{eye}} = \left(\frac{24}{0.8}\right)^2 = (30)^2 = 900 \text{ times more light}$$

The resolving power of a telescope is the angular distance between two stars that are just barely visible through the telescope as two separate images. For optical telescopes, the re-

solving power α, in seconds of arc, equals 11.6 divided by the diameter of the telescope in centimeters:

$$\alpha = \frac{11.6}{D}$$

Example C: What is the resolving power of a 25-cm telescope? *Solution:* The resolving power is 11.6 divided by 25, or 0.46 seconds of arc:

$$\alpha = \frac{11.6}{25} = 0.46 \text{ seconds of arc}$$

If the lenses are of good quality and if the seeing is good, we should be able to distinguish as separate points of light any pair of stars farther apart than 0.46 seconds of arc. If the stars are any closer together, diffraction fringes blur the stars together into a single image (see Figure 5-15).

The magnification of a telescope is the ratio of the focal length of the objective lens or mirror F_o divided by the focal length of the eyepiece F_e.

$$M = \frac{F_o}{F_e}$$

Example D: What is the magnification of a telescope whose focal length is 80 cm used with an eyepiece whose focal length is 0.5 cm? *Solution:* The magnification is 80 divided by 0.5, or 160 times.

place their largest telescopes on high mountaintops as far as possible from civilization (Figure 5-17).

Now that we have described the three powers of the telescope in general terms (Box 5-2 gives them mathematical form), we are ready to study the special instruments that analyze the light gathered by optical telescopes.

5-3 SPECIAL INSTRUMENTS

Merely looking through a telescope tells us little about celestial objects. Our eyes are not accurate measuring instruments and our memories are not perfectly dependable. Also, most celestial objects are faint even when viewed through a large telescope. Consequently, astronomers have developed special instruments that, attached to telescopes, can dig out the clues hidden in starlight.

Photography The photographic plate is a particularly valuable special instrument. A photograph, especially one recorded on a glass photographic plate, can be measured precisely, and it can record a complex image and not "forget" the details, as people are apt to do. More important, time exposures can reveal features too faint to be visible to a human observer peering through an eyepiece. Ever since the first astronomical photograph was taken in 1840 (a daguerreotype of the moon), astronomers have made photography a part of their tool kit.

In the 1980s, electronic cameras began to replace the photographic plate. High-sensitivity television cameras were common a few years ago but have been replaced by small arrays of diodes called **charge-coupled devices (CCDs)** (Figure 5-18a). These can record the faint image produced by a telescope in long time exposures and then read the image directly into computer memory. CCDs are more sensitive and more accurate than photographic plates.

Note that astronomical photographs, however they are recorded, are often reproduced in negative form. That is, bright and dark are reversed, so stars against a black sky look like black spots against a white background (Figure 5-18b). Such negative images reveal subtle brightness differences better than do positive images.

You may not own a CCD, but you probably have a camera, so photography is a fairly familiar process. Most astronomical instruments, however, are less common. We will discuss two kinds—spectrographs and photometers.

The Spectrograph A **spectrograph** is a device that separates starlight according to wavelength to produce a spectrum. White light is a mixture of light of various wavelengths, a fact that we can demonstrate by shining a beam of white light through a glass prism. The angle through which the prism bends the light depends on wavelength—violet light bends most, and red least (Figure 5-19). Thus a prism can separate a beam of light into a band of its component colors, that is, a spectrum. The astronomer can build a spectrograph with a single

a

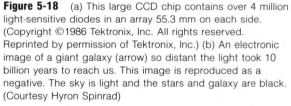

b

Figure 5-18 (a) This large CCD chip contains over 4 million light-sensitive diodes in an array 55.3 mm on each side. (Copyright ©1986 Tektronix, Inc. All rights reserved. Reprinted by permission of Tektronix, Inc.) (b) An electronic image of a giant galaxy (arrow) so distant the light took 10 billion years to reach us. This image is reproduced as a negative. The sky is light and the stars and galaxy are black. (Courtesy Hyron Spinrad)

prism or with a series of prisms to spread the wavelengths farther.

A spectrograph can be built using a grating in place of a prism. A **grating** is a piece of glass with thousands of microscopic parallel lines scribed into its surface. Different wavelengths of light reflect from the grating surface at slightly different angles, so white light striking a grating produces a spectrum. We see this effect in the iridescent colors reflected from the fine grooves in long-playing phonograph records.

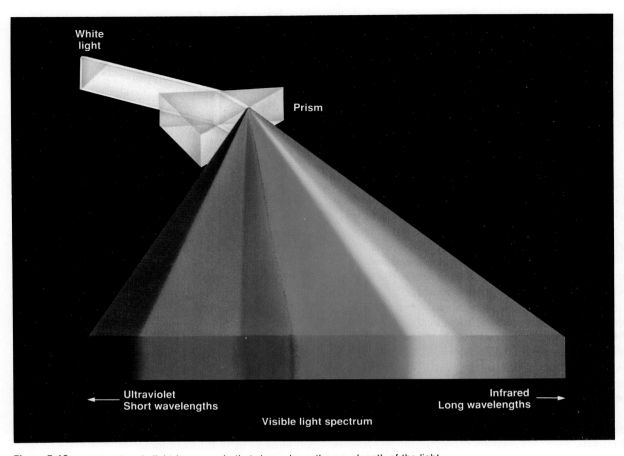

White light

Prism

Ultraviolet
Short wavelengths

Infrared
Long wavelengths

Visible light spectrum

Figure 5-19 A prism bends light by an angle that depends on the wavelength of the light. Short wavelengths bend most, and long wavelengths least. Thus white light passing through a prism is spread into a spectrum. (From James W. Kalat, *Biological Psychology,* Wadsworth Publishing)

Prism and grating spectrographs both work the same way. The telescope focuses starlight on a thin slit that admits light to the spectrograph. To keep the telescope properly pointed, the observer looks through a guiding eyepiece and keeps the image of the star on the slit. After passing through the slit, the light strikes a lens (or mirror) that guides it to the prism (or grating). A camera lens (or mirror) then focuses the various colors onto a photographic plate or CCD, where the spectrum is recorded.

To help identify the lines in a stellar spectrum, the astronomer often adds a **comparison spectrum.** A mirror directs light from an iron arc or other source of light into the spectrograph, producing an emission spectrum above and below the star's spectrum. Because the wavelengths of the lines in the comparison spectrum are precisely known, the astronomer can use it as a road map to identify wavelengths in the stellar spectrum (Figure 5-20).

The Photometer Another way to analyze starlight is to measure its intensity and color with a sensitive light meter called a **photometer.** A photometer contains a sensitive detector that produces an electric current when struck by light. The strength of this current is proportional to the intensity of the light, so a measurement of the current determines the brightness of the star.

In addition to brightness, a photometer can measure a star's color. A filter transmits only those wavelengths in a certain range—blue light between 400 nm and 480 nm, for instance. The

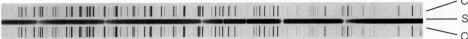

Comparison spectrum
Stellar spectrum
Comparison spectrum

Figure 5-20 In this negative reproduction of a stellar absorption spectrum, we see the spectrum of the star as the horizontal stripe in the middle. The comparison (emission) spectrum lies above and below the stellar spectrum, and contains many individual lines of known wavelength, which allow the identification and measurement of the lines in the stellar spectrum. (Palomar Observatory photograph)

difference between the brightness of the star measured through a blue filter and its brightness measured through a red filter gives a numerical index of the star's color. A blue star, for example, looks brighter through a blue filter than through a red filter. Such color measurements are important because, as we will see in the next chapter, the color of the star is related to its temperature.

The telescope and special instruments discussed so far are important tools in the search for clues in starlight, but they tell only part of the story because they are limited to the visible spectrum. To get the information in the rest of the electromagnetic spectrum, we must use radio telescopes and space telescopes.

5-4 RADIO TELESCOPES

Operation of a Radio Telescope A radio telescope usually consists of four parts: a dish reflector, an antenna, an amplifier, and a recorder (Figure 5-21). The components, working together, make it possible for astronomers to detect radio radiation from celestial objects.

The dish reflector of a radio telescope, like the mirror of a reflecting telescope, collects and focuses radiation. Because radio waves are much longer than light waves, the dish need not be as smooth as a mirror. Wire mesh works quite well as a reflector of all but the shortest radio waves (Figure 5-21). In some radio telescopes, the reflector may not even be dish-shaped, or the telescope may contain no reflector at all.

Though a radio telescope's dish may be many meters in diameter, the antenna may be as small as your hand. Like the antenna on a TV set, its only

function is to absorb the radio energy and direct it along a cable to an amplifier. After amplification, the signal goes to some kind of recording instrument. Most radio observatories record data on magnetic tape or feed it directly to a computer. However it is recorded, an observation with a radio telescope measures the amount of radio energy coming from a specific point on the sky.

Because humans can't see radio waves, astronomers must convert them into something perceptible. One way is to measure the strength of the radio signal at various places in the sky and draw a map in which contours mark areas of uniform radio intensity. We might compare such a map to a seating diagram for a baseball stadium in which the contours mark areas in which the seats have the same price (Figure 5-22a). Contour maps are very common in radio astronomy.

Radio astronomers often color code their maps to produce **false-color images.** In a false-color image, the different colors stand for different levels of intensity (Figure 5-22b). Note that not all false-color images are radio maps. X-ray, ultraviolet, visible, and infrared images can also be displayed in false-color-image format to reveal subtle variation in intensity.

Limitations of the Radio Telescope A radio astronomer works under three handicaps: poor resolution, low intensity, and interference. We saw that the resolving power of an optical telescope depends on the diameter of the objective lens or mirror. It also depends on the wavelength of the radiation. At very long wavelengths, like those of radio waves, images become fuzzy because of the large diffraction fringes. As in the case of an optical telescope, the only way to improve the resolving power is to build a bigger telescope. Consequently, radio telescopes must be quite large.

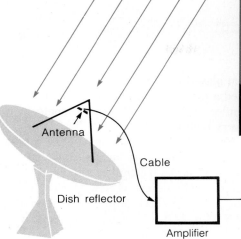

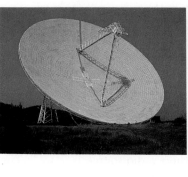

Figure 5-21 In most radio telescopes, a dish reflector concentrates the radio signal on the antenna. The signal is then amplified and recorded. For all but the shortest radio waves, wire mesh is an adequate reflector (inset). (Courtesy NRAO/AUI)

Antenna

Cable

Dish reflector

Amplifier

Computer

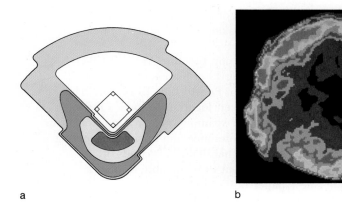

a

b

Figure 5-22 (a) A contour map of a baseball stadium shows regions of similar admission prices. The most expensive seats are those behind home plate. (b) A false-color-image radio map of Tycho's supernova remnant, the expanding shell of gas produced by the explosion of a star in 1572. The radio contour map has been color-coded to show intensity. Red is the strongest radio intensity, and violet the weakest. (Courtesy NRAO/AUI)

Even so, the resolving power of a radio telescope is not good. A dish 30 m in diameter receiving radiation with a wavelength of 21 cm has a resolving power of about 0.5°. Such a radio telescope would be unable to show us any details in the sky smaller than the moon. Fortunately, radio astronomers can combine two or more radio telescopes to improve the resolving power. (See Box 5-3.)

The second handicap radio astronomers face is the low intensity of the radio signals. We saw earlier that the energy of a photon depends on its wavelength. Photons of radio energy have such long wavelengths that their individual energies are quite low. In order to get strong signals focused on the antenna, the radio astronomer must build large collecting dishes.

The largest radio dish in the world is 300 m (1000 ft) in diameter. So large a dish can't be supported in the usual way, so it is built into a mountain valley in Arecibo, Puerto Rico (Figure 5-24). A thin metallic surface supported above the valley floor by cables serves as the reflecting dish, and the antenna hangs on cables from three towers built on the three mountains that surround the valley. Although this telescope can look only overhead, the operators can change its aim slightly by

BOX 5-3

The Radio Interferometer

Because of the poor resolving power of radio telescopes, radio astronomers can see little detail. Consequently, they cannot pinpoint the location of radio objects. They can draw a box on a star chart and say that their radio source is somewhere in the box, but there may be hundreds of stars and galaxies in the box. Which one is emitting radio signals?

To improve the resolving power of their telescopes, radio astronomers have devised the radio interferometer. A **radio interferometer** consists of two or more radio telescopes that combine their signals as if the two signals were coming from different parts of one big telescope (Figure 5-23a). The system has the resolving power of a telescope whose diameter is equal to the distance between the two dishes. Since we can locate radio telescopes miles apart, we can simulate a telescope much bigger than one we could actually build.

The smallest radio sources are only a fraction of a second of arc in diameter. To study such sources, radio astronomers connected radio telescopes in Europe, the United States, Canada, and Australia into a planetwide radio interferometer. Because it was impractical to connect the telescopes with cables, they recorded the signals on magnetic tape together with time signals from atomic clocks. Tapes made simultaneously at different locations were later synchronized according to the time signals and then played together. The combined signal simulated a radio telescope nearly 12,800 km (8000 miles) in diameter (Earth's diameter), giving very high resolution.

The National Radio Astronomy Observatory, with most of its telescopes located in Green Bank, West Virginia, has built a radio observatory in the desert of New Mexico. The Very Large Array (VLA) radio interferometer (Figure 5-23b), as it is known, consists of 27 radio dishes, each 25 m (82 ft) in diameter, located in a Y-shaped pattern about 20 km (13 miles) on each leg. The signals from the dishes, combined by computer, simulate a radio telescope 40 km (25 miles) in diameter. It can produce radio maps with a resolution better than 1 second of arc—as good as the best photographs taken by the largest optical telescopes on earth.

The Very Long Baseline Array (VLBA) consists of a series of matched radio dishes stretching from Puerto Rico to Hawaii. The dishes are linked together to form a radio interferometer spanning nearly a quarter of the Earth's circumference and having a resolving power better than 0.0003 seconds of arc—equivalent to being able to read a newspaper 600 miles away. A new proposal to place a 15-m radio dish in orbit around the earth and link it to the VLBA would produce a resolution of 0.000025 seconds of arc.

moving the antenna and by waiting for the rotation of Earth to point the telescope in the proper direction. This may sound clumsy, but the telescope's ability to detect weak radio sources, together with its good resolution, makes it one of the most important radio observatories in the world.

The third handicap the radio astronomer faces is interference. A radio telescope is an extremely sensitive radio receiver listening to radio signals thousands of times weaker than artificial radio and TV transmissions. Such weak signals are easily drowned out by interference. Sources of such

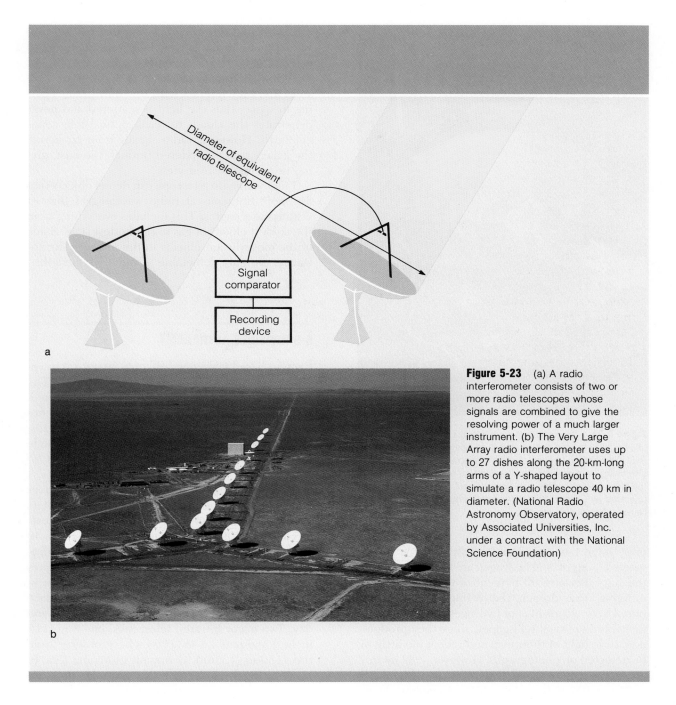

a

b

Figure 5-23 (a) A radio interferometer consists of two or more radio telescopes whose signals are combined to give the resolving power of a much larger instrument. (b) The Very Large Array radio interferometer uses up to 27 dishes along the 20-km-long arms of a Y-shaped layout to simulate a radio telescope 40 km in diameter. (National Radio Astronomy Observatory, operated by Associated Universities, Inc. under a contract with the National Science Foundation)

interference include everything from poorly designed transmitters in Earth satellites to automobiles with faulty ignition systems. To avoid this kind of interference, radio astronomers locate their telescopes as far from civilization as possible. Hidden deep in mountain valleys, they are able to listen to the sky protected from human-made radio noise.

Advantages of Radio Telescopes Building large radio telescopes in isolated locations is expensive, but three factors make it all worthwhile. First, and

Figure 5-24 The largest radio telescope in the world is the 300-m (1000-ft) dish suspended in a valley in Arecibo, Puerto Rico. The antenna hangs above the dish on cables stretching from towers. The Arecibo Observatory is part of the National Astronomy and Ionosphere Center, which is operated by Cornell University under contract with the National Science Foundation.

most important, a radio telescope can show us where clouds of cool hydrogen are located between the stars. Because 90 percent of the atoms in the universe are hydrogen, that is important information. Large clouds of cool hydrogen are completely invisible to normal telescopes, since they produce no visible light of their own and reflect too little to be detected on photographs. However, cool hydrogen emits a radio signal at the specific wavelength of 21 cm. (We will see how the hydrogen produces this radiation when we discuss atoms in the next chapter.) The only way we can detect these clouds of gas is with a radio telescope that receives the 21-cm radiation. These hydrogen clouds are the places where stars are born, and that is one reason that radio telescopes are so important.

Nevertheless, there is a second reason. Because radio signals have relatively long wavelengths, they can penetrate the vast clouds of dust that obscure our view at visual wavelengths. Light waves are short and they interact with tiny dust grains floating in space; thus the light is scattered and never penetrates the clouds to reach optical telescopes on Earth. However, radio signals from far across the galaxy pass unhindered through the dust, giving us an unobscured view.

Finally, a radio telescope can detect objects that are more luminous at radio wavelengths than at visible wavelengths. This includes everything from the coldest clouds of gas to the hottest stars. Some of the most distant objects in the universe, for instance, are only detectable at radio wavelengths.

5-5 INVISIBLE ASTRONOMY

Ground-based telescopes can only operate at wavelengths in the optical and radio "windows." The rest of the electromagnetic radiation—infrared, ultraviolet, X ray, and gamma ray—never reaches Earth's surface. To observe at these wavelengths, telescopes must fly above the atmosphere in high-flying aircraft, rockets, balloons, and satellites.

Infrared Astronomy Some infrared radiation does leak through our atmosphere. This radiation enters narrow, partially open atmospheric windows scattered from 1200 nm to about 40,000 nm (1.2 to 40 micrometers). In this range, called the near infrared, much of the radiation is absorbed by water vapor, carbon dioxide, and oxygen molecules in Earth's atmosphere, so it is an advantage to place telescopes on mountains where the air is thin and dry. Two major infrared telescopes, one 3 m (118 inches) and one 3.8 m (150 inches), observe from the 4150-m (13,600-ft) summit of Mauna Kea, in Hawaii (Figure 5-25). At this altitude, they are above much of the water vapor, which is the main absorber of infrared.

The far infrared range, which includes wavelengths longer than 40 micrometers, can tell us about planets, comets, forming stars, and other cool objects, but these wavelengths are absorbed high in the atmosphere. To observe in the far

Figure 5-25 The thin, dry air atop the 13,600-ft volcano Mauna Kea in Hawaii is ideal for infrared astronomy. From left to right, the large domes shown here house the 10-m Keck Telescope, the 3-m NASA Infrared Telescope Facility, the 3.6-m Canada–France–Hawaii Telescope, and the 3.6-m United Kingdom Infrared Telescope. (California Association for Research in Astronomy)

infrared, telescopes must venture to high altitudes. Remotely operated infrared telescopes suspended under balloons have reached altitudes as high as 41 km (25 miles).

Another solution is to fly the infrared telescope to high altitude in an airplane. NASA has modified a Lockheed C-141 jet transport to carry a 91-cm infrared telescope and a crew of a dozen astronomers to altitudes of 12,000 m (40,000 ft) to get above 99 percent of the water vapor in Earth's atmosphere (Figure 5-26). The rings of Uranus, for example, were discovered in 1977 by astronomers flying in this aircraft. American and German astronomers now propose to build the Stratospheric Observatory for Infrared Astronomy (SOFIA). SOFIA would consist of a Boeing 747-SP air freighter that would carry a 2.5-m telescope to the fringes of the atmosphere.

The ultimate solution to atmospheric absorption is to place the telescope in orbit above the atmosphere. The Infrared Astronomical Satellite (IRAS) (Figure 5-27a) carried a 56-cm (22-inch) telescope cooled to nearly absolute zero by liquid helium. It observed from an orbit 900 km high throughout most of 1983 before its helium coolant was exhausted. Among its many discoveries, IRAS found a

disk or shell of cold matter around the bright star Vega. This is the first clear evidence of planetlike material orbiting a star other than our sun. (See Chapter 16.) NASA is now planning the Space Infrared Telescope Facility (SIRTF) (Figure 5-27b), which would become a major infrared observatory in space.

IRAS, like all infrared telescopes, had to be cooled to very low temperatures. Infrared radiation is heat, and if the telescope is warm it will emit many times more infrared radiation than that coming from a distant object. Imagine trying to look at a dim, moonlit scene through binoculars that are glowing brightly. In the near infrared, only the detector, the element on which the infrared radiation is focused, must be cooled. To observe in the far infrared, however, as IRAS did, the entire telescope must be cooled.

Ultraviolet Astronomy Some infrared telescopes can observe from mountaintops, but ultraviolet radiation beyond about 290 nm, the far ultraviolet, is completely absorbed by the ozone layer extending from 20 km up to about 40 km in our atmosphere. To get above this layer, ultraviolet telescopes must go into space.

Figure 5-26 The Gerard P. Kuiper Airborne Observatory carries a 91-cm infrared telescope, astronomers, instruments, and computers to altitudes of 40,000 ft to get above 99 percent of the water vapor in Earth's atmosphere. Note the open telescope port just behind the cockpit. (Photos courtesy of NASA Ames Research Center)

a   b

Figure 5-27 (a) The Infrared Astronomical Satellite mapped the entire sky at wavelengths of 10, 25, 50, and 100 micrometers. (b) The Space Infrared Telescope Facility, now in the planning stage, would observe in the far infrared. (NASA)

The first far-ultraviolet observations of a celestial body were made in 1946 when a captured German V-2 rocket lifted instruments to an altitude of 100 km and recorded the spectrum of the sun. Since then, many rockets have carried ultraviolet instruments on short trips into space, but extended studies are possible only from satellites.

One of the most successful ultraviolet observatories is the International Ultraviolet Explorer (IUE) (Figure 5-28). Launched in January 1978, it

Figure 5-28 The International Ultraviolet Explorer, commemorated on this stamp from Sierra Leone, was launched in January 1978 with an expected life of 3 years. It is still operating as an observatory in space run remotely by astronomers working in control rooms on Earth.

carries a 45-cm (18-inch) telescope with two attached spectrographs using TV systems to record the spectra from 320 nm to 115 nm. Images are later transmitted to an Earth-based computer system that controls the satellite.

IUE has long outlived its projected lifespan and has become an important observatory available to any astronomer with a good research proposal. However, the satellite has had a number of mechanical and electronic failures over the years. Only the creative genius of NASA engineers and programmers has kept it operating.

X-Ray Astronomy Beyond the far ultraviolet, at wavelengths from 10 nm to 0.01 nm, lie the X rays. These very-high-energy photons can be produced only by violent, high-energy events, so they tell us about high-energy violence in the universe. We detect X rays coming from exploding stars and colliding galaxies and from matter smashing onto the surface of a neutron star or falling into a black hole—phenomena we will be examining in greater detail in later chapters.

Although early X-ray observations of the sky were made from balloons and small rockets in the 1960s, the age of X-ray astronomy did not really begin until 1970, when an X-ray telescope named Uhuru (Swahili for "freedom") was put into orbit. Uhuru detected nearly 170 separate sources of celestial X rays. In the late 1970s, the three High Energy Astronomy Observatories (HEAO) satel-

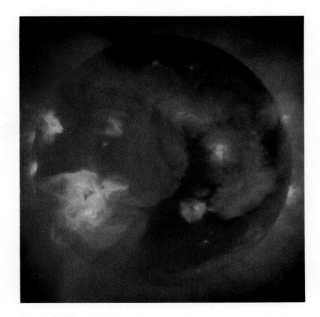

Figure 5-29 This X-ray image of the sun was recorded by the Yohkoh solar research spacecraft of the Japanese Institute of Space and Astronautical Science. The image reveals intensely hot regions in the solar atmosphere linked by magnetic fields. (Courtesy Japan/US Yohkoh Team)

lites, carrying more sensitive and more sophisticated equipment, pushed the total to many hundreds. The second HEAO satellite, named the Einstein Observatory, used special optics to produce X-ray images. A few other orbiting X-ray telescopes have imaged the sun (Figure 5-29).

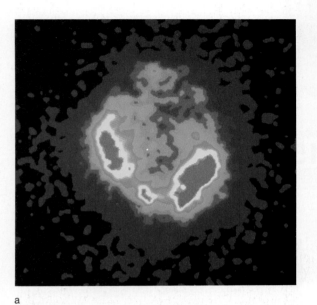

a   b

Figure 5-30 (a) An X-ray image of the Cassiopeia A supernova remnant obtained by the Exosat orbiting observatory shows a shell of hot gas expanding away from the site of a stellar explosion that occurred about 310 years ago. (Courtesy of J. Trümpter, Max Planck Institute) (b) The Advanced X-ray Astrophysics Facility will carry a 1.2-m X-ray telescope. Designed to be launched and maintained by a space shuttle, it is expected to reach orbit in the 1990s. (Courtesy TRW and Harvey Tananbaum)

A number of smaller X-ray telescopes have ventured into space. The European Space Agency satellite Exosat observed from 1983 to 1986, and Japanese and Soviet satellites have also observed at X-ray wavelengths. The German–U.S.–British satellite ROSAT was launched in 1990, first to survey the entire sky and then to study selected objects. NASA is now planning a new X-ray observatory called the Advanced X-ray Astrophysics Facility (AXAF) (Figure 5-30). With a 1.2-m (47-inch) telescope operating from 6 nm to 0.15 nm, AXAF will have about 100 times the sensitivity of the best X-ray telescopes yet launched.

The Hubble Space Telescope Astronomers have dreamed of having a giant telescope in orbit since the 1920s, but work on the Hubble Space Telescope did not begin until 1978, and problems with

the space shuttle forced delay after delay. The telescope was finally released into Earth orbit on April 24, 1990.

Named after Edwin P. Hubble, who discovered the expansion of the universe (Chapter 15), the Hubble Space Telescope (HST) is the largest orbiting telescope ever built (Figure 5-31). As big as a bus, it carries a 2.4-m (96-inch) mirror and five instruments each the size of a telephone booth. One spectrograph is designed for high resolution, while another is designed to observe faint objects such as galaxies. A high-speed photometer contains over 100 filters ranging from the ultraviolet to the infrared. Two cameras are on board, one for faint objects and one for planets and wide-field images. When fully operational, the telescope will be able to see objects 50 times fainter than can any telescope on Earth. It will be able to see objects of a

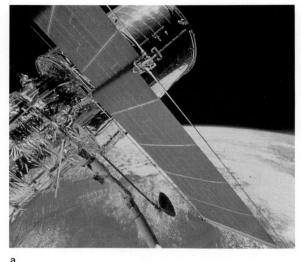

a

b

c

Figure 5-31 (a) The Hubble Space Telescope (HST) was carried into orbit by the space shuttle and released at an altitude of 610 km (380 miles). The shuttle manipulator arm extends from bottom center. (ESA-NASA) (b) The photo of the stellar object R136 taken from Earth shows only an irregular blob. (Meylan-ESO) (c) But computer-processed images from the HST reveal many stars in the cluster never seen separately before. (NASA)

given brightness up to 7 times farther away, and without the turbulence of Earth's atmosphere, it will reveal details 10 times smaller than Earth-based telescopes can.

HST should be able to study an impressive list of astronomical mysteries. Improved measurements of distances to nearby galaxies will help pin down the age and size of the universe. Photometric and spectroscopic observations will illuminate the birth and death of stars, and camera images will monitor weather on the planets, the shape of comets, and the distribution of asteroids.

After years of disappointments, astronomers everywhere were elated by the launch of HST. But in the weeks after the launch, engineers discovered a tiny error in the giant mirror. It is 2 micrometers too shallow, and consequently it cannot bring all of the light into focus at the same time. The original design called for 70 percent of a star's light to fall into an image only 0.2 seconds of arc in diameter. In fact, only 15 percent of the light falls into such a small region. The rest is scattered into a fuzzy halo around the image.

The error in the mirror affects the wide-field and planetary camera and the faint-object camera because the images cannot be focused. The spectrographs also suffer because it is not possible to focus all of the light from a star or galaxy into the spectrographs' entrance slit. Thus spectra can be recorded, but much longer exposures are needed.

Plans call for shuttle astronauts to install corrective optics during a 1994 spacewalk, but other engineering problems have developed. The solar panels vibrate when they warm in sunlight after emerging from earth's shadow, and some of the gyroscopes have failed. Thus new panels and gyroscopes must be installed in 1994 as well, and trouble with a power supply may indicate a need for yet more repairs.

Although the press has harped on the problems, astronomers have been able to computer-process images to correct for some of the optical problems, and engineers have found ways to work around problems with the solar panels and gyroscopes. The telescope is producing valuable results (Figure 5-31b and c), but some of the most significant observations will have to be postponed until after the repair mission.

Space telescopes, radio telescopes, and Earth-based telescopes are the astronomer's link to the sky. They gather electromagnetic radiation for analysis. Now that we have described these tools, we are ready to begin analyzing starlight. In Unit 2, we seek an answer to a simple question: What are the stars?

SUMMARY

Electromagnetic radiation is an electric and magnetic disturbance that transports energy at the speed of light. The electromagnetic spectrum includes radio waves, infrared radiation, visible light, ultraviolet radiation, X rays, and gamma rays.

We can think of a "particle" of light, or a photon, as a bundle of waves that sometimes acts as a particle and sometimes as a wave. The energy a photon carries depends on its wavelength. The wavelength of visible light, usually measured in nanometers (10^{-9} m), ranges from 400 nm to 700 nm. Infrared and radio photons have longer wavelengths and carry less energy. Ultraviolet, X-ray, and gamma ray photons have shorter wavelengths and carry more energy.

To obtain data, astronomers use telescopes to gather light, see fine detail, and magnify the image. The first two of these three powers of the telescope depend on the telescope's diameter; thus astronomical telescopes often have large diameters.

Astronomical telescopes are of two types, refractor and reflector. A refractor uses a lens to bend the light and focus it into an image. Because of chromatic aberration, refracting telescopes cannot bring all colors to the same focus, resulting in color fringes around the images. An achromatic lens partially corrects for this, but such lenses are expensive and cannot be made larger than about 1 m in diameter.

Reflecting telescopes use a mirror to focus the light and are less expensive than refracting telescopes of the same diameter. In addition, reflecting telescopes do not suffer from chromatic aberration. Thus most recently built large telescopes are reflectors.

The light gathered by an astronomical telescope can be recorded and analyzed by special instruments attached to the telescope. For many decades, astronomers have used photographic plates to record images at the telescope, but modern electronic systems such as CCDs have now replaced photographic plates in many applications. Spectrographs can spread starlight out according to wavelength to form a spectrum, and a photometer can measure the brightness and color of stars viewed through a telescope.

To observe radio signals from celestial objects, we need a radio telescope, which usually consists of a dish reflector, an antenna, an amplifier, and a recorder. Such an instrument can measure the intensity of radio signals over the sky and construct radio maps. The poor resolution of the radio telescope can be improved by combining it with another radio telescope to make a radio interferometer. Radio telescopes have three important features—they can detect cool hydrogen, they can see through dust clouds in space, and they can detect certain objects invisible at other wavelengths.

Earth's atmosphere admits radiation primarily through two wavelength intervals, or windows—the visual window and the radio window. At other wavelengths, our atmosphere absorbs radiation. To observe in the far infrared, astronomers must fly telescopes high in balloons or aircraft, though they can work at some wavelengths in the near infrared from high mountaintops. To observe in the ultraviolet and X-ray range and some parts of the infrared, they must send their telescopes into space to get above our atmosphere.

NEW TERMS

electromagnetic
    radiation
wavelength
photon
angstrom (Å)
atmospheric window
refracting telescope
focal length
objective lens
eyepiece
chromatic aberration
achromatic lens
reflecting telescope
objective mirror
prime focus
secondary mirror
Cassegrain telescope
Newtonian focus
coudé focus
Schmidt camera

Schmidt-Cassegrain
    telescope
active otpics
sidereal drive
equatorial mounting
alt-azimuth mounting
light-gathering power
resolving power
diffraction fringe
seeing
magnifying power
light pollution
charge-coupled device
    (CCD)
spectrograph
grating
comparison spectrum
photometer
false-color image
radio interferometer

REVIEW QUESTIONS

1. If you visited another planet, its atmosphere might contain gases different from those in Earth's atmosphere. Would you expect its atmospheric windows to be the same as those on Earth?

2. Identify by name and approximate wavelength the different regions of the electromagnetic spectrum. How do astronomers observe in each region?

3. What are the three powers of a telescope? Which is most important, which depends mostly on the atmosphere, and which can be changed easily?

4. Why are large, modern telescopes reflectors rather than refractors?

5. How has the development of computers changed the design of large astronomical telescopes and their mountings?

6. Optical astronomers place their telescopes on mountaintops, but radio astronomers often place their telescopes in mountain valleys. Why?

7. What are some of the advantages and disadvantages of radio telescopes?

8. Why is it that we can observe in the infrared from aircraft, but we must go into space to observe in the ultraviolet?

9. Why is it necessary to cool a telescope observing in the far infrared, but not necessary to cool a telescope observing in the far ultraviolet?

10. What are the advantages of placing a telescope in orbit?

DISCUSSION QUESTIONS

1. Why does the wavelength response of the human eye match so well the visual window of Earth's atmosphere?

2. Basic research in chemistry, physics, biology, and similar sciences is supported in part by industry. How is astronomy different? Who funds the major observatories?

PROBLEMS

1. The wavelength of red light is about 650 nm. How many of these waves would be needed to stretch 1 mm?

2. What is the approximate wavelength of blue light in nanometers and in meters? (HINT: See Figure 5-2.)

3. Use a ruler to measure the actual wavelength of the wave in Figure 5-1. In what portion of the electromagnetic spectrum would it belong?

4. How much more energy is carried by a photon of wavelength 1 nm (an X ray) than by a photon of wavelength 500 nm (visible light)?

5. Compare the light-gathering powers of a 5-m telescope and a 0.5-m telescope.

6. One proposed new telescope would place two 8-m mirrors on one mounting and combine their light. How large would a single mirror have to be to match the telescope's light-gathering power?

7. What is the resolving power of a 25-cm telescope? What do two stars, 1.5 seconds of arc apart, look like through this telescope?

8. If we built a telescope with a focal length of 50 cm, what focal length would the eyepiece have to be to give a magnification of 100 times?

9. Use a ruler to measure the focal lengths of the objective and the eyepiece of the telescope in Figure 5-5. What is the magnification?

10. Compare the radio-wave-gathering power of a radio telescope 100 ft across with that of the Arecibo telescope.

RECOMMENDED READING

BEATTY, J. KELLY. "Rosat and the X-Ray Universe." *Sky and Telescope* 80 (August 1990), p. 128.

BERRY, RICHARD. "The Telescope That Defies Gravity." *Astronomy* 16 (July 1988), p. 42.

———."Thinking Telescopes." *Astronomy* 16 (August 1988), p. 42.

———."Seeing Sharp." *Astronomy* 18 (July 1990), p. 38.

BISHOP, ROY L. "Newton's Telescope Revealed." *Sky and Telescope* 59 (March 1980), p. 207.

CORDOVA, FRANCE A., and KEITH O. MASON. "Exosat: Europe's New X-Ray Satellite." *Sky and Telescope* 67 (May 1984), p. 397.

CRAWFORD, DAVID L. "The Battle Against Light Pollution." *Sky and Telescope* 80 (July 1990), p. 23.

DAVIES, JOHN K. "The Extreme Ultraviolet: A Promising New Window on the Universe." *Astronomy* 15 (July 1987), p. 82.

FIELD, GEORGE B., and ERIC J. CHAISSON. *The Invisible Universe.* Cambridge, Mass.: Birkhauser Boston, 1985.

FIELD, GEORGE B., and DONALD GOLDSMITH. *Space Telescope: Eyes Above the Atmosphere.* Chicago: Contemporary Books, 1990.

FIENBERG, RICHARD TRESCH "HST: Astronomy's Discovery Machine." *Sky and Telescope* 79 (April 1990), p. 366.

———."HST Update: Science amid Setbacks." *Sky and Telescope* 82 (September 1991), p. 242.

FISCHER, DANIEL. "A Telescope for Tomorrow." *Sky and Telescope* 78 (September 1989), p. 248.

GENET, RUSSELL, KENNETH KISSELL, and GEORGE ROBERTS. "Our Turn at Kitt Peak." *Sky and Telescope* 63 (March 1982), p. 240.

GILLETT, FREDERICK C., IAN GATLEY, and DAVID HOLLENBACH. "Infrared Astronomy Takes Center Stage." *Sky and Telescope* 82 (August 1991), p. 148.

GULL, THEODORE R. "Astro: Observatory in a Shuttle." *Sky and Telescope* 79 (June 1990), p. 591.

JANESICK, JAMES, and MORELEY BLOUKE. "Sky on a Chip: The Fabulous CCD." *Sky and Telescope* 74 (September 1987), p. 238.

KELLERMANN, KENNETH I. "Radio Astronomy: The Next Decade." *Sky and Telescope* 82 (September 1991), p. 247.

KING, G. C. *The History of the Telescope.* Cambridge, Mass.: Sky, 1955.

LONGAIR, MALCOLM. *Alice and the Space Telescope.* Baltimore: Johns Hopkins University Press, 1989.

MAMMANA, DENNIS L. "The Incredible Spinning Oven." *Sky and Telescope* 70 (July 1985), p. 7.

ROBINSON, LEIF J. "Monster Telescopes for the 1990's." *Sky and Telescope* 73 (May 1987), p. 495.

SCHIELDS, JOHN POTTER. "Backyard Radio Astronomy." *Astronomy* 11 (March 1983), p. 75.

SHORE, LYS ANN. "IUE: Nine Years of Astronomy." *Astronomy* 15 (April 1987), p. 14.

———."The Telescope That Never Sleeps." *Astronomy* 15 (August 1987), p. 14.

SINNOTT, ROGER W. "The Keck Telescope's Giant Eye." *Sky and Telescope* 80 (July 1990), p. 15.

SMITH, DAVID H. "MERLIN: A Wizard of a Telescope." *Sky and Telescope* 67 (January 1984), p. 31.

SMITH, ROBERT W. *The Space Telescope: A Study of NASA, Science, Technology, and Politics.* New York: Cambridge University Press, 1989.

SOIFER, BARUCH T., CHARLES A. BEICHMAN, and DAVID B. SANDERS. "An Infrared View of the Universe." *American Scientist* 77 (January-February 1989), p. 46.

SPRADLEY, JOSEPH L. "The First True Radio Telescope." *Sky and Telescope* 76 (July 1988), p. 28.

STROM, STEPHEN. "New Frontiers in Ground-Based Optical Astronomy." *Sky and Telescope* 82 (July 1991), p. 18.

TUCKER, WALLACE, and RICCARDO GIACCONI. *The X-Ray Universe.* Cambridge, Mass.: Harvard University Press, 1985.

TUCKER, WALLACE, and KAREN TUCKER. *The Cosmic Inquirers.* Cambridge, Mass.: Harvard University Press, 1986.

VERSCHUUR, GARRIT. *The Invisible Universe Revealed.* New York: Springer-Verlag, 1987.

WOLFF, S. "The Search for Aperture: A Selective History of Telescopes." *Mercury* 14 (September-October 1985), p. 139.

The
Stars

Atoms and Starlight

No laboratory jar on Earth holds a sample labeled "star stuff," and no instrument has ever probed inside a star. The stars are far beyond our reach, and the only information we can obtain about them comes to us hidden in light (Figure 6-1). Whatever we want to know about the stars we must catch in a noose of light.

We earth-bound humans knew almost nothing about stars until the early 19th century, when the Munich optician Joseph von Fraunhofer studied the solar spectrum and found it crossed by some 600 dark lines. As scientists realized that the lines were related to the various atoms in the sun and found that stellar spectra had similar patterns of lines, the door to an understanding of stars finally opened.

In this chapter, we will go through that door by considering how atoms interact with light to produce spectral lines. We begin

Awake! for Morning in the Bowl of Night Has flung the stone that put the Stars to Flight! And lo! the Hunter of the East has caught The Sultan's Turret in a Noose of Light.

The Rubaiyat of Omar Khayyám
Edward FitzGerald, translator

with the hydrogen atom because it is the most common atom in the universe, as well as the simplest. Other atoms are larger and more complicated, but in many ways their properties resemble those of hydrogen. Another reason for studying hydrogen is to consider how the structure of its atoms can give rise to the 21-cm wavelength radiation that is so important to radio astronomers (see Chapter 5).

Once we understand how an atom's structure can interact with light to produce spectral lines, we will recognize certain patterns in stellar spectra. By classifying the spectra according to these patterns, we can arrange the stars in a sequence according to temperature. One of the most important pieces of information revealed in a star's spectrum is its temperature.

But, properly analyzed, a stellar spectrum can tell us much more. The strength of different

Figure 6-1 So vast that light takes 20 years to cross its diameter, the Eagle Nebula glows deep red. At a distance of 6000 ly, the nebula and the associated star cluster NGC 6611 are far beyond our reach. But by analyzing the light from the nebula and the cluster, astronomers conclude that the nebula is mostly hydrogen gas and that the star cluster is a young group of stars that have formed recently from the nebula. (National Optical Astronomy Observatories)

features in the spectrum indicates the chemical composition of the star, and the wavelengths of such features can tell us about the star's motion relative to Earth.

6-1 ATOMS

A Model Atom   In Chapter 2, we devised a model of the sky, the celestial sphere, to help us think about the nature and motion of the heavens. In the case of the atom, we again need a model.

Our model of the atom consists of a small central **nucleus** surrounded by a cloud of whirling electrons. The nucleus has a diameter of about 0.0000016 nm, and the cloud of electrons has a diameter of about 0.1–0.5 nm. (Recall from Chapter 5 that 1 nm is 10^{-9} m.) Household plastic wrap is about 100,000 atoms thick. This makes the atom seem very small, but the nucleus is 100,000 times smaller still. (Box 6-1 describes a scale model of an atom.)

The nucleus of a typical atom consists of two different kinds of particles—protons and neutrons. **Protons** carry a positive electrical charge, and **neutrons** have no charge. Consequently, an atomic nucleus, made of protons and neutrons, has a net positive charge.

The **electrons** surrounding the nucleus are low-mass particles carrying a charge equal to but opposite from that of the protons. In a normal atom, the number of electrons equals the number of protons. Thus the positive charge of each proton is balanced by the negative charge of an electron, and the atom is electrically neutral.

There are over a hundred kinds of atoms, called chemical elements. The kind of element an atom represents depends only on the number of protons in the nucleus. For example, carbon has six protons and six neutrons in its nucleus. Adding a proton produces nitrogen, and subtracting a proton produces boron.

We can, however, change the number of neutrons in an atom's nucleus without changing the atom significantly. For instance, if we add a neutron to the carbon nucleus, we still have carbon, but it is slightly heavier than normal carbon. Atoms that have the same number of protons but a different number of neutrons are **isotopes.** Carbon has two stable isotopes. One form contains six protons and six neutrons for a total of 12 particles, and is thus called carbon-12. Carbon-13 has six protons and seven neutrons in its nucleus.

Protons and neutrons are bound tightly into the nucleus, but the electrons are held loosely in the electron cloud. Running a comb through your hair creates a static charge by removing a few electrons from their atoms. This process is called **ionization,** and the atom that has lost one or more electrons is an **ion.** The neutral carbon atom, with six protons and six neutrons in its nucleus, has six electrons, which balance the positive charge of the nucleus. If we ionize the atom by removing one or more electrons, the atom is left with a net positive charge. Under some circumstances, an atom may capture one or more extra electrons, giving it more negative charges than positive. Such a negatively charged atom is also considered an ion.

Atoms form bonds with each other by exchanging or sharing electrons. Two or more atoms bonded together form a **molecule.** Few atoms can form chemical bonds in stars, where the high temperatures produce such violent collisions between atoms that most molecules break up. Only in the coolest stars are the collisions gentle enough to permit chemical bonds. We will see later that the presence of molecules such as titanium oxide (TiO) in a star is a clue that the star is cool. In later chapters, we will see that molecules can form in cool gas clouds in space and in the atmospheres of planets.

Electron Shells   So far we have described the electron cloud only in a general way, but the specific way electrons behave within the cloud is very important in astronomy.

The electrons are bound to the atom by the attraction between their negative charge and the positive charge on the nucleus. This force is known as the **Coulomb force,** after the French physicist Charles-Augustin de Coulomb (1736–1806). If we wish to ionize the atom, we need a certain amount of energy to pull an electron away from its nucleus. This energy is the electron's **binding energy,** the energy that holds it to the atom.

An electron may orbit the nucleus at various distances. If the orbit is small, the electron is close to the nucleus, and a large amount of energy is needed to pull it away. Therefore its binding ener-

BOX 6-1

A Scale Model of the Atom

Suppose we could make a hydrogen atom bigger by a factor of 10^{12} (1 million million). Only then would it be big enough to examine.

The nucleus of a hydrogen atom is a proton whose diameter is about 0.0000016 nm, or 1.6 $\times$ 10^{-15} m. Multiplying by a factor of 10^{12} magnifies it to 0.16 cm, about the size of a grape seed. The electron cloud* has a diameter of about 0.4 nm, or 4×10^{-10} m. When we magnify the atom by 10^{12}, this becomes 400 m, or about $4\frac{1}{2}$ football fields laid end to end (Figure 6-2). When you imagine a grape seed in the midst of $4\frac{1}{2}$ football fields, orbited by one magnified electron still too small to be visible, you see that an atom is mostly empty space.

Looking at our model atom, we find that the nucleus contains nearly all of the mass. Protons and neutrons have almost equal mass—about 1836 times the mass of an electron. Thus the electrons in an atom never represent more than about 0.05 percent of the atom's mass.

Magnifying the diameter of an atom by 10^{12} makes it large enough to see, but multiplying the mass of a typical atom by 10^{12} would still leave it with a mass less than 2×10^{-15} kg. This is too small to be meaningful. We would have to multiply the mass by another factor of 10^{12} just

*For a representative diameter, we take the size of the atom's second orbit (see Figure 6-3).

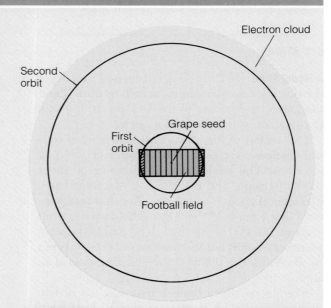

Figure 6-2 Magnifying a hydrogen atom by 10^{12} makes the nucleus the size of a grape seed and the outer electron cloud about $4\frac{1}{2}$ times bigger than a football field. The electron itself is still too small to see.

to get a mass we could imagine. Individual atoms have such small masses that only by assembling vast numbers can nature build such massive objects as stars. The sun, for instance, contains about 10^{57} atoms. Even a single grain of table salt contains over 10^{18} atoms—enough to give a billion atoms to every person on Earth.

---

gy is large. An electron orbiting farther from the nucleus is held more loosely, and less energy will pull it away. It therefore has less binding energy. The size of an electron's orbit is related to the energy that binds it to the atom.

Nature permits atoms only certain amounts (quanta) of binding energy, and the laws that describe how atoms behave are called the laws of **quantum mechanics.** Much of our discussion of

atoms is based on the laws of quantum mechanics.

Because atoms can have only certain amounts of binding energy, our model atoms can have orbits of only certain sizes called **permitted orbits** (Figure 6-3). These are like steps in a staircase: you can stand on the number-one step or the number-two step, but not on the number-one-and-one-quarter step. The electron can occupy any permitted orbit but not orbits in between.

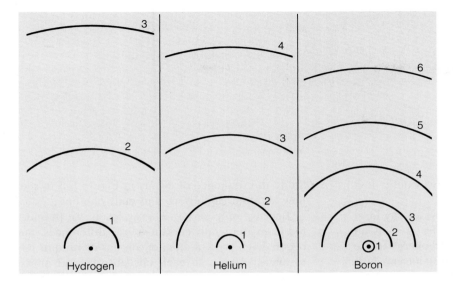

Figure 6-3 The electrons in atoms may occupy only certain, permitted orbits. Since they have different charges on the nucleus, different elements have different patterns of permitted orbits.

The arrangement of permitted orbits depends primarily on the charge of the nucleus, which in turn depends on the number of protons. Thus each kind of element has its own pattern of permitted orbits (Figure 6-3). Isotopes of the same elements have nearly the same pattern because they have the same number of protons. However, ionized atoms have orbital patterns that differ from their unionized forms. Thus the arrangement of permitted orbits differs for every kind of atom and ion.

The properties of electron orbits are important to astronomy because the electrons can interact with light, our major clue from afar. By understanding this interaction, we can interpret the lines in spectra and learn such things as the composition and temperature of stars, gas clouds in space, and atmospheres of planets.

6-2 THE INTERACTION OF LIGHT AND MATTER

We begin our study of light and matter by considering the hydrogen atom. As we noted earlier, hydrogen is both simple and common. Roughly 90 percent of all atoms in the universe are hydrogen.

The Excitation of Atoms Each orbit in an atom represents a specific amount of binding energy, so

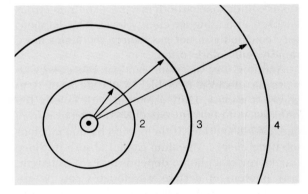

Figure 6-4 A hydrogen atom can absorb one of a number of different wavelength photons and move the electron to one of its higher orbits.

physicists commonly refer to the orbits as **energy levels.** Using this terminology, we can say of the hydrogen atom in Figure 6-4 that the electron is in the lowest permitted energy level, where it is tightly bound to the atom.

We can move the electron from one energy level to another by supplying enough energy to make up the difference between the two energy levels. It is like moving a flowerpot from a low shelf to a high shelf; the greater the distance between the shelves, the more energy we need to raise the pot. The amount of energy needed to move the electron is

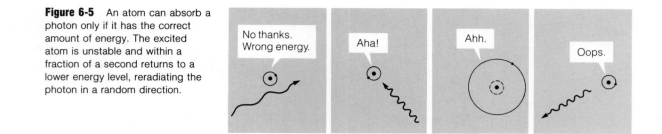

Figure 6-5 An atom can absorb a photon only if it has the correct amount of energy. The excited atom is unstable and within a fraction of a second returns to a lower energy level, reradiating the photon in a random direction.

the energy difference between the two energy levels.

If we move the electron from a low energy level to a higher energy level, we say the atom is an **excited atom.** That is, we have added energy to the atom in moving its electron. If the electron falls back to the lower energy level, that energy is released.

An atom can become excited by collision. If two atoms collide, one or both may have their electrons knocked into a higher energy level. This happens very commonly in hot gas, where the atoms move rapidly and collide often.

Another way an atom can get the energy to move an electron to a higher energy level is to absorb a photon. Only a photon with exactly the right amount of energy can move the electron from one level to another. If the photon has too much or too little energy, the atom cannot absorb it. Since the energy of a photon depends on its wavelength, only photons of certain wavelengths can be absorbed by a given kind of atom. The atom in Figure 6-4 can absorb any of three wavelengths, moving its electron up to any of three permitted energy levels. Any other wavelength photon has too much or too little energy to be absorbed.

Atoms, like humans, cannot exist in an excited state forever. The excited atom is unstable and must eventually (usually within 10^{-6} to 10^{-9} sec) give up the energy it has absorbed and return its electron to the lowest energy level. Since the electrons eventually tumble down to this bottom level, physicists call it the **ground state.**

When the electron drops from a higher to a lower energy level, it moves from a loosely bound level to one more tightly bound. The atom then has a surplus of energy—the energy difference between the levels—which it can emit as a photon.

Study the sequence of events in Figure 6-5 to see how an atom can absorb and emit photons.

Because only certain energy levels are permitted in an atom, only certain energy differences can occur. Each type of atom or ion has its unique set of energy levels, so each one absorbs and emits photons with a unique set of wavelengths. Thus we can identify the elements in a gas by studying the characteristic wavelengths of light absorbed or emitted.

This process of excitation and emission is a common sight. The gas in a neon sign glows because a high voltage forces electrons to flow through the gas, exciting the atoms by collisions. Almost as soon as an atom is excited, its electron drops back to a lower energy level, emitting the surplus energy as a photon of a certain wavelength. Neon is a popular gas for signs because the pattern of its energy levels makes it emit a rich reddish-orange light when it is excited. So-called neon signs of other colors contain other gases or mixtures of gases instead of pure neon.

The Formation of a Spectrum To see how an astronomical object like a star can produce a spectrum, imagine a cloud of hydrogen gas floating in space with an incandescent light bulb glowing behind it. This imaginary experiment will help us understand how heated bodies produce light, how light interacts with a gas, and how stars produce spectra.

The bulb glows because its filament is hot, and it emits what physicists call **black-body radiation.** (See Box 6-2). An idealized perfect radiator emits radiation at all wavelengths to produce a **continuous spectrum,** a spectrum with no gaps (Figure 6-7). According to the theory, a perfect radiator would also be a perfect absorber, capable of absorbing radiation at all wavelengths. Such an object

BOX 6-2

Radiation from a Heated Object

Whenever we produce a changing electric field, we produce electromagnetic radiation. If we run a comb through our hair, we disturb electrons (negatively charged subatomic particles) in both hair and comb, producing static electricity. Since each electron is surrounded by an electric field, any sudden change in the electron's motion gives rise to electromagnetic radiation. Running a comb through your hair while standing near an AM radio produces radio static. This illustrates an important principle: whenever we change the motion of an electron, we generate electromagnetic waves.

To see what this has to do with a heated object, think of what we mean by heat. When we say an object is hot, we mean that its atoms are vibrating rapidly. The hotter the object is, the more motion among the atoms. The vibrating atoms collide with electrons in the material, and each time the motion of one of these electrons gets disturbed, it emits a photon. Consequently, we should expect a heated object to emit electromagnetic radiation. Such radiation is called black-body radiation and is quite common. In fact, it is responsible for the light emitted by the hot filament in an incandescent light bulb.

In the heated filament, gentle collisions produce low-energy photons with long wavelengths, and violent collisions produce high-energy photons with short wavelengths. If we graph the energy emitted at different wavelengths, we get a curve like those shown in Figure 6-6. The curve shows that gentle collisions and violent collisions are rare. Most collisions are intermediate in violence, producing photons of intermediate wavelength.

The wavelength at which an object emits the maximum amount of energy, called the **wavelength of maximum intensity (λ_{max})**, depends on the object's temperature. If we heat the object, the average collision should *(continued)*

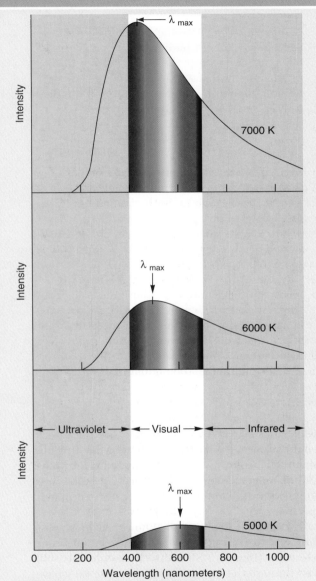

Figure 6-6 Black-body radiation from three objects of slightly different temperature demonstrates how a hotter object (top) emits more total energy than a cooler object (bottom). Also, the hotter body emits more short-wavelength radiation and thus looks blue, while the cooler object emits more long-wavelength radiation and thus looks red. Note that the wavelength of maximum intensity, λ_{max}, shifts to longer wavelengths as temperature falls.

Radiation from a Heated Object
(continued)

be more violent, producing higher-energy, shorter-wavelength photons. The hotter the object is, the shorter the λ_{max} (Figure 6-6). Basic physics shows that λ_{max} in nanometers equals 3 million divided by the temperature in degrees Kelvin (see Appendix C):

$$\lambda_{max} = \frac{3,000,000}{T}$$

This equation, commonly known as Wien's law, is a powerful tool in astronomy because it means we can determine the temperature of a star from its light.

In fact, we can estimate a star's temperature from its color. For a hot star, λ_{max} lies in the ultraviolet and we cannot see most of the radiation, but in the visible range the star emits more blue light than red. Thus a hot star looks blue. In contrast, a cool star radiates its maximum energy in the infrared. In the visible part of the spectrum, however, it radiates more red than blue and thus looks red.

The total amount of radiation emitted at all wavelengths depends on the number of collisions per second. If an object's temperature is high, there are many collisions and it emits more light than a cooler object of the same size. We measure energy in units called **joules**; 1 J is about the energy of an apple falling from a table to the floor. The total radiation given off by 1 m² of the object in joules per second equals a constant number, represented by σ, times the temperature raised to the fourth power.* This is called the Stefan–Boltzmann law:

$$E = \sigma T^4 \ (\text{J/sec/m}^2)$$

If we doubled an object's temperature, for instance, it would radiate 2^4, or 16, times more energy. Thus we can expect hot stars to radiate large amounts of energy from each square meter, and most of their radiation is at short wavelengths.

For the sake of completeness, we should note that the constant σ equals 5.67×10^{-8} J/m²·sec·degree⁴.

would look black at room temperature because it would absorb all of the radiation that hits it. Thus physicists refer to an ideal radiator as a black body. The filament in our incandescent light bulb emits black-body radiation and produces a continuous spectrum.

The light from this bulb, however, must pass through the hydrogen gas before it can reach our telescope (Figure 6-8). Most of the photons will pass through the gas unaffected because they have wavelengths the hydrogen atoms cannot absorb, but a few photons will have the right wavelengths. These photons cannot pass through the gas because they are absorbed by the first atom they meet. The atom is excited for a fraction of a second, and the electron then drops back to a lower energy level and a new photon is emitted. The original photon was traveling through the gas toward our telescope, but the new photon is emitted in some random direction. Very few of these new photons leave the cloud in the direction of our telescope, so the light that finally enters the telescope has very few photons at the wavelengths the atoms can absorb. When we form a spectrum from this light, photons of these wavelengths are missing and the spectrum has dark lines at the positions these photons would have occupied. These dark lines, like those the Munich optician Fraunhofer saw in the solar spectrum in 1814 and 1815, are called **absorption lines** because the atoms absorbed the photons. A spectrum containing absorption lines is an **absorption spectrum** (also called a **dark-line spectrum**).

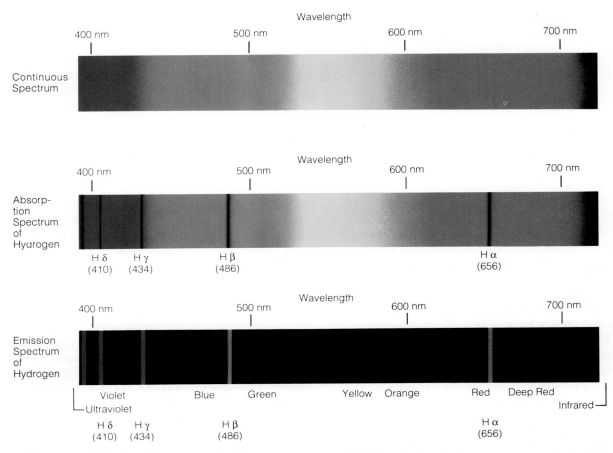

Figure 6-7 The three types of spectra. A continuous spectrum (top) contains no bright or dark lines, but an absorption spectrum (middle) is interrupted by dark absorption lines. An emission spectrum (bottom) is dark except at certain wavelengths where emission lines occur. Note that the lines in the absorption spectrum of hydrogen have the same wavelength as the lines in the emission spectrum of hydrogen.

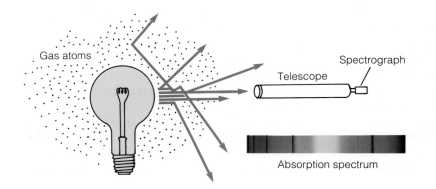

Figure 6-8 Photons of the proper wavelengths can be absorbed by the gas atoms and reradiated in random directions. Since these photons do not reach the telescope, their wavelengths are dark, producing absorption lines in the spectrum.

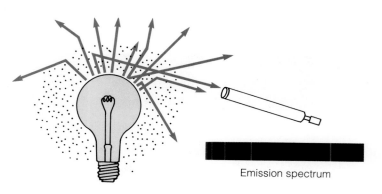

Figure 6-9 Pointing the telescope away from the bulb, we can receive only those photons the atoms can absorb and reradiate, producing emission lines in the spectrum.

Emission spectrum

What happens to the photons that were absorbed? They bounce from atom to atom, being absorbed and emitted over and over until they escape from the cloud. If, instead of aiming our telescope at the bulb, we swing it to one side so that no light from the bulb enters the telescope, we can photograph a spectrum of the light emitted by the gas atoms (Figure 6-9). In this case, the only photons entering the telescope are photons that were absorbed and reemitted. A spectrum of this light is almost entirely dark except for the wavelengths corresponding to the photons the gas can absorb and reemit. Thus we will see a spectrum containing only bright lines on a dark background. These bright lines are called **emission lines,** and a spectrum with emission lines is an **emission spectrum** (also called a **bright-line spectrum**). The spectrum of a neon sign, for instance, is an emission spectrum. Also, the bluish-purple color of mercury-vapor streetlights and the pink-orange color of sodium-vapor streetlights are produced by the emission lines of these elements.

The properties of the three spectra—continuous, absorption, and emission—can be described by the three rules known as **Kirchhoff's laws.** Law I states that a heated solid, liquid, or dense gas will produce a continuous spectrum. We see this in the filament of a light bulb. Law II states that a low-density gas excited to emit light will produce an emission spectrum. We see this in neon signs. And Law III states that if light comprising a continuous spectrum passes through a cool, low-density gas, the result will be an absorption spectrum. We see this in stars.

Stellar spectra are absorption spectra formed as light from the bright surface, the photosphere, travels outward through the stellar atmosphere (Figure 6-10). The bright surface of the star is sufficiently dense and hot that collisions between gas atoms and electrons produce a continuous spectrum just as does the filament of the light bulb in Figure 6-8. The lower-density gases in the stellar atmosphere absorb certain wavelength photons, and thus the light that finally reaches our telescope is missing those specific wavelengths, and we see an absorption spectrum.

The absorption lines in stellar spectra provide a windfall of data about the star's atmosphere. By studying the spectral lines we can identify the elements in the stellar atmosphere and find the temperature of the atoms. To see how to get all this information, we need to look carefully at the way the hydrogen atom produces lines in a star's spectrum.

The Hydrogen Spectrum As you must have gathered by now, each element has its own spectrum, unique as a human fingerprint, and it can be recognized by its spectrum across trillions of miles. To see how hydrogen produces its unique spectrum, we must draw a detailed diagram of its permitted orbits, making the size of each orbit proportional to its energy (Figure 6-11). Then we can examine the way such atoms interact with light.

A transition occurs in an atom when an electron changes energy levels. In our diagram of a hydrogen atom, we can represent transitions by arrows pointing from one level to another. If the arrow points upward, the atom must absorb energy, and if the arrow points downward, the atom must emit a photon.

If the transition results in the absorption or emission of a photon, the length of the arrow tells us its energy. Long arrows represent large amounts

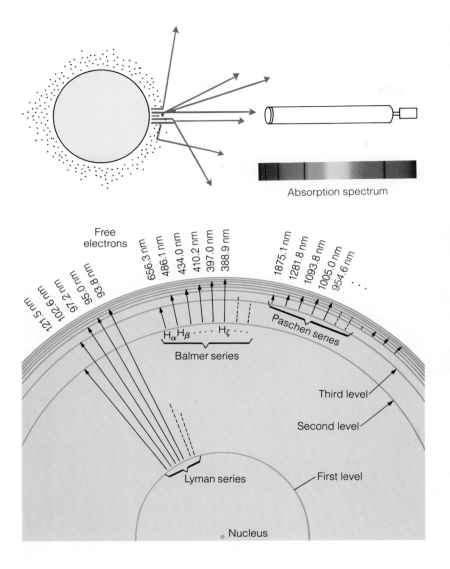

Figure 6-10 A star produces an absorption spectrum because its atmosphere absorbs certain wavelengths in the spectrum.

Absorption spectrum

Figure 6-11 The electron orbits in this diagram of the hydrogen atom are spaced to represent the energy the electron would have in each orbit. Thus we can refer to the orbits as energy levels. The transitions, drawn as arrows, can be grouped into series according to their lowest energy level. This drawing shows only a few of the infinite number of transitions and series possible.

of energy and thus short-wavelength photons. Short arrows represent smaller amounts of energy and longer-wavelength photons.

We can divide the possible transitions in a hydrogen atom into groups called series, according to their lowest energy level. Those arrows whose lower ends rest on the ground state represent the **Lyman series;** those resting on the second energy level, the **Balmer series;** and those resting on the third, the **Paschen series.** In principle, each series contains an infinite number of transitions, and there are an infinite number of series. Figure 6-11 shows the first few transitions in the first few series.

The Lyman series transitions involve large energies, as shown by the long arrows in Figure 6-11. These energetic transitions produce lines in the ultraviolet part of the spectrum, where they are invisible to the human eye. The Paschen series also lies outside the visible part of the spectrum. These transitions involve small energies, and thus produce spectral lines in the infrared.

Balmer series transitions produce the only spectral lines of hydrogen in the visible part of the spectrum. Figure 6-11 shows that the first few Balmer series transitions are intermediate between the energetic Lyman transitions and the low-

Figure 6-12 The Balmer lines photographed in the near ultraviolet. (Mount Wilson and Las Campanas observatories, Carnegie Institution of Washington)

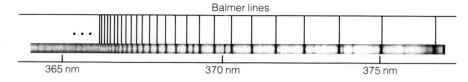

Balmer lines

365 nm   370 nm   375 nm

Figure 6-13 The Lagoon Nebula in Sagittarius is a cloud of gas and dust about 60 ly in diameter. Its gases are excited by the ultraviolet radiation of the hot, young stars within, and it glows in the pink color produced by the mixture of the red, blue, and violet Balmer lines. (National Optical Astronomy Observatories)

energy Paschen transitions. These Balmer lines are labeled by Greek letters for easy identification. H_{α} is a red line, H_{β} is blue, and H_{γ} and H_{δ} are violet. The remaining Balmer lines have wavelengths too short to see, though they can be photographed easily (Figure 6-12). In fact, these four lines create the purple-pink color characteristic of glowing clouds of hydrogen in space (Figure 6-13).

Before we leave the subject of hydrogen and its energy levels, we should note that the 21-cm radiation so important to radio astronomers is produced in hydrogen atoms by a special kind of transition that can occur when the electron in the ground state of a hydrogen atom changes the direction of its spin. (See Box 6-3.) Thus only cold, unexcited hydrogen gas such as that in interstellar space can emit 21-cm photons.

In frigid space or torrid stars, hydrogen atoms and their spectra give away nature's secrets. Balmer series lines are important because, as we have seen, they are the only hydrogen lines in the visible part of the spectrum. In the next section, we will see how the Balmer series lines can tell us a star's surface temperature.

6-3 STELLAR SPECTRA

A spectrum can tell us about such things as temperature, composition, and even motion. In later chapters, we will use spectra to study galaxies and planets, but we begin by studying the spectra of stars, including the sun. Stellar spectra are the easiest to understand, and the nature of stars is central to our study of all celestial objects.

The Balmer Thermometer We can use the Balmer lines as a thermometer to find the temperatures of stellar surfaces. From our discussion of black-body radiation (see Box 6-2), we know that we can estimate stellar temperatures from color—red stars are cool and blue stars are hot. But the Balmer lines give us much greater accuracy.

To refer to stellar temperatures, we will use the Kelvin temperature scale. On this scale, zero degrees Kelvin (written as 0 K) (−459.7°F) is absolute zero, the theoretical temperature at which there is the least possible motion among an object's atoms. Nothing can become that cold. Water freezes at 273 K and boils at 373 K (see Appendix C).

Note that when we discuss the temperature of a star, we mean the surface temperature—typically 40,000–2000 K. The centers of stars are much hotter—many millions of degrees—but the spectra tell us about the surface layers from which the light originates.

The Balmer thermometer works because the Balmer absorption lines are produced by hydrogen atoms whose electrons are in the second energy level (see Figure 6-11). If the surface of a star is as cool as the sun or cooler, there are few violent collisions between atoms to excite the electrons, and most atoms have their electrons in the ground state. These atoms can't absorb photons in the Balmer series. As a result, we should expect to find weak Balmer lines in the spectra of cool stars.

In the surface layers of stars hotter than about 20,000 K, however, there are many violent collisions between atoms, exciting electrons to high energy levels or knocking the electrons completely out of some atoms. That is, some atoms become ionized. Therefore, few atoms have electrons in the second energy level to form Balmer absorption lines, and we should expect hot stars, like cool stars, to have weak Balmer absorption lines.

At an intermediate temperature, roughly 10,000 K, the collisions have the correct amount of energy to excite large numbers of electrons into the second energy level. With many atoms excited to the second level, the gas absorbs Balmer-wavelength photons well and thus produces strong Balmer lines.

To summarize, the strength of the Balmer lines depends on the temperature of the star's surface layers. Both hot and cool stars have weak Balmer lines, but medium-temperature stars have strong Balmer lines.

Theoretical calculations can predict just how strong the Balmer lines should be for stars of various temperatures. Such calculations are the key to finding temperatures from stellar spectra. Figure 6-16 shows the calculated strength of the Balmer lines for various stellar temperatures. We could use this as a temperature indicator, except that the curve gives us two answers. A star with Balmer lines of a certain strength might have either of two temperatures, one high and one low. How do we know the true temperature of the star? We must examine other spectral lines to arrive at the correct temperature.

We have seen how the strength of the Balmer lines depends on temperature. The same process affects the spectral lines of other elements. That is, their lines are weak at high and low temperatures and strong at some intermediate temperature. But the temperature at which their lines reach maximum strength is different for each element. If we add these elements to our graph, we get a powerful tool for finding the temperature of stars (Figure 6-17).

We can determine a star's temperature by comparing the strengths of its spectral lines with our graph. For instance, if we photographed a spectrum of a star and found medium-strength Balmer lines and strong helium lines, we could conclude it had a temperature of about 20,000 K. But if the star had weak hydrogen lines and strong lines of ionized iron, we would assign it a temperature of about 5800 K, similar to the sun.

BOX 6-3

The 21-cm Radiation

As mentioned in Chapter 5, radio telescopes are important because they can detect the 21-cm-wavelength radiation emitted by clouds of cool hydrogen in space. This radiation is emitted when a hydrogen atom's electron changes its energy by changing the direction of its spin.

Quantum mechanics tells us that particles like protons and electrons have a property akin to the spinning of a top. This spin creates a magnetic field around the particle like that produced when electricity flows through a coil of wire in an electromagnet. These magnetic fields can alter the binding energy of the hydrogen atom (Figure 6-14), and that is what produces the 21-cm radiation.

The proton and electron in a hydrogen atom can spin in only two ways: in the same direction or in the opposite direction. If they are spinning such that their magnetic fields help hold the atom together, then the electron is more tightly bound. If they spin the other way, the magnetic field slightly opposes the binding of the atom.

This means that the ground state of the hydrogen atom is really two states that differ

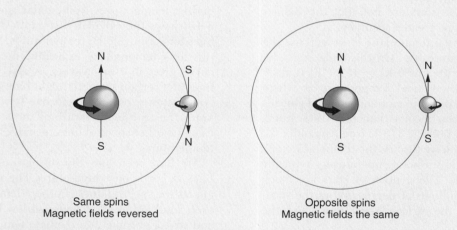

Same spins
Magnetic fields reversed

Opposite spins
Magnetic fields the same

Figure 6-14   The proton (red) and electron (blue) in a hydrogen atom spin and so produce small magnetic fields. If they spin in the same direction, the magnetic fields are reversed. If they spin in the opposite direction, the fields are the same. Because the fields repel or attract, the electron is slightly more tightly bound to the proton in one case and slightly less tightly bound in the other case.

The spectra of stars cooler than about 3000 K contain dark bands produced by molecules such as titanium oxide (TiO). Because of their structure, molecules can absorb photons at many wavelengths, producing numerous, closely spaced spectral lines that blend together to form bands. These molecular bands appear only in the spectra of the coolest stars because, as mentioned before, only there can molecules avoid the collisions that would break them up in hotter stars. Thus the presence of dark bands in a star's spectrum indicates that the star is very cool.

From stellar spectra, astronomers have found that the hottest stars have surface temperatures

from each other according to the spin of the proton and electron. If we could look at the energy levels of a hydrogen atom with a powerful magnifying glass (Figure 6-15), we would find the ground state split into two levels that differ by a very small amount of energy. If the electron is in the higher level, it can spontaneously flip over and spin the other way, changing its magnetic field to the lower energy state and thus dropping to the lower energy level. The atom gives up this extra energy by emitting a photon. The two energy levels are so close together that the photon must have a very

low energy—corresponding, in fact, to a wavelength of 21 cm.

Only cold, low-density clouds of hydrogen produce 21-cm radiation. If the gas is warm and dense, the atoms collide so often that the electrons are never in the ground state long enough to flip their direction of spin and emit a 21-cm photon. Thus radio telescopes are not confused by hot hydrogen in stars and can use the 21-cm radiation to map the clouds of cold hydrogen between the stars.

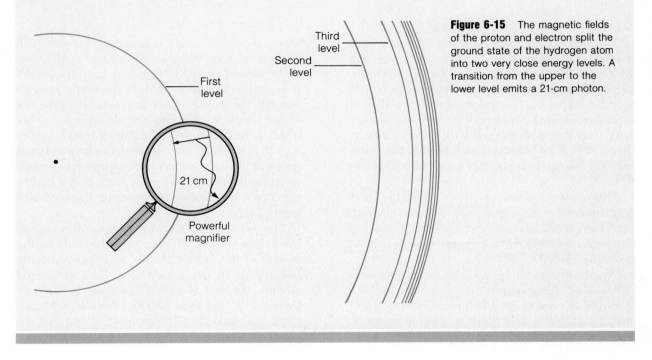

Figure 6-15 The magnetic fields of the proton and electron split the ground state of the hydrogen atom into two very close energy levels. A transition from the upper to the lower level emits a 21-cm photon.

above 40,000 K and the coolest about 2000 K. Compare these with the surface temperature of the sun, which is about 5800 K.

Spectral Classification We have seen that the strengths of spectral lines depend on the surface temperature of the star. From this we can predict

that all stars of a given temperature should have similar spectra. If we learn to recognize the pattern of spectral lines produced by a 6000-K star, for instance, we need not use Figure 6-17 every time we see that kind of spectrum. We can save time by classifying stellar spectra rather than analyzing each one individually.

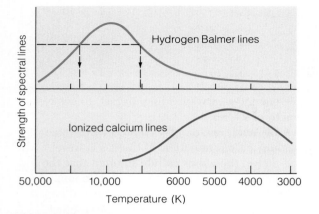

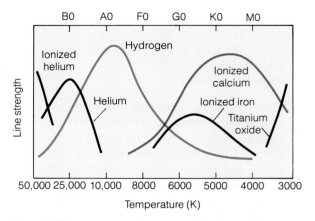

Figure 6-16 The strength of the Balmer lines in a stellar spectrum depends on the temperature of the star. A star with medium-strength Balmer lines can have one of two possible temperatures. Other elements, such as once ionized calcium, behave similarly, but reach maximum strength at different temperatures. This helps us choose the correct temperature.

Figure 6-17 If we plot the curves for hydrogen and ionized calcium shown in Figure 6-16 along with similar curves for other atoms and molecules, we get a diagram that can tell us the temperature of a star. This relationship is the basis for the spectral classification system.

The first widely used classification system was devised by astronomers at Harvard during the 1890s and 1900s. One of them, Annie J. Cannon, personally inspected and classified the spectra of over 250,000 stars. The spectra were first classified in groups labeled A through Q, but some groups were later dropped, merged with others, or reordered. The final classification includes the seven **spectral classes,** or **types,** still used today: O, B, A, F, G, K, M.*

This sequence of spectral types, called the **spectral sequence,** is important because it is a temperature sequence. The O stars are the hottest, and the temperature continues to decrease down to the M stars, the coolest of all.

We classify a star by features in its spectrum (Table 6-1). For example, if it has weak Balmer lines and lines of ionized helium, it must be an O star. This table is based on the same information we used in Figure 6-17.

The spectra shown in Figure 6-18 illustrate how spectral features change from class to class. Note that the Balmer lines are strongest in A stars,

where the temperature is moderate but still high enough to excite the electrons in hydrogen atoms to the second energy level, where they can absorb Balmer-wavelength photons. In the hotter stars (O and B), the Balmer lines are weak because the higher temperature excites the electrons to energy levels above the second. The Balmer lines in cooler stars (F through M) are also weak but for a different reason. The lower temperature cannot excite many electrons to the second energy level, so few hydrogen atoms are capable of absorbing Balmer-wavelength photons.

The spectral lines of other atoms also change from class to class. Helium is visible only in the spectra of the hottest classes, and titanium oxide bands, only in the coolest. Two lines of ionized calcium, labeled H and K, increase in strength from A to K and then decrease from K to M. Because the strength of these spectral features depends on temperature, it requires only a few minutes to compare a star's spectrum with Table 6-1 or Figure 6-18 and determine its temperature.

To be more precise, astronomers divide each spectral class into 10 subclasses. For example, spectral class A consists of the subclasses A0, A1, A2, . . . A8, A9. Next comes F0, F1, F2, and so on. This finer division, of course, demands that we

Generations of astronomy students have remembered the spectral sequence using the mnemonic "Oh, Be A Fine Girl, Kiss Me." Nationwide contests to find a less sexist mnemonic have failed to displace this traditional sentence.

TABLE 6-1
Spectral Classes

Spectral Class	Approximate Temperature (K)	Hydrogen Balmer Lines	Other Spectral Features
O	40,000	Weak	Ionized helium
B	20,000	Medium	Neutral helium
A	10,000	Strong	Ionized calcium weak
F	7500	Medium	Ionized calcium weak
G	5500	Weak	Ionized calcium medium
K	4500	Very weak	Ionized calcium strong
M	3000	Very weak	Titanium oxide strong

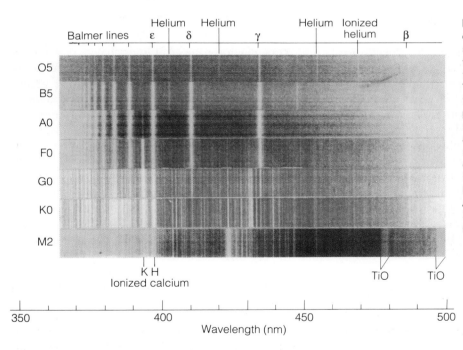

Figure 6-18 The spectra of stars of various classes illustrate how the strengths of lines change with temperature. Hot stars are at the top and cool stars at the bottom. These spectra are reproduced as the astronomer sees them on the photographic plates—that is, as negatives. Thus the dark absorption lines in the bright spectra appear as light lines on a darker background. (Adapted from H. A. Abt, A. B. Meinel, W. W. Morgan, and J. W. Tapscott, *An Atlas of Low-Dispersion Grating Stellar Spectra,* Kitt Peak National Observatory, 1968)

look carefully at a spectrum. But it is worth the effort, for the subclasses give us a star's temperature to an accuracy within about 5 percent. The sun, for example, is not just a G star, but a G2 star, with a temperature of about 5800 K.

A stellar spectrum can tell us many things besides temperature. Box 6-4 describes how a spectrum can reveal a star's motion. Lines in the spectrum of a star coming toward Earth are shifted slightly toward the blue end of the spectrum, and lines in the spectrum of a star moving away from Earth are shifted slightly toward the red. This effect, called the **Doppler effect,** can tell us the star's velocity with respect to Earth.

The Doppler Effect

The Doppler effect is the change in wavelength of radiation due to the relative motion of the source and observer. To see how this works, imagine standing on a railroad track as a train approaches with the engine bell ringing once each second (Figure 6-19). When the bell rings, the sound travels ahead of the engine to reach your ears. One second later the bell rings again, but not at the same place. During that one second, the engine moved closer to you, so the bell is closer at its second ring. Now the sound has a shorter distance to travel and reaches your ears a little sooner than it would have if the engine had not moved. The third time the bell rings, it is even closer. By timing the bell, you would observe that the clangs are slightly less than 1 second apart, all because the engine is approaching.

Standing behind the engine would give the opposite effect. You would find that each successive clang takes place farther away from you, and the clangs would sound more than 1 second apart. This apparent change in the rate of the ringing bell is an example of the Doppler effect.

We can think of the peaks of the electromagnetic waves leaving a star as a series of clangs from a bell. If a star is moving toward us, we see the peaks of the light waves closer together than expected, making the wavelengths slightly shorter than they would be if the star were not moving. If the star is moving away from us, the peaks of the light waves are slightly farther apart and the wavelengths are longer. Thus the lines in a star's spectrum are shifted slightly toward the blue if the star is approaching, and toward the red if it is receding.

For convenience, we have assumed that Earth is standing still and the star is moving, but the Doppler effect depends only on relative motion. Thus we cannot say that either Earth or the star is stationary—only that there is relative motion between them. In addition, the Doppler effect depends only on the **radial velocity** V_r, that part of the velocity directed away from or toward Earth. The Doppler effect cannot reveal relative motion to right or left.

How much the spectral lines change depends on the radial velocity. This can be expressed as a simple ratio relating the radial velocity V_r divided by the speed of light c, to the change in wavelength $\Delta\lambda$ divided by the unshifted wavelength λ_0.

$$\frac{V_r}{c} = \frac{\Delta\lambda}{\lambda_0}$$

This expression is quite accurate for the low velocities of stars, but we will need a better version later when we discuss objects moving with very high velocities.

Chemical Composition The stellar spectrum can also be used to pinpoint the chemical composition. Identifying the elements in a star by identifying the lines in the star's spectrum is a relatively straightforward procedure. For example, two dark absorption lines appear in the yellow region of the solar spectrum at the wavelengths 589 nm and 589.6 nm. The only atom that can produce this pair of lines is sodium, so we must conclude that the sun contains sodium. Over 90 elements in the sun have been identified this way.

However, just because the spectral lines characteristic of an element are missing, we cannot conclude that the element itself is absent. For example, the hydrogen Balmer lines are weak in the sun's spectrum, yet 90 percent of the atoms in the sun are hydrogen. The reason for this apparent paradox is that the sun is too cool to produce strong Balmer

Example: We observe a line in a star's spectrum with a wavelength of 600.1 nm. Laboratory measurements show that the line should have a wavelength of 600.0 nm; that is, its unshifted wavelength is 600.0 nm. What is the star's radial velocity? *Solution:* First, we note that the change in wavelength is 0.1 nm:

$$\frac{V_r}{c} = \frac{0.1}{600} = 0.000167$$

Multiplying by the speed of light, 3×10^5 km/sec, gives the radial velocity, 50 km/sec. Since the wavelength is shifted to the red (lengthened), the star must be receding from us.

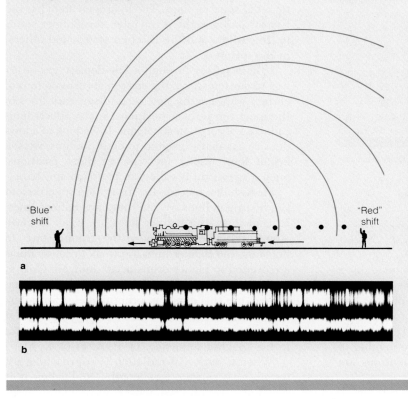

a

b

Figure 6-19 (a) Successive clangs of the engine bell (marked by dots) occur closer to the observer ahead, decreasing the distance the sound must travel. Thus the observer hears the bell ring at shorter time intervals than it really does. The observer behind the train hears the bell ring at longer time intervals. This is an example of the Doppler effect. (b) The upper spectrum of Arcturus was taken when Earth's orbital motion carried it toward the star. The lower spectrum was taken 6 months later, when Earth was receding from Arcturus. The difference in the wavelengths of the lines is due to the Doppler shift. (Mount Wilson and Las Campanas observatories, Carnegie Institution of Washington)

lines. Similarly, an element's spectral lines may be absent from a star's spectrum because the star is too hot or too cool to excite those atoms to the energy levels that produce visible spectral lines.

Detailed spectral analysis taking the star's temperature into consideration can reveal the abundance of the chemical elements in the star. The results of such studies show that nearly all stars have compositions similar to the sun's— about 92 percent of the atoms are hydrogen, and 7.8 percent are helium, with small traces of heavier elements (see Table 6-2).

The spectrum of the sun, like the spectra of all stars, is filled with clues to the nature of the gas emitting the light. Much of astronomy is based on unraveling these clues. We will begin in the following chapter by studying the nearest star—the sun.

TABLE 6-2

The Most Abundant Elements in the Sun

Element	Percentage by Number of Atoms	Percentage by Mass
Hydrogen	92.0	73.4
Helium	7.8	25.0
Carbon	0.03	0.3
Nitrogen	0.008	0.1
Oxygen	0.06	0.8
Neon	0.008	0.1
Magnesium	0.002	0.05
Silicon	0.003	0.07
Sulfur	0.002	0.04
Iron	0.004	0.2

Source: Adapted from C. W. Allen, *Astrophysical Quantities*. London: Athlone Press, 1976.

SUMMARY

An atom consists of a nucleus surrounded by a cloud of electrons. The nucleus is made up of two kinds of particles: positively charged protons and uncharged neutrons. The number of protons in an atom determines which element it is. Atoms of the same element (that is, having the same number of protons) with different numbers of neutrons are called isotopes.

The motion among the atoms in a solid, liquid, or dense gas causes the emission of black-body radiation. The hotter the object, the more radiation it emits. This produces a continuous spectrum, and the energy is most intense at the wavelength of maximum intensity, λ_{max}, which depends on the body's temperature. Hot objects emit mostly short-wavelength radiation, while cool objects emit mostly long-wavelength radiation. This effect gives us clues to the temperatures of stars—hot stars are blue and cool stars are red.

The negatively charged electrons surrounding an atomic nucleus may occupy various permitted orbits. An electron may be excited to a higher orbit during a collision between atoms, or it may move from one orbit to another by absorbing or emitting a photon of the proper energy. If the energy of an absorbed photon is too large, the atom may lose an electron and become ionized.

Because only certain orbits are permitted, only photons of certain wavelengths can be absorbed or emitted. Each kind of atom has its own characteristic set of spectral lines. The hydrogen atom has the Lyman series in the ultraviolet, the Balmer series in the visible, and the Paschen series (and others) in the infrared.

If light passes through a low-density gas on its way to our telescope, the gas can absorb photons of certain wavelengths, and we will see dark lines in the spectrum at those positions. Such a spectrum is called an absorption spectrum. If we look at a low-density gas that is excited to emit photons, we see bright lines in the spectrum at those positions. Such a spectrum is called an emission spectrum.

Most stellar spectra are absorption spectra, and the hydrogen lines we see there are Balmer lines. In cool stars, the Balmer lines are weak because atoms are not excited out of the ground state. In hot stars, the Balmer lines are weak because atoms are excited to higher orbits or are ionized. Only at medium temperatures are the Balmer lines strong. We can use this effect as a thermometer for determining the temperature of a star. In its simplest form, this amounts to classifying the star's spectra in the spectral sequence: O, B, A, F, G, K, M.

When a source of radiation is approaching us, we observe shorter wavelengths, and when it is receding, we observe longer wavelengths. This Doppler effect makes it possible for the astronomer to measure a star's radial velocity, that part of its velocity directed toward or away from Earth.

We can also find the composition of a gas from its spectrum. Such studies of the sun show that it is mostly hydrogen, contains only about 7.8 percent helium atoms, and includes only traces of heavier elements.

NEW TERMS

nucleus

proton

neutron

electron

isotope

ionization

ion

molecule

Coulomb force

binding energy

quantum mechanics

permitted orbit

energy level

excited atom

ground state

black-body radiation

continuous spectrum

wavelength of maximum
    intensity (λ_{max})

joule

absorption line

absorption spectrum
    (dark-line spectrum)

emission line

emission spectrum
    (bright-line spectrum)

Kirchhoff's laws

transition

Lyman series

Balmer series

Paschen series

spectral class or type

spectral sequence

Doppler effect

radial velocity (V_r)

REVIEW QUESTIONS

1. Why can a good blacksmith judge the temperature of a piece of heated iron by its color?

2. Why does the amount of black-body radiation emitted depend on the temperature of the object?

3. Why is the binding energy of an electron related to the size of its orbit?

4. Describe two ways an atom can become excited.

5. Why do different atoms have different lines in their spectra?

6. Why are stellar spectra usually absorption spectra and not continuous or emission spectra?

7. Why does the strength of the Balmer lines depend on temperature?

8. Explain the similarities between Figure 6-17 and Table 6-1.

9. Why do we expect TiO bands to be weak in the spectra of all but the coolest stars?

10. If spectral lines of a certain element are absent from a star's spectrum, may we conclude that the element is absent? Why or why not?

DISCUSSION QUESTIONS

1. In what ways is our model of an atom a scientific model? How can we use it when it is not a completely correct description of an atom?

2. Can you think of classification systems we use commonly to simplify what would otherwise be very complex measurements? Consider foods, movies, cars, grades, and clothes.

PROBLEMS

1. If the average atom is 0.5 nm in diameter, how many atoms are needed to reach 1 cm?

2. Human body temperature is about 310 K (98.6°F). At what wavelength do humans radiate the most energy? What kind of radiation do we emit?

3. If a star is five times hotter than the sun, how much more energy per second will it radiate from each square meter of its surface? At what wavelength will it radiate the most energy?

4. Transition A produces light with a wavelength of 500 nm. Transition B involves twice as much energy as A. What wavelength light does it produce?

5. Where would the arrow for the 21-cm transition be located in Figure 6-11?

6. Estimate the temperatures of the following stars based on their spectra. Use Figure 6-17.
    a. medium-strength Balmer lines, strong helium lines
    b. medium-strength Balmer lines, weak ionized calcium lines
    c. strong TiO bands
    d. very weak Balmer lines, strong ionized calcium lines

7. To which spectral classes do the stars in Problem 6 belong?

8. In a laboratory, the Balmer alpha line has a wavelength of 656.3 nm. If the line appears in a star's spectrum at 656.5 nm, what is the star's radial velocity? Is it approaching or receding?

RECOMMENDED READING

ALLER, L. H. *Atoms, Stars, and Nebulae.* Cambridge, Mass.: Harvard University Press, 1971.

BOOM, V. *The Structure of Atoms.* New York: Macmillan, 1964.

BOORSE, H., and L. MOTZ. *The World of the Atom.* New York: Basic Books, 1966.

FEINBERG, G. *What Is the World Made Of?* Garden City, N.Y.: Anchor/Doubleday, 1977.

GOLDBERG, I. "Ultraviolet Astronomy." *Scientific American* 220 (June 1969), p. 92.

GOLUB, LEON. "X-Rays from Stars." *Astronomy* 11 (September 1983), p. 66.

HOLZINGER, J. R., and M. A. SEEDS. *Laboratory Exercises in Astronomy.* Ex. 25, 28, 29. New York: Macmillan, 1976.

KALER, JAMES B. "Origins of the Spectral Sequence." *Sky and Telescope* 71 (February 1986), p. 129.

———. *Stars and Their Spectra.* New York: Cambridge University Press, 1989.

KIDWELL, PEGGY ALDRICH. "Three Women of American Astronomy." *American Scientist* 78 (May/June 1990), p. 244.

SNEDEN, CHRISTOPHER. "Reading the Colors of the Stars." *Astronomy* 17 (April 1989), p. 36

WALKER, JEARL. "The Amateur Scientist, the Physics and Chemistry Underlying the Infinite Charm of a Candle Flame." *Scientific American* 238 (April 1978), p. 154.

———. "The Amateur Scientist, the Spectra of Streetlights Illuminate Basic Principles of Quantum Mechanics." *Scientific American* 250 (January 1984), p. 138.

WELTHER, BARBARA. "Annie Jump Cannon: Classifier of the Stars." *Mercury* 13 (January/February 1984), p. 28.

The Sun

All cannot live on the piazza, but everyone may enjoy the sun.

Italian proverb

The sun is just a typical star. Most stars are a bit smaller, a bit cooler, and a bit less powerful, but our sun is typical of stars in general. The universe is filled with billions of stars like the sun, and our sun is special in only one way—it is ours.

Our Earth, revolving in an orbit around the sun, is much closer to the sun than to any other star, and this makes the sun important to us. First, the sun is the source of heat, light, energy, and life for our planet. Not only does the sun warm and illuminate our planet, but we use stored sunlight in the form of petroleum, coal, gas, wind, hydroelectric power, and the like to supply nearly all of our energy. Even the food we eat is nothing more than sunlight transformed into bread, meat, produce, and fish.

Just as the sun belongs to us, we belong to the sun. We exist because of the sun's energy, and small changes in the sun's output

caused by the sun's periodic cycles of activity could alter conditions on Earth dramatically. The nature of the sun, its past behavior, and its future dependability as an energy source for our planet are critically important to us, and we should study the sun as a matter of self-preservation and self-understanding.

But the sun is also important to astronomers. The sun is a typical star, but we can see events and processes on the sun that are invisible on any other star. One way to understand stars is to try to understand the sun.

As we study the sun in this chapter, we will be using the tools discussed in the previous chapter. The sun and stars are great globes of hot gas held together by their own gravity and powered by nuclear reactions near their centers. In this chapter we will study the outer layers of the sun—the solar atmosphere. We will reserve our study of the sun's interior for Chapter 9. The layers of the solar atmosphere are of interest at this point because they are the layers from which light can escape. This light carries with it clues to the temperature, pressure, density, magnetic field, and motion of the gas. To read those clues, we must remember how atoms interact with light to produce spectra.

In addition, we should study the sun for its own sake, and the more we study the sun, the more intricate and beautiful it becomes. Thus we should study the sun not only because it is *our star*, not only because it is *a star*, but because it is the sun.

7-1 THE SOLAR ATMOSPHERE

With a radius 109.1 times Earth's (109.1 R$_{\oplus}$)*, the sun is gaseous throughout. (See Data File 1.) The layers we call its atmosphere extend from the visible surface out to about 5×10^6 km (7 R$_{\odot}$) and consist of the photosphere, the chromosphere, and the corona. If we include the low-density gas flowing away from the sun, the atmosphere extends to envelope the earth.

The Photosphere The visible surface of the sun, the **photosphere,** is not a solid surface. In fact, the sun is gaseous from its outer atmosphere right down to its center. The photosphere is the thin layer of gas from which we receive most of the sun's light. It is less than 500 km deep and has an average temperature of about 6000 K. If the sun magically shrank to the size of a bowling ball, the photosphere would be no thicker than a layer of tissue paper wrapped around the ball (Figure 7-1a).

Below the photosphere, the gas is denser and hotter, and therefore radiates plenty of light, but that light cannot escape from the sun because of the outer layers of gas. Thus we cannot detect light from these deeper layers. Above the photosphere, the gas is less dense and thus is unable to radiate much light. The photosphere is the layer in the sun's atmosphere that is dense enough to emit

In astronomy, the symbols ⊙ and ⊕ represent the sun and the earth, respectively. Thus 7 R$_{\odot}$ equals 7 solar radii.

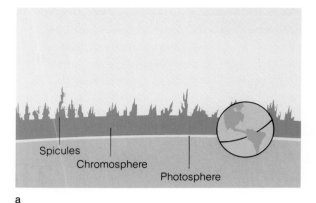
a

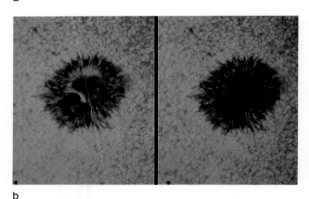
b

Figure 7-1 (a) A cross section at the edge of the sun shows the relative thickness of the photosphere and chromosphere. Earth shown for scale. (b) A visible-light photograph of a sunspot (right) shows the surrounding surface maked by granulation. An enhanced image (left) shows detail within the sunspot. (Courtesy William Livingston)

plenty of light but not so dense that the light can't escape.

Although the photosphere appears to be substantial, it is really a very low-density gas. Even in the deepest and densest layers visible, the photosphere is 3400 times less dense than the air we breathe. To find gases as dense as the air we breathe, we would have to descend about 7×10^4 km below the photosphere, about 10 percent of the way to the sun's center. With a fantastically efficient insulation system, we could fly a spaceship right through the photosphere.

Spectra of the photosphere reveal an absorption spectrum. The deeper layers are dense enough to radiate as a black body (see Box 6-2) and thus produce a continuous spectrum. But the higher

The Sun

Average distance from Earth	1.00 AU (1.495979 × 10^8 km)
Maximum distance from Earth	1.0167 AU (1.5210 × 10^8 km)
Minimum distance from Earth	0.9833 AU (1.4710 × 10^8 km)
Average diameter seen from Earth	0.53° (32 minutes of arc)
Period of rotation	25 days at equator
Radius	6.9599 × 10^5 km
Mass	1.989 × 10^{30} kg
Average density	1.409 g/cm^3
Escape velocity at surface	617.7 km/sec
Luminosity	3.826 × 10^{26} J/sec
Surface temperature	5800 K
Central temperature	15 × 10^6 K
Spectral type	G2 V
Apparent visual magnitude	−26.74
Absolute visual magnitude	4.83

Earth   Moon

a. An image of the sun in visible light shows a few sunspots. The spiral sunspot at the right is very unusual. The Earth–moon system is added for scale. (National Optical Astronomy Observatories)

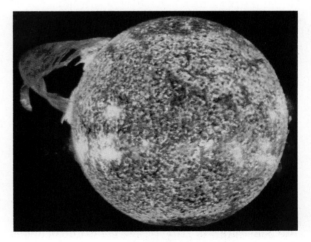

b. The sun photographed in the light emitted by ionized helium (30 nm) reveals a mottled surface and a major arch prominence erupting from the surface. (NASA Skylab)

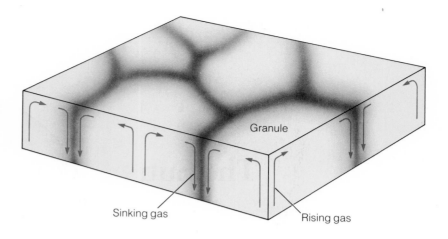

Figure 7-2 Rising convection currents in the sun heat areas of the surface, producing bright granules. Cooler material sinks at the darker edges of the granule.

Granule

Sinking gas

Rising gas

layers contain lower-density gas that absorbs photons of certain wavelengths, producing absorption lines.

Good photographs of the photosphere show that it is not uniform but is mottled by a pattern of bright cells called **granulation** (Figure 7-1b). Each granule is about 1500 km in diameter—slightly larger than Texas—and is separated from its neighbors by a dark boundary. A granule lasts for about 10 minutes before it dissipates or merges with neighboring granules.

The granulation is the surface effect of convection currents just below the photosphere (Figure 7-2). Convection occurs when hot gas rises and cool gas sinks, as when, for example, a convection current of hot gas rises above a candle flame. You can create convection in a liquid by adding a bit of nondairy creamer to an unstirred cup of hot coffee. The cool creamer sinks, warms, rises, cools, sinks again, and so on, creating small convection currents. The tops of these currents look much like solar granules. Doppler shifts show that the centers of granules are indeed rising and the edges are sinking. Thus the granulation reveals motions below the photosphere.

The Chromosphere Above the photosphere lies a nearly invisible layer of gas about 10,000 km deep, the chromosphere (see Figure 7-1a). It is about 1000 times fainter than the photosphere, so we can see it only during a total solar eclipse when the moon covers the brilliant photosphere. Then, for a few seconds, the chromosphere flashes into view as a thin line of pink just above the photosphere. The

term *chromosphere* comes from the Greek word *chroma*, meaning "color."

The pink color of the chromosphere is related to its spectrum. The hot, low-density gas produces a brilliant emission spectrum, and the red, blue, and violet Balmer lines of hydrogen plus the lines of other elements blend to produce pink. This emission spectrum tells us that the chromosphere is much hotter and much less dense than the photosphere (Figure 7-3a).

Although the chromosphere is not visible to the naked eye except during solar eclipses, it can be photographed if special filters are used to admit only those photons easily absorbed by certain atoms and ions. Such photographs are called **filtergrams.** For instance, an H$_a$ filtergram (Figure 7-3b) is formed by photons with wavelengths in the Balmer alpha (red) line of hydrogen. Because hydrogen can absorb these photons so readily, they cannot have come from deep in the chromosphere. Photons with these wavelengths could only have escaped from the upper layers of the chromosphere. Thus the H$_a$ filtergram shows detail in the uppermost layers of the chromosphere.

Filtergrams such as the one shown in Figure 7-3b reveal **spicules,** flamelike structures 100–1000 km in diameter, extending up to 10,000 km above the photosphere and lasting from 5 to 15 minutes. These spicules appear to be cool regions (about 10,000 K) extending up into the much hotter corona (about 500,000 K). Seen at the edge of the solar disk, these spicules blend together and look like flames covering a burning prairie, but filtergrams of spicules located near the center of

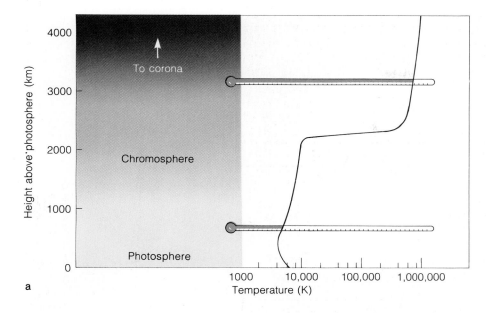

Figure 7-3 The chromosphere. (a) If we could place thermometers in the sun's atmosphere, we would discover that the temperature increases from 6000 K at the photosphere to 10^6 K at the top of the chromosphere. (b) An H_α filtergram shows that the chromosphere is not uniform. Spicules appear like black grass blades at the edges of supergranules over twice the diameter of Earth. (© 1971 National Optical Astronomy Observatories/NSO)

the solar disk show that they spring up around the edges of **supergranules** like weeds around flagstones. Over twice the diameter of the earth, supergranules appear to be related to magnetic fields coupled to gas motions below the photosphere. Spicules are apparently created by the magnetic fields at the edges of supergranules.

The Corona The sun's atmosphere extending above the chromosphere is termed the **corona**, after the Greek word for "crown." The corona is visible to the naked eye only during total solar eclipses, when the moon covers the bright photosphere (see Figure 3-17). Then the corona shines with a milky glow not quite as bright as the full moon. Eclipse photographs taken from the ground can trace the corona out to a distance of a few solar radii, and photographs taken from high-flying balloons or aircraft can trace the corona out to 12 solar radii or more (Figure 7-4). Ground-based telescopes of special design can record the inner corona of the uneclipsed sun, but the best studies of the corona have been made from spacecraft above Earth's atmosphere.

The corona is actually made up of a number of different components. The corona we see during total solar eclipses is mostly sunlight scattering off of interplanetary dust near the sun. Additional light scatters off of free electrons in the coronal gas.

We might expect this light to produce a solar absorption spectrum, but because the gas is very hot (about 1,000,000 K), the electrons travel very fast and the resulting Doppler shifts smear out any absorption lines. Thus we see a continuous spectrum.

Another component of the corona is produced by photons emitted by the low-density, highly ionized gases. In the lower corona, where the solar

Figure 7-4 Streamers in the solar corona. An image from the Solar Maximum Mission satellite (inset) has been computer-enhanced to produce a false-color image that reveals subtle variations in brightness. (NASA/JPL)

magnetic field confines the gas in long radial streamers, this emission can be strong. Thus the spectrum of the corona is continuous with superimposed emission lines.

The temperature in the corona rises as we travel outward. Just above the chromosphere, in the region called the transition region, the temperature climbs 500,000 K in only 300 km. In the lower corona, the temperature is about 500,000 K, and in the outer corona, it can be as high as 2,000,000 K. The density of this gas must be very low indeed or it would emit a great deal of light. In fact, the density of the outer corona is only about 1–10 atoms/cm³. Many solar astronomers now believe that the corona is heated by interaction with the sun's rotation and its magnetic field.

The outer corona is so hot the sun is unable to hold it. The high-velocity gas atoms stream away from the sun in a continuous breeze called the **solar wind.** It contains mostly protons and electrons (ionized hydrogen) but also carries heavier particles. With an average density of a few atoms per cubic centimeter, this solar wind blows past Earth at 300–800 km/sec with irregular gusts that can reach 1000 km/sec. Continuing out past the planets, the solar wind eventually mixes with the gases between the stars. Thus Earth is bathed in the corona's hot breath.

Do other stars have chromospheres and coronas like the sun's? Ultraviolet spectra taken by orbiting space telescopes such as the IUE (see Figure 5-28) suggest that the answer is yes. The spectra of many stars contain emission lines in the far ultraviolet that could have been formed only in the low-density, high-temperature gases of the upper chromosphere and lower corona. Thus the sun, for all its complexity, seems to be a normal star.

7-2 SOLAR ACTIVITY

So far we have described the sun as if it were a static, unchanging object. But, in fact, the sun is a complex variable body whose face is constantly changing. The most obvious evidence of this solar activity is the sunspots.

Sunspots A sunspot is a cool, dark area of the solar surface (see Figure 7-1b). The center of the spot, called the umbra, is darker than the outer border, the penumbra. The average spot is about twice the diameter of the earth and may last for a week or so. Sunspots tend to form in groups, and a large group may contain up to 100 individual spots and may last as long as 2 months.

Sunspots look dark because they are cooler than the photosphere. The center of a large sunspot is about 4240 K. The temperature of the photosphere is about 6000 K, so the cooler spot looks dark in contrast. In fact, a sunspot emits quite a bit of radiation. If the sun were magically removed and only an average-size sunspot were left behind, it would glow a brilliant orange-red and would be brighter than the full moon.

In 1908, George Ellery Hale discovered that sunspots contain powerful magnetic fields. He made this discovery by studying the **Zeeman effect,** the splitting of single spectral lines into multiple components through the influence of a magnetic field (Figure 7-5). When an atom is in a magnetic field, the electron orbits are altered, and the atom is able to absorb a number of different-wavelength photons where it was originally limited to a single wavelength. In the spectrum, we see single lines split into multiple components, with the separation between components proportional to the strength of the magnetic field.

Using the Zeeman effect, Hale found that the field in a typical sunspot is about 1000 times stronger than the sun's average field (Figure 7-6).

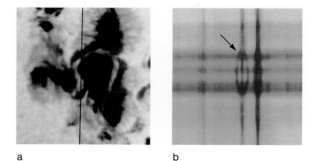

a   b

Figure 7-5 (a) The slit of a spectrograph has been placed across a complex sunspot. Thus light from the sunspot enters the spectrograph. (b) The resulting spectrum shows a typical absorption line split into three components (arrow) by the magnetic field in the sunspot. (National Optical Astronomy Observatories)

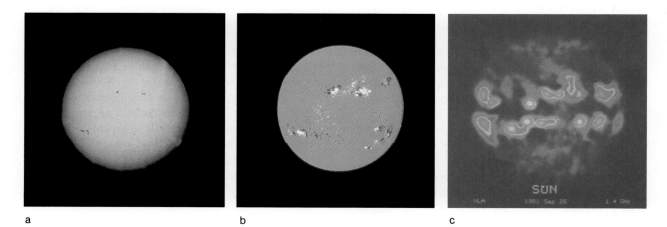

a   b   c

Figure 7-6 (a) A white-light image of the sun shows the location of sunspots. (b) On the same day the magnetic map, a magnetogram constructed using the Zeeman effect, shows the location of strong magnetic fields associated with the sunspots in (a). (c) A radio image of the sun (made on a different day) reveals a similar pattern of high-temperature coronal gas trapped by the magnetic fields above sunspots. (a., b. National Optical Astronomy Observatories; c. National Radio Astronomy Observatory, operated by Associated Universities, Inc. under contract with the National Science Foundation)

Figure 7-7 (a) The number of sunspots varies over an 11-year period. Note that between about 1645 and 1715, the Maunder minimum, there were very few sunspots. (b) The Maunder butterfly diagram shows that the spots first occur at higher latitudes and later nearer the solar equator.

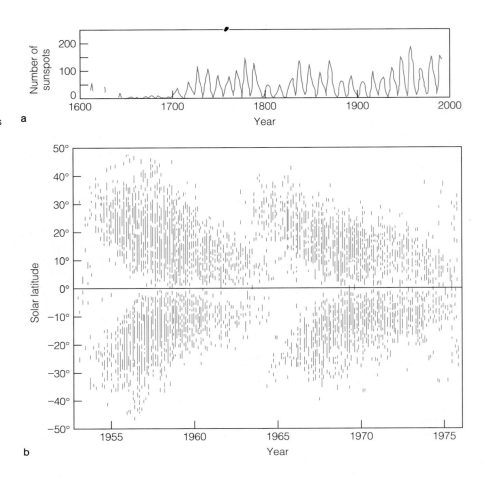

Apparently, this powerful magnetic field inhibits gas motions just below the photosphere, and rising currents of hot gas cannot deliver their heat to the surface. Thus the area cools slightly, and we see it as a sunspot.

The total number of sunspots visible on the sun is not constant. In 1843, the German amateur astronomer Heinrich Schwabe noticed that the number of sunspots varies with a period of about 11 years. This is now known as the sunspot cycle (Figure 7-7a). At sunspot maximum there are often as many as 100 spots visible at any one time, but at sunspot minimum there are only a few small spots. The last sunspot maximum occurred about 1990.

At the beginning of each sunspot cycle, the spots begin to appear in the sun's middle latitudes about 35° above and below the sun's equator. As the cycle proceeds, the spots appear at lower latitudes, until, near the end of the cycle, they are appearing within 5° of the sun's equator. If we plot the latitude of the appearance of sunspots over time, the diagram takes on the appearance of butterfly wings (Figure 7-7b). Such diagrams are now known as **Maunder butterfly diagrams,** named after E. Walter Maunder of the Greenwich Observatory.

Maunder also noticed that records show very few sunspots from 1645 to 1715. Modern studies show that this **Maunder minimum** (Figure 7-7a) coincided with a period of reduced solar activity and may be linked to the "little ice age," a period of unusually cool weather in Europe and America from about 1430 to 1850.

The Magnetic Cycle   The key to the sunspot cycle is the sun's rotation, which is complicated by the fact

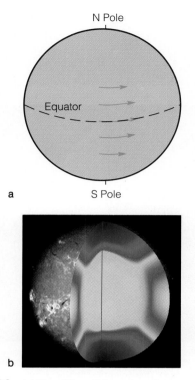

N Pole

Equator

a   S Pole

b

Figure 7-8 (a) The differential rotation of the sun's surface has been known for many years. The equatorial region rotates faster than do those regions at higher latitudes. (b) Helioseismology shows how the interior rotates relative to the surface. In this cutaway diagram, the equator rotates fastest, once in 25 days, but regions shaded red rotate more slowly. Regions near the poles, shaded blue, rotate most slowly with a period of 35 days. (Courtesy Kenneth Libbrecht, Caltech)

that the equatorial region rotates faster than do regions at the higher latitudes (Figure 7-8). At the equator the sun rotates once every 25 days, but at a latitude of 45° one rotation takes 27.8 days. The rotation of the sun also varies with depth, with the interior rotating more slowly than the surface. (Box 7-1 explains how solar astronomers can explore the interior of the sun.) This **differential rotation** is clearly linked to the magnetic cycle.

The sunspot cycle is controlled by the magnetic cycle. Sunspots tend to occur in pairs, and the magnetic field around the pair resembles that around a bar magnet with one end magnetic north and the other end magnetic south (Figure 7-10). During alternate sunspot cycles, the polarity of these magnetic fields is reversed. Thinking of the rotation of the sun, we can refer to one spot as the leading spot and the other as the trailing spot. We

might notice that the leading spot of each pair in the sun's northern hemisphere is a north pole and that each trailing spot is a south pole. South of the sun's equator we would find the polarities reversed. If we watched for an entire sunspot cycle, we would discover that the overall polarity reverses cycle to cycle.

This magnetic cycle is not fully understood, but the **Babcock model** (named after its inventor) explains the magnetic cycle as a progressive tangling of the solar magnetic field. Because the electrons in an ionized gas are free to move, the gas is a very good conductor of electricity, and any magnetic field in the gas is "frozen" into the gas. If the gas moves, the magnetic field must move with it. Thus the sun's magnetic field is frozen into its gases, and the differential rotation wraps this field around the sun like a long string caught on a hubcap. Rising and sinking gas currents twist the field into ropelike tubes, which tend to float upward. Where these magnetic tubes burst through the sun's surface, sunspot pairs occur (Figure 7-11).

The Babcock model explains the reversal of the sun's magnetic field from cycle to cycle. As the magnetic field becomes tangled, adjacent regions of the sun's surface are dominated by magnetic fields that point in different directions. After about 11 years of tangling, the field becomes so complex that adjacent regions of the solar surface begin changing their magnetic field to agree with neighboring regions. Quickly the entire field rearranges itself into a simpler pattern, and differential rotation begins winding it up to start a new cycle. But the newly organized field is reversed, and the next sunspot cycle begins with magnetic north replaced by magnetic south. Thus the complete magnetic cycle is 22 years long and the sunspot cycle is 11 years long.

Notice the power of a scientific model. The Babcock model may in fact be incorrect in some or all details, but it gives us a framework on which to organize all of the complex solar activity. Even though our models of the sky (see Chapter 2) and the atom (see Chapter 6) were only partially correct, they served as organizing themes to guide our thinking. Similarly, although the precise details of the solar magnetic cycle are not yet understood, the Babcock model gives us a general picture of the behavior of the sun's magnetic field.

Solar astronomers are now exploring the sun's interior using **helioseismology,** the study of the modes of vibration of the sun. Just as geologists can study the earth's interior by observing how sound waves produced by earthquakes are reflected and transmitted by the layers of the earth's interior, so too can solar astronomers explore the sun's interior by studying how vibrations (sound waves) travel through the sun. As these vibrations rise from the interior and reach the surface of the sun, we see the photosphere moving up and down in a complicated pattern that depends on the frequency (Figure 7-9). The Doppler shifts caused by the moving surface, although very small, can tell solar astronomers which frequency vibrations are present. Though 10 million different frequencies are possible, some penetrate deeper than others, and conditions in the sun's interior layers can weaken or strengthen the waves. By observing which of the millions of frequencies actually occur, solar astronomers can determine the temperature, density, pressure, composition, and motion of the sun's internal layers.

Because there are so many possible frequencies of vibration, solar astronomers need long series of data to separate the different pulsations. Observations have been made from the South Pole where the sun does not set for 6 months at a time, but now the Global Oscillation

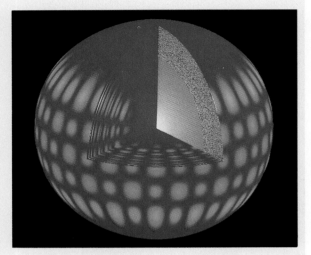

Figure 7-9 Helioseismology is the study of the modes of vibration of the sun. This computer image shows one of nearly 10 million possible modes of oscillation. Red regions are receding, and blue approaching. Such studies can reveal details about the sun's interior. (National Optical Astronomy Observatories)

Network Group (GONG) is building a chain of at least six robotic telescopes to encircle the earth. The network will observe the sun 24 hours a day for as long as 3 years. Needless to say, supercomputers will be used to analyze the data and find the internal structure and motion of the sun's interior (see Figure 7-8b).

If the sun is truly a representative star, we might expect to find similar magnetic cycles on other stars, but stars other than the sun are too distant for spots to be directly visible. Some stars, however, vary in brightness over a period of days in a way that reveals they are marked with dark spots believed to resemble sunspots (Figure 7-12). Other stars have spectral features that vary over periods

of years, suggesting that they are subject to magnetic cycles much like the sun's. In fact, some stars are subject to sudden flares that may resemble solar eruptions.

Prominences and Flares Prominences are visible during total solar eclipses as red protrusions at the edge of the solar disk (see Figure 3-17). The red

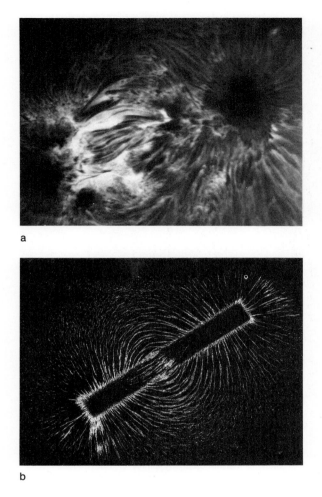

a

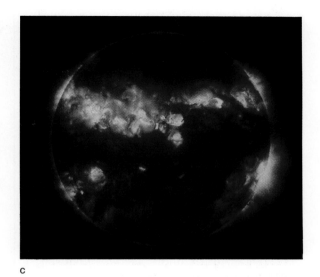

c

b

Figure 7-10 An H$_\alpha$ filtergram of a sunspot group (a) shows that structures in the chromosphere over sunspots follow a pattern much like iron filings sprinkled over a bar magnet (b). The magnetic fields on the sun are responsible for much of the activity we see in the chromosphere and corona (c), as shown in this X-ray image obtained during a rocket flight. (a. ©Association of Universities for Research in Astronomy, Inc., Sacramento Peak Observatory; b. Grundy Observatory; c. IBM Research and Smithsonian Astrophysical Observatory)

color is the same as the red color of the chromosphere and comes from the emission lines of hydrogen.

Prominences seem to be controlled by the magnetic fields. Many are arch-shaped, looking much like the patterns made by iron filings sprinkled over a magnet (see Data File 1, Figure b). Eruptive prominences burst out of the confused magnetic fields around sunspots and may shoot upward 500,000 km in a few hours. (Note that prominences are 50 times bigger than spicules.) Quiescent prominences (Figure 7-13) may develop as graceful arches over sunspot groups and, supported by magnetic fields, can last weeks or even months. Whether eruptive or quiescent, prominences are clearly ionized gases trapped in the twisted magnetic fields of active regions.

Flares are much more violent than prominences (Figure 7-14). A **flare** is an eruption on the solar surface that rises to maximum in a few minutes and decays in an hour or less. During that time it emits vast amounts of X-ray, ultraviolet, and visible radiation, and streams of high-energy protons and electrons. A large flare can release 10^{25} J, the equivalent of 2 billion megatons of TNT. The temperature in a flare can reach 500,000 K, and X-ray observations suggest that some nuclear reactions occur in solar flares.

Solar flares seem clearly linked to the magnetic field. They almost always occur near sunspot groups and may recur over and over at the same place. A large spot group may experience 100 flares a day. Although solar astronomers do not agree on the cause of flares, the outbursts are

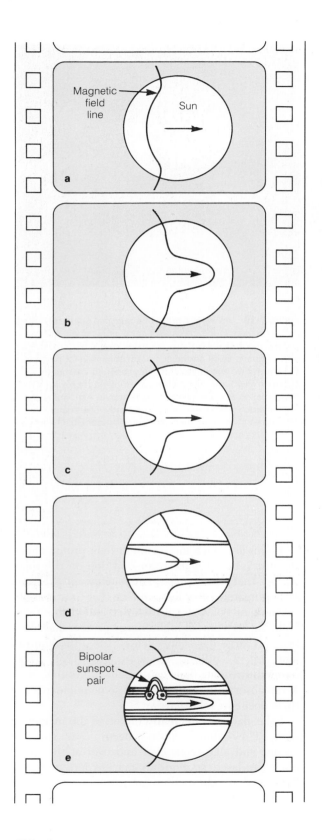

a

b

c

d

Bipolar
sunspot
pair

e

Figure 7-11 The Babcock model suggests that the differential rotation of the sun winds up the magnetic field. When the field becomes tangled and bursts through the surface, it forms sunspot pairs. For the sake of clarity, only one line in the sun's magnetic field is shown here.

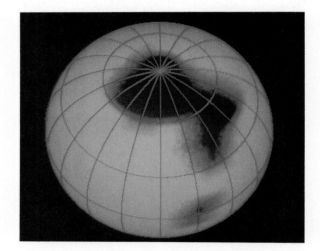

Figure 7-12 This computer-generated map shows the distribution of dark spots over the surface of the star HR1099. The map was constructed by analyzing the Doppler shifts in the spectrum of the rotating star. Although more extensive than sunspots, these dark regions are believed to be related to magnetic activity similar to that responsible for sunspots. (Courtesy Steven Vogt, Artie Hatzes, and Don Penrod)

known to be associated with the sudden release of energy stored in the magnetic field above the photosphere. A large flare can release energy equivalent to 10 billion megatons of TNT. (Large hydrogen bombs on Earth rate about 100 megatons.)

During a major solar flare, gas trapped in the magnetic field can be heated to over 5,000,000 K, and can emit large amounts of X rays and ultraviolet. Although flares are common on the sun, most are detected only in filtergrams. So-called whitelight flares, which are detectable in visible-light photographs, occur only about once a year.

Solar flares can have important effects on Earth. The X-ray and ultraviolet radiation reaches Earth in only 8 minutes, and increases the ionization in Earth's upper atmosphere. This alters the reflec-

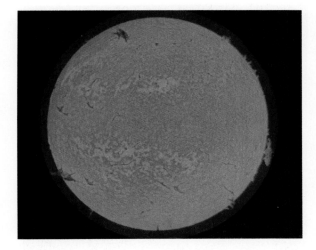

Figure 7-13 The sun photographed in the red light of the hydrogen Balmer alpha line reveals bright emission from the regions above and below the solar equator occupied by sunspots. Prominences are visible at the edge of the disk. A large quiescent prominence lies at lower right. (National Optical Astronomy Observatories)

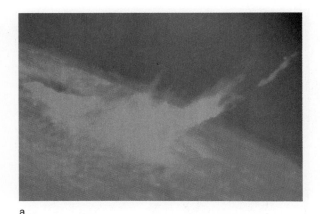

a

b

Figure 7-14 Solar flares are violent eruptions involving an area of the solar surface that is usually larger than the earth. (a) An H_α filtergram of a solar flare shows a surge of heated gas being ejected from the surface. (National Optical Astronomy Observatories) (b) Seen near the edge of the solar disk (in a computer-enhanced image), a solar flare ejects material up into the corona. (National Oceanic and Atmospheric Administration, Space Environment Laboratory, and USAF Air Weather Service)

tion of shortwave radio signals and can absorb them completely, thereby interfering with communications. Flares can eject high-energy particles at a third the speed of light, but most particles ejected from flares have lower velocities and reach Earth hours or days after the flare as gusts in the solar wind. Such gusts interact with Earth's magnetic field and generate tremendous electrical currents (as much as a million megawatts), which flow down into Earth's atmosphere near the magnetic poles. There they can excite the atoms of the upper atmosphere to glow in displays called **auroras** (Figure 7-15) at altitudes of 100–400 km. These currents can also disturb Earth's magnetic field and cause magnetic storms in which compasses behave erratically.

Other effects of solar flares include surges in high-voltage power lines and radiation hazards to passengers in supersonic transports and in spacecraft. The U.S. Air Force watches for flares from observatories around the world, and solar astronomers have developed ways of predicting which flares will affect Earth.

Coronal Activity Even the corona of the sun takes part in the solar activity cycle. At sunspot minimum, observers of total solar eclipses see a small,

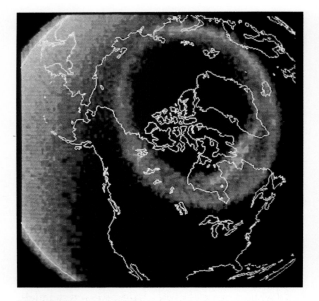

Figure 7-15 Auroral displays are triggered by the arrival of disturbances in the solar wind. This ultraviolet image of Earth was obtained by the Dynamics Explorer 1 satellite. The oval shape of the display encircles Earth's north magnetic pole and shows the influence of Earth's magnetic field. (L. A. Frank, University of Iowa/NASA)

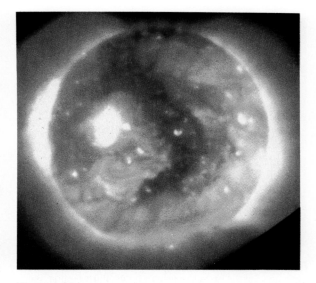

Figure 7-17 A coronal hole shows as a dark area near the center of the solar disk in this Skylab photo of the sun made at X-ray wavelengths. (American Science and Engineering)

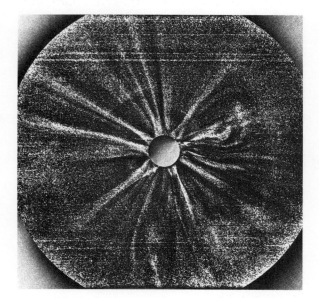

Figure 7-16 The solar eclipse of February 16, 1980, was photographed by astronomers in a jet aircraft flying 11 km (36,000 ft) above the Indian Ocean. Thirty-two separate images have been added together and computer-processed to reveal streamers in the solar corona extending out 12 solar radii. (Courtesy C. Keller, Los Alamos National Laboratory)

slightly flattened corona. But at sunspot maximum, eclipse observers are treated to a blazing corona that is nearly circular. Using special telescopes, solar astronomers study the corona daily from high mountaintops where the air is clear and steady, from high-flying aircraft, or from orbiting space satellites.

The results of such studies show that the corona is not the uniform halo of gas that earlier astronomers imagined. It is actually composed of streamers shaped by the solar magnetic field (Figure 7-16). The corona looks uniform during some eclipses because we see the streamers in projection at the sun's edge.

The streamers seem to draw their bulbous shapes from loops of magnetic fields that extend to a few solar radii, where the charged particles trapped in the fields are able to escape in long, thin streams. In some regions of the solar surface, however, the magnetic fields do not loop back, and the particles stream away from the sun unimpeded. These regions appear in X-ray images of the corona as cooler, lower-density regions called **coronal holes** (Figure 7-17). Although there are permanent coronal holes at the sun's north and south poles, the distribution of coronal holes over the rest of the sun depends on the solar activity cycle, further evidence that it is a magnetic phenomenon. Coronal activity is important because it affects the solar wind and thus affects Earth. Many solar astronomers believe that the solar wind is composed of particles streaming away from coronal holes. Understanding the solar wind, therefore, seems to depend on our understanding coronal activity.

The solar activity we see is fascinating in its power and complexity, but it is also important to us for what it tells us about other stars. The sun is a typical star, so similar processes must be occurring on the surfaces of many other stars. Now that we have observed the sun in detail, we can turn our attention to the more distant stars. In the next chapter, we will begin our study of the family of stars.

SUMMARY

The atmosphere of the sun consists of three layers: the photosphere, chromosphere, and corona. The photosphere, or visible surface, is a thin layer of low-density gas that is the level in the sun from which visible photons most easily escape. It is marked by granulation, a pattern produced by gas currents rising from below the photosphere.

The chromosphere is most easily visible during total solar eclipses, when it flashes into view for a few seconds. It is a thin, hot layer of gas just above the photosphere, and its brilliant pink color is caused by the emission lines in its spectrum. Filtergrams of the chromosphere reveal large jets called spicules extending up into the corona.

The corona is the sun's outermost atmospheric layer. It is composed of a very low-density, very hot gas extending at least 30 $R_\odot$ from the sun. Its high temperature—up to 2×10^6 K—is believed to be maintained by the magnetic field and the rotation of the sun. The outer parts of the corona merge with the solar wind, a breeze of low-density ionized gas streaming away from the sun. Thus Earth, which is bathed in the solar wind, is orbiting within in the solar atmosphere.

Sunspots are the most prominent example of solar activity. A sunspot seems dark because it is slightly cooler than the rest of the photosphere. The average sunspot is about twice the size of Earth and contains magnetic fields about 1000 times stronger than the sun's average field. Sunspots are thought to form because the magnetic field inhibits rising currents of hot gas and allows the surface to cool.

The average number of sunspots varies over a period of about 11 years and appears to be related to a magnetic cycle. The sunspot cycle does not repeat exactly each cycle, and the period from 1645 to 1715, known as the Maunder minimum, seems to have been a period when solar activity was very low.

Alternate sunspot cycles have reversed magnetic polarity, and this has been explained by the Babcock model of the magnetic cycle. In this theory the differential rotation of the sun winds up the magnetic field. Tangles in the field rise to the surface and cause sunspot pairs. When the field becomes strongly tangled, it reorders itself into a simpler but reversed field, and the cycle starts over.

Prominences and flares are other examples of solar activity. Prominences occur in the chromosphere; their arch shapes show that they are formed of ionized gas trapped in the magnetic field. Flares, too, seem to be related to the magnetic field. They are sudden eruptions of X-ray, ultraviolet, and visible radiation, and high-energy particles that occur among the twisted magnetic fields around sunspot groups. Flares are important because they can have dramatic effects on Earth such as communications blackouts and auroras.

Activity in the corona is also guided by the magnetic field. The corona seems to be composed of streamers of thin, hot gas escaping from the magnetic field. In some regions of the corona, the magnetic field does not loop back to the sun, and the gas escapes unimpeded. These regions are called coronal holes and are believed to be the source of the solar wind.

Helioseismology, the study of the oscillations of the solar surface, will eventually allow solar astronomers to study the layers below the photosphere.

NEW TERMS

photosphere

granulation

filtergram

spicule

supergranule

solar wind

sunspot

Zeeman effect

Maunder butterfly
   diagram

Maunder minimum

differential rotation

helioseismology

Babcock model

flare

aurora

coronal hole

REVIEW QUESTIONS

1. Why can't we see deeper into the sun than the photosphere?

2. What does granulation tell us about the sun?

3. What kinds of spectra do the photosphere, chromosphere, and corona produce? Why?

4. Why are sunspots dark?

5. Why do we think the sunspot cycle is controlled by the magnetic cycle of the sun?

6. What evidence shows that prominences and flares are magnetic phenomena?

7. How can solar activity affect Earth?

8. What evidence do we have that other stars have stellar winds, chromospheres, coronas, spots, and magnetic cycles?

DISCUSSION QUESTIONS

1. What energy sources on Earth cannot be thought of as stored sunlight?

2. How are helioseismology and the study of the sun similar to seismology and the study of the Earth?

3. What would the spectrum of an auroral display look like? Why?

PROBLEMS

1. The radius of the sun is 0.7 million km. What percent of the radius is taken up by the chromosphere?

2. If a sunspot has a temperature of 4200 K and the solar surface has a temperature of 6000 K, how many times brighter is the surface than the sunspot? (HINT: See Box 6-2.)

3. What is the angular diameter of a star the same size as the sun but located 5 ly away? Will the Hubble Space Telescope be able to detect detail on the surface of such a star? (HINT: See Box 3-2.)

4. If particles in the solar wind move at 400 km/sec, how long do they take to travel from the sun to Earth?

5. The United States consumes about 2.5×10^{19} J of energy in all forms in a year. If a solar flare releases 10^{25} J, how many years could we run the United States on the energy from the flare?

RECOMMENDED READING

AKASOFU, S. "The Aurora: New Light on an Old Subject." *Sky and Telescope* 64 (December 1982), p. 534.

BOHM-VITENSE, ERIKA. "Chromospheres, Transition Regions, and Coronas." *Science* 223 (24 February 1984), p. 777.

DELANCY, MARY MARTIN. "The Case of the Missing Sunspots." *Astronomy* 9 (February 1981), p. 66.

EMSLIE, A. GORDON. "Explosions in the Solar Atmosphere." *Astronomy* 15 (November 1987), p. 18.

GIAMPAPA, MARK S. "The Solar-Stellar Connection." *Sky and Telescope* 74 (August 1987), p. 142.

GIOVANELLI, R. *Secrets of the Sun.* New York: Cambridge University Press, 1984.

GOLUB, LEON. "Solar Magnetism: A New Look." *Astronomy* 9 (March 1981), p. 66.

———. "What Heats the Solar Corona?" *Astronomy* 10 (September 1982), p. 74.

HALL, DOUG. "Starspots." *Astronomy* 11 (February 1983), p. 66.

HARVEY, JOHN W., JAMES R. KENNEDY, and JOHN W. LEIBACHER. "GONG: To See Inside Our Sun." *Sky and Telescope* 74 (November 1987), p. 470.

HARVEY, J., M. POMERANTX, and T. DUVALL. "Astronomy on Ice." *Sky and Telescope* 64 (December 1982), p. 520.

MARAN, STEPHEN P. "Coronal Revisionism." *Natural History* 92 (January 1983), p. 74.

———. "Flareup on the Sun." *Natural History* 93 (April 1984), p. 90.

———. "The Jewel in the Satellite." *Natural History* 95 (January 1986), p. 84.

MAXWELL, ALAN. "Solar Flares and Shock Waves." *Sky and Telescope* 66 (October 1983), p. 285.

MIMS, FORREST M. "Sunspots and How to Observe Them Safely." *Scientific American* 262 (June 1990), p. 130.

NEIDIG, DONALD F., and JACQUES M. BECKERS. "Observing White-Light Flares." *Sky and Telescope* 65 (March 1983), p. 226.

O'LEARY, BRIAN. "The Stormy Sun." *Sky and Telescope* 60 (September 1980), p. 199.

REDDY, FRANCIS. "Polar Lights." *Astronomy* 11 (August 1983), p. 6.

ROBINSON, LEIF J. "The Sunspot Cycle: Tip of the Iceberg." *Sky and Telescope* 73 (June 1987), p. 589.

RUST, DAVID M. "Solar Flares, Proton Showers, and the Space Shuttle." *Science* 216 (28 May 1982), p. 939.

SCHORN, RONALD A. "Stellar Granulation." *Sky and Telescope* 74 (September 1987), p. 247.

SOFIA, SABATINO, PIERRE DEMARQUE, and ANDREW ENDAL. "From Solar Dynamo to Terrestrial Climate." *American Scientist* 73 (July/August 1985), p. 326.

STAHL, PHILIP A. "Prominences." *Astronomy* 11 (January 1983), p. 66.

WALLENHORST, STEVEN G. "Sunspot Numbers and Solar Cycles." *Sky and Telescope* 64 (September 1982), p. 234.

WILLIAMS, GEORGE E. "The Solar Cycle in Precambrian Time." *Scientific American* 255 (August 1986), p. 88.

WILSON, OLIN C., ARTHUR H. VAUGHAN, and DIMITRI MIHALAS. "The Activity Cycles of Stars." *Scientific American* 244 (February 1981), p. 104.

WILSON, RICHARD C., HUGH HUDSON, and MARTIN WOODARD. "The Inconstant Solar Constant." *Sky and Telescope* 67 (June 1984), p. 501.

WOLFSON, RICHARD. "The Active Solar Corona." *Scientific American* 248 (February 1983), p. 104.

C H A P T E R
8

The Properties of Stars

The stars are unimaginably remote. The nearest star is the sun, only 150 million km (93 million miles) away, so close that light takes only 8 minutes to reach Earth. The next nearest star is nearly 300,000 times farther away, a distance so great that the starlight takes over 4 years to reach Earth. These distances are so large we can express them in light-years. A light-year is the distance light travels in 1 year—about 9.5 trillion kilometers or about 5.9 trillion miles.

In spite of their great distances from us, the stars are the key to the secrets of the universe. The universe is filled with stars, and if we are to understand the universe, we must discover how stars are born, live, and die. We begin our study in this chapter by gathering data about the intrinsic properties of the stars—those properties inherent in the nature of the stars (Figure 8-1). In later chapters, we will use these data to

If it can't be expressed in figures, it is not science; it is opinion.

Robert Heinlein,
The Notebooks of Lazarus Long

deduce the life stories of different kinds of stars.

Unfortunately, determining a star's intrinsic properties is quite difficult. When we look at a star through a telescope, we see only a point of light that tells us nothing about the star's energy production, temperature, diameter, or mass. Because we cannot visit stars, we can only observe from Earth and unravel their properties through the analysis of starlight. One of the reasons astronomy is interesting is that it contains so many such puzzles, each demanding a different method of solution.

To simplify our task in this chapter, we will concentrate on three intrinsic stellar properties. Our goals will be to find out how much energy stars emit, how large stars are, and how much mass they contain. These three parameters, combined with stellar temperatures—an intrinsic stellar property already discussed in

Figure 8-1 The central question of modern astronomy is manifest in this photo of the Cone Nebula, where clouds of gas and dust are actively forming new stars. How do stars form, evolve, and die? To answer that question, we must know the luminosities, diameters, and masses of the stars—the fundamental properties of stars. (Copyright Anglo-Australian Telescope Board)

Chapter 6—will give us an overview of the nature of stars and provide us with the data we need to consider the lives of stars.

Although we begin with three goals firmly in mind, we immediately meet a short detour. To find out how much energy a star emits, we must know how far away it is. If at night we see bright lights approaching on the highway, we cannot tell whether the lights are the intrinsically bright headlights of a distant truck or the intrinsically faint lights on a pair of nearby bicycles. Only when we know the distance to the lights can we judge their intrinsic brightness. In the same way, to find the intrinsic brightness of a star, and thus the amount of energy it emits, we must know its distance from us. Our short detour will provide us with a method of measuring stellar distances.

Having reached our three goals, we will pause to consider the densities of stars, an intrinsic property that is easily determined once we know a star's size and mass. The densities of different kinds of stars will be helpful when we consider the internal structure of stars in the next two chapters.

We will conclude this chapter by considering the frequency of stellar types—that is, which kinds of stars are common and which are rare. This information, like all of the data in this chapter, is aimed at helping us understand what stars are.

8-1 MEASURING THE DISTANCES TO STARS

Determining the distance to a star is difficult because astronomers cannot actually journey to the star. They must instead measure the distance indirectly, much as surveyors measure the distance across a river they cannot cross. We will begin by reviewing this method, and then we will apply it to stars.

The Surveyor's Method To measure the distance across a river, a team of surveyors begins by driving two stakes into the ground. The distance between the stakes is the baseline of the measurement. The surveyors then choose a landmark on the opposite side of the river, a tree perhaps, thus establishing a large triangle marked by the two stakes and the tree. Using their surveyor's instruments, they sight the tree from the two ends of the baseline and measure the two angles on their side of the river (Figure 8-2).

Knowing two angles of this large triangle and the length of the side between them, the surveyors can then find the distance across the river by simple trigonometry. Another way to find the distance is to construct a scale drawing. For example, if the baseline were 50 m and the angles were 66° and 71°, they could draw a line 50 mm long to represent the baseline. Using a protractor, they could construct angles of 66° and 71° at each end of the baseline, and then extend the two sides until they met at C, the location of the tree. Measuring the height of the triangle in the drawing, they would find it to be 64 mm high and would thus conclude that the distance across the river to the tree is 64 m.

Figure 8-2 Surveyors can find the distance d across the river by measuring the baseline and the angles A and B and then constructing a scale drawing of the triangle.

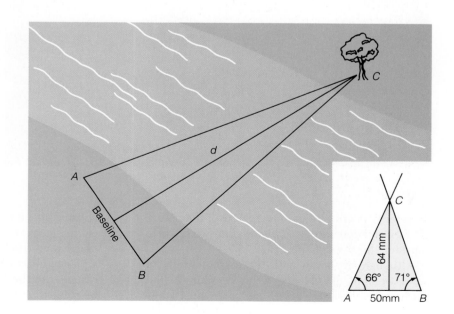

The Astronomer's Method To find the distance to a star, we must use a very long baseline, the diameter of Earth's orbit. If we took a photograph of a nearby star and then waited 6 months, Earth would have moved halfway around its orbit. We could then take another photograph of the star at a point in space 2 AU (astronomical units) from the point where the first photograph was taken. Thus our baseline would equal the diameter of Earth's orbit, or 2 AU.

We would then have two photographs of the same part of the sky taken from slightly different locations in space. If we examined the photographs, we would discover that the star was not in exactly the same place in the two photographs. This apparent shift in the position of the star is called *parallax* (Figure 8-3).

Parallax is the apparent change in the position of an object due to a change in the location of the observer. We saw in Chapter 4 that parallax is an everyday experience. Our thumb, held at arm's length, appears to shift position against a distant background when we look with first one eye and then with the other (see Figure 4-3). In this case, the baseline is the distance between our eyes, and the parallax is the angle through which our thumb appears to move when we change eyes. The farther away we hold our thumb, the smaller the parallax. If we know the length of the baseline and measure the parallax, we can calculate the distance from our eyes to our thumb.

Because the stars are so distant, their parallaxes are very small angles, usually expressed in seconds of arc. The quantity that astronomers call **stellar parallax (*p*)** is half the total shift of the star, as shown in Figure 8-3. Astronomers measure the parallax and surveyors measure the angles at the ends of the baseline, but both measurements tell us the same thing—the shape of the triangle and thus the distance to the object in question.

Measuring the small angle p is very difficult. The nearest star, α Centauri, has a parallax of only 0.76 seconds of arc, and the more distant stars have even smaller parallaxes. To see how small these angles are, hold a piece of paper edgewise at arm's length. The thickness of the paper covers an angle of about 30 seconds of arc.

We cannot use scale drawings to find the distances to stars because the distances are so large and the angles so small. Even for the nearest star, the triangle would have to be 300,000 times longer than it was wide. If the baseline in our drawing

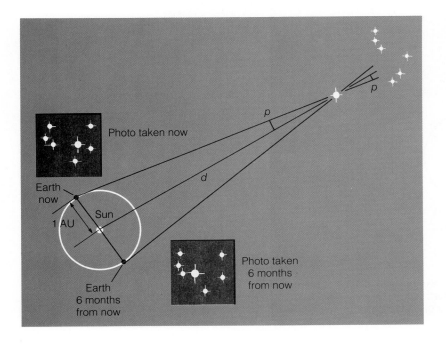

Photo taken now

Earth now

1 AU

Sun

Earth 6 months from now

Photo taken 6 months from now

p

d

p

Figure 8-3 We can measure the parallax of a nearby star by photographing it from two points along Earth's orbit. For example, we might photograph it now, and again in 6 months. Half of the star's total change in position from one photograph to the other is its stellar parallax p.

BOX 8-1

Parallax and Distance

We wish to find the distance to a star from its measured parallax. To see how this is done, imagine that we observe the earth from the star. Figure 8-4 shows that the angular distance from the sun to the earth would equal the star's parallax p. To find the distance, we recall that the small-angle formula (see Box 3-2) relates an object's angular diameter, its linear diameter, and its distance. In this case, the angular diameter is p and the linear diameter is 1 AU. Then the small-angle formula, rearranged slightly, tells us that the distance to the star in AU is equal to 206,265 divided by the parallax in seconds of arc:

$$d = \frac{206{,}265}{p}$$

Because the parallaxes of even the nearest stars are less than 1 second of arc, the distances in AU are inconveniently large numbers. To keep the numbers manageable, astronomers have defined the parsec as their unit of distance in a way that simplifies the arithmetic. One parsec equals 206,265 AU, so the equation becomes:

$$d = \frac{1}{p}$$

Thus a parsec is the distance to an imaginary star whose parallax is 1 second of arc.

Example: The star Altair has a parallax of 0.20 seconds of arc. How far away is it? *Solution:* The distance in parsecs equals 1 divided by 0.2, or 5 pc:

$$d = \frac{1}{0.2} = 5 \text{ pc}$$

Since 1 pc equals about 3.26 ly, Altair is about 16.3 ly away.

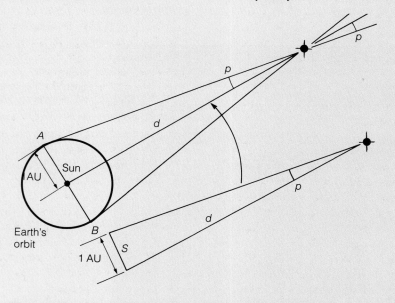

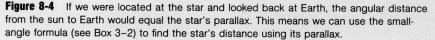

Figure 8-4 If we were located at the star and looked back at Earth, the angular distance from the sun to Earth would equal the star's parallax. This means we can use the small-angle formula (see Box 3-2) to find the star's distance using its parallax.

were 1 cm, the triangle would have to be about 3 km long. (Box 8-1 describes how we could find the distance from the parallax without drawing scale triangles.)

The distances to the stars are so large it is not convenient to use astronomical units. As Box 8-1 explains, when we measure distance via parallax, it is convenient to use the unit of distance called a **parsec (pc).** One parsec is 206,265 AU, roughly 3.26 ly (light-years).*

Using ground-based telescopes, we can't measure parallax when p is smaller than about 0.02 seconds of arc, so we can't measure parallax with any accuracy beyond about 50 pc. Although some 500,000 stars are within this distance, less than 10,000 have measured parallaxes, and a much smaller number have accurate parallaxes. Thus our knowledge of stars is based on a small sample with well-known distances.

Recently it has become possible to measure parallaxes from space, far beyond Earth's turbulent atmosphere. In 1989, the European Space Agency launched the satellite Hipparcos, which is currently measuring the parallaxes of 120,000 stars to better than 0.002 seconds of arc. The Hubble Space Telescope, though not intended for such tasks, will be able to make a few high-precision parallax observations. When these accurate distances become available, it will give astronomers a new window on the nature of the local stars.

8-2 INTRINSIC BRIGHTNESS

Our eyes tell us that some stars look brighter than others, but unless we know how distant a star is, we cannot judge its intrinsic brightness. It might be a very distant star emitting tremendous amounts of light or a nearby star emitting much less light. To discover the true brightnesses of the stars, we must know how brightness depends on distance.

Brightness and Distance When we look at a bright light, our eyes respond to the visual-wavelength

---

The parsec is used throughout astronomy because it simplifies the calculation of distance. However, there are instances in which the light-year is also convenient. Consequently, the chapters that follow use either parsecs or light-years as convenience and custom dictate.

energy falling on the eye's retina, which tells us how bright the object looks. Thus brightness is related to the flux of energy entering our eye. **Flux is the energy in joules (J) per second falling on one square meter.** (We commonly refer to 1 J/sec as 1 watt.)

If we placed a screen 1m² near a light bulb, a certain amount of flux would fall on the screen (Figure 8-5). If we moved the screen twice as far from the bulb, the light that previously fell on the screen would be spread to cover an area four times larger, and the screen would receive only one-fourth as much light. If we tripled the distance to the screen, it would receive only one-ninth as much light. Thus the flux we receive from a light source depends inversely on the square of the distance to the source. This is known as the inverse square law. (We first encountered the inverse square law in Chapter 4 where it was applied to the strength of gravity.)

Absolute Visual Magnitude If all the stars were the same distance away, we could compare one with another and decide which was emitting more light and which less. Of course, we can't move the stars, but if we know the distance to a star, we can use the inverse square law to calculate the brightness the star would have at some standard distance. Astronomers take 10 pc as the standard distance and refer to the intrinsic brightness of the star as its **absolute visual magnitude (M_v),** which is the apparent visual magnitude the star would have if it were 10 pc away.

For example, suppose we know that a star is 100 pc away and has an apparent magnitude of 8.5. Because it is 10 times farther than the standard distance, it must look 100 times fainter than it would at the standard distance. We recall from our discussion of the magnitude system (Chapter 2) that a factor of 100 times in brightness is the same as a difference of 5 magnitudes, so the star must look 5 magnitudes fainter than its absolute magnitude. Thus the absolute magnitude of the star must be 8.5 minus 5, or 3.5. (Box 8-2 expresses this relation in a simple equation.)

The symbol for absolute visual magnitude is a capital M with a subscript v. The subscript tells us it is a visual magnitude based only on the wavelengths of light we can see. Other magnitude systems are based on other parts of the electromag-

Figure 8-5 The inverse square law. A light source is surrounded by spheres with radii of 1 unit and 2 units. The light falling on an area of 1 m² on the inner sphere spreads to illuminate an area of 4 m² on the outer sphere. Thus the brightness of the light source is inversely proportional to the square of the distance.

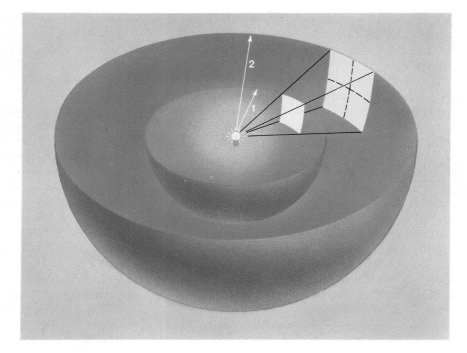

netic spectrum such as the infrared, ultraviolet, and so on.

The intrinsically brightest stars known have absolute magnitudes of about −8, and the faintest about +15 or fainter. The nearest star to the sun, α Centauri, is only 1.4 pc away, and its apparent magnitude is 0.0, indicating that it looks bright in the sky. Its absolute magnitude, however, is 4.39, telling us it is not intrinsically very bright. Because we know the distance to the sun and can measure its apparent magnitude, we can find its absolute magnitude—about 4.78. If the sun were only 10 pc away from us, it would look no brighter than the faintest star in the handle of the Little Dipper.

Luminosity Our goal is to find the **luminosities (L)** of the stars, the total amount of energy they emit per second. We can do that by comparing them with the sun.

We know the sun has an absolute magnitude of 4.78, and we know how much energy the sun emits because we can measure it as it falls on Earth's surface. The luminosity of the sun, $L_\odot$, is 4 × 10²⁶ J/sec. If a star has an absolute magnitude that is 5 magnitudes brighter than the sun, then

that star must emit 100 times more energy per second.

However, one correction is necessary. The absolute visual magnitude of a star refers only to visible light, but luminosity includes energy of all wavelengths. Hot stars emit a great deal of ultraviolet radiation we cannot see, and cool stars emit infrared. Thus astronomers must make a small correction that depends on the star's temperature. With that correction, we can use the absolute magnitudes of the stars to calculate their luminosities.

We can express luminosity in two ways. For example, we can say that the star Canopus, which is 100 times more luminous than the sun, has a luminosity of 100 $L_\odot$, or we can multiply by the luminosity of the sun and write the luminosity of Canopus as 4 × 10²⁸ J/sec.

The range of luminosities is very large. The most luminous stars emit more than 10⁵ $L_\odot$, and the least luminous stars roughly 10⁻⁴ $L_\odot$.

Both absolute magnitude and luminosity are measures of the intrinsic brightness of a star. Thus we have reached our first goal, determining the energy output of stars. Our second goal, finding the diameters of stars, is related to stellar luminosity.

Absolute Magnitude

Apparent magnitude tells us how bright a star looks (see Box 2-2), but absolute magnitude tells us how bright the star really is. The absolute visual magnitude M_v of a star is the apparent visual magnitude of the star if it were 10 pc away. If we know a star's apparent magnitude and its distance away, we can calculate its absolute magnitude. The equation that allows this calculation relates apparent magnitude m, distance in parsecs d, and absolute magnitude M_v:

$$m - M_v = -5 + 5 \log_{10}(d)$$

Sometimes it is convenient to rearrange the equation and write it in this form:

$$d = 10^{\frac{m - M_v + 5}{5}}$$

It is the same equation, so we can use whichever form is most convenient in a given problem. This equation shows that the difference between apparent and absolute magnitude depends only on the distance to the star.

The quantity $m - M_v$ is called the **distance modulus**, a measure of how far away the star is. If the star is very far away, the distance modulus is large; if the star is close, the distance modulus is small. We could use the preceding equation to make a table of distance moduli (Table 8-1).

If we know the distance to a star, we can find its distance modulus from the table. If we subtract the distance modulus from the apparent magnitude, we get the star's absolute magnitude.

Example: Deneb is 490 pc away and has an apparent magnitude of 1.26. What is its absolute magnitude? *Solution:* From the table we find that Deneb's distance modulus is about

T A B L E 8 - 1 Distance Moduli	
$m - M_v$	d (in pc)
0	10
1	16
2	25
3	40
4	63
5	100
6	160
7	250
8	400
9	630
10	1000
.	.
.	.
.	.
15	10,000
.	.
.	.
.	.
20	100,000
.	.
.	.

8.5. Thus its absolute magnitude is about 1.26 minus 8.5, or about −7.2. Deneb is intrinsically a very bright star. If it were only 10 pc away, it would dominate the night sky, shining 5.8 magnitudes or over 200 times brighter than Sirius.

BOX 8-3

Luminosity, Radius, and Temperature

The luminosity L of a star depends on two things—its size and its temperature. If the star has a large surface area from which to radiate, it can radiate a great deal. Recall from our discussion of black-body radiation in Box 6-2 that the amount of energy emitted per second from each square meter of the star's surface is σT^4. Thus the star's luminosity can be written as its surface area in square meters times the amount it radiates per square meter:

$$L = \text{area} \times \sigma T^4$$

Because a star is a sphere, we can use the formula area $= 4\pi R^2$. Then the luminosity is:

$$L = 4\pi R^2 \sigma T^4$$

This seems complicated, but if we divide by the same quantities for the sun we can cancel out the constants:

$$\frac{L}{L_\odot} = \left(\frac{R}{R_\odot}\right)^2 \left(\frac{T}{T_\odot}\right)^4$$

Recall that the symbol $\odot$ stands for the sun.

Example A: Suppose a star is 10 times the sun's radius but only half as hot. How luminous would it be? *Solution:*

$$\frac{L}{L_\odot} = \left(\frac{10}{1}\right)^2 \left(\frac{1}{2}\right)^4 = \frac{100}{1} \times \frac{1}{16} = 6.25$$

The star would be 6.25 times the sun's luminosity.

We can also use this formula to find diameters.

Example B: Suppose we found a star whose absolute magnitude is +1 and whose spectrum shows it is twice the sun's temperature. What is the diameter of the star? *Solution:* The star's absolute magnitude is 4 magnitudes brighter than the sun, and we recall from Box 2-2 that 4 magnitudes is a factor of 2.512^4, or about 40. The star's luminosity is therefore about 40 $L_\odot$. With the luminosity and temperature, we can find the radius:

$$\frac{40}{1} = \left(\frac{R}{R_\odot}\right)^2 \left(\frac{2}{1}\right)^4$$

Solving for the radius we get:

$$\left(\frac{R}{R_\odot}\right)^2 = \frac{40}{2^4} = \frac{40}{16} = 2.5$$

So the radius is:

$$\frac{R}{R_\odot} = \sqrt{2.5} = 1.58$$

The star is 58 percent larger in radius than the sun.

8-3 THE H–R DIAGRAM

Our second goal in this chapter is to find the diameters of stars. Are all stars similar in size to our sun, or are some much bigger and some much smaller? We can answer that question by drawing what is surely the most important diagram in astronomy, a diagram that relates stellar luminosity and temperature.

Luminosity, Temperature, and Diameter The luminosity of a star depends on its temperature and its diameter. To understand this relationship, imagine a candle. You can eat dinner by candlelight because the candle flame has a small surface area and, consequently, a small luminosity. If the flame were 12 ft tall, however, it would have a very large surface area from which to radiate, and even if it were no hotter than a normal candle flame, its luminosity would drive you away from the table.

In a similar way, a star's luminosity is proportional to its surface area. A hot star may not be very luminous if it has a small surface area, although it could be highly luminous if it were larger. Even a cool star could be luminous if it had a large surface area. (See Box 8-3.) Because of this dependence on both temperature and surface area, we can use stellar luminosities to determine the diameters of stars if we can separate the effects of temperature and surface area.

The **Hertzsprung–Russell (H–R) diagram,** named after its originators, Ejnar Hertzsprung and Henry Norris Russell, is a graph that separates the effects of temperature and surface area on stellar luminosities, and enables us to sort the stars according to their diameters. Before we discuss the details of the H–R diagram (as it is often called), let us look at a similar diagram we might use to sort automobiles.

We can plot a diagram such as Figure 8-6 to show horsepower versus weight for various makes of cars. And in so doing, we find that in general the more a car weighs, the more horsepower it has. Most cars fall somewhere along the sequence of cars, running from heavy, high-powered cars to light, low-powered models. We might call this the main sequence of cars. But some cars have much more horsepower than normal for their weight— the sport or racing models—and the economy models have less power than normal for cars of the same weight. Just as this diagram helps us understand the different kinds of autos, the H–R diagram helps us understand the kinds of stars.

The H–R diagram relates the intrinsic brightness of stars to their surface temperatures (Figure 8-7). We may plot either absolute magnitude or luminosity on the vertical axis of the graph, since both refer to intrinsic brightness. As you will remember from Chapter 6, spectral type is related to temperature, so we may plot either spectral type or temperature on the horizontal axis. Technically, only graphs of absolute magnitude versus spectral type are H–R diagrams. However, we will refer to

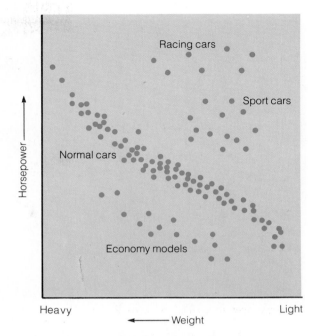

Figure 8-6 We could analyze automobiles by plotting their horsepower versus their weight and thus reveal relationships between various models. Most would lie somewhere along the main sequence of "normal" cars.

plots of luminosity versus either spectral type or surface temperature by the generic term *H–R diagram*.

A point on an H–R diagram shows a star's luminosity and surface temperature. Points near the top of the diagram represent very luminous stars, and points near the bottom represent very faint stars. Points on the left represent hot stars, and points on the right represent cool stars. Notice that the location of a star in the H–R diagram has nothing to do with its location in space. Also, as we will see in later chapters, a star's luminosity and surface temperature change as it ages and moves in the H–R diagram, but this has nothing to do with the star's actual motion through space.

Dwarfs, Giants, and Supergiants The **main sequence** is the region of the H–R diagram running from upper left to lower right, which includes roughly 90 percent of all stars. These are the "ordinary" stars. As we might expect, the hot main-sequence stars are brighter than the cool main-sequence

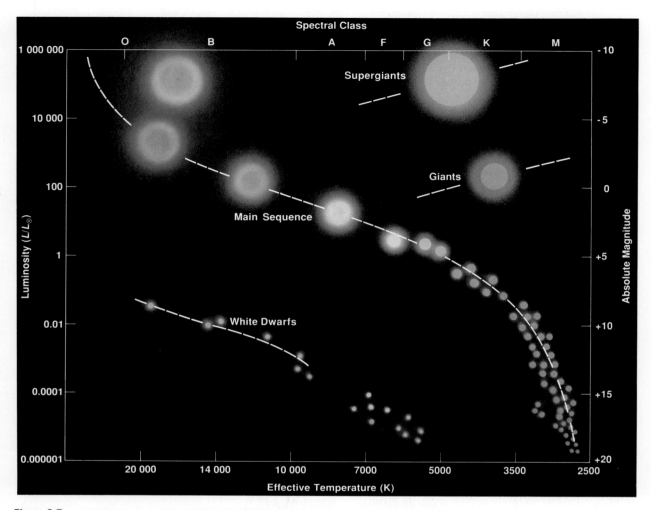

Figure 8-7 An H–R diagram. Roughly 90 percent of all stars are main-sequence stars, including the sun—a G2 star with an absolute magnitude of about +5. Star diameters not to scale. (Courtesy William K. Hartmann)

stars. The sun is a medium-temperature main-sequence star.

Just as sports cars do not fit in with the normal cars in Figure 8-6, some stars do not fit in with the main-sequence stars in Figure 8-7. The **giant stars** lie at the upper right of the H–R diagram. These stars are cool, radiating little energy per square centimeter. Nevertheless, they are highly luminous because they have enormous surface areas, hence the name *giant stars*. In fact, we can estimate the size of these giants with a simple calculation. Notice that giants are about 100 times more lumi-

nous than the sun even though they have about the same surface temperature. Thus they must have about 100 times more surface area than the sun. This means they must be about 10 times larger in diameter than the sun (Figure 8-8).

Near the top of the H–R diagram we find the **supergiants.** These exceptionally luminous stars are 10 to 1000 times the diameter of the sun. Betelgeuse in Orion is a supergiant (Figure 8-8). If it magically replaced the sun at the center of our solar system, it would swallow up Mercury, Venus, Earth, and Mars, and would just reach Jupiter.

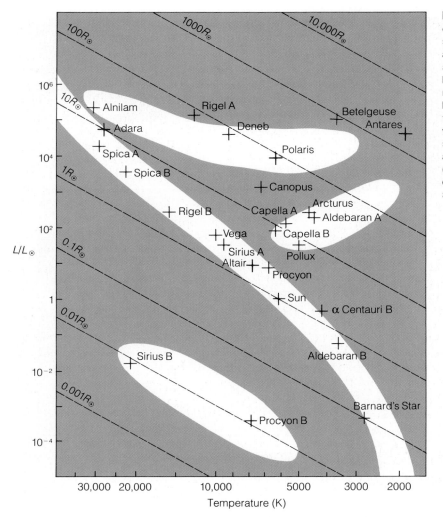

Figure 8-8 An H–R diagram drawn with luminosity versus surface temperature. Diagonal lines are lines of constant radius. White areas mark the approximate location of supergiants, giants, the main sequence, and white dwarfs. Note that giant stars are 10 to 100 times larger than the sun and that white dwarfs are about 100 times smaller than the sun—that is, about the size of the earth. (Individual stars that orbit each other are designated by A and B, as in Spica A and Spica B.)

μ Cephei is believed to be over three times larger than Betelgeuse. It would reach nearly to the orbit of Uranus.

In the lower left of the H–R diagram are the economy models, stars that are very faint even though they are hot. Clearly such stars must be small. These are the **white dwarf stars,** about the size of Earth (Figure 8-8).

Luminosity Classification We can tell from a star's spectrum what kind of star it is. Main-sequence stars are rather small and have dense atmospheres.

The gas atoms collide with one another often, distorting the energy levels and smearing the spectral lines, making them broader. Giant stars are larger, their atmospheres are less dense, and the atoms disturb each other relatively little (Figure 8-9). The lines in the spectra of giant stars are sharp; the lines in the spectra of supergiants are even sharper.

Thus we can look at a star's spectrum and classify its luminosity. Although these are **luminosity classes,** the names refer to the sizes of stars because size is the dominating factor determining

Figure 8-9 Differences in widths and strengths of spectral lines distinguish the spectra of supergiants, giants, and main-sequence stars, thus making the luminosity classification possible. (Adapted from H. A. Abt, A. B. Meinel, W. W. Morgan, and J. W. Tapscott, *An Atlas of Low-Dispersion Grating Stellar Spectra,* Kitt Peak National Observatory, 1968)

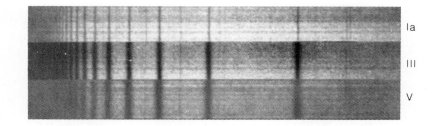

Figure 8-10 The approximate location of the luminosity classes on the H–R diagram.

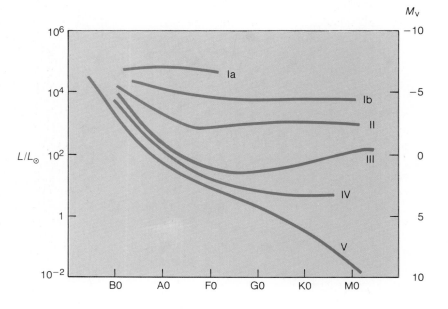

luminosity. Supergiants, for example, are very luminous because they are very large.

The luminosity classes are represented by the roman numerals I through V, with supergiants further subdivided into types Ia and Ib, as follows:

Luminosity Classes

Ia	Bright supergiant
Ib	Supergiant
II	Bright giant
III	Giant
IV	Subgiant
V	Main-sequence star

Using letters for subclasses, we can distinguish between the bright supergiants (Ia) such as Rigel (β Orionis) and the regular supergiants (Ib) such as Polaris, the North Star. The star Adhara (ϵ Canis Majoris) is a bright giant (II), Capella (α Aurigae) is a giant (III), and Altair (α Aquilae) is a subgiant (IV). The sun is a main-sequence star (V). The luminosity class appears after the spectral type, as in G2 V for the sun. White dwarfs don't enter into this classification because their spectra are peculiar.

We can plot the positions of the luminosity classes on the H–R diagram (Figure 8-10). Remember that these are rather broad classifications. A star of luminosity class III may lie slightly above or below the line labeled III. The lines are only approximate.

The luminosity classes are an important tool because the luminosity of a star can give us a clue to its distance. This method of finding distance is called **spectroscopic parallax,** and we can use it to estimate the distance to stars that are too far away

Spectroscopic Parallax

Driving along the highway at night, we often see lights dotting the countryside. We cannot immediately tell how far away these lights are unless we know how bright they really are. If we know that one of the lights is an airport searchlight and another is the headlight on a bicycle, we can judge their distances. The same is true of stars. If we can discover the luminosity of a star, we can use its apparent brightness to estimate its distance.

The method of spectroscopic parallax lets us find the distance to a star by classifying its spectrum according to spectral type and luminosity class. We can then look it up on an H–R diagram such as Figure 8-10 and read off its absolute magnitude M_v. Once we measure its apparent magnitude m, we can calculate the distance using either the equation from Box 8-2 or a distance modulus table such as Table 8-1.

Example: Spica is classified B1 V, and its apparent magnitude is +1. How far away is it?
Solution: From Figure 8-10 we can estimate that a B1 V star should have an absolute magnitude of about −3. Therefore its distance modulus is 1 − (−3), or 4, and the distance (taken from Table 8-1) is about 63 pc.

This method is not very accurate because there is some uncertainty in Figure 8-10 due to individual differences between stars. Consequently, when we classify a star's spectrum, we can't be sure of its exact absolute magnitude. It might be a little brighter or fainter than the diagram predicts. If the star is just 1 magnitude fainter than we expect, the distance we calculate is 37 percent too small. Although this method is not very accurate, spectroscopic parallax is often the only method available to measure distance.

---

to have measurable parallaxes. As long as we can record a star's spectrum, we can estimate its distance. (Box 8-4 explains how spectroscopic parallax works.)

In reaching our second goal, finding the diameters of stars, we discovered four different kinds of stars: main-sequence stars, giants, supergiants, and white dwarfs. This raises the question: Are giant stars large because they contain more mass, or do all stars have about the same mass? This question is related to our third goal, finding the masses of stars.

8-4 THE MASSES OF STARS

A star's mass is difficult to determine. Observing a single star through a telescope, we see nothing but a point of light that tells us nothing about the star's mass. To measure the masses of stars, we must

discuss the motions of **binary stars,** pairs of stars that orbit each other.

Binary Stars in General   The key to finding the mass of a binary star is understanding orbital motion— which provides clues to mass.

In Chapter 4, we illustrated orbital motion by imagining a cannon firing a cannonball from a high mountain (see Figure 4-24). If the cannonball traveled fast enough, it could orbit Earth. Newton's laws explain that the cannonball follows a curved path around Earth only because gravity pulls it away from its straight-line motion.

Similarly, the two stars in Figure 8-11 would move at constant speed in the straight-line paths shown by the dashed lines if some force did not act on them. In a binary star system, the stars move around each other in orbits because their mutual gravitational attraction pulls them away from straight-line motion.

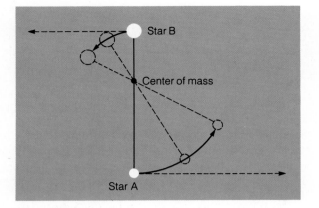

Figure 8-11 Without gravity, the two stars would follow the straight-line paths shown by dashed lines. Gravity pulls the stars into curved paths about their common center of mass.

Each star in a binary system moves in its own orbit around the system's **center of mass,** the balance point of the system. If the stars were connected by a massless rod and placed in a uniform gravitational field such as that near Earth's surface, the system would balance at its center of mass like a child's seesaw (Figure 8-12a). If one star were more massive than its companion, then the massive star would be closer to the center of mass and would travel in a smaller orbit, while the lower-mass star would whip around in a larger orbit (Figure 8-12b). By observing the relative sizes of the two orbits, we could determine the ratio of the masses. That is, we could say which was more massive and by how much. Unfortunately, this does not tell us the individual masses of the stars, which is what we really want to know. That requires further analysis.

To find the mass of a binary star system, we must know the size of the orbits and the orbital period —the length of time the stars take to complete one orbit. The smaller the orbits are and the shorter the orbital period is, the stronger the stars' gravity must be to hold each other in orbit. For example, if two stars whirl rapidly around each other in small orbits, then their gravity must be very strong to prevent their flying apart. Such stars would have to be very massive. By knowing the size of the orbits and the orbital period, we could figure out how much mass the stars contain. (See Box 8-5.) Calculations such as those shown in Box 8-5 will yield the total mass, which, combined with the ratio of the masses found from the relative sizes of the orbits, can tell us the individual masses of the stars.

Actually, figuring out the mass of a binary star system is not as easy as it might seem from the examples in Box 8-5. The orbits of the stars are often elliptical and tipped at an unknown angle to our line of sight. For some kinds of binary star systems, astronomers can overcome these problems, analyze the system, and find the masses of the stars.

Although there are many different kinds of binary stars, three types are especially important for determining stellar masses. We will discuss these separately in the next sections.

Visual Binaries In a **visual binary,** such as Sirius and its companion, the two stars are separately visible in the telescope (Figure 8-13a). Only a pair with large orbits can be separated visually, for if the orbits are small, the star images blend together in the telescope and we see only a single point of light. Since visual binaries have such large orbits,

Figure 8-12 (a) If the two stars in a binary system were connected by a massless rod and placed in a uniform gravitational field like that at Earth's surface, the system would balance like a seesaw at its center of mass. (b) The stars orbit around their center of mass, and thus the more massive star follows a smaller orbit than its lower-mass companion.

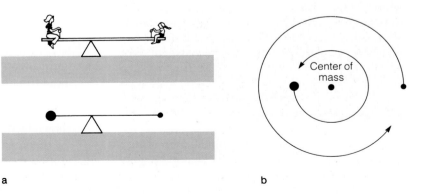

a

b

The Masses of Binary Stars

According to Newton's laws of motion and gravity, the total mass of two stars orbiting each other is related to the average distance a between them and their orbital period P. If the masses are M_A and M_B, then:

$$M_A + M_B = \frac{a^3}{P^2}$$

In this formula, we will measure a in AU, P in years, and the mass in solar masses.

Notice that this formula is related to Kepler's third law of planetary motion. Almost all of the mass of the solar system is in the sun. If we apply this formula to any planet in our solar system, the total mass is 1 solar mass. Then the formula becomes $P^2 = a^3$, which is Kepler's third law.

In other star systems, the total mass is not necessarily 1 solar mass, and this gives us a way to find the masses of binary stars. If we can find the average distance between the two stars in AU and their orbital period in years, the sum of the masses of the two stars is just a^3/P^2.

Example A: If we observe a visual binary with a period of 32 years and an average separation of 16 AU, what is the total mass? *Solution:* The total mass equals $16^3/32^2$, or 4 solar masses.

Example B: In the previous example, suppose star A is 12 AU away from the center of mass and star B is 4 AU away. What are the individual masses? *Solution:* The ratio of the masses must be 12:4. What two numbers add up to 4 and have the ratio 12:4? Star B must be 3 solar masses, and star A must be 1 solar mass.

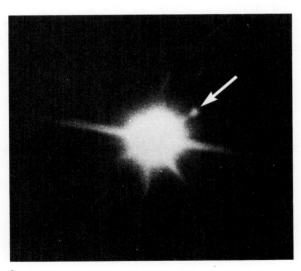

a

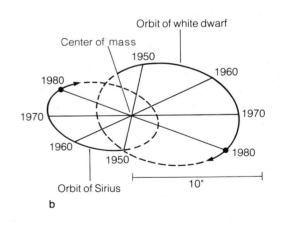

b

Figure 8-13   (a) Sirius and its white-dwarf companion (arrow), a visual binary system, photographed in 1960. The size of the images is related to the brightness of the stars and not to their diameters. The spikes on the image of Sirius are caused by diffraction in the telescope. (Lick Observatory photograph) (b) Sirius and its companion move in elliptical orbits around a common center of mass.

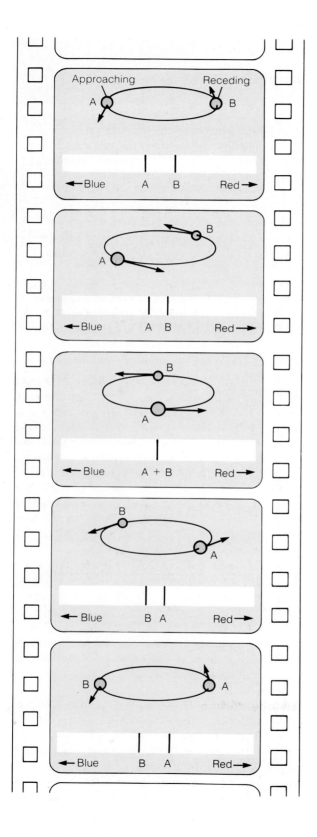

they also have long orbital periods. Some take hundreds or even thousands of years to complete a single revolution.

Astronomers study visual binaries by measuring the position of the two stars directly at the telescope or on photographic plates. In either case, they need measurements over many years to map the orbits. Figure 8-13b shows the orbits of the visual binary Sirius and its white dwarf companion. The dates show the observed location of the stars since 1950 and the predicted locations in the future. The orbital period of this system is about 50 years.

Cygnus (the swan) is an interesting constellation in terms of binary stars. Albireo (β Cygni) is a beautiful sight through a small telescope, appearing as a golden-yellow star of spectral type K3 and a sapphire-blue B8 star. For a long time, astronomers considered Albireo a very long-period binary star, but it shows no orbital motion, and a recent study suggests that the stars lie at different distances from the earth and thus are not a real binary system.

Though less beautiful than Albireo, the star system 61 Cygni is a true visual binary. It has a period of only 653 years, which is still too long for convenient analysis, but it is of special interest for a number of reasons. It is only 3.4 pc from the sun, and in 1838 it became the first star to have its parallax measured.

More than half of all stars are members of multiple systems, so binaries are common. Few, however, can be analyzed completely. Many are so far apart that their periods are much too long for practical mapping of their orbits. Others are so close together they are not visible as separate stars.

Spectroscopic Binaries   If the stars of a binary system are close together, the telescope, limited by diffraction and seeing, shows us a single point of light. Only by taking a spectrum, which is formed by light from both stars and contains spectral lines from both, can we tell that there are two stars present and not one. Such a system is called a **spectroscopic binary.**

Figure 8-14   As the stars of a spectroscopic binary revolve around their common center of mass, they alternately approach and recede from Earth. The Doppler shifts cause their spectral lines to move back and forth across each other.

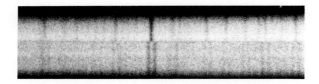

Figure 8-15 Mizar, at the bend of the handle of the Big Dipper, is a spectroscopic binary, as these two spectra demonstrate. The upper spectrum was taken when the stars were moving perpendicular to the line of sight and the lines were single. The lower spectrum was taken when one star was approaching and the other receding; thus the lines are double. (Mount Wilson and Las Campanas observatories, Carnegie Institution of Washington)

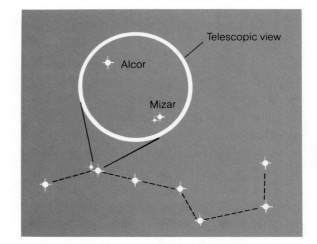

Figure 8-16 At the bend in the handle of the Big Dipper lies Mizar, a visual binary. It and its companion, visible through small telescopes, plus the nearby star Alcor, are all spectroscopic binaries (see Figure 8-14).

Because the stars in a spectroscopic binary orbit each other, they alternately approach and recede from us, as in Figure 8-14. As one star comes toward us, its spectral lines are Doppler-shifted toward the blue. The other star is moving away from us, and its spectral lines are shifted toward the red. Half an orbit later, the star that was approaching is receding. As we watch the spectrum of the binary system, we see the spectral lines split into two parts that move apart and then back together as the stars follow their orbits (Figures 8-14 and 8-15).

We can find the orbital period of a spectroscopic binary by observing its changing radial velocities. We can also determine the velocity of each star. If we know the velocity and the length of time it takes for the star to complete one orbit, we can multiply to find the circumference of the orbit. From that we can find the radius of the orbit. Thus we would know most of what we need to find the masses of the stars—the size and period of the orbits. One important detail is still missing, however. We don't know how the orbits are inclined to our line of sight.

We can find the inclination of a visual binary because we can see the two stars moving along their orbits. In a spectroscopic binary, however, we cannot see the individual stars, find the inclination, or untip the orbits. The velocities we observe are not the true orbital velocities but only the part of that velocity directed radially toward or away from earth. Because we cannot find the inclination, we cannot correct these radial velocities to their true orbital velocities. Therefore we cannot find the true masses. All we can find from a spectroscopic binary is a lower limit to the masses.

Spectroscopic binaries are common. Capella, for instance, is such a binary. A small telescope reveals that Mizar in the Big Dipper is a visual binary (Figure 8-16). Spectra show that both of the stars in the Mizar visual binary are themselves spectroscopic binaries, making Mizar a "double double star." Near Mizar is Alcor, just visible to the naked eye. It is 22 pc to Alcor and only 21 pc to Mizar. Apparently Mizar and Alcor are an optical double—that is, they appear close as seen from Earth but do not orbit each other. But Alcor, like many stars, is a spectroscopic binary.

So far we have discussed an ideal spectroscopic binary in whose spectrum lines of both stars appear. In many spectroscopic binaries, however, one of the stars is significantly fainter than its companion, and the lines of the fainter star are invisible in the spectrum. These systems are even more difficult to analyze and provide even less information about stellar masses.

Eclipsing Binaries Rare among binary stars are those with orbits tipped so the stars cross in front of each other as seen from Earth. Imagine a model of a binary star system in which a cardboard disk represents the orbital plane, as in Figure 8-17. If the orbits are seen edge-on from Earth, then the two stars cross in front of each other. The small star

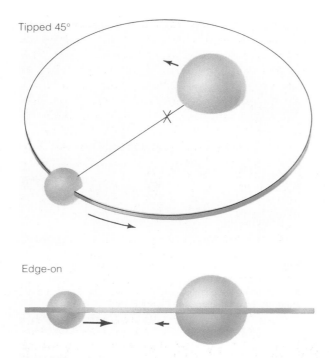

Tipped 45°

Edge-on

Figure 8-17 Imagine a model of a binary system with balls for stars and a disk of cardboard for the plane of the orbits. Only if we view the system edge-on do we see the stars cross in front of each other. These eclipsing binary systems are rare.

crosses in front of the large star and then, half an orbit later, the larger star crosses in front of the small star. We say the stars are eclipsing each other, and we call the system an **eclipsing binary.**

Seen from Earth, the two stars are not resolvable—that is, they are not visible separately. The system looks like a single point of light. But when one star moves in front of the other star, part of the light is blocked, and the total brightness of the system decreases. Figure 8-18 shows a smaller star moving in an orbit around a larger star, first eclipsing the larger star as it crosses in front and then being eclipsed as it moves behind. The resulting variation in brightness is recorded in a graph

Figure 8-18 As the stars in an eclipsing binary cross in front of each other, the total brightness of the system changes. In this example, a small hot star orbits a giant cool star.

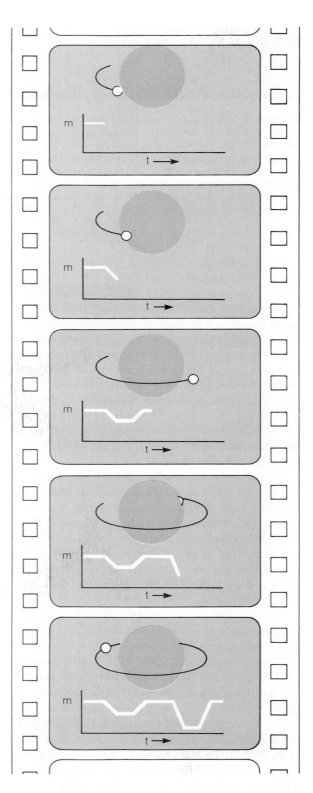

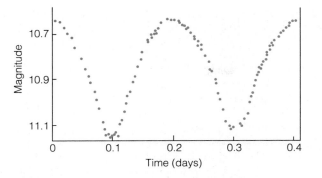

Figure 8-19 A light curve of a typical eclipsing binary shows how the brightness of the system falls when first one and then the other star crosses in front of its companion. In this system, the stars are very close to each other and are distorted into egg shapes. This causes the rounded shape of the curve. Compare with Figure 8-18, where the stars are not distorted. (Grundy Observatory)

called the **light curve** of the system (Figure 8-19).

It is often possible to find the masses of eclipsing binaries. We could find the orbital period easily, and if we could get spectra showing the Doppler shifts of the two stars, we could find the orbital velocity. Then we could find the size of the orbits and the masses of the stars. The inclination of the orbits poses no problem because we know that we must be observing the orbits nearly edge on. Otherwise, we would not see the stars eclipse each other.

Eclipsing binaries are especially important because they enable us to measure the diameters of the stars. From the light curve we can tell how long it took for the small star to cross the large star. Multiplying this time interval by the orbital velocity of the small star gives us the diameter of the larger star. We can also determine the diameter of the small star by noting how long it took to disappear behind the edge of the large star. For example, if it took 300 seconds for the small star to disappear while traveling 500 km/sec relative to the large star, then it must be 150,000 km in diameter.

Of course there are complications due to the inclination and eccentricity of orbits, but often these effects can be taken into account, and the system can tell us not only the masses of its stars but also their diameters.

Algol (β Persei) is one of the best-known eclipsing binaries because its eclipses are visible to the naked eye. Normally, its magnitude is about 2.15, but its brightness drops to 3.4 in eclipses that occur every 68.8 hours. Although the nature of the star was not recognized until 1783, its periodic dimming was probably known to the ancients. *Algol* comes from the Arabic for "the demon's head," and it is associated in constellation mythology with the severed head of Medusa, the sight of whose serpentine locks turned mortals to stone (Figure 8-20). Indeed, in some accounts, Algol is the winking eye of the demon.

Mass, Luminosity, and Density Although binary systems are common, masses can be accurately derived for fewer than 100. Nevertheless, this handful of stars reveals important relationships among a star's mass, luminosity, and density.

The most massive stars known are about 55 to 100 times the mass of the sun, and the least massive are slightly less than 0.1 solar masses. Among main-sequence stars, the most massive are the O and B stars, the so-called upper-main-sequence stars. As we run our eye down the main sequence, we find lower-mass stars. The lowest-mass are the K and M stars, the lower-main-sequence stars. Thus a star's location along the main sequence depends on its mass (Figure 8-21).

Stars that do not lie on the main sequence do not appear to be arranged in any particular pattern according to mass. Some giants and supergiants are quite massive, while others are no more massive than our sun. White dwarfs are about the mass of the sun or slightly less.

Because of the systematic ordering of mass along the main sequence, these stars obey a **mass–luminosity relation**—the more massive a star is, the more luminous it is (Figure 8-22). In fact, the mass–luminosity relation can be expressed as a simple formula. (See Box 8-6.) Giants and supergiants do not follow the mass–luminosity relation very closely, and white dwarfs not at all. In the next chapters, the mass–luminosity relation will help us understand how stars generate their energy.

Though mass alone does not reveal any pattern among giants, supergiants, and white dwarfs, density does. Once we know a star's mass and diameter, we can calculate its average density by dividing its mass by its volume. Stars are not uniform in

Figure 8-20 The eclipsing binary Algol is the star on the demon's forehead in this 1837 engraving of Perseus and the head of the gorgon Medusa. *Algol* comes from the Arabic for "the demon's head." (From Duncan Bradford, *Wonders of the Heavens,* Boston: John B. Russell, 1837)

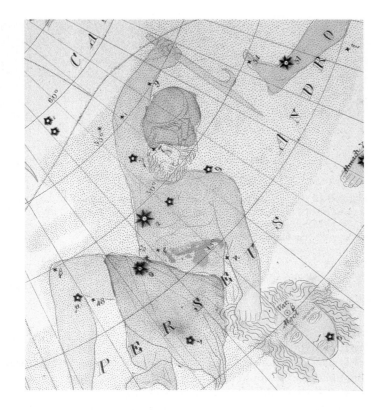

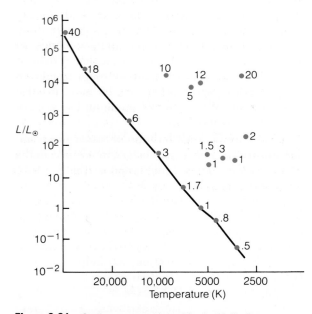

Figure 8-21 Stellar masses plotted in the H–R diagram. Giants and supergiants consist of stars of various masses, but masses along the main sequence are ordered. The most massive stars are the most luminous.

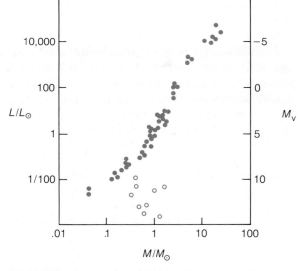

Figure 8-22 The mass–luminosity relation shows that the more massive a star is, the more luminous it is. The open circles represent white dwarfs, which do not obey the relation.

BOX 8-6
The Mass–Luminosity Relation

We can calculate the approximate luminosity of a star from a simple equation. A star's luminosity in terms of the sun's luminosity equals its mass in solar masses raised to the 3.5 power:

$$L = M^{3.5}$$

This is the mathematical form of the mass–luminosity relation.

We can do simple calculations with this equation if we remember that raising a number to the 3.5 power is the same as cubing it and then multiplying by its square root.

Example: What is the luminosity of a star four times the mass of the sun? *Solution:* The star must be 128 times more luminous than the sun because:

$$L = M^{3.5} = 4^{3.5} = 4 \cdot 4 \cdot 4 \sqrt{4} = 64 \cdot 2 = 128$$

density, but are most dense at their centers and least dense near their surface. The center of the sun, for instance, is about 100 times as dense as water; its density near the visible surface is about 3400 times less dense than Earth's atmosphere at sea level. A star's average density is intermediate between its central and surface densities. The sun's average density is approximately 1 gm/cm^3—about the density of water.

Main-sequence stars have average densities similar to the sun's, but giant stars, being large, have low average densities, ranging from 0.1 to 0.01 gm/cm^3. The enormous supergiants have still lower densities, ranging from 0.001 to 0.000001 gm/cm^3. These densities are thinner than the air we breathe, and if we could insulate ourselves from the heat, we could fly an airplane through these stars. Only near the center would we be in any danger, for there the material is very dense—about 3,000,000 gm/cm^3.

The white dwarfs have masses of about 1 $M_\odot$ but are very small, only about the size of Earth. Thus the matter is compressed to densities of 10,000,000 gm/cm^3 or more. On Earth, 1 cm^3 of this material would weigh about 20 tons.

Density divides stars into three groups. Most stars are main-sequence stars with densities like the sun's. Giants and supergiants are very low-density stars, and white dwarfs are high-density objects. We will see in later chapters that these densities reflect different stages in the evolution of stars.

In this chapter, we have tried to describe the properties of stars. We have found ways of determining their luminosity (or absolute magnitude), their diameter, and their mass. Along the way we have discovered relationships between mass and luminosity and between density and position in the H–R diagram. Thus it might appear that we have sufficient data to begin the study of the birth, evolution, and death of stars. Before we begin, however, we must answer one last question: Among the different kinds of stars, which are common and which are rare? The only way we can find the answer is by taking a survey.

A Neighborhood Survey

Surveying a Representative Sample   Suppose you took a survey in your neighborhood to find how many people had gray eyes. If you knew the area of your neighborhood in square miles, you could then say, "In my neighborhood, x people per square mile have gray eyes." Next you might think of extending your conclusion to the country as a whole: "In America x people per square mile have gray eyes." Of course, if you did not live in an average neighborhood, your result would be wrong, but if you had sampled a truly representative neighborhood, your survey would give valid results about the entire population.

We can do the same thing with the stars. We can ask how many stars of each spectral type are whirling through each million cubic parsecs of space. We can't count every star in our galaxy, but we can take a survey in the region of space near the sun. Because we think the solar neighborhood is a fairly average sort of place, we can use this local survey to reach conclusions about the entire population of stars.

Units of Stellar Density   The result of such a survey of stars is called a **stellar density function,** a description of the abundance of different types of stars in space. A simple form of the stellar density function appears in Figure 8-23, giving the abundance of the stars of each spectral type in terms of the number of each type we would expect to find in $1,000,000$ pc^3.

The stellar density function does not tell us how many stars are giants and how many are main-sequence stars. To distinguish among the luminosity classes, we would have to take a survey to find the number of stars of a given luminosity per million cubic parsecs.

Three Problems in Counting Stars   An astronomer could make these surveys in the neighborhood of the sun by counting all the stars within a given distance. A sphere of radius 62 pc contains $1,000,000$ pc^3. Thus if we could count all the M stars, for example, within 62 pc of the sun, we

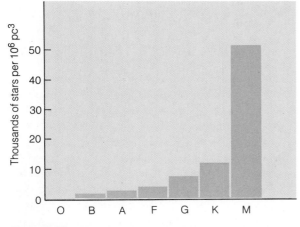

Figure 8-23 The stellar density function shows that M stars are the most common stars in space and that O stars are very rare.

would have the frequency of M stars per million cubic parsecs.

But this survey of the stars poses three problems. First, to determine which stars are within 62 pc of the sun, we must measure their distances. However, stars near the outer edge of a sphere 62 pc in radius have small parallaxes, which are difficult to measure accurately, and the method of spectroscopic parallax is not accurate enough for this purpose. We could count stars in a smaller sphere, but some stars are so rare that we might not find any in such a small volume of space.

A second problem for the stellar surveyor is the intrinsic faintness of stars such as M stars and white dwarfs. These are so faint that they are very hard to see if they are only a few dozen parsecs away. For example, a white dwarf 62 pc away is over 1500 times fainter than the faintest star visible to the naked eye—very hard to find, indeed.

The third problem with these surveys is that the hottest stars are rare. There are no O stars at all within 62 pc of the sun. We must extend our survey to great distances before we find many of

these hot, luminous stars. At such distances the parallaxes are too small to be measured, and we have to find distance in other, less accurate ways, such as spectroscopic parallax.

The Frequency of Stellar Types In spite of these difficulties, astronomers can use statistical methods to study the abundance of stars within a few hundred parsecs of the sun. These studies determine not just the abundance of stars of a given spectral type but also how many stars of that type are likely to be giants, how many are likely to be supergiants, and how many white dwarfs. This detailed version of the stellar density function is shown in Figure 8-24, an H–R diagram with bars added to represent the frequency of stars in space. Compare this figure with the H–R diagram in Figure 8-7 and locate the main-sequence, giants, white dwarfs, and supergiants.

Notice how common the main-sequence M stars are. There are about 65,000 of these red dwarfs in every million cubic parsecs. White dwarfs are also very common. Fortunately for us, these stars are very faint, for if they were as luminous as supergiants, the sky would be filled with their glare, and we would see hardly anything else. As it is, the most common stars are so faint they are hard to find even with large telescopes.

Notice that the more luminous stars are rare. The stars become less common as we go up the main sequence; A stars are rare, and type O and B stars are so rare we cannot plot them on this diagram. There are only about 0.036 of the type O stars per million cubic parsecs. That is, we would have to search through about 30,000,000 pc^3 to find an O star. Put yet another way, only 1 star in 4 million is an O star. Giants and supergiants are also very rare. Every million cubic parsecs contains only about 100 giants and only 0.07 supergiants. Luckily, these very rare kinds of stars are also very luminous—we can see them from great distances. If they were as faint as

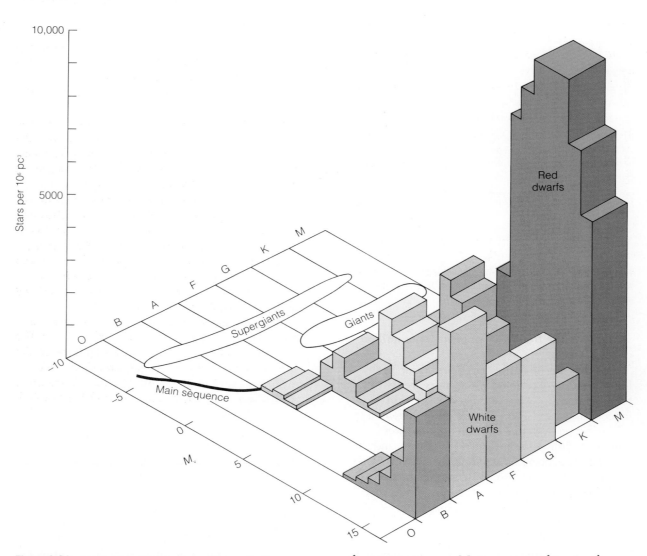

Figure 8-24 In this histogram, bars rise from an H–R diagram to represent the number of stars of given spectral type and absolute magnitude found in 1 million cubic parsecs. The main-sequence M stars are the most common stars, followed closely by the white dwarfs. The upper-main-sequence stars and the giants and supergiants are so rare their bars are not visible in this diagram.

the main-sequence M stars, we might never know they existed.

The relative numbers of stars give us clues to how stars are born and how they die. In the next two chapters, we will discover why O stars are rare and why red dwarfs are common, why giants are rare and why white dwarfs are common.

SUMMARY

We can measure the distance to the nearer stars by observing their parallaxes. The more distant stars are so far away that their parallaxes are unmeasurably small. To find the distances to these stars, we must use spectroscopic parallax. Stellar distances are commonly expressed in parsecs. One parsec is 206,265 AU—the distance to an imaginary star whose parallax is 1 second of arc.

Once we know the distance to a star, we can find its intrinsic brightness, expressed as its absolute magnitude or its luminosity. A star's absolute magnitude is the apparent magnitude we would see if the star were only 10 pc away. The luminosity of the star is the total energy radiated in 1 second, usually expressed in terms of the luminosity of the sun.

The H–R diagram plots stars according to their intrinsic brightness and their surface temperature. In the diagram, roughly 90 percent of all stars fall on the main sequence, the more massive being hotter, larger, and more luminous. The giants and supergiants, however, are much larger and lie above the main sequence, more luminous than main-sequence stars of the same temperature. The white dwarfs are hot stars, but they fall below the main sequence because they are so small.

The large size of the giants and supergiants means their atmospheres have low densities and their spectra have sharper spectral lines than the spectra of main-sequence stars. In fact, it is possible to assign stars to luminosity classes by the widths of their spectral lines. Class V stars are main-sequence stars with broad spectral lines. Giant stars (III) have sharper lines, and supergiants (I) have extremely sharp spectral lines.

The only direct way we can find the mass of a star is by studying binary stars. When two stars orbit a common center of mass, we can find their masses by observing the period and sizes of their orbits.

Given the mass and diameter of a star, we can find its average density. On the main sequence, the stars are about as dense as the sun, but the giants and supergiants are very low-density stars. Some are much thinner than air. The white dwarfs, lying below the main sequence, are tremendously dense.

The mass–luminosity relation says that the more massive a star is, the more luminous it is. Main-sequence stars follow this rule closely, the most massive being the upper-main-sequence stars and the least massive the lower-main-sequence stars. Giants and supergiants do not follow the relation precisely, and white dwarfs not at all.

A survey in the neighborhood of the sun shows us that the most common kind of stars are the lower-main-sequence stars. The hot stars of the upper-main-sequence are very rare. Giants and supergiants are also rare, but white dwarfs are quite common, although they are faint and hard to find.

NEW TERMS

stellar parallax (p)

parsec (pc)

flux

absolute visual magnitude (M_v)

distance modulus ($m - M_v$)

luminosity (L)

Hertzsprung–Russell (H–R) diagram

main sequence

giant star

supergiant star

white dwarf star

luminosity class

spectroscopic parallax

binary stars

center of mass

visual binary

spectroscopic binary

eclipsing binary

light curve

mass–luminosity relation

stellar density function

REVIEW QUESTIONS

1. Why are parallax measurements limited to the nearest stars?

2. Describe two ways of specifying a star's intrinsic brightness.

3. Draw and label an H–R diagram. Show the proper locations of the main-sequence stars, giants, white dwarfs, and the sun.

4. Describe the steps in using the method of spectroscopic parallax. Do we really measure a parallax?

5. Give the approximate radii in terms of the sun of stars in the following classes: G2 V, G2 III, G2 Ia. Which of these is most dense, and which is least dense?

6. What is the most common type of star?

DISCUSSION QUESTIONS

1. How would having an observatory on Mars help astronomers measure parallax more accurately? How would having an observatory in orbit around Earth help?

2. How could a cool star be more luminous than a hot star?

3. How can we be certain the giant stars are actually larger than the sun?

4. Why is it impossible to determine the masses of binary stars that are very far apart?

5. The sun is sometimes described as an average star. What is the average star really like?

PROBLEMS

1. If a star has a parallax of 0.050 seconds of arc, what is its distance in pc? in ly? in AU?

2. If you place a screen of area 1 m² at a distance of 2.8 m from a 100-watt light bulb, the light flux falling on the screen will be 1 J/sec. To what distance must you move the screen to make the flux striking it equal 0.01 J/sec? (This assumes the light bulb emits all of its energy as light.)

3. If a star has a parallax of 0.016 seconds of arc and an apparent magnitude of 6, how far away is it and what is its absolute magnitude?

4. Complete the following table.

m	M$_v$	d (pc)	P (sec of arc)
___	7	10	___
11	___	1000	___
___	−2	___	0.025
4	___	___	0.040

5. The unaided human eye can see stars no fainter than those with an apparent magnitude of 6. If you can see a bright firefly blinking up to 0.5 km away, what is the absolute magnitude of the firefly? (HINT: Convert the distance to parsecs and use the formula in Box 8-2.)

6. If a main-sequence star has a luminosity of 400 L$_\odot$, what is its spectral type? (HINT: see Figure 8-7.)

7. If a star is 10 times the radius of the sun and half as hot, what will its luminosity be? (HINT: See Box 8-3.)

8. An O8 V star has an apparent magnitude of +1. Use the method of spectroscopic parallax to find the distance to the star. Why might this distance be inaccurate?

9. Find the luminosity and spectral type of a 5-M$_\odot$ main-sequence star.

10. In the following table, which star is brightest in apparent magnitude? Most luminous in absolute magnitude? largest? least dense? farthest away?

Star	Spectral Type	m
a	G2 V	5
b	B1 V	8
c	G2 Ib	10
d	M5 III	19
e	White dwarf	15

11. If two stars orbit each other with a period of 6 years and a separation of 4 AU, what is their total mass? (HINT: See Box 8-5.)

OBSERVATIONAL ACTIVITY: OBSERVING ALGOL

Most eclipsing binary stars are too faint to see with the naked eye, but a few are quite bright. Algol, the winking eye of the demon, goes through eclipses every 2.8673 days, and its changes in brightness are easy to detect.

Observations   Use the star charts at the back of the book to locate the constellation Perseus. It is visible in the evening sky from September through March. Then use the star chart of Figure 8-25 to find Algol (β Persei).

Estimate the brightness of Algol to a tenth of a magnitude by comparing it to other stars in the constellation.

Eclipses   Normally, Algol's magnitude is about 2.1, but once every 2.8673 days it fades to a magnitude of 3.4. These eclipses last only about 10 hours, so you may have to look at Algol on a number of evenings before you catch it in eclipse. Dates and times of mideclipse are published regularly in *Sky and Telescope* magazine.

If you see Algol entering eclipse, estimate its brightness every half hour or so, making careful note of the time of each observation. Later, you can plot these data on graph paper to produce a light curve.

If you are observing Algol with binoculars or a small telescope, don't fail to look at the double cluster in Perseus, as shown on the chart.

RECOMMENDED READING

ASHBROOK, JOSEPH. "Visual Double Stars for the Amateur." *Sky and Telescope* 60 (November 1980), p. 379.

BATTEN, ALAN H. "The Story of OE 341." *Sky and Telescope* 59 (March 1980), p. 200.

CROSWELL, KEN. "When Stars Coalesce." *Astronomy* 13 (May 1985), p. 67.

DARLING, DAVID. "Mystery Star." *Astronomy* 12 (August 1983), p. 66.

DEVORKIN, D. H. "Steps Toward the Hertzsprung–Russell Diagram." *Physics Today* 31 (March 1978), p. 32.

EVANS, ANEURIN. "Laboratory Exercises in Astronomy —The Orbit of a Visual Binary." *Sky and Telescope* 60 (September 1980), p. 195.

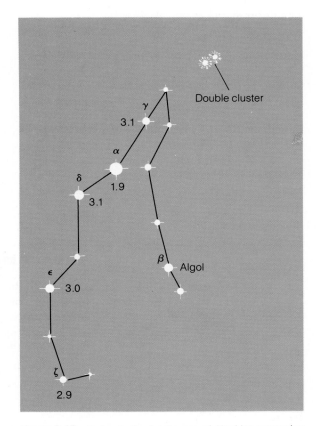

Figure 8-25   Estimate the brightness of Algol by comparing it with the other stars in the constellation of Perseus.

EVANS, D. S., T. G. BARNES, and C. H. LACY. "Measuring Diameters of Stars." *Sky and Telescope* 58 (August 1979), p. 130.

GETTS, JUDY A. "Decoding the Hertzsprung–Russell Diagram." *Astronomy* 11 (October 1983), p. 16.

HACK, M. "The Hertzsprung–Russell Diagram Today." *Sky and Telescope* 31 (May/June 1966), pp. 260, 333.

HODGE, PAUL. "How Far Are the Hyades?" *Sky and Telescope* 75 (February 1988), p. 138.

HOLZINGER, J. R., and M. A. SEEDS. *Laboratory Exercises in Astronomy*. Ex. 23, 24, 30, 31. New York: Macmillan, 1976.

HOPKINS, JEFFREY L., and ROBERT E. STENCEL. "Epsilon Aurigae." *Astronomy* 14 (February 1986), p. 6.

IRWIN, J. B. "The Case of the Degenerate Dwarf." *Mercury* 7 (November/December 1978), p. 125.

KALER, JAMES. "Journeys on the H–R Diagram." *Sky and Telescope* 75 (May 1988), p. 482.

———. *The Stars and Their Spectra.* New York: Cambridge University Press, 1989.

MacRobert, Alan. "Epsilon Aurigae: Puzzle Solved?" *Sky and Telescope* 75 (January 1988), p. 15.

Moore, Patrick. *Astronomer's Stars.* New York: Norton, 1987.

Philip, A. G. D., and L. C. Green. "Henry Norris Russell and the H–R Diagram." *Sky and Telescope* 55 (April 1978), p. 306.

———. "The H–R Diagram as an Astronomical Tool." *Sky and Telescope* 55 (May 1978), p. 395.

Reddy, Francis. "How Far the Stars." *Astronomy* 11 (June 1983), p. 6.

Seeds, Michael A. "The Wink in the Demon's Eye." *Astronomy* 8 (December 1980), p. 66.

Sinnott, Roger W. "Double Stars Waiting to Be Measured." *Sky and Telescope* 76 (November 1988), p. 487.

Soderblom, David R. "The Alpha Centauri System." *Mercury* 16 (September/October 1987), p. 138.

Tomkin, Jocelyn, and David L. Lambert. "The Strange Case of Beta Lyrae." *Sky and Telescope* 74 (October 1987), p. 354.

Trimble, Virginia. "A Field Guide to Close Binary Stars." *Sky and Telescope* 68 (October 1984), p. 306.

Verschuur, G. L. "Measuring Star Diameters." *Astronomy* 2 (December 1974), p. 36.

FURTHER EXPLORATION

VOYAGER, the Interactive Desktop Planetarium, can be used for further exploration of the topics in this chapter. This chapter corresponds with the following projects in *Voyages Through Space and Time:* Project 22, Magnitude Systems; Project 23, Spectral Classes and the H–R Diagram.

The Formation and Structure of Stars

Stars exist because of gravity. They form because gravity makes clouds of gas contract, and they spend long, stable lives generating nuclear energy at their centers and balancing their own gravity with the pressure of the hot gas in their interiors. In the end, stars die because they exhaust their fuel supply and can no longer withstand the force of their own gravity.

A star can remain stable only by maintaining great gas pressure in its interior. Gravity tries to make the stellar gas contract, but if the internal temperature is high enough, the pressure of the gas pushes outward just enough to balance gravity. Thus a star is a battlefield where gas pressure and gravity struggle for dominance. If gas pressure wins the star must expand, but if gravity wins the star must contract.

Only by generating tremendous amounts of nuclear energy can a star keep its interior hot

Jim he allowed [the stars] was made, but I allowed they happened. Jim said the moon could'a laid them; well, that looked kind of reasonable, so I didn't say nothing against it, because I've seen a frog lay most as many, so of course it could be done.

Mark Twain
The Adventures of Huckleberry Finn

enough to maintain the gravity–pressure balance. The sun, for example, generates enough energy in 1 second to power Earth's civilization at its current level for 2 million years.

To understand how stars form, we must understand how they generate their energy and how that energy eventually escapes from the star. Thus we must explore the nuclear reactions that generate energy in the cores of stars, and we must explore the internal structure of stars, which resists the inward pull of the star's gravity. These explorations will allow us to imagine the slow evolution of stable stars over billions of years.

We cannot see inside stars, and we do not live long enough to see stars evolve, but that is the subject of this chapter. Here we unravel two of the great secrets of nature —the origin and structure of the stars.

9-1 THE BIRTH OF STARS

The key to understanding star formation is the correlation between young stars and clouds of gas and dust. Where we find the youngest groups of stars, we also find large clouds of gas illuminated by the hottest and brightest of the new stars (Figure 9-1). This leads us to suspect that stars form from such clouds, just as raindrops condense from the water vapor in a thundercloud. To study the formation of stars, we must examine these cold clouds of gas that float between the stars and understand how they contract, heat up, and become stars.

The Interstellar Medium   The gas and dust distributed between the stars are called the **interstellar medium.** About 75 percent of the mass of the gas is hydrogen, and 25 percent helium plus traces of carbon, nitrogen, oxygen, calcium, sodium, and heavier atoms. Roughly 1 percent of the mass of the interstellar medium is made up of microscopic dust about the size of the particles in cigarette smoke. The dust is believed to be carbon, iron, and silicates (rocklike minerals) mixed with or coated with frozen water. The average distance between dust grains is about 150 m.

The interstellar medium is not uniformly distributed through space; it consists of a complex tangle of cool, dense clouds pushed and twisted by currents of hot, low-density gas. Although the cool clouds contain only 10–1000 atoms/cm^3 (fewer than any vacuum on Earth), we refer to them as dense clouds in contrast with the hot, low-density gas that contains only 0.1 atoms/cm^3.

These clouds make their presence known by affecting starlight passing through them. If a cloud is unusually dense and dusty, it totally obscures our view of the stars beyond, and we see it as a dark cloud. Photographs of such dark clouds reveal that they are generally not spherical, but torn and twisted like wisps of smoke distorted by a breeze. If a cloud is less dense, starlight may be able to penetrate it, but the stars look dimmer because the cloud absorbs some of the light.

Besides dimming stars, the dust in the clouds makes stars look redder. Because the dust grains have diameters comparable to the wavelength of light, they can interact with photons and deflect them. Red photons, due to their longer wave-

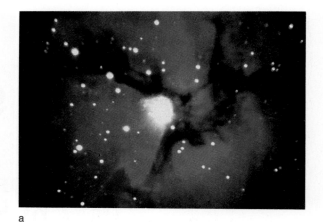
a

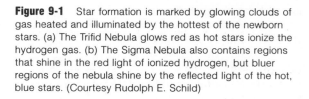
b

Figure 9-1   Star formation is marked by glowing clouds of gas heated and illuminated by the hottest of the newborn stars. (a) The Trifid Nebula glows red as hot stars ionize the hydrogen gas. (b) The Sigma Nebula also contains regions that shine in the red light of ionized hydrogen, but bluer regions of the nebula shine by the reflected light of the hot, blue stars. (Courtesy Rudolph E. Schild)

lengths, are less likely to be deflected than shorter-wavelength, blue photons (Figure 9-2). Thus the light that reaches us is rich in long-wavelength photons, making it seem redder. This **interstellar reddening** is an important source of information about the nature and distribution of the clouds and about the size and composition of the grains.

The clouds also reveal their presence by forming interstellar absorption lines in the spectra of distant stars (Figure 9-3a). As starlight passes through a cloud, gas atoms of elements such as calcium and sodium absorb certain wavelength photons, producing narrow interstellar absorption lines. We can

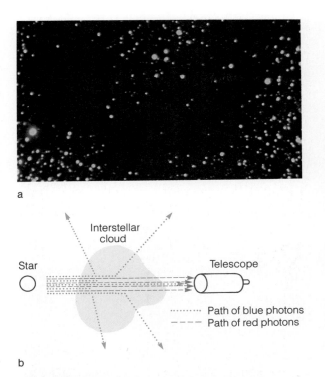

a

a

Interstellar cloud

Star

Telescope

········· Path of blue photons
- - - - Path of red photons

b

Figure 9-2 Some interstellar clouds of gas and dust are so dense we cannot see through them and they appear as regions nearly free of stars. (a) In the case of Barnard's S Nebula, only a few stars are visible through the dark cloud. (b) Stars that are visible through interstellar dust will look redder than normal because blue photons, having shorter wavelengths, are more likely to be scattered by the dust grains. Thus more red photons reach our telescope and such stars look redder. Note the reddened stars visible through the dusty nebula in part a. (a. Courtesy Rudolph E. Schild)

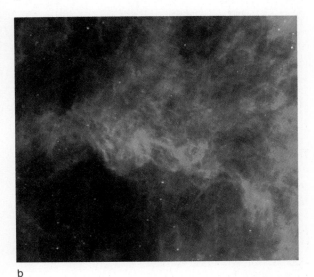

b

Figure 9-3 Evidence of gas and dust between the stars. (a) Hot stars do not produce calcium lines in their spectra, but light from such stars traveling toward the earth can pass through clouds of gas that produce narrow calcium lines. Multiple lines indicate multiple clouds with different velocities. (Mt. Wilson and Las Campanas Observatories, Carnegie Institution of Washington) (b) The Infrared Astronomy Satellite discovered the infrared cirrus, a wispy distribution of dust slightly warmed by starlight. The cirrus is visible only in the infrared. (NASA/IPAC Courtesy Deborah Levine)

be sure these lines originate in the interstellar medium because they appear in the spectra of O and B stars—stars that are too hot to form their own calcium and sodium absorption lines. The narrowness of the lines tells us the clouds are very cool, about 10–50 K. If they were warmer, the lines would be broadened by the motions of the individual atoms. It is not unusual to find interstellar absorption lines split into several components, indicating that the starlight has passed through more than one cloud. Small differences in the velocities of the clouds Doppler-shift the components and make them appear at slightly different wavelengths, as in Figure 9-3a.

The Infrared Astronomy Satellite discovered a faint, wispy network of dust clouds covering the entire sky (Figure 9-3b). The material is very cold,

15–30 K, and seems to be associated with the coldest parts of the interstellar medium. Called the **infrared cirrus** because it resembles cirrus clouds in Earth's atmosphere, it illustrates how the cold, dark, tenuous clouds of gas and dust fill interstellar space.

Astronomers commonly refer to any cloud of gas and dust as a **nebula** (from the Latin word for "cloud"). Such nebulae (plural) give us clear evidence of an interstellar medium. If a star is hot enough, it can ionize the gas of the nebula and produce an **emission nebula**—a nebula whose spectrum is an emission spectrum. In some nebulae, however, the spectrum produced by the nebula is merely the reflected spectrum of the stars, showing that the nebula is rich in dust. Such a

Figure 9-4 The Pleiades star cluster, just visible to the naked eye in Taurus, is surrounded by a faint nebula caused by the reflection of starlight from dust in the nebula. In this case, the stars do not appear to have formed from this gas and dust but are merely passing through the nebula. (Copyright California Institute of Technology and Carnegie Institution of Washington, by permission from Hale Observatories)

nebula is known as a **reflection nebula.** The Trifid Nebula in Figure 9-1 is an emission nebula, while the nebula around the stars of the Pleiades star cluster (Figure 9-4) is a reflection nebula.

In many cases, the hot stars exciting emission nebulae appear to have formed recently from the nebulae. But how can a very low-density, very cold cloud of gas form a dense, hot star? The answer lies in the force of the cloud's gravity.

The Formation of Stars from the Interstellar Medium

The combined gravitational attraction of the atoms in a cloud of gas squeezes the cloud, pulling every atom toward the center. Thus we might expect that every cloud would eventually collapse and become a star; however, the heat in the cloud resists collapse. Heat is related to the motion of the atoms

or molecules in a material—in a hot gas the atoms move more rapidly than do those in a cool gas. The interstellar clouds are very cold, but even at a temperature of only 10 K, the average hydrogen atom moves about 0.5 km/sec (1100 mph). This thermal motion would make the cloud drift apart if gravity were too weak to hold it together.

The densest interstellar clouds contain about 1000 atoms/cm³, include from a few hundred to a billion solar masses, and have temperatures as low as 10 K. In such clouds hydrogen can exist as molecules (H_2) rather than as atoms. Although hydrogen molecules cannot be detected by radio telescopes, the clouds can be mapped by the emission of carbon monoxide molecules (CO) present in small amounts in the gas. Consequently, these clouds are known as **molecular clouds,** and the largest ones are called giant molecular clouds. Stars form in these clouds when the densest parts of the clouds become unstable and contract under the influence of their own gravity.

Most clouds do not appear to be gravitationally unstable, but such a cloud colliding with a **shock wave** (the astronomical equivalent of a sonic boom) can be compressed and disrupted into fragments (Figure 9-5). Theoretical calculations show that some of these fragments can become dense enough to collapse and form stars.

Thus most interstellar clouds may not collapse and form stars until they are triggered by the compression of a shock wave. Happily for this hypothesis, space is filled with shock waves. Supernova explosions (exploding stars described in the next chapter) produce shock waves that compress the interstellar medium, and recent observations show young stars forming at the edges of such shock waves. Another source of shock waves may be the ignition of very hot stars. If one part of a cloud forms a massive star, the sudden flare of radiation as the star "turns on" could push against other parts of the cloud, compress the gas, and trigger more star formation. Even the collision of two clouds can produce a shock wave and trigger star formation.

Although these are important sources of shock waves, the dominant trigger of star formation in our galaxy may be the spiral pattern itself. In Chapter 1, we noted that our galaxy contains spiral arms, and many astronomers believe the arms are marked by shock waves traveling through the interstellar medium. If they are, then the collision of an interstellar cloud with a spiral arm could trigger the formation of new stars.

Astronomers have found a number of giant molecular clouds where stars are forming in a repeating cycle. Both high-mass and low-mass stars form in such a cloud, but when the massive stars form, their intense radiation or eventual supernovae explosions push back the surrounding gas and compress it. This compression in turn can trigger the

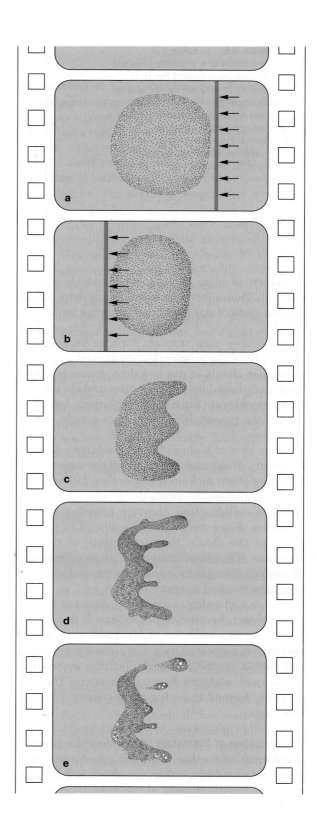

Figure 9-5 An interstellar gas cloud has such a low density it is unlikely to form stars without an outside stimulus. Computer models show that a passing shock wave (red line) can compress and fragment a cloud, driving some regions to high enough densities to trigger star formation (bottom). From first frame to last, this figure spans about 6 million years.

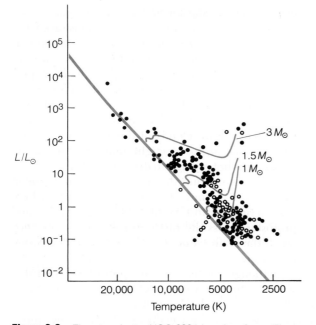

Figure 9-8 The star cluster NGC 2264 is only a few million years old, and its H–R diagram shows that many of its lower-mass stars have not yet reached the main sequence. T Tauri stars (open circles) are common in the cluster. Note that the T Tauri stars lie between the birth line and the main sequence (see Figure 9-7). Blue evolutionary tracks are derived from theoretical models of contracting protostars. (Adapted from data by M. Walker)

can change their brightness and shape over only a few years. Apparently, Herbig–Haro objects occur where jets of gas flowing away from young stars smash into clouds of interstellar matter. The excited clouds glow, and as the jets of gas fluctuate, the clouds change in brightness and shape. Herbig–Haro objects HH34S and HH34N are clearly related to a young star located halfway between them (Figure 9-9), and the hottest part of the jet is visible pointing toward HH34S. Another example with a visible jet is HH46 and HH47 (Figure 9-10).

When a young star blasts gas away in two jets shooting in opposite directions, astronomers refer to it as a **bipolar flow.** According to recent theoretical work, such beams of gas appear to be caused by the formation of a spinning disk of gas around a forming star. As additional gas falls into the disk, it drags in a magnetic field from the interstellar

medium, and the resulting interaction between the magnetic field and the rotating disk blows beams of gas outward along the axis of rotation (Figure 9-11). We will discover in later chapters that both disks and bipolar flows are common—they occur in dying stars and exploding galaxies as well as in young stars.

We know that many infrared objects, Herbig–Haro objects, and T Tauri stars are young because they are located in regions where hot, blue stars are found. Since these stars have such short lives, they too must be young, a fact that implies that the entire region is a site of active star formation.

Where we find the youngest stars we often find objects called **Bok globules** (named after the astronomer Bart Bok)—small clouds only about 1 ly in diameter that contain 10 to 1000 solar masses of gas and dust. Since these clouds produce no light of their own, they are only seen silhouetted against bright nebulae (Figure 9-12). Some astronomers believe the globules are collapsing clouds that will become stars. Infrared observations, including those made by the Infrared Astronomy Satellite, show that some globules do contain infrared sources that appear to be contracting protostars. The globule Barnard 5, for instance, contains at least four protostars.

All of these observations confirm our theoretical model of contracting protostars. Although star formation still holds many mysteries, the general process seems clear. In at least some cases, interstellar gas clouds are compressed by passing shock waves, and the clouds' gravity, acting unopposed, draws the matter inward to form protostars.

Most astronomers now believe that the planets of our solar system formed with the sun over an interval of no more than a few million years. At some point during the sun's contraction, when it was surrounded by swirling clouds of gas and dust, solid bodies condensed and grew into planets. Infrared observations have found a number of nearby stars surrounded by such clouds, so such clouds are not unusual. We will discuss planet building in Chapter 16, but it is important to note here because it implies that stars may form planets as they contract.

Only when the star is hot enough to ignite its thermonuclear fires can it halt the collapse and reach stability. Clearly, internal nuclear reactions

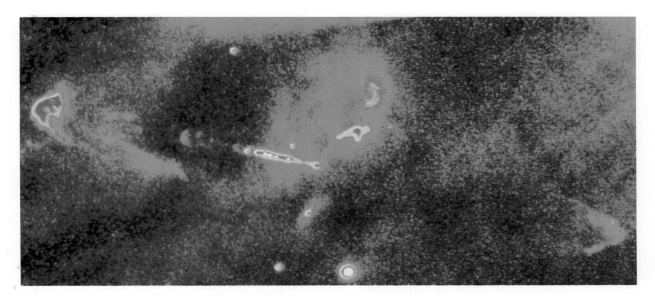

Figure 9-9 Originally recorded in the light emitted by once-ionized sulfur, this CCD image has been computer-enhanced to reveal Herbig–Haro object HH34S (left) and the fainter HH34N (right). At the center, a newly formed star emits jets of gas to the left and right, which excite the Herbig–Haro objects. These objects are located only 1° south of the Orion Nebula. (Obtained by Reinhard Mundt at the Calar Alto 3.5-m telescope)

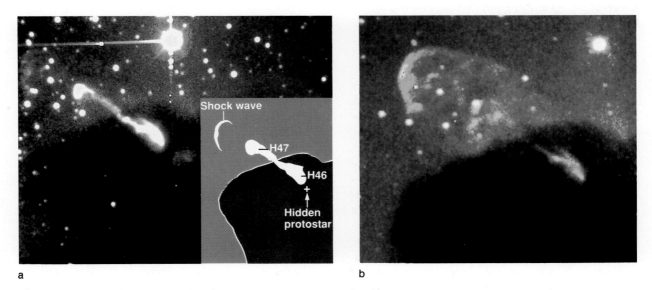

a

b

Figure 9-10 Herbig–Haro objects HH46 and HH47 are part of a jet moving toward us from a young protostar buried inside a dark globule. Image (a) was recorded in the light of once-ionized sulfur and shows the location of the jet as it bursts out of the globule. Image (b) was recorded in the light of once-ionized oxygen and shows the shock wave formed where the jet collides with the interstellar medium. (Courtesy National Optical Astronomy Observatories/Cerro Tololo Inter-American Observatory and Patrick Hartigan)

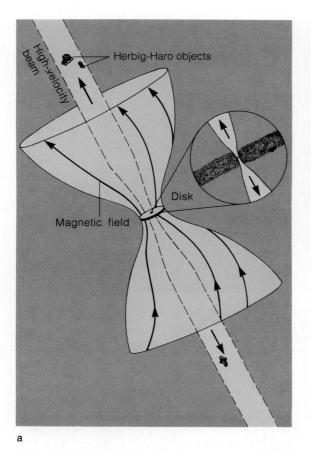

a

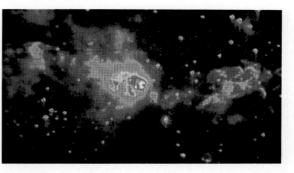

b

Figure 9-11 Bipolar flow. (a) A theoretical model shows how gas falling into a spinning disk around a star can drag in a magnetic field and generate flows along the axis of rotation. Herbig–Haro objects can form where the beams excite small gas clouds. (Adapted in part from diagrams by Pudritz and Norman) (b) This infrared CCD image of Cepheus A shows a cluster of young stars hidden in the gas and dust at the center. One of the stars is emitting beams of gas to right and left that interact with the surrounding gas to produce the complex glowing filaments and Herbig–Haro objects. (Courtesy Adair P. Lane and John Bally)

are as important in a star's life as gravity is in its formation. Before we can follow the evolution of stars further, we must examine these reactions—the sources of stellar energy.

9-2 HOW STARS GENERATE NUCLEAR ENERGY

Stars generate energy through nuclear reactions, which keep the interior of the star hot and the gas pressure high. That pressure balances the force of the star's gravity and prevents further contraction. In a sense, the matter in a star is supported against its own gravity by the nuclear reactions.

The high temperatures inside stars keep the gas entirely ionized. That is, the violent collisions between the atoms strip the electrons away, so the gas is a hot mixture of positively charged nuclei and negatively charged electrons. Thus, when we discuss fusion reactions in stars, we refer to atomic nuclei and not to atoms.

But how exactly can the nucleus of an atom yield energy? The answer lies in the forces that hold the nuclei together.

Nuclear Binding Energy Stars generate their energy by breaking and reconnecting the bonds between the particles inside atomic nuclei. This is quite different from the way we generate energy by burning wood in our fireplaces. The process of burning wood extracts energy by breaking and reconnecting chemical bonds between atoms in the wood. Chemical bonds are formed by the electrons in atoms, and we saw in Chapter 6 that the electrons are bound to the atoms by the electromagnetic force. Thus chemical energy originates in the force that binds the electrons.

There are only four known forces in nature: the

Figure 9-12 The Rosette Nebula in Monoceros lies about 3000 ly away. Evidently, hot young stars in the nebula are driving nebular gas outward and compressing small, dense dust clouds to form dark globules. (One is marked at upper left.) Such globules may be the precursors of protostars. (Copyright © Royal Observatory, Edinburgh, and Anglo-Australian Telescope Board)

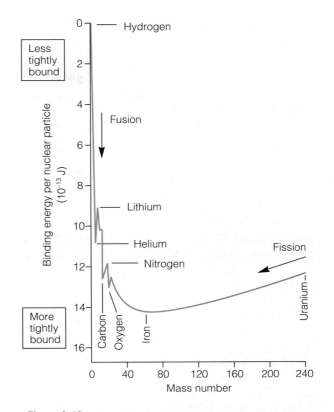

Figure 9-13 The binding energy is the energy that holds an atomic nucleus together. Fission reactions can break heavy nuclei such as uranium into lighter fragments and release energy, and fusion reactions can combine low-mass nuclei to form more massive nuclei and release energy. Iron has the most tightly bound nucleus, so it and more massive nuclei cannot release energy by fusion.

force of gravity (see Chapter 4), the electromagnetic force (see Chapter 6), the **strong force,** and the **weak force.** The weak force is involved in the radioactive decay of certain kinds of nuclear particles, and the strong force binds together atomic nuclei. Thus nuclear energy has its origin in the strong force.

Nuclear power plants on Earth generate energy through **nuclear fission** reactions that split uranium nuclei into less massive fragments. A uranium nucleus contains a total of 235 protons and neutrons, and it splits into a range of fragments containing roughly half as many particles. Because uranium nuclei are less tightly bound than the

fragments, energy is released during uranium fission (Figure 9-13).

Stars generate their energy through **nuclear fusion** reactions that combine very light nuclei into slightly heavier nuclei. The most common reaction fuses hydrogen nuclei (single protons) into helium nuclei (two protons and two neutrons). Because the products of the fusion are more tightly bound, energy is released during these fusion reactions (Figure 9-13).

Note from Figure 9-13 that iron has the most tightly bound nucleus. Fusion reactions can release energy only when the nuclei are less massive than iron. This ultimate limitation will be important in the next chapter, when we discuss the deaths of stars.

Hydrogen Fusion

The nuclear reactions that fuse hydrogen in stars join four hydrogen nuclei to make one helium nucleus. Because one helium nucleus has 0.7 percent less mass than four hydrogen nuclei, it seems that some mass vanishes in the process:

$$
\begin{array}{ll}
\text{4 hydrogen nuclei} = 6.693 \times 10^{-27} \text{ kg} \\
\text{1 helium nucleus} \;\; = 6.645 \times 10^{-27} \text{ kg} \\
\hline
\text{difference in mass} = 0.048 \times 10^{-27} \text{ kg}
\end{array}
$$

The difference in mass is the mass equivalent to the difference in the binding energy of the nuclei. Recall from Einstein's equation $E = mc^2$ that energy and mass are related. Thus the 0.048×10^{-27} kg is the binding energy released in the fusion reaction:

$$
\begin{aligned}
E &= mc^2 \\
&= (0.048 \times 10^{-27} \text{ kg})(3 \times 10^8 \text{ m/sec})^2 \\
&= 0.43 \times 10^{-11} \text{ J}
\end{aligned}
$$

This is a very small amount of energy, hardly enough to raise a housefly 0.001 inch into the air. Only by concentrating many reactions in a small area can nature produce significant results. A single kilogram (2.2 lb) of hydrogen, for instance, converted entirely to energy, would produce enough power to raise an average-sized mountain 10 km (6 miles) into the air! The sun has a voracious energy appetite and needs 10^{38} reactions per second, transforming 5 million tons of mass into energy every second, just to balance its own gravity. This sounds like a lot of mass, but in its entire 10-billion-year lifetime, the sun will convert less than 0.07 percent of its mass into energy.

These nuclear reactions can only occur when the nuclei of two atoms get very close to each other. Since atomic nuclei carry positive charges, they repel each other with an electrostatic force called the Coulomb force (see Chapter 6). Physicists commonly refer to this repulsion between nuclei as the **Coulomb barrier.** To overcome this barrier, atomic nuclei must collide violently. Violent collisions are rare unless the gas is very hot, in which case the nuclei move at high speeds and collide violently. (Remember, an object's tempera-

ture is just a measure of the speed with which its particles move.)

Thus nuclear reactions can only occur near the centers of stars, where they are hot and dense. A high temperature ensures that collisions between nuclei are violent enough to overcome the Coulomb barrier, and a high density ensures that there are enough collisions, and thus enough reactions, to meet the star's energy needs.

We can symbolize this process with a simple nuclear reaction:

$$
4 \;^1\text{H} \rightarrow {}^4\text{He} + \text{energy}
$$

In this equation, ^{1}H represents a proton, the nucleus of the hydrogen atom, and ^{4}He represents the nucleus of a helium atom. The superscripts indicate the approximate weight of the nuclei. The actual steps in the process are more complicated than this convenient summary suggests. Instead of waiting for four hydrogen nuclei to collide simultaneously, a highly unlikely event, the process can proceed step by step in a chain of reactions that can occur in either of two ways—the proton–proton chain or the CNO cycle.

The **proton–proton chain** is a series of three nuclear reactions that builds a helium nucleus by adding together protons. This process is efficient at temperatures above 10,000,000 K. The sun, for example, manufactures over 90 percent of its energy in this way.

The three steps in the proton–proton chain entail these reactions:

$$
\begin{aligned}
{}^1\text{H} + {}^1\text{H} &\rightarrow {}^2\text{H} + e^+ + \nu \\
{}^2\text{H} + {}^1\text{H} &\rightarrow {}^3\text{He} + \gamma \\
{}^3\text{He} + {}^3\text{He} &\rightarrow {}^4\text{He} + {}^1\text{H} + {}^1\text{H}
\end{aligned}
$$

In the first step, two hydrogen nuclei (two protons) combine to form a heavy hydrogen nucleus, emitting a particle called a positron (a positively charged electron) and another called a neutrino. (See Box 9-1 for more about neutrinos.) In the second reaction, the heavy hydrogen nucleus absorbs another proton and, with the emission of a gamma ray, becomes a lightweight helium nucleus. Finally, two light helium nuclei combine to form a normal helium nucleus and two hydrogen nuclei. Since the last reaction needs two ^{3}He nuclei, the first and second reactions must occur twice (Figure 9-14). The net result of this chain reaction is the

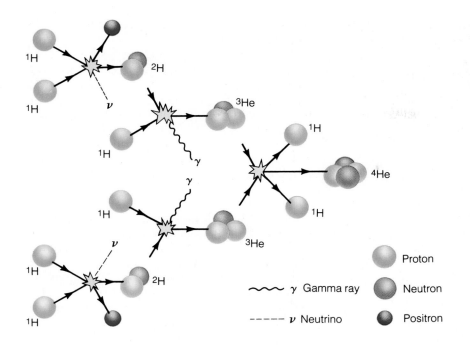

¹H

¹H

ν

2H

¹H

3He

¹H

γ

γ

¹H

¹H

4He

¹H

3He

¹H

ν

¹H

2H

¹H

~~~~ *γ* Gamma ray

- - - - *ν* Neutrino

Proton

Neutron

Positron

transformation of four hydrogen nuclei into one helium nucleus plus energy.

The energy appears in the form of gamma rays, positrons, and neutrinos. The gamma rays are photons that are absorbed by the surrounding gas before they can travel more than a few centimeters. This heats the gas. The positrons produced in the first reaction combine with free electrons, and both particles vanish, converting their mass into gamma rays. Thus the positrons also help keep the center of the star hot. The neutrinos, however, are particles that resemble photons except that they almost never interact with other particles. The average neutrino could pass unhindered through a lead wall 1 ly thick. Thus the neutrinos do not help heat the gas, but race out of the star at the speed of light, carrying away roughly 2 percent of the energy produced.

Like the proton–proton chain, the **CNO cycle** is a series of nuclear reactions that combines four hydrogen nuclei to make one helium nucleus plus energy. The CNO cycle, however, needs carbon as a catalyst—a substance that aids a reaction but is not altered by it. This process is most efficient at temperatures above 16,000,000 K. Below this

temperature, the efficiency of the CNO cycle drops rapidly. Since the sun's central temperature is only about 15,000,000 K, the CNO cycle creates slightly less than 10 percent of the sun's energy.

The steps in the CNO cycle (Figure 9-16) begin with carbon-12 (^{12}C); pass through stages involving isotopes of carbon, nitrogen (N), and oxygen (O); and finish with the reappearance of a carbon-12 nucleus just like the one that started the process. Thus we say the carbon is a catalyst; it makes the reactions possible but is not altered in the end.

Counting protons in the CNO cycle reactions in Figure 9-16 shows that the net result of the CNO cycle is four hydrogen nuclei combined to make one helium nucleus plus energy.

Because the carbon nucleus has a charge six times that of hydrogen, the Coulomb barrier is high, and much hotter temperatures are necessary to force the proton and carbon nucleus together. Thus the CNO cycle is important only in stars more massive than about 1.1 $M_\odot$. These stars have central temperatures hotter than about 16,000,000 K. Stars less massive than 1.1 $M_\odot$, such as the sun, are not hot enough at their centers to generate much energy on the CNO cycle.

BOX 9-1

The Mystery of the Solar Neutrinos

Because the study of stellar structure is highly theoretical, astronomers are ever alert for ways to confirm their theories. One method measures the rate at which the sun produces energy by trapping **neutrinos,** massless atomic particles that travel at the speed of light.

The nuclear reactions in the sun's core produce a flood of neutrinos that rush outward into space. During the moment it takes to read this sentence, approximately 10^{12} solar neutrinos pass through your body. If we could catch and study these particles, we could learn about conditions at the sun's center. But neutrinos almost never react with other particles, so they are very difficult to detect.

In 1970, an experiment to trap solar neutrinos began. Because a few of the many solar neutrinos do interact with chlorine atoms, Raymond Davis, Jr., a research chemist from the Brookhaven National Laboratory, filled a 100,000-gallon tank with a type of cleaning fluid that contains a large percentage of chlorine atoms. This neutrino trap had to be buried nearly a mile underground in a South Dakota gold mine to shield it from cosmic rays from space, which could react with the chlorine and imitate the behavior of neutrinos (Figure 9-15).

Of the trillions of solar neutrinos that pass through the tank each second, one should occasionally interact with a chlorine atom, transforming it into a radioactive argon atom.

The argon atoms can be counted via their radioactivity. If we know the probability that a neutrino will interact with a chlorine atom, the number of argon atoms that appear in the tank can tell us how rapid the nuclear reactions are at the sun's core.

The experiment has been in operation for two decades and has been carefully tested and calibrated. Yet it detects only about one-third as many neutrinos as theory predicts. Can there be something wrong with our theory of stellar structure, or is the experiment not counting all of the neutrinos it should?

Explanations have taken one of three forms. One idea is that the sun has temporarily stopped producing energy at its core, but no theory suggests why energy production should vary. A second theory is that the sun contains exotic particles such as **WIMPs** (weakly interacting massive particles) that distribute energy through the core and help generate energy at a slightly lower temperature. However, WIMPs have never been detected in the laboratory. A third suggestion is that neutrinos oscillate naturally between three different types and that Davis's tank can detect only one type. If the neutrinos changed types during their 8-minute flight from the sun to the earth, we would count only one-third the number that began the journey—just what is observed.

One problem is that the Davis experiment

Heavy-Element Fusion At later stages in the life of a star, when it has exhausted its hydrogen fuel, it may fuse other nuclear fuels. As we learned in the last section, the ignition of these fuels requires high temperatures. Helium fusion occurs at temperatures above 100,000,000 K, and carbon fusion does not begin until the temperature exceeds 600,000,000 K.

We can summarize the helium-fusion process in two steps:

$$^{4}\text{He} + {}^{4}\text{He} \rightarrow {}^{8}\text{Be} + \gamma$$
$$^{8}\text{Be} + {}^{4}\text{He} \rightarrow {}^{12}\text{C} + \gamma$$

Because a helium nucleus is called an alpha particle, these reactions are commonly known as the **triple-alpha process.** Helium fusion is complicated

cannot detect neutrinos produced directly in the proton–proton chain reactions. Rather, it counts neutrinos from a side reaction that occurs less frequently. The Soviet-American Gallium Experiment (SAGE) solves this problem by using 60 tons of the rare metal Gallium to trap neutrinos produced by the proton–proton reactions. Early results confirm Davis's findings. So few neutrinos have been counted that the statistical uncertainties are still high, but the SAGE measurement can be no more than 60 percent of the neutrinos expected from the sun, and the most likely value is only 15 percent. This result strongly supports the theory of neutrino oscillation, but further observations and theoretical analysis are needed.

Still more detectors are being developed. A second gallium detector is starting operation, and the Sudbury Neutrino Observatory (SNO) will put 1000 tons of heavy water (deuterium oxide) 2000 m deep in a mine shaft in Canada. Phototubes will watch for light flashes from neutrinos passing through the heavy water. Another team of scientists is planning to put photocells in the Antarctic ice cap to detect neutrinos.

These new detectors are very expensive and difficult to build, but they are important experiments because the mystery of the solar neutrinos is a fundamental problem in modern astronomy. The mystery represents a direct conflict between observation and theory.

Figure 9-15 The solar neutrino experiment consists of 100,000 gallons of cleaning fluid held in a tank nearly a mile underground. Solar neutrinos trapped in the cleaning fluid convert chlorine atoms into argon atoms that can be counted by their radioactivity. (Brookhaven National Laboratory)

by the fact that beryllium-8, produced in the first reaction of the process, is very unstable and may break up into two helium nuclei before it can absorb another helium nucleus. Three helium nuclei can also form carbon directly, but such a triple collision is unlikely.

At temperatures above 600,000,000 K, carbon fuses rapidly in a complex network of reactions illustrated in Figure 9-17, where each arrow represents a different nuclear reaction. The process is complicated because nuclei can react by adding a proton, a neutron, or a helium nucleus, or by combining directly with other nuclei. Unstable nuclei can decay by ejecting an electron, a positron, or a helium nucleus, or by splitting into fragments. The complexity of this process makes it

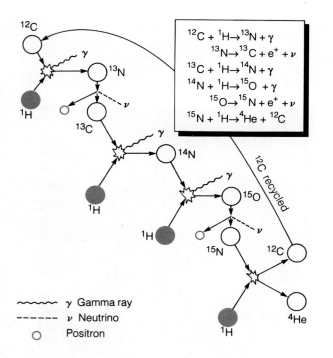

$$^{12}C + {}^{1}H \rightarrow {}^{13}N + \gamma$$
$$^{13}N \rightarrow {}^{13}C + e^+ + \nu$$
$$^{13}C + {}^{1}H \rightarrow {}^{14}N + \gamma$$
$$^{14}N + {}^{1}H \rightarrow {}^{15}O + \gamma$$
$$^{15}O \rightarrow {}^{15}N + e^+ + \nu$$
$$^{15}N + {}^{1}H \rightarrow {}^{4}He + {}^{12}C$$

~~~~~ γ Gamma ray

------ ν Neutrino

○ Positron

Figure 9-16 The CNO cycle uses ^{12}C as a catalyst to combine four hydrogen atoms (^{1}H) to make one helium atom (^{4}He) plus energy. The carbon atom reappears at the end of the process ready to start the cycle over.

difficult to determine exactly how much energy will be generated and how many heavy atoms will be produced.

Reactions at still higher temperatures can convert magnesium, aluminum, and silicon into yet heavier atoms. These reactions involving heavy elements will be important in the study of the deaths of massive stars in the next chapter.

The Pressure–Temperature Thermostat Nuclear reactions in stars manufacture energy and heavy atoms under the supervision of a built-in thermostat that keeps the reactions from erupting out of control. That thermostat is the relation between gas pressure and temperature.

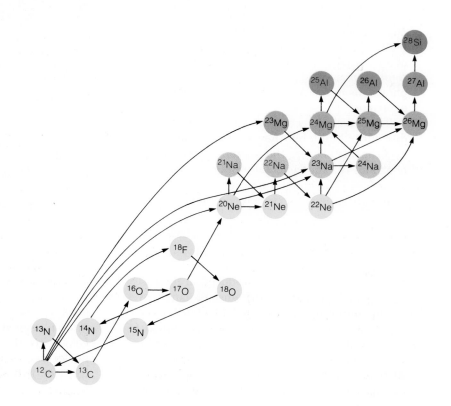

Figure 9-17 Carbon fusion involves many possible reactions that build numerous heavy atoms.

In a star, the nuclear reactions generate just enough energy to balance the inward pull of gravity. Consider what would happen if the reactions began to produce too much energy. The extra energy would raise the internal temperature of the star, and since the pressure of the gas depends on its temperature, the pressure would also rise. The increased pressure would make the star expand. Expansion of the gas would cool the star slightly, slowing the nuclear reactions. Thus the star has a built-in regulator that keeps the nuclear reactions from occurring too rapidly.

The same thermostat also keeps the reactions from dying down. Suppose the nuclear reactions began to produce too little energy. Then the inner temperature would decrease, lowering the pressure and allowing gravity to compress the star slightly. As the gas was compressed, it would heat up, increasing the nuclear energy generation until the star regained stability.

The stability of a star depends on this relation between gas pressure and temperature. If an increase or decrease in temperature produces a corresponding change in pressure, then the thermostat functions correctly and the star is stable. We will see in this chapter how the thermostat accounts for the mass–luminosity relation. In the next chapter, we will see what happens to a star when the thermostat breaks down completely and the nuclear fires rage unregulated.

9-3 STELLAR STRUCTURE

The nuclear fusion in the centers of stars heats their interiors, creates high gas pressure, and thus balances the inward force of gravity. If there is a single idea in modern astronomy that can be called critical, it is that balance. Stars are simple, elegant, power sources held together by their own gravity and supported by their nuclear fusion.

Having discussed the birth of stars and their nuclear power sources, we can now consider the structure of a star—that is, the variation in temperature, density, pressure, and so on from the surface of the star to its center. A star's structure depends on how it generates its energy and on four simple laws that give us a deep insight into how

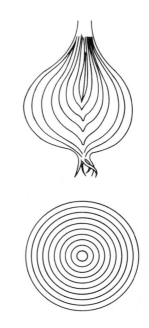

Figure 9-18 The layers of the model star resemble the layers of the common onion.

stars work. Each is a basic law of physics applied to the matter and energy inside a star.

It will be easier to think about stellar structure if we imagine that the star is divided into concentric shells like those in an onion (Figure 9-18). We can then discuss the temperature, density, pressure, and so on in each shell. Keep in mind, however, that these helpful shells do not really exist; stars have no separable layers.

The Laws of Mass and Energy The first two laws of stellar structure have something in common—they are both laws of continuity. They tell us that the distribution of matter and energy inside the star must vary smoothly from surface to center. No gaps are allowed.

The **continuity of mass** law says that the total mass of the star must equal the sum of the masses of its shells and that the mass must be distributed smoothly throughout the star. In a sense, the continuity of mass law is the law of conservation of mass with the added proviso that the mass be distributed smoothly.

The **continuity of energy** law says that the amount of energy flowing out the top of a shell

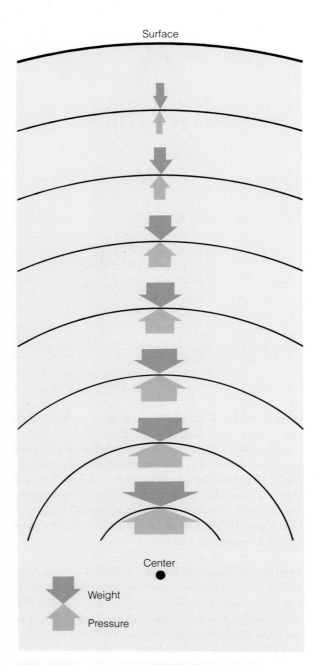

Surface

Center

Weight

Pressure

Figure 9-19 Hydrostatic equilibrium says the pressure in each layer must balance the weight on that layer. As a result, pressure and temperature must increase from the surface of a star to its center.

must equal the amount coming in at the bottom plus whatever energy is generated within the shell. Further, it says that the energy leaving the surface of the star—the luminosity—must equal the sum of the energies generated in the shells. Thus the continuity of energy law is really a version of the law of conservation of energy.

In fact, both laws may seem familiar, since we have all heard of the conservation of mass and energy. The third law of stellar structure, the law of hydrostatic equilibrium, is also familiar because we have been using it under a different name in the previous sections.

Hydrostatic Equilibrium The law of **hydrostatic equilibrium** says that, in a stable star, the weight of the material pressing downward on a layer must be balanced by the pressure of the gas in that layer. Hydrostatic equilibrium is just a new name for the gravity–pressure balance discussed earlier. *Hydro* (from the Greek word for "water") implies we are discussing a fluid, the gases of a star, and *static* implies that the fluid is stable, neither expanding nor contracting.

The law of hydrostatic equilibrium can prove to us that the temperature must increase as we descend into a star. Near the surface, there is little weight pressing down on the gas, so the pressure must be low, implying a low temperature. But as we go deeper into the star, the weight becomes greater, so the pressure, and therefore the temperature, must also increase (Figure 9-19).

Although the law of hydrostatic equilibrium can tell us some things about the inner structure of stars, we need one more law to completely describe a star. We need a law that describes the flow of energy from the center to the surface.

Energy Transport The surface of a star radiates light and heat into space, and would quickly cool if that energy were not replaced. Because the inside of the star is hotter than the surface, energy must flow outward from the core, where it is generated, to the surface, where it radiates away. This flow of energy through the shells determines their temperature, which, as we saw previously, determines how much weight each shell can balance. To understand the structure of a star, we must understand how energy moves from the center through the shells to the surface.

The law of **energy transport** says that energy must flow from hot regions to cooler regions by conduction, convection, or radiation.

Conduction is the most familiar form of heat flow. If you hold the bowl of a spoon in a candle flame, the handle of the spoon grows warmer. Heat, in the form of motion among the molecules of the spoon, is conducted from molecule to molecule up the handle, until the molecules of metal under your fingers begin to move faster and you sense heat (Figure 9-20). Thus conduction requires close contact between the molecules. Since matter in most stars is gaseous, conduction is unimportant. Conduction is significant in white dwarfs, which have tremendous internal densities.

The transport of energy by radiation is another familiar experience. Put your hand beside a candle flame, and you can feel the heat. What you actually feel are infrared photons radiated by the flame (Figure 9-20). Because photons are packets of energy, your hand grows warm as it absorbs them.

Radiation is the principal means of energy transport in the sun's interior. Photons are absorbed and reemitted in random directions over and over as they work their way outward. The process of absorption and reemission breaks the high-energy photons common near the sun's center into large numbers of low-energy photons in the cooler outer layers. Thus the energy of a single high-energy photon at the center of the sun takes about 1 million years to reach the sun's surface, where it emerges as roughly 1600 photons of visible light.

The flow of energy by radiation depends on how difficult it is for the photons to move through the gas. If the gas is cool and dense, the photons are more likely to be absorbed or scattered, and thus the radiation does not get through easily. We would call such a gas opaque. In a hot, thin gas, the photons can get through more easily; such a gas is less opaque. The **opacity** of the gas, its resistance to the flow of radiation, depends strongly on its temperature.

If the opacity is high, radiation cannot flow through the gas easily, and it backs up like water behind a dam. When enough heat builds up, the gas begins to churn as hot gas rises upward and cool gas sinks downward. This is convection, the third way energy can move in a star. Convection is a common experience; the wisp of smoke rising

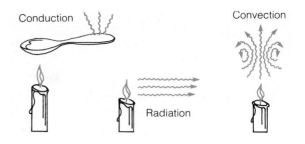

Figure 9-20 Conduction, radiation, and convection are the only modes of energy transport within a star.

above a candle flame travels upward in a small convection current (Figure 9-20). If you hold your hand above the flame, you can feel the rising current of hot gas. In stars, energy may be carried upward by rising currents of hot gas hundreds or thousands of miles in diameter.

Convection is important in stars because it carries energy and because it mixes the gas. Convection currents flowing through the layers of a star tend to homogenize the gas, giving it a uniform composition throughout the convective zone. As you might expect, this mixing affects the fuel supply of the nuclear reactions, just as the stirring of a campfire makes it burn more efficiently.

The four laws of stellar structure are much more precise than the four statements in Table 9-1. In fact, the four laws can be expressed as the four equations shown in Figure 9-21. Those equations permit astronomers to build mathematical models of stars that can tell us how stars work, how they are born, and how they die.

Stellar Models If we wanted to build a model of a star, we would have to divide the star into about 100 concentric shells, and then write down the four equations of stellar structure for each shell. We would then have 400 equations that would have 400 unknowns, namely, the temperature, density, mass, and energy flow in each shell. Solving 400 equations simultaneously is not easy, and the first such solutions, done by hand before the invention of the electronic computer, took months of work. Now a properly programmed computer can solve the equations in a few seconds and print a table of numbers that represent the conditions in each shell of the star. Such a table, shown in Figure 9-21, is a **stellar model**.

TABLE 9-1

The Four Laws of Stellar Structure

I.	Continuity of mass	Total mass equals sum of shell masses.
II.	Continuity of energy	Total luminosity equals sum of energies generated in each shell.
III.	Hydrostatic equilibrium	The weight on each layer is balanced by the pressure in that layer.
IV.	Energy transport	Energy moves from hot to cool by conduction, radiation, or convection.

The model shown in Figure 9-21 represents our sun. As we scan the table from top to bottom, we descend from the surface of the sun to its center. The temperature increases rapidly as we move downward, reaching a maximum of about 15,000,000 K at the center. At this temperature, the gas is not very opaque and the energy can flow outward as radiation. In the cooler outer layers, the gas is more opaque and the outward-flowing energy forces these layers to churn in convection. This model of the sun, like all stellar models, lets us study the otherwise inaccessible layers inside a star. In addition, the proximity of the sun allows us to check the results of stellar models by direct observation. In Chapter 7, we learned that the solar surface is marked by granulation caused by rising convection currents just below the photosphere (see Figures 7-1b and 7-2). We can now recognize these convection currents as the uppermost layer of the sun's convection zone, a region that extends to a depth of about 0.2 $R_\odot$.

Stellar models also let us look into the star's past and future. In fact, we can use models as time machines to follow the evolution of stars over billions of years. To look into a star's future, for

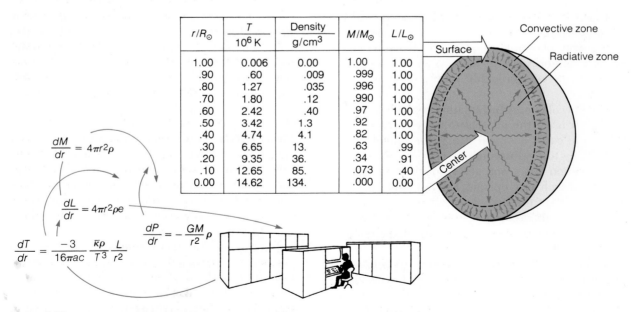

$r/R_\odot$	$\dfrac{T}{10^6 \, K}$	Density g/cm^3	$M/M_\odot$	$L/L_\odot$
1.00	0.006	0.00	1.00	1.00
.90	.60	.009	.999	1.00
.80	1.27	.035	.996	1.00
.70	1.80	.12	.990	1.00
.60	2.42	.40	.97	1.00
.50	3.42	1.3	.92	1.00
.40	4.74	4.1	.82	1.00
.30	6.65	13.	.63	.99
.20	9.35	36.	.34	.91
.10	12.65	85.	.073	.40
0.00	14.62	134.	.000	0.00

$$\frac{dM}{dr} = 4\pi r^2 \rho$$

$$\frac{dL}{dr} = 4\pi r^2 \rho e$$

$$\frac{dP}{dr} = -\frac{GM}{r^2}\rho$$

$$\frac{dT}{dr} = \frac{-3}{16\pi ac}\frac{\bar{\kappa}\rho}{T^3}\frac{L}{r^2}$$

Figure 9-21 A stellar model is a table of numbers that represents conditions inside a star. Such tables can be computed using the four laws of stellar structure, shown here in mathematical form. The table in this figure describes the interior of the sun. (Adapted from Michael A. Seeds, "Stellar Evolution," *Astronomy,* February 1979)

instance, we use a stellar model to determine how fast the star uses its fuel in each shell. As the fuel is consumed, the chemical composition of the gas changes and the amount of energy generated declines. By calculating the rate of these changes, we can predict what the star will look like at any point in the future.

Although this sounds simple, it is actually a highly challenging problem involving nuclear and atomic physics, thermodynamics, and sophisticated computational methods. Only since the 1950s have electronic computers made the rapid calculation of stellar models possible, and the advance of astronomy since then has been heavily influenced by the use of such models to study the structure and evolution of stars. Our summary of star formation in this chapter is based on thousands of stellar models. We will continue to rely on theoretical models as we study main-sequence stars in the next section and the deaths of stars in the next chapter.

9-4 MAIN-SEQUENCE STARS

When a contracting protostar begins to fuse hydrogen, it stops contracting and becomes stable. The location of these stable stars in the H–R diagram is marked by the main sequence. Since 90 percent of all stars lie on the main sequence, it is important that we understand their structure.

The Mass–Luminosity Relation   Main-sequence stars obey a simple rule—the more massive a star is, the more luminous it is. That rule, the mass–luminosity relation discussed in the previous chapter, is the key to understanding the stability of stars. In addition, the mass–luminosity relation can be explained by the theory of stellar structure, giving us direct observational confirmation of the theory.

To understand the mass–luminosity relation, we must consider the law of hydrostatic equilibrium, which says that pressure must balance weight and the pressure–temperature thermostat, which regulates energy production. We have seen that a star's internal pressure stays high because generation of thermonuclear energy keeps its interior hot. Because more massive stars have more weight pressing down on the inner layers, their interiors must have

high pressures and thus must be hot. For example, the temperature at the center of a 15 $M_\odot$ star is about 34,000,000 K, more than twice the central temperature of the sun.

Because massive stars have hotter cores, their nuclear reactions rage more fiercely. That is, their pressure–temperature thermostat is set higher. One gram of material at the center of a 15 $M_\odot$ star fuses over 3000 times more rapidly than 1 gm of material at the center of the sun. The rapid reactions in massive stars make them more luminous than the lower-mass stars. Thus the mass–luminosity relation results from the requirement that a star support its weight by generating nuclear energy.

This also tells us why the main sequence must have a lower end. Masses below about 0.08 $M_\odot$ cannot raise their central temperatures high enough to ignite hydrogen fusion. A mass of 0.08 $M_\odot$ corresponds to about 80 Jupiter masses, and astronomers have speculated that nature must make such failed stars. They would be small, not much larger than Jupiter, and warm from their contraction. Such objects should emit copious infrared radiation and consequently have been called **brown dwarfs.** In detailed searches, however, astronomers have found no conclusive example of a brown dwarf. While some astronomers continue to search, others are asking how nature avoids making brown dwarfs.

In spite of the mystery of the brown dwarfs, the mass–luminosity relation combined with our knowledge of stellar structure can tell us the life story of main-sequence stars.

The Life of a Main-Sequence Star   While a star is on the main sequence it is stable, so we might think its life would be totally uneventful. But a main-sequence star balances its gravity by fusing hydrogen, and as the star gradually uses up its fuel, that balance must change. Thus even the stable main-sequence stars are changing as they consume their hydrogen fuel.

Recall that hydrogen fusion combines four nuclei into one. Thus, as a main-sequence star fuses its hydrogen, the total number of particles in its interior decreases. Each newly made helium nucleus can exert the same pressure as a hydrogen nucleus, but because the gas has fewer nuclei, its total pressure is less. This unbalances the gravity–pressure stability, and gravity squeezes the core of

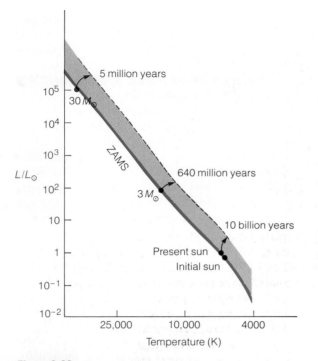

Figure 9-22 The main sequence is not a line but a band (shaded). Stars begin their main-sequence lives on the lower edge, which is called the zero-age main sequence (ZAMS). As hydrogen fusion changes their composition, the stars slowly move across the band.

the star more tightly. As the core contracts, its temperature increases and the nuclear reactions run faster, releasing more energy. This additional energy flowing outward through the envelope forces the outer layers to expand. As the star becomes larger, it becomes more luminous, and eventually the expansion begins to cool the surface.

As a result of these gradual changes in main-sequence stars, the main sequence is not a sharp line across the H-R diagram but rather a band (shaded in Figure 9-22). Stars begin their stable lives fusing hydrogen on the lower edge of this band, the **zero-age main sequence (ZAMS),** but gradual changes in luminosity and surface temperature move the stars upward and to the right. By the time they reach the upper edge of the main sequence (the dashed line in Figure 9-22), they have exhausted nearly all the hydrogen in their centers. Thus we find main-sequence stars scattered throughout this band at various stages of their main-sequence lives.

TABLE 9-2
Main-Sequence Stars

Spectral Type	Mass (Sun = 1)	Luminosity (Sun = 1)	Years on Main Sequence
O5	40	405,000	1×10^6
B0	15	13,000	11×10^6
A0	3.5	80	440×10^6
F0	1.7	6.4	3×10^9
G0	1.1	1.4	8×10^9
K0	0.8	0.46	17×10^9
M0	0.5	0.08	56×10^9

The sun is a typical main-sequence star, and as it undergoes these gradual changes, the earth will suffer. When the sun began its main-sequence life about 5 billion years ago, it was only about 30 percent as luminous as it is now, and by the time it leaves the main sequence in another 5 billion years, the sun will have twice its present luminosity. This will raise the average temperature on Earth by at least 19°C (34°F). As this happens over the next few billion years, the polar caps will melt, and Earth's climate will change dramatically. Life on Earth may not survive this change in the sun.

The average star spends 90 percent of its life on the main sequence. This explains why 90 percent of all stars are main-sequence stars—we are most likely to see a star during that long, stable period while it is on the main sequence. To illustrate: imagine that you photograph, at a single instant, a crowd of 20,000 people. We all sneeze now and then, but the act of sneezing is very short compared to a human lifetime, so we would expect to find very few people in our photograph in the act of sneezing. By contrast, because the main-sequence phase is very long compared to a stellar lifetime, we can expect to see many stars on the main sequence.

The number of years a star spends on the main sequence depends on its mass (Table 9-2). Massive stars consume fuel rapidly and live short lives, but

BOX 9-2

The Life Expectancies of Stars

We can estimate the amount of time a star spends on the main sequence—its life expectancy T—by estimating the amount of fuel it has and the rate at which it consumes that fuel:

$$T = \frac{\text{fuel}}{\text{rate of consumption}}$$

The amount of fuel a star has is proportional to its mass, and the rate at which it uses up its fuel is proportional to its luminosity. Thus its life expectancy must be proportional to M/L. But we can simplify this equation further because, as we saw in the last chapter, the luminosity of a star depends on its mass raised to the 3.5 power ($L = M^{3.5}$). Thus the life expectancy is:

$$T = \frac{M}{M^{3.5}}$$

or

$$T = \frac{1}{M^{2.5}}$$

If we express the mass in solar masses, the lifetime will be in solar lifetimes.

Example: How long can a 4-solar-mass star live? *Solution:*

$$T = \frac{1}{4^{2.5}} = \frac{1}{4 \cdot 4\sqrt{4}}$$
$$= \frac{1}{32} \text{ solar lifetimes}$$

Studies of solar models show that the sun, presently 5 billion years old, can last another 5 billion years. Thus a solar lifetime is approximately 10 billion years, and a 4-solar-mass star will last for about:

$$T = \frac{1}{32} \times (10 \times 10^9 \text{ yr})$$
$$= 310 \times 10^6 \text{ years}$$

low-mass stars conserve their fuel and shine for billions of years. For example, a 25 $M_\odot$ star will exhaust its hydrogen and die in only about 7 million years. Very low-mass stars, the red dwarfs, use up their fuel so slowly that they last for 200–300 billion years. Since the universe seems to be only 10 to 20 billion years old, red dwarfs must still be in their infancy. (Box 9-2 explains how we can quickly estimate the life expectancies of stars from their masses.)

Nature makes more low-mass stars than high-mass stars, but this fact is not sufficient to explain the vast numbers of low-mass stars that fill the sky. An additional factor is the stellar lifetimes. Because low-mass stars live long lives, there are more

of them in the sky than massive stars. Look at Figure 8-24 and notice how much more common the lower-main-sequence stars are than the massive O and B stars. The main-sequence K and M stars are so faint they are difficult to locate, but they are very common. The O and B stars are luminous and easy to locate, but, because of their fleeting lives, there are never more than a few on the main sequence at any one time.

When a star finally exhausts its hydrogen fuel, it can no longer resist the pull of its own gravity. Contraction resumes and the star collapses. As the star dies, it can delay its end by fusing other fuels, but as we will discover in the next chapter, nothing can steal gravity's final victory.

The Orion Nebula

The fuzzy wisp of nebula visible to the naked eye in Orion's sword has attracted the attention of astronomers and casual stargazers throughout history. Commonly referred to as the Great Nebula in Orion, it is a striking sight through binoculars or a small telescope, and through a large telescope it is breathtaking. At the center lie four brilliant blue-white stars known as the Trapezium, and surrounding them are the glowing filaments of a nebula more than 8 pc across (Figure 9-23a). Like a great thundercloud illuminated from within, the churning currents of gas and dust testify to the violence of the mechanisms that created them. However, a deeper significance lies hidden, figuratively and literally, behind the Great Nebula, for radio and infrared astronomers have discovered a vast dark cloud lying just beyond the visible nebula—a cloud in which stars are now being created.

Star Formation in Orion It should not surprise us that Orion is a site of star formation. The stars that make up the constellation are mostly hot, bright, main-sequence stars. These massive stars have short lifetimes and therefore must have formed recently. In addition, the region contains associations of T Tauri stars, which are probably stars in the later stages of contraction to the main sequence. Both ground-based telescopes and the Hubble Space Telescope have detected jets of gas leading away from young stars in and near the Orion Nebula (Figure 9-23b; see also Figure 9-9). These new stars most likely formed from the large gas and dust clouds that fill the Orion region.

The Great Nebula represents a late stage of star formation. The stars in the Trapezium reached the main sequence no more than a million years ago. The most massive, a star of about 40 solar masses, must consume its fuel at a tremendous rate to support its large mass. Thus it is hot and luminous. Its surface temperature is about 30,000 K, and it is about 300,000 times

a

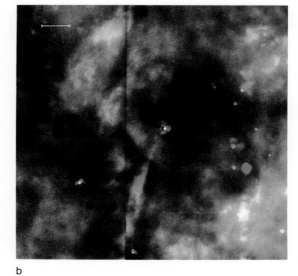

b

Figure 9-23 (a) The Great Nebula in Orion is a glowing cloud of excited gas and dust over 8 pc in diameter. It is excited by a cluster of young, hot stars, the Trapezium, visible near its center. Compare with Figure 9-25. (Copyright Anglo-Australian Telescope Board) (b) Evidence of recent star formation within the nebula appears in this Hubble Space Telescope photo as a jet of gas flowing away from a star. Such bipolar flows are typical of young stars. (NASA)

more luminous than the sun. The other stars are too cool to affect the surrounding gas very much, but the 40-solar-mass star is so hot it radiates large amounts of ultraviolet radiation (Figure 9-24). These ultraviolet photons have enough energy to ionize the hydrogen gas in the region near the Trapezium, creating the glowing clouds we see as the Great Nebula.

Although the nebula looks impressive, it is nearly a vacuum, containing a mere 600 atoms/ cm³. For comparison, the interstellar medium has an average density of about 1 atom/cm³, and the density of air at sea level is about 10^{19} atoms/ cm³. Infrared observations show that the thin, ionized gas is mixed with sparsely scattered dust, heated by the central stars to a temperature of about 70 K.

The Molecular Cloud The importance of the Orion region was established when observations at wavelengths longer than visible light revealed a dense cloud of gas lying just beyond the Great Nebula. These observations spanned the region of the spectrum that includes infrared and short-wavelength radio waves. At these wavelengths, hot stars and ionized gas are invisible, but cool, dense gas is detectable. The gas cloud beyond the Great Nebula is so dense that molecules such as carbon monoxide have formed where the dense dust protects them from ultraviolet radiation. These molecules emit the radio signals that make the cloud detectable. As shown in

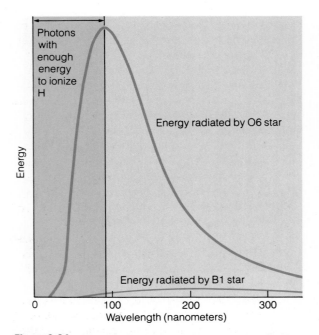

Figure 9-24 Photons with wavelengths shorter than 91.2 nm have enough energy to ionize hydrogen. The O6 star is the only star in the Trapezium hot enough to produce appreciable ionization.

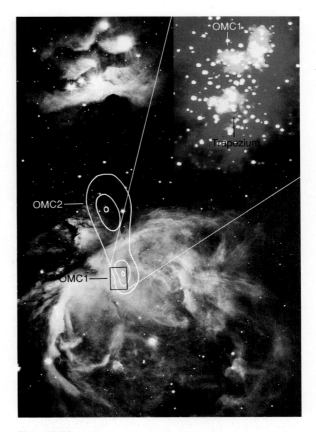

Figure 9-25 The Great Nebula in Orion is only the visible portion of a much larger cloud of gas and dust. The Orion molecular clouds (OMC1 and OMC2) can be mapped at radio and infrared wavelengths. An infrared image of the region shows the Trapezium cluster of stars (lower left of inset), which is visible in small telescopes, and OMC1 containing the BN object and KL nebula at the upper right. (Lick Observatory photograph; inset courtesy Ian McLean, Joint Astronomy Centre, Hawaii)

Figure 9-25, the cloud contains two dense regions, named the Orion molecular clouds 1 and 2 (OMC1 and OMC2).

These clouds are significantly different from the Great Nebula. The molecular clouds contain at least 10^6 atoms/cm³ and large amounts of dust. In addition, observations at different wavelengths show that the clouds are warmest near their centers. Astronomers conclude that the clouds are growing hotter at their centers because they are contracting.

The Infrared Clusters Infrared observations reveal that clusters of warm objects lie at the centers of both OMC1 and OMC2. Invisible at optical wavelengths, these objects are evidently stars in pre-main-sequence stages, wrapped deep in dust cocoons. Two especially interesting infrared objects lie near the center of OMC1.

The Becklin–Neugebauer object (BN object), named after its discoverers, was first thought to

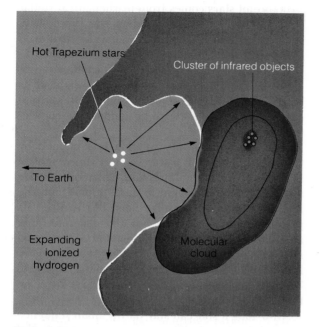

Figure 9-26 A side view of the Great Nebula would show the newly formed Trapezium stars heating and driving away the surrounding gas as new stars form in OMC1.

The Future of the Orion Cloud All evidence points to star formation within the molecular cloud just behind the Great Nebula in Orion. To predict the future of this cloud, we have only to look at the ionized nebula around the Trapezium stars.

About 6 million years ago, the Trapezium stars were contracting protostars buried within a molecular cloud. As they approached the main sequence, their temperatures increased and their radiation drove away their cocoons. But not until the 40 $M_\odot$ star turned on was there sufficient ultraviolet radiation to ionize the gas. The ionization transformed the cloud from a cold, dusty cocoon into a hot, transparent bubble of gas expanding away from the Trapezium. We can see evidence of this expansion in the Doppler shifts of spectral lines and in the twisted filaments of gas within the nebula.

Once the hot star ionized the gas, star formation in the Trapezium region stopped. In more distant parts of the cloud, the contraction of the gas into protostars continues. Indeed, the ionized gases pushing against the remains of the cloud may have triggered the collapse of more protostars. We now see the front of the cloud torn by the expanding ionized gases around the Trapezium cluster, while deeper within the cloud more protostars are forming (Figure 9-26).

In the next few thousand years, the familiar outline of the Great Nebula will change, and a new nebula may form as the protostars in the molecular clouds reach the main sequence and the most massive become hot enough to ionize the surrounding gas. Thus the Great Nebula in Orion and its parent molecular cloud show how the formation of stars continues even today.

be a protostar, but infrared observations show that the gas near its center is ionized. Thus it is probably a young B0 star that has just reached the main sequence and has not yet cleared away its cocoon nebula.

The second infrared object in OMC1 was originally called the Kleinmann–Low Nebula, but recent observations show that it is a cluster of infrared objects, at least five of which are protostars. Radio observations show that one of these objects, IRc2, is surrounded by an expanding doughnut of dense gas and is ejecting gas in a powerful bipolar flow along the axis of the doughnut. The object cannot sustain such mass loss for more than about 1000 years, so it must be a very young protostar.

photons survive a trip of 2000 pc.)
2-2.)

tific American 251 (January 1985), p. 40.

CASH, WEBSTER, and PHILIP CHARLES. "Stalking the Cygnus Superbubble." *Sky and Telescope* 59 (June 1980), p. 455.

CHURCHWELL, ED, and KEVIN J. ANDERSON. "The Anatomy of a Nebula." *Astronomy* 13 (June 1985), p. 66.

CROSWELL, KEN. "Stars Too Small to Burn." *Astronomy* 12 (April 1984), p. 16.

FRANCO, ANNA, and DAVID H. SMITH. "Vanishing Solar Neutrinos." *Sky and Telescope* 73 (February 1987), p. 149.

GEHRZ, ROBERT D., DAVID C. BLACK, and PHILIP M. SOLOMON. "The Formation of Stellar Systems from Interstellar Molecular Clouds." *Science* 224 (25 May 1984), p. 823.

HARPAZ, AMOS. "How Much Energy Does a Star Radiate?" *Physics Today* 28 (November 1990), p. 526.

HERBST, WILLIAM, and GEORGE E. ASSOUSA. "Supernova and Star Formation." *Scientific American* 241 (August 1979), p. 138.

KANIPE, RICHARD. "Inside Orion's Stellar Nursery." *Astronomy* 17 (August 1989), p. 40.

LADA, CHARLES J. "Energetic Outflows from Young Stars." *Scientific American* 247 (July 1982), p. 82.

———. "Star in the Making." *Sky and Telescope* 72 (October 1986), p. 334.

LITTLE, LESLIE. "How Astronomers Watch the Birth of Stars." *New Scientist* 101 (March 1, 1984), p. 12.

MARAN, STEPHEN P. "Debut of an Interstellar Bubble." *Natural History* 90 (August 1981), p. 28.

———. "The Impossible Star." *Natural History* 91 (June 1982), p. 64.

MARSCHALL, LAURENCE. "Secrets of Interstellar Clouds." *Astronomy* 10 (March 1982), p. 6.

PARESCE, FRANCESCO, and STUART BOWYER. "The Sun and the Interstellar Medium." *Scientific American* 255 (September 1986), p. 93.

REIPURTH, BO. "Bok Globules." *Mercury* 13 (March/April 1984), p. 50.

ROBINSON, LEIF J. "Orion's Stellar Nursery." *Sky and Telescope* 64 (November 1982), p. 430.

RODRIGUEZ, L. "Searching for the Energy Source of the Herbig–Haro Objects." *Mercury* 10 (March/April 1981), p. 34.

———. "Cosmic Jets: Bipolar Outflows in the Universe." *Astronomy* 12 (June 1984), p. 66.

SCHNEPS, MATTHEW H., and NEO WRIGHT. "A Bubble in Space: The Shell of NGC 2359." *Sky and Telescope* 59 (March 1980), p. 195.

SCHWARZSCHILD, BERTRAM. "Conversion in Matter May Account for Missing Solar Neutrinos." *Physics Today* 39 (June 1986), p. 17.

SEEDS, M. "Stellar Evolution." *Astronomy* 7 (February 1979), p. 6. Reprinted in *Astronomy: Selected Readings*, ed. M. A. Seeds. Menlo Park, Calif.: Benjamin/Cummings, 1980.

SHORE, LYS ANN, and STEVEN N. SHORE. "The Chaotic Material Between the Stars." *Astronomy* 16 (January 1988), p. 6.

SPITZER, L. "Interstellar Matter and the Birth and Death of Stars." *Mercury* 12 (September/October 1983), p. 22.

STAHLER, STEVEN W. "The Early Life of Stars." *Scientific American* 265 (July 1991), p. 48.

VERSCHUUR, GERRITL. "Molecules Between the Stars." *Mercury* 14 (May/June 1987), p. 66.

WELCH, W. J., S. N. VOGEL, R. L. PIAMBECK, M. C. H. WRIGHT, and J. H. BIEGING. "Gas Jets Associated with Star Formation." *Science* 228 (June 21, 1985), p. 1389.

CHAPTER

10

The Deaths of Stars

Natural laws have no pity.

Robert Heinlein
The Notebooks of Lazarus Long

The stars are dying. In the previous chapter, we saw that stars can resist their own gravity by generating energy through nuclear fusion. The energy keeps their interiors hot, and the resulting high pressure balances gravity and prevents the stars from collapsing. But stars have limited fuel. When they exhaust that fuel, gravity wins, and the stars die.

Perhaps the most important principle of stellar evolution is that the properties of stars depend mainly on their mass. For one thing, high-mass stars evolve faster and die sooner than low-mass stars. The most massive stars may live only a few million years, while the lowest-mass stars live hundreds of billions of years. In addition, stars of different masses die in different ways. Low-mass stars die quietly, but high-mass stars die in tremendous explosions. To follow the evolution

of stars to their final collapse, we must consider stars of various masses: low-mass red dwarfs, medium-mass sunlike stars, and massive upper-main-sequence stars (Figure 10-1).

The deaths of stars lead invariably to one of three final states. Most stars, including stars like the sun, become white dwarfs, stars about the size of the earth with no usable fuels. The most massive stars explode and leave behind either a neutron star or a black hole, exotic objects we will study in detail in the next chapter.

The final states of stellar evolution—white dwarfs, neutron stars, and black holes—represent the final victory of gravity. Understanding stellar evolution and death means we must first understand how the star delays its inevitable collapse. We begin this chapter at the beginning of the end, the evolution of giant stars.

BOX 10-1
Degenerate Matter

Normally, the pressure in a gas depends on its temperature, but under certain circumstances nature may break that rule, often with catastrophic consequences for the star. To see how this works, we must consider the energy of the free electrons in an ionized gas.

The gas inside a star is completely ionized. That is, it consists of the bare nuclei of the atoms mixed with a large number of free electrons. At high densities and temperatures, the pressure of the gas is due almost entirely to the numerous electrons.

If the density is very high, the particles of the gas are forced close together, and two laws of quantum mechanics become important. First, quantum mechanics says that the moving electrons confined in the star's core can have only certain amounts of energy of motion, just as the electron in an atom can occupy only certain energy levels (see Chapter 6). We can think of these permitted energies as the rungs in a ladder. An electron can occupy any rung, but not the spaces in between.

The second quantum-mechanical law (called the Pauli Exclusion Principle) says that two identical electrons may not occupy the same energy level. Because electrons spin in one direction or the other, two electrons can occupy an energy level if they spin in opposite directions. Such a level is completely filled, and a third electron could not enter because, whichever way it spun, it would be identical to one or the other of the two electrons already in the level. Therefore no more than two electrons can occupy the same energy level. That is, no more than two electrons in a given volume can have the same velocities.

In a low-density gas, there are few electrons per cubic centimeter, so there are plenty of empty energy levels (Figure 10-4). If a gas becomes very dense, however, nearly all of the lower energy levels may be occupied, and the gas is termed **degenerate matter.** In such matter, an electron cannot slow down because there are no open energy levels for it to drop down to. It can speed up only if it can absorb enough energy to leap to the top of the energy ladder, where there are empty energy levels.

Because degenerate matter is ruled by quantum mechanics, such matter has two properties that are important to dying stars. First, the pressure of degenerate gas resists compression. To compress the gas, we must push against the moving electrons, and changing their motion

of the eruption. The helium core is quite small (Figure 10-5), and all of the energy of the explosion is absorbed by the distended envelope. In addition, the helium flash is a very short-lived event in the life of the star. In a matter of hours, the core of the star becomes so hot it is no longer degenerate, the pressure–temperature thermostat brings the helium fusion under control, and the star proceeds to fuse helium steadily in its core (Figure 10-6).

When helium fusion begins in the core of a star, its envelope contracts slightly and becomes hotter. This makes the star a bit less luminous and a bit bluer. In the H–R diagram, the star moves down and then to the left toward the hot side of the diagram.

Not all stars experience a helium flash. Stars less massive than about $0.4\ M_\odot$ can never get hot enough to ignite helium, and stars more massive than about $3\ M_\odot$ ignite helium before their cores become degenerate. In such stars, pressure depends on temperature, so the pressure–temperature thermostat keeps the helium fusion under control.

Helium fusion produces carbon and oxygen, atoms that do not fuse at the low temperatures of

means changing their energy. That requires tremendous effort because we must boost them to the top of the energy ladder. Thus degenerate matter, though still a gas, takes on the consistency of hardened steel.

In addition, the pressure of degenerate gas does not depend on temperature. The pressure depends on the speed of the electrons, which cannot be changed without tremendous effort. The temperature, however, depends on the motion of all of the particles in the gas, both electrons and nuclei. If we add heat, most of it goes to speed up the motions of the nuclei, and only a few electrons can absorb enough energy to reach the empty energy levels at the top of the energy ladder. Thus changing the temperature of the gas has almost no effect on the pressure.

These two properties of degenerate matter become important when stars leave the main sequence and approach their final collapse. The independence of pressure and temperature causes the helium flash and other explosive effects such as supernova explosions, and the hardened steel effect supports white dwarfs and neutron stars.

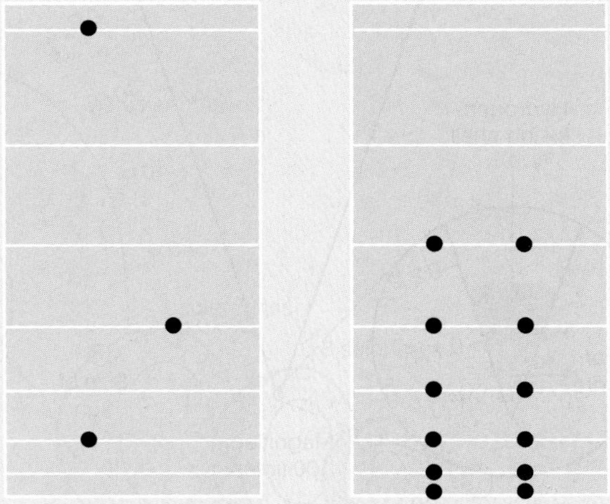

Figure 10-4 Electron energy levels are arranged like rungs on a ladder. In a low-density gas many levels are open, but in a degenerate gas all lower energy levels are filled.

the helium core. Thus, when the helium in the core is used up, the core contracts, grows hotter, and ignites a helium-fusion shell. The star expands and moves back toward the right in the H–R diagram, completing a loop (Figure 10-6).

The evolution of a star after helium exhaustion is uncertain, but the general course is clear. The inert carbon–oxygen core contracts and becomes hotter. Stars more massive than about 3 $M_\odot$ can reach temperatures of 600,000,000 K and ignite carbon fusion. Subsequent contraction may be able to fuse oxygen and silicon as well as other heavy elements.

Eventually, the star must exhaust all of its fuels and die in a final collapse. Exactly how the star collapses depends on which fuels it has consumed and on the strength of its gravity. Both of these factors depend on the star's mass, so we must divide the stars into two groups, low-mass stars and high-mass stars.

Now we can understand why giant stars are so rare (see Figure 8-24). A star spends about 90 percent of its lifetime on the main sequence and only 10 percent as a giant star. During any particular human lifetime, only a fraction of the visible stars will be passing through the giant stage.

a

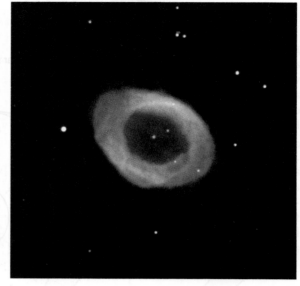

b

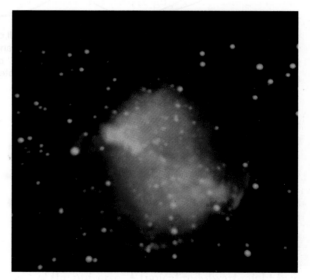

c

Figure 10-8 (a) A young planetary nebula, NGC 1535, surrounds its dying parent star. (b) The Ring Nebula has expanded to become a hollow shell and is beginning to lose its spherical shape. (c) M27, an older planetary nebula, is asymmetrical and is mixing into the interstellar medium. (a. and c. Courtesy Rudolph E. Schild; b. National Optical Astronomy Observatories)

As the star loses its surface gases, other effects accelerate the process. Elements such as magnesium, silicon, calcium, and iron can condense into dust, and the pressure of starlight can push the dust outward. The dust can collide with gas atoms and thus speed up the stellar wind. Deep inside the star, the helium-fusion shell can become unstable and experience bursts of energy production that increase its output by a factor of 1000. These, and other processes not yet understood, can drive the surface of the star outward at a speed of 20–30 km/sec. This ejected shell of gas can sweep up the gas in the stellar wind to form a hollow shell of gas.

In many cases, we can see the remains of the star at the center. The ejected layers were the insulating blanket that confined the star's internal heat. When the surface puffs into space, the white-hot interior is exposed, and the star emits intense

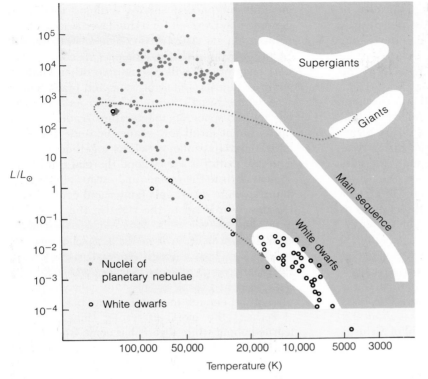

Figure 10-9 Our customary H–R diagram (shaded) must be expanded to show the evolution of a star after it ejects a planetary nebula. Central stars of planetary nebulae (filled circles) are hotter and more luminous than most white dwarfs (open circles). The dotted line shows the evolution of a 0.8 $M_\odot$ star as it collapses toward the white-dwarf region. (Adapted from diagrams by C. R. O'Dell and S. C. Vila)

ultraviolet radiation. The ultraviolet radiation can ionize the gas in the expanding shell, and we see it glow like a great doughnut of fog up to 2 ly in diameter.

Although the planetary nebulae are actually hollow spheres, they look like rings of gas. When we look at the center of the nebula, we look through two thicknesses—the front and the back. But when we look at the edge of the nebula, we look through a much greater thickness of gas, and the nebula looks brighter. Thus we see a doughnut (Figure 10-8).

About 1500 of these nebulae are known, which is not many compared with the billions of stars visible in telescopes. The planetary nebula is a short-lived phenomenon. The expanding shell of gas mixes with the interstellar medium and is gone within 50,000 years. Thus we do not catch many stars in the act of producing such nebulae.

Most medium-mass stars end by producing planetary nebulae. The sun will do so within about 6 billion years. Stars more massive than the sun can also produce planetary nebulae because they have strong stellar winds and lose much of their mass

before they reach the planetary nebula stage. Stars as massive as 6 $M_\odot$ and perhaps 8 $M_\odot$ eventually produce planetary nebulae, though our growing understanding of these nebulae suggests that the more massive of these stars produce planetary nebulae that are slightly different in shape and composition.

Once the star blows its surface layers into space, the remaining hot interior is exposed. It consists of the carbon core surrounded by the helium- and hydrogen-fusion shells and topped by a shallow atmosphere of hydrogen. Such a planetary nebula nucleus is very hot at first—in some cases over 100,000 K—and we find it far to the upper left in the H–R diagram (Figure 10-9). As the nucleus cools it moves toward the lower right in the diagram and eventually reaches the region of the white dwarfs.

White Dwarfs Both low-mass red dwarfs and medium-mass stars will eventually become white dwarfs. Our survey of neighboring stars (see Chapter 8) showed that most stars have masses less than that of the sun, but these are very long-lived stars.

Our galaxy is probably not old enough for many of these red dwarfs to have become white dwarfs. White dwarfs, however, are very numerous—our galaxy probably contains billions—so they must be the remains of stars with masses similar to the sun's.

The first white dwarf discovered was the faint companion to Sirius. In that visual binary system, the bright star is Sirius A. The white dwarf, Sirius B, is over 11,000 times fainter than Sirius A at visual wavelengths. The orbital motions of the stars (shown in Figure 8-13b) tell us that the white dwarf's mass is 0.98 $M_\odot$, and its blue-white color tells us that its surface is very hot. Observations in the ultraviolet, where it radiates most of its energy, show that the temperature is between 27,000 K and 32,000 K. Because it has a low luminosity and a high temperature, we can conclude that it must have a small surface area (see Box 8-3); in fact, it is about 85 percent of Earth's diameter. The mass and size imply that its average density is about 2.8 × 10⁶ gm/cm³. On Earth, a teaspoonful of Sirius B material would weigh over 15 tons.

A normal star is supported by energy flowing outward from its core, but a white dwarf has no internal energy source, so there is nothing to oppose its gravity, and the gas becomes degenerate (see Box 10-1). Thus a white dwarf is supported not by energy flowing outward but by the refusal of its electrons to pack themselves into a smaller volume.

The interior of a white dwarf is a degenerate gas—the ashes of nuclear fusion. Many white dwarfs are composed of degenerate helium, but others may contain degenerate carbon and oxygen produced by helium fusion. If further heavy-element fusion occurred in the star, the resulting white dwarf might be composed of nuclei as massive as iron.

Although the interior of a white dwarf is degenerate, computer models predict that normal ionized gas can exist in a surface layer about 50 km (30 miles) deep. These layers blend into an intensely hot atmosphere of hydrogen and helium gas. Because the surface gravity of a white dwarf is over 200,000 times that on Earth, the white dwarf's atmosphere is pulled down into a very shallow layer. If Earth's atmosphere were equally shallow, people on the top floors of skyscrapers would have to wear space suits.

Clearly, a white dwarf is not a normal star. It generates no nuclear energy, is almost totally degenerate, and, except for a thin layer at its surface, contains no gas. Instead of calling a white dwarf a "star," we can call it a "compact object." In the next chapter, we will discuss two other kinds of compact objects—neutron stars and black holes.

A white dwarf's future is bleak. As it radiates energy into space, its temperature gradually falls, but it cannot shrink any smaller because its degenerate electrons cannot get closer together. This degenerate matter is a very good thermal conductor, so heat flows to the surface and escapes into space, and the white dwarf gets fainter and cooler, moving downward and to the right in the H–R diagram. Because the white dwarf contains a tremendous amount of heat, it needs billions of years to radiate that heat through its small surface area. Eventually, such objects may become cold and dark, so-called **black dwarfs.** Our galaxy is probably not old enough to contain many black dwarfs.

Perhaps the most interesting thing we have learned about white dwarfs has come from mathematical models. The equations predict that if we added mass to a white dwarf, its radius would *shrink* because added mass would increase its gravity and squeeze it tighter. If we added enough to raise its total mass to about 1.4 $M_\odot$, its radius would shrink to zero (Figure 10-10). This is called the **Chandrasekhar limit** after the astronomer who discovered it. It seems to imply that a star more massive than 1.4 $M_\odot$ could not become a white dwarf unless it shed mass in some way.

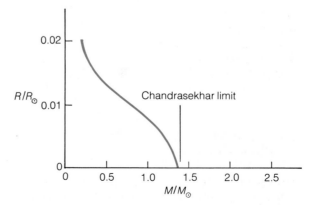

Figure 10-10 The more massive a white dwarf is, the smaller its radius. Stars more massive than the Chandrasekhar limit of 1.4 $M_\odot$ cannot be white dwarfs.

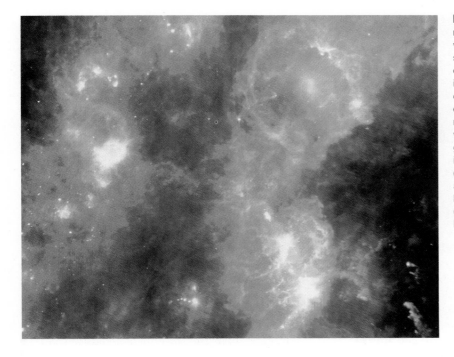

Figure 10-11 The IRAS satellite mapped the sky at far infrared wavelengths and found regions of star formation filled with twisting currents of gas and dust. This image includes most of the constellation of Orion. The hot stars of the Great Nebula in Orion (lower right) are only one of the regions where young, hot stars drive away gas and dust and stir the interstellar medium. The large ring (upper right) is composed of gas and dust expanding away from Lambda Orionis (an 08 giant). (NASA/IPAC Courtesy Deborah Levine)

Stars do lose mass. We know that some stars have stellar winds similar to but stronger than the sun's solar wind, and such flows of gas can reduce the mass of a star dramatically over the star's lifetime. The Infrared Astronomy Satellite found that such mass loss is common (Figure 10-11).

By ejecting mass in this way, a star as massive as 7 $M_\odot$ could lose enough mass to die as a white dwarf. Typical mass loss, however, could not reduce a star with an initial mass larger than about 7 $M_\odot$ to less than the Chandrasekhar limit. Some stars must approach their final end with more than 1.4 $M_\odot$, and that will lead us to one of the most exciting advances in modern astronomy—the discovery of neutron stars and black holes.

10-3 THE DEATHS OF MASSIVE STARS

We have seen that low- and medium-mass stars die relatively quietly as they exhaust their nuclear fuels. The most violent event in the lives of these stars is the ejection of their surface layers to form planetary nebulae. Massive stars live much more violent lives, however, and they die in spectacular supernova explosions that destroy the star.

Nuclear Fusion in Massive Stars The evolution of a massive star begins in the same way as the evolution of a medium-mass star. The star exhausts the hydrogen in its core, and when the core contracts, a hydrogen-fusion shell ignites, swelling the star into a giant. As a giant, it fuses helium in its core and then in a shell, leaving behind a carbon–oxygen core that contracts and grows hotter.

If a star approaches this stage with a mass of about 3–9 $M_\odot$, the carbon–oxygen core becomes degenerate before it gets hot enough to ignite carbon. Because the pressure–temperature thermostat is turned off in degenerate matter, the carbon–oxygen core is a bomb destined to explode when the temperature reaches 600,000,000 K, the ignition temperature for carbon fusion. This explosion, called the **carbon detonation,** is more violent than the helium flash because it occurs at higher temperatures.

The carbon detonation may be powerful enough to blow some stars apart, and it could be responsible for some of the violent stellar explosions known as supernovae. As we will see later, supernovae are

TABLE 10-1

Heavy-Element Fusion in a 25 $M_{\odot}$ Star

Fuel	Time	Percentage of Lifetime
H	7,000,000 years	93.3
He	500,000 years	6.7
C	600 years	0.008
O	0.5 years	0.000007
Si	1 day	0.00000004

rare and poorly understood. Most stars are able to survive carbon detonation, since the rising temperature eventually forces the core to expand and the gas to stop being degenerate.

Giant stars more massive than about 9 $M_{\odot}$ do not face the carbon detonation because their cores are so hot carbon ignites before the gas becomes degenerate. Thus the pressure–temperature thermostat is in working order and carbon fusion turns on gradually. Stars that survive carbon ignition continue their lives for a short while as they burn heavier elements (Table 10-1), but they, too, eventually face an explosive end. The development of an iron core spells their absolute and final collapse.

The Iron Core The evolution of stars that survive carbon ignition is uncertain because of the complexity of the nuclear reactions involving carbon and heavier elements (see Figure 9-17). Nevertheless, the general outline is clear. As the star burns heavy elements, each new fuel ignites first in the core and then in a shell, building layer after layer of heavy elements (Figure 10-12).

This heavy-element fusion must stop when the star develops an iron core because the iron nucleus is the most tightly bound of all nuclei (see Figure 9-13). Reactions that fuse iron atoms into even heavier atoms must absorb rather than release energy.

As a star develops an iron core, energy production begins to decline, and the core contracts. For less massive atoms than iron, this contraction would heat the gas and ignite new fusion reactions, but any nuclear reactions involving iron will absorb energy in either of two ways. For stars between 8 and 12 solar masses, the heavy nuclei in the core capture electrons, thus removing energy from the gas. In yet more massive stars, temperatures are so high that many photons have wavelengths in the gamma ray part of the spectrum. These gamma rays can break up the iron nuclei and cause the core to collapse while releasing a burst of neutrinos.

Although a massive star may live for millions of years, its inner core collapses in only a few thousandths of a second. This collapse happens so rapidly that our most powerful computers are not capable of predicting the details. Thus models of the final instants in the existence of a massive star are only approximations. One thing is clear, however. The collapse of the core of a massive star triggers a star-destroying explosion—called a supernova.

We can't be sure how a supernova explodes, but recent advances in mathematical modeling techniques and increases in computer power and speed are giving us some clues. The collapse of the innermost part of the degenerate core allows the rest of the core to fall inward, and this creates a tremendous traffic jam as all of the nuclei fall toward the center. It is as if all of the car owners in Indiana suddenly tried to drive their cars into downtown Indianapolis. Not only would there be a traffic jam downtown but also in the suburbs, and as more cars arrived, the traffic jam would spread outward. Similarly, as the inner core falls inward, a shock wave (a "traffic jam") develops and begins to move outward. Such a shock wave can blast the outer part of the star into space and produce the explosion we see as a supernova.

Theorists like to refer to this outward traveling ejection as a *bounce*. The in-falling core is said to bounce off the inner core now compressed to the density of an atomic nucleus. Recent studies suggest that under certain circumstances the star may not bounce. The shock wave may stall as the matter falls inward faster than the shock wave can travel outward. This may lead to the total collapse of the star into a black hole.

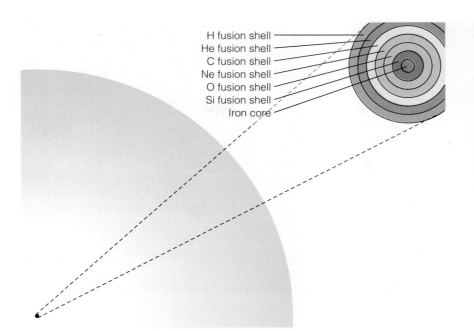

H fusion shell
He fusion shell
C fusion shell
Ne fusion shell
O fusion shell
Si fusion shell
Iron core

Figure 10-12 A star on the verge of a supernova explosion is usually a cool, red supergiant almost as large as the orbit of Jupiter. The core (magnified 100,000 times here) is about the size of Earth and contains concentric layers fusing heavier fuels around the growing core of iron. Because iron cannot produce energy through fusion, the iron core must eventually collapse and trigger a supernova explosion.

The supernova we see is the brightening of the star as its distended envelope is blasted outward by the shock wave rising from the interior. As the cloud of gas expands and thins, it begins to fade, but the way some supernovae fade suggests to some astronomers that the gas is enriched with short-lived radioactive nuclei such as nickel-56. The gradual decay of these nuclei can keep the gas hot and prevent it from fading rapidly.

New supercomputers may eventually tell us more about how supernova explosions occur, but whatever the mechanism, it must generate tremendous energy to account for supernovae.

Observations of Supernovae In A.D. 1054, Chinese astronomers saw a "guest star" appear in the constellation we know as Taurus the Bull. The star quickly became so bright it was visible in the daytime, and then, after a month, it slowly faded, taking almost 2 years to vanish from sight. When modern astronomers turned their telescopes to the location of the "guest star," they found a cloud of gas about 1.35 pc in radius expanding at 1400 km/sec. Projecting the expansion back in time, they concluded the expansion must have begun about 900 years ago, just when the "guest star" made its visit. Thus we think the nebula, now called the Crab Nebula because of its shape (Fig-

ure 10-13a), marks the site of the A.D. 1054 supernova.

Supernovae are rare. Only a few have been seen with the naked eye in recorded history. Arab astronomers saw one in A.D. 1006 and the Chinese saw one in A.D. 1054. European astronomers observed two—one in A.D. 1572 (Tycho's supernova) and one in A.D. 1604 (Kepler's supernova). Also, the guest stars of A.D. 185, 386, 393, and 1181 may have been supernovae.

In the centuries following the invention of the astronomical telescope in 1609, no supernova seen was bright enough to be visible to the naked eye. The ones seen were in distant galaxies and were thus fainter and harder to study. Then, in the early hours of February 24, 1987, astronomer Ian Shelton discovered a 5th-magnitude object in the southern sky—a supernova still growing in brightness (Figure 10-14a). The supernova, known officially as SN 1987A, is located only 53,000 pc away in the Large Magellanic Cloud, a small satellite galaxy to our own Milky Way. This first naked-eye supernova in 383 years will give astronomers a ringside seat at the most spectacular event in stellar evolution.

For example, at 2:35 A.M. EST on February 23, 1987, some 20 hours before the supernova was discovered, a blast wave of neutrinos swept through

a

a

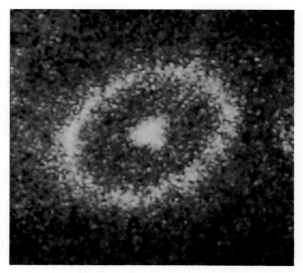

b

b

Figure 10-13 The Crab Nebula is a supernova remnant, the remains of a supernova observed by the Chinese in A.D. 1054. (a) At visual wavelengths, the glow of the nebula is woven through a network of expanding filaments. (Copyright California Institute of Technology and Carnegie Institution of Washington; by permission from Hale observatories) (b) A VLA radio map shows that the nebula is filled with hot gas emitting energy by synchrotron radiation. (National Radio Astronomy Observatory)

Figure 10-14 Supernova 1987A in the Large Magellanic Cloud was the first supernova visible to the naked eye since 1604. (a) The Tarantula Nebula is at upper left and the supernova is at the right. (National Optical Astronomy Observatories) (b) A Hubble Space Telescope image reveals a ring of gas only 1.66 seconds of arc in diameter around the supernova. The gas was expelled by normal mass loss about 5000 years ago when the star was a cool, red supergiant. (European Space Agency/NASA)

the earth.* Instruments buried in a salt mine near Cleveland and similar detectors in Japan and Europe, though designed for another purpose, recorded the passage of the neutrinos coming from the direction of SN 1987A and thus confirmed the theory that core collapse triggers some supernovae.

Within a few seconds of that moment, roughly 20 trillion neutrinos from the supernova passed harmlessly through your body.

Studies of many supernovae seen in other galaxies (Figure 10-15) show that there are at least two kinds. Type I supernovae become about 4 billion times more luminous than the sun ($M_v = -19$) and then decline rapidly at first and then more slowly. Type II supernovae become only about 0.6 billion times as bright as the sun ($M_v = -17$) and decline in a more irregular way (Figure 10-16). Spectra of Type II supernovae show hydrogen lines; spectra of Type I do not. The differences are not well under-

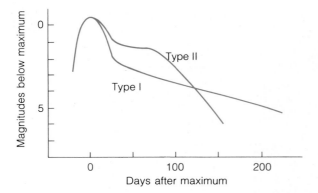

Figure 10-16 Type I supernovae decline rapidly at first and then more slowly, but Type II supernovae pause for about 100 days before beginning a steep decline. These curves have been adjusted to the same maximum brightness. Generally, Type II supernovae are about 2 magnitudes fainter than Type I.

stood, but many astronomers believe that the Type II supernovae are produced by the collapse of the iron cores in massive stars.

Although SN 1987A originated in the explosion of a massive star, it was not a normal Type II supernova. It was caused by the explosion of a hot, blue supergiant rather than the usual cool, red supergiant (see Figure 10-1). In fact, the Hubble Space Telescope detected a ring around the supernova made up of gas lost from the star by normal mass loss during its red-supergiant phase about 5000 years ago (see Figure 10-14b). Subsequently, the star contracted slightly, heating up and becoming bluer and hotter until it exploded. Normal Type II supernovae are believed to occur during the red-supergiant phase.

Type I supernovae seem to be related to white dwarfs. If a white dwarf were to gain mass from a companion in a binary system (a process we will discuss in detail later in this chapter), then the white dwarf might exceed the Chandrasekhar limit. This could cause the little star to collapse, and the sudden fusion of degenerate fuels such as carbon could trigger an explosion that would destroy the star. Spectra of Type I supernovae may lack hydrogen lines because white dwarfs contain only a thin layer of hydrogen at their surfaces.

Although the supernova explosion fades to obscurity in a year or two, an expanding shell of gas marks the site of the explosion. The gas, originally expelled at 10,000–20,000 km/sec, may carry away a fifth of the mass of the star. The collision of that expanding gas with the surrounding interstellar medium can sweep up even more gas and excite it to produce a **supernova remnant,** the nebulous remains of a supernova explosion (Figure 10-17a).

Supernova remnants look quite delicate and do not survive very long—a few tens of thousands of years—before they gradually mix with the interstellar medium and vanish. The Crab Nebula is a young remnant, only about 900 years old and about 2.7 pc in diameter. The Cygnus Loop (see Figure 10-17a) is an estimated 20,000 years old, with a diameter of 40 pc. Some supernova remnants are visible only at radio and X-ray wavelengths (Figure 10-17b; see also Figure 5-30a). They have become too tenuous to emit detectable light, but the collision of the expanding gas shell with the interstellar medium can generate radio and X-ray radiation. We saw in Chapter 9 that the compression of the interstellar medium by supernova remnants can also trigger star formation.

The radio emission coming from the Crab Nebula is **synchrotron radiation** (Figure 10-17c). This form of radiation is produced by rapidly moving electrons spiraling through magnetic fields, and in the case of the Crab Nebula, the electrons are so energetic that they also emit visible light. This leaves us with a puzzle. The Crab Nebula is 900 years old, so the electrons should have radiated away their energy long ago. The Crab Nebula must contain a powerful energy source to maintain the synchrotron radiation.

The energy source in the Crab Nebula is related to another puzzle. What happens when a supernova leaves behind an object with more than the Chandrasekhar limit of 1.4 $M_\odot$? Such a collapsing object could not become a white dwarf. Then what could it become? The answer is a neutron star or a black hole, objects we will discuss in the next chapter.

Evolution and Star Clusters In this chapter, we have concentrated on the effects of evolution on individual stars, but we should also consider its effects on groups of stars. We saw in Chapter 9 that some stars form in clusters. By studying the evolution of stars in clusters, we can confirm our theory of stellar evolution.

The stars in a cluster form over a range of

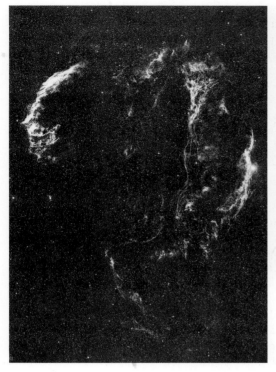

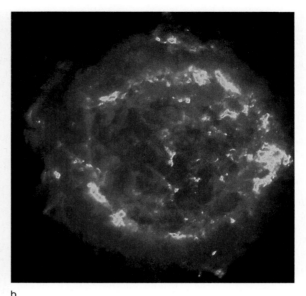

b

Figure 10-17 (a) The Cygnus Loop is a supernova remnant created when a massive star exploded 20,000 years ago. (Hale Observatories) (b) This radio map of the 300-year-old remnant Cas A was made with the VLA radio telescope. Cas A is also detectable in X rays (see Figure 5-30a). (National Radio Astronomy Observatory/Association of Universities, Inc.) (c) Some of the energy emitted by supernova remnants is synchrotron radiation, produced when an electron spirals around a magnetic field and radiates its energy away as photons. Usually synchrotron radiation is detected at radio wavelengths, but the Crab Nebula also produces synchrotron radiation at visible wavelengths.

a

c

Magnetic line of force

Photons

Path of electron

masses, and we have seen that the life history of a star depends on its mass (Figure 10-18). The least massive stars, red dwarfs, will evolve directly to white dwarfs after very long lifetimes on the main sequence. Stars more massive than about 0.4 $M_\odot$ can become giant stars. Mass loss can reduce the mass of such a star so that even a star with an initial mass of 8 $M_\odot$ may be able to die by ejecting a planetary nebula and collapsing into a white dwarf. Mass loss, however, does not seem adequate to save the most massive stars from an explosive fate. These stars become giants and supergiants, and fuse heavier elements until they form an iron core. When the iron core collapses, the star may explode as a supernova and leave behind a neutron star or a black hole.

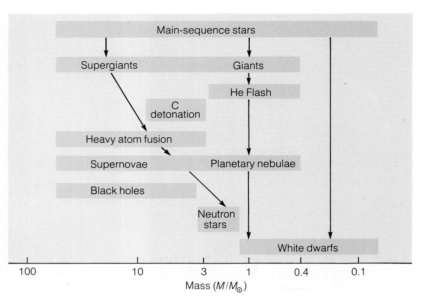

Figure 10-18 How a star evolves depends on its mass. The lowest-mass stars become white dwarfs, while stars like the sun become giants and then white dwarfs. Mass loss can move a star to the right in this diagram, as shown for the massive star. (Adapted from Michael A. Seeds, "Stellar Evolution," *Astronomy*, February 1979)

Because all of the stars in a cluster begin their lives at about the same time, we should see the most massive stars in a cluster die first, followed by progressively less massive stars. A detailed study of the H–R diagrams of star clusters reveals that this is indeed the case. (See Box 10-2.) This not only makes the evolution of stars visible, but gives us a way to find the ages of clusters.

The theory of stellar evolution is one of the great accomplishments of 20th-century astronomy. It gives us insights into how individual stars and clusters of stars evolve. But our discussion is incomplete until we study the way close pairs of stars interact as they evolve.

10-4 THE EVOLUTION OF BINARY STARS

So far we have discussed the deaths of stars as if they were all single objects that never interact. But more than half of all stars are members of binary star systems. Most such binaries are far apart (thousands of AU or more), and one of the stars can swell into a giant and eventually collapse without affecting the companion star. Some systems, however, are close together (as close as 0.01 AU). When the more massive star begins to expand, it interacts with its companion star in peculiar ways.

These interacting binary stars are interesting objects in their own right, but they are also important because they help us explain observed phenomena such as nova explosions. In the next chapter, we will use them to help us find black holes.

Mass Transfer Binary stars can sometimes interact by transferring mass from one star to the other. Of course, the gravitational field of each star holds its mass together, but the gravitational fields of the two stars, combined with the rotation of the binary system, define a teardrop-shaped surface around each star called the **Roche surface** (Figure 10-21). Matter inside a star's Roche surface is gravitationally bound to the star. The size of the Roche surface depends on the mass of the star and on the distance between the stars.

Like the lobes of a giant dumbell, the Roche surfaces in a binary star meet at the **Lagrangian point** somewhere between the stars (Figure 10-21). The Lagrangian point connects the two Roche surfaces (often termed Roche lobes) and allows matter to flow from one lobe to the other.

In general, there are only two ways matter can escape from a star and reach the Lagrangian point. First, if a star has a strong stellar wind, some of the gas blowing away from the star can pass through the Lagrangian point and be captured by the other star. Second, if an evolving star expands so far that

Cluster H–R Diagrams

The theory of stellar evolution is so complex and involves so many assumptions that astronomers would have little confidence in it were it not for H–R diagrams of star clusters. Because the stars in a cluster begin their lives at about the same time, an H–R diagram of the cluster makes the slow evolution of the stars visible.

Suppose we follow the evolution of a star cluster by making H–R diagrams like frames in a film (Figure 10-19). Our first frame shows the cluster only 10^6 years after it began forming, and already the most massive stars have reached the main sequence, consumed their fuel, and moved off to become supergiants. The medium- to low-mass stars have not yet reached the main sequence, however.

Because evolution is such a slow process, we cannot make the time step between frames equal or we would fill over 1000 pages with nearly identical diagrams. Instead, we increase the time step by a factor of 10 with each frame. Thus the second frame shows the cluster after 10^7 years, and the third after 10^8 years.

By the third frame, all massive stars have died, and stars slightly more massive than the sun are beginning to leave the main sequence. Notice that the lowest-mass stars have finally begun to fuse hydrogen. Only after 10^{10} years will the sun begin to swell into a giant.

These five frames were made from theoretical models of stellar evolution, but they compare very well with H–R diagrams of real star clusters. NGC 2264 is only a few million years old and still has many of its lower-mass stars contracting toward the main sequence

(continued)

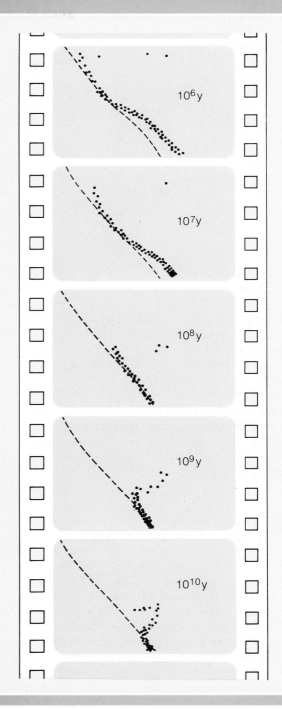

Figure 10-19 A series of theoretical H–R diagrams, like frames in a film, illustrates the evolution of a cluster of stars. Massive stars approach the main sequence faster, live shorter lives, and die sooner than lower-mass stars. Compare with Figure 10-20.

BOX 10-2

Cluster H–R Diagrams *(continued)*

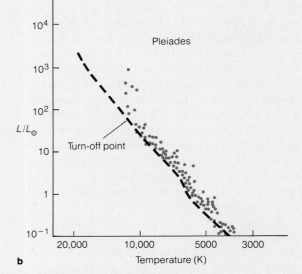

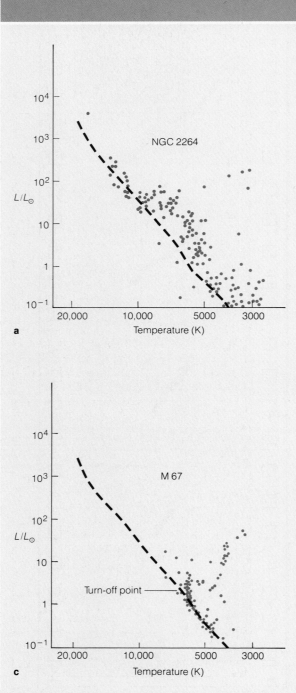

Figure 10-20 (a) NGC 2264 is a cluster only a few million years old. Its lower-mass stars are still approaching the main sequence. (b) The Pleiades is an older star cluster, but most of its stars are still on the main sequence. (c) M 67 is about 5 billion years old, and all of its more massive stars have died.

(Figure 10-20a). The Pleiades, a cluster visible to the naked eye in Taurus, is older than NGC 2264, dating back about 10^8 years (Figure 10-20b). Compare these younger clusters with M 67, a faint cluster of stars about 5×10^9 years old (Figure 10-20c).

We can estimate the age of a star cluster by noting the point at which its stars turn off the main sequence and move toward the red-giant region. The masses of the stars at this **turn-off point** will tell us the age of the cluster, because those stars are on the verge of exhausting their hydrogen fusion cores. Thus the life expectancy of the stars at the turn-off point equals the age of the cluster (see Box 9-2).

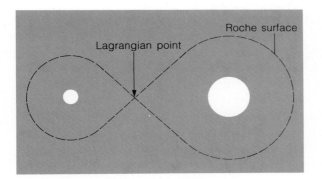

Figure 10-21 The Roche surface around a pair of binary stars outlines the two volumes of space that the stars control gravitationally. The Roche surfaces touch at the Lagrangian point, where matter can flow from one star to the other.

it fills its Roche lobe, then matter will flow through the Lagrangian point onto the other star. Mass transfer driven by a stellar wind tends to be slow, but mass transfer driven by an expanding star can occur rapidly.

Evolution with Mass Transfer Mass transfer between stars can affect the evolution of the stars in surprising ways. In fact, this is the explanation of a problem that puzzled astronomers for many years.

In some binary systems, the less massive star has become a giant, while the more massive star is still on the main sequence. If higher-mass stars evolve faster than lower-mass stars, how do the lower-mass stars in such binaries manage to leave the main sequence first? This is called the Algol paradox, after the binary system Algol.

Mass transfer explains how this could happen. Imagine a binary system that contains a 5 $M_\odot$ star and a 1 $M_\odot$ companion (Figure 10-22). The two stars formed at the same time, so the higher-mass star will evolve faster and leave the main sequence first. When it expands into a giant, it can fill its

Figure 10-22 The evolution of a close binary system. As the more massive star evolves (a), it fills its Roche lobe (b) and begins to transfer mass through the Lagrangian point to its companion. The companion grows more massive (c), and the star losing mass collapses (in this example) into a white dwarf (d). The companion, now a massive star, evolves into a giant and begins transferring mass back to the white dwarf (e).

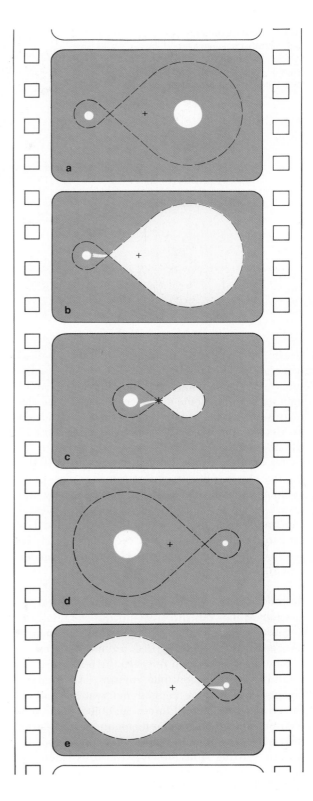

Figure 10-23 Skaters demonstrate conservation of angular momentum when they spin faster by drawing their arms and legs closer to their axes of rotation.

Roche lobe and transfer matter to the low-mass companion. Thus the higher-mass star could evolve into a lower-mass star, and the companion could gain mass and become a high-mass star still on the main sequence. Thus we might find a system such as Algol containing a 5 $M_\odot$ main-sequence star and a 1 $M_\odot$ giant.

Yet another exotic possibility can arise if the massive star in a binary system transfers mass to its companion and then collapses to form a white dwarf. Such a system can later become a site of tremendous explosions when the companion begins to transfer mass back to the white dwarf. To see how these explosions occur, we must consider how mass falls into a star.

Accretion Disks Matter flowing from one star to another cannot fall directly into the star. Rather, because of conservation of angular momentum, it must flow into a whirling disk around the star.

Angular momentum refers to the tendency of a rotating object to continue rotating. All rotating objects possess some angular momentum, and in the absence of external forces, an object maintains (conserves) its total angular momentum. Ice skaters take advantage of conservation of angular momentum by starting spins slowly with arms extended and then drawing their arms closer to their axes of rotation. As their mass is concentrated closer to the axis of rotation, they spin faster (Figure 10-23). The same effect causes the slowly circulating water in a bathtub to spin in a whirlpool as it approaches the drain.

Mass transferred through the Lagrangian point in a binary system toward a white dwarf must conserve its angular momentum. Thus it must flow into a rapidly rotating whirlpool called an **accretion disk** around the white dwarf (Figure 10-24).

Two important things happen in an accretion disk. First, the gas in the disk grows very hot due to friction and tidal forces. The disk also acts as a brake, ridding the gas of its angular momentum and allowing it to fall into the white dwarf. The temperature of the gas in the inner parts of an accretion disk can exceed 1,000,000 K and emit intense X rays. In addition, the matter falling inward from the accretion disk can cause a violent explosion if it accumulates on a white dwarf.

Novae The word **nova** (plural *novae*) is Latin for "new," and in astronomy it refers to the appearance of what seems to be a new star. A nova can appear in the sky and brighten in a few days, and then fade back to obscurity during the next few months (Figure 10-25). A nova, however, is not a new star, but the eruption of an old star, a white dwarf.

Nova explosions appear to be caused by the

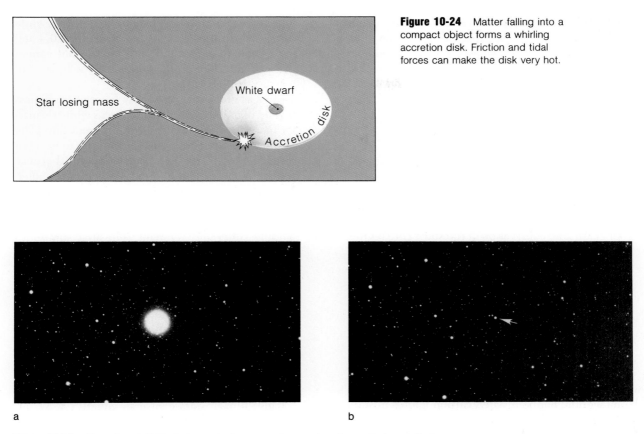

Figure 10-24 Matter falling into a compact object forms a whirling accretion disk. Friction and tidal forces can make the disk very hot.

a

b

Figure 10-25 Nova Cygni 1975 photographed near maximum when it was 2nd magnitude (a) and later when it had declined to about 11th (b). (Lick Observatory)

transfer of matter from a normal star, through an accretion disk, onto the surface of a white dwarf. Because the matter comes from the surface of a normal star, it is rich in unfused fuel, mostly hydrogen, and when it accumulates on the surface of the white dwarf, it forms a layer of unprocessed fuel. As the layer deepens, it becomes denser and hotter until the hydrogen fuses in a sudden explosion.

The nova explosion blows off the surface of the white dwarf in a shell of gas traveling thousands of kilometers per second. Although it contains only about 0.0001 $M_\odot$, this expanding shell can become 100,000 times more luminous than the sun. As the shell grows larger and less dense, it cools, and the nova fades.

The explosion hardly disturbs the white dwarf and its companion star. Mass transfer quickly resumes, and a new layer of fuel begins to accumu-

late. How fast the fuel builds up depends on the rate of mass transfer. According to this theory, some novae might need 1000–100,000 years to accumulate another explosive layer. Others might need only a few weeks.

Almost all novae are known to be members of binary systems, but it is possible for a single white dwarf to explode as a nova. For example, a star shrinking toward the white dwarf stage would heat up as it shrank. If its surface layers were rich in unfused fuel and if it heated rapidly enough, its surface layers might explode. Another way a single white dwarf might become a nova is by slowly accumulating matter from the interstellar medium. This would gradually build a surface layer of fuels.

In this chapter, we have traced the evolution of stars, both as single objects and as members of binary systems. We have found that all stars end in one of three final states—white dwarfs, neutron stars, and black holes. We have considered white dwarfs, but we have not discussed neutron stars and black holes. Those objects will be the subject of the next chapter.

SUMMARY

When a star's central hydrogen-fusion reactions cease, its core contracts and heats up, igniting a hydrogen-fusion shell and swelling the star into a cool giant. The contraction of the star's core ignites helium first in the core and later in a shell. If the star is massive enough, it can eventually fuse carbon and other elements.

If a star's mass lies between about 0.4 and 3 $M_\odot$, its helium core becomes degenerate before the helium ignites. In degenerate gas, pressure does not depend on temperature, so there is no pressure–temperature thermostat to control the reactions. As a result, the core explodes in a helium flash. All of the energy produced is absorbed by the star. A similar thing happens when carbon ignites in stars between about 3 and 9 $M_\odot$, except that it is much more powerful. This carbon detonation is one possible cause of supernovae.

How a star evolves depends on its mass. Stars less massive than about 0.4 $M_\odot$ are completely mixed and will have very little hydrogen left when they die. They cannot ignite a hydrogen shell or a helium core, so they will eventually become white dwarfs. Medium-mass stars between about 0.4 and 3 $M_\odot$ become giants and fuse helium but cannot fuse carbon. They produce planetary nebulae and become white dwarfs.

Stars as massive as 8 $M_\odot$ may loose enough mass to eject planetary nebulae and die as white dwarfs, but more massive stars suffer a different fate. The most massive stars fuse nuclear fuels up to iron but cannot generate further nuclear energy because iron is the most tightly bound of all atomic nuclei. When an iron core forms in a massive star, the core collapses and triggers a supernova explosion that expels the outer layers of the star to form an expanding supernova remnant. The first supernova visible to the naked eye since 1604 was discovered in February 1987.

We can see evidence of stellar evolution in the H–R diagrams of clusters of stars. Beginning their evolution at about the same time, the stars evolve in different ways, depending on their masses. The most massive evolve off of the main sequence first and are followed later by progressively less massive stars. Thus we can estimate the age of a star cluster from the turn-off point in its H–R diagram.

Close binary stars evolve in complex ways because they can transfer mass from one star to the other as they evolve. This can explain why some binary systems contain a main-sequence star more massive than its giant companion—the Algol paradox. Also, mass transfer into an accretion disk around a white dwarf can produce X rays from the hot disk and can trigger nova explosions.

NEW TERMS

helium flash	synchrotron radiation
degenerate matter	turn-off point
planetary nebula	Roche surface
black dwarf	Lagrangian point
Chandrasekhar limit	angular momentum
carbon detonation	accretion disk
supernova remnant	nova

REVIEW QUESTIONS

1. How does the contraction of a star's helium core make the star swell into a giant?

2. Why does the expansion of the star's envelope make the star cooler and more luminous?

3. Describe the two properties of degenerate matter that are important in stars.

4. Why don't red dwarfs become giants?

5. Why can't a white dwarf contract as it cools? What is its fate?

6. Describe step by step how the sun will die.

7. Give four possible causes of a supernova explosion.

8. How can we estimate the age of a star cluster?

9. What is the Algol paradox? How might we explain it?

10. Why do we expect novae to be binary systems?

DISCUSSION QUESTION

1. How do we know the helium flash occurs if it cannot be observed? Can we accept something as real if we can never observe it?

PROBLEMS

1. About how long would a 0.4 $M_\odot$ star spend on the main sequence? (HINT: See Box 9-2.)

2. The Ring Nebula in Lyrae is a planetary nebula with an angular diameter of 72 seconds of arc and a distance of 5000 ly. What is its linear diameter? (HINT: See Box 3-2.)

3. If the Ring Nebula is expanding at a velocity of 15 km/sec, typical of planetary nebulae, how old is it?

4. Suppose a planetary nebula is 1 pc in radius. If the Doppler shifts in its spectrum show it is expanding at 30 km/sec, how old is it? (HINTS: 1 pc equals 3×10^{13} km, and 1 year equals 3.15×10^7 seconds.)

5. If a star the size of the sun expands to form a giant 20 times larger in radius, by what factor will its average density decrease? (HINT: The volume of a sphere is $(\frac{4}{3})\pi r^3$.)

6. If a star the size of the sun collapses to form a white dwarf the size of the earth, by what factor will its density increase? (HINTS: The volume of a sphere is $(\frac{4}{3})\pi r^3$. See Appendix D for radii of sun and Earth.)

7. The Crab Nebula is now 1.35 pc in radius and is expanding at 1400 km/sec. About when did the supernova occur? (HINT: 1 pc equals 3×10^{13} km.)

8. If the Cygnus Loop is 40 pc in diameter and is 40,000 years old, with what average velocity has it been expanding? (HINTS: 1 pc equals 3×10^{13} km, and 1 year equals 3.15×10^7 seconds.)

9. Observations show that the gas ejected from SN 1987A is moving at about 10,000 km/sec. How long will it take to travel one astronomical unit? one parsec? (HINTS: 1 AU equals 1.5×10^8 km, and 1 pc equals 3×10^{13} km.)

10. If the stars at the turn-off point in a star cluster have masses of about 4 $M_\odot$, how old is the cluster?

RECOMMENDED READING

BALIK, BRUCE. "The Shaping of Planetary Nebulae." *Sky and Telescope* 73 (February 1987), p. 125.

CANNIZZO, JOHN K., and RONALD H. KAITCHUCK. "Accretion Disks in Interacting Binary Stars," *Scientific American* 266 (January 1992), p. 92.

DE VAUCAULEURS, G. "The Supernova of 1885 in Messier 31." *Sky and Telescope* 70 (August 1985), p. 115.

FELTEN, JAMES. "Light Echoes of Nova Persei 1901." *Sky and Telescope* 81 (February 1991), p. 153.

HARPAZ, AMOS. "The Formation of a Planetary Nebula." *The Physics Teacher* 29 (May 1991), p. 268.

KAFATOS, MINAS, and ANDREW G. MICHALITSIAHOS. "Symbiotic Stars." *Scientific American* 25 (July 1984), p. 84.

KAHN, RONALD N. "Desperately Seeking Supernovae." *Sky and Telescope* 73 (June 1987), p. 594.

KALER, JAMES B. "Planetary Nebulae and Stellar Evolution." *Mercury* 10 (July/August 1981), p. 114.

———. "Bubbles from Dying Stars." *Sky and Telescope* 63 (February 1982), p. 129.

———. "Planetary Nebulae and the Death of Stars." *American Scientist* 74 (May/June 1986), p. 244.

———. "Realm of the Hottest Stars." *Astronomy* 18 (February 1990), p. 22.

KAWALER, STEVEN D., and DONALD E. WINGET. "White Dwarfs: Fossil Stars." *Sky and Telescope* 74 (August 1987), p. 132.

LATTIMER, JAMES M., and ADAM S. BURROWS. "Neutrinos from Supernova 1987A." *Sky and Telescope* 76 (October 1988), p. 348.

MALIN, DAVID, and DAVID ALLEN. "Echoes of the Supernova." *Sky and Telescope* 79 (January 1990), p. 22.

MARSCHALL, LAURENCE A. *The Supernova Story*. New York: Plenum Press, 1988.

MORRISON, NANCY D., and STEPHEN GREGORY. "What Massive Stars Explode?" *Mercury* 15 (May/June 1986), p. 77.

NICASTRO, A. J. "White Dwarfs: Big Things in Small Packages." *Astronomy* 12 (July 1984), p. 6.

PACZYNSKI, BOHDAN. "Binary Stars." *Science* 225 (20 July 1984), p. 275.

PLAVEC, M. J. "IUE Looks at the Algol Paradox." *Sky and Telescope* 65 (May 1983), p. 413.

SCHORN, RONALD A. "Supernova in Our Backyard." *Sky and Telescope* 73 (April 1987), p. 382.

———. "Neutrinos from Hell." *Sky and Telescope* 73 (May 1987), p. 477.

———. "Supernova 1987A After 200 Days." *Sky and Telescope* 74 (November 1987), p. 477.

———. "Happy Birthday, Supernova." *Sky and Telescope* 75 (February 1988), p. 134.

———. "Supernova 1987A's Changing Face." *Sky and Telescope* 76 (July 1988), p. 32.

SEEDS, M. A. "Stellar Evolution." *Astronomy* 7 (February 1979), p. 6.

———. "The Wink of the Demon's Eye." *Astronomy* 8 (December 1980), p. 66.

SEWARD, FREDERICK D., PAUL GORENSTEIN, and WALLACE H. TUCKER. "Young Supernova Remnants." *Scientific American* 253 (August 1985), p. 88.

STENCEL, ROBERT E. "Mass Loss from Stars." *Astronomy* 7 (November 1979), p. 78.

STEPHENSON, F. R., and D. H. CLARK. "Historical Supernovas." *Scientific American* 234 (June 1976), p. 100.

STRAKA, W. "The Cygnus Loop: An Older Supernova Remnant." *Mercury* 16 (September/October 1987), p. 150.

TALCOTT, RICHARD. "Insight into Star Death." *Astronomy* 16 (February 1988), p. 6.

TRIMBLE, V. "How to Survive the Cataclysmic Binaries." *Mercury* 9 (January/February 1980), p. 8.

———. "Exploding Stars, Superbubbles, and the HEAO Observations." *Mercury* 13 (September/October 1984), p. 130.

TUCKER, W. "Supernovae, Dinosaurs, and Us." *Mercury* 9 (July/August 1980), p. 95.

WALLERSTEIN, G., and S. WOLFF. "The Next Supernova." *Mercury* 10 (March 1981), p. 44.

WHEELER, CRAIG, and ROBERT P. HARKNESS. "Helium-Rich Supernovas." *Scientific American* 257 (November 1987), p. 50.

WOOSLEY, STAN, and TOM WEAVER. "The Great Supernova of 1987." *Scientific American* 261 (August 1989), p. 32.

FURTHER EXPLORATION

VOYAGER, the Interactive Desktop Planetarium, can be used for further exploration of the topics in this chapter. This chapter corresponds with the following projects in *Voyages Through Space and Time:* Project 22, Magnitude Systems; Project 23, Spectral Classes and the H–R Diagram.

CHAPTER
11

Neutron Stars and Black Holes

Almost anything is easier to get into than out of.

Agnes Allen

However a star dies, gravity assures us that its last remains must eventually reach one of three final states—white dwarf, neutron star, or black hole. These objects, often called compact objects, are small, high-density monuments to the power of gravity. Every star faces an ultimate collapse into such an object.

We discussed white dwarfs in the preceding chapter. In this chapter, we will compare the theoretical predictions of the existence of neutron stars and black holes to observations. The question is, Do neutron stars and black holes really exist? The search for neutron stars and black holes has been one of the greatest adventures of modern astronomy, and it isn't over yet.

11-1 NEUTRON STARS

A neutron star is the core of a star that has collapsed to a radius of

10–15 km and to a density so high only neutrons can exist. Although neutron stars were originally predicted from theoretical physics, astronomers have found evidence that such objects do really exist.

Theoretical Prediction of Neutron Stars The neutron was discovered in the laboratory in 1932, and the following year Walter Baade and Fritz Zwicky predicted that supernova explosions might leave behind neutron stars.

During the collapse of a massive star in a supernova explosion, the core of the star is compressed to tremendous density. A white dwarf is very dense, but it is supported by the pressure of its degenerate electrons. Inside the collapsing star, however, the density of the gas becomes so great that the degenerate electrons cannot stop the collapse, the atomic nuclei break up, and protons and electrons are forced to combine to form a gas of neutrons. Such a neutron gas can become degenerate itself, and if the remaining core of the star is not too massive, the degenerate neutrons can support the weight of the core. Thus a supernova explosion might leave behind a self-supporting star made entirely of neutrons—a neutron star.

Theoretical calculations predict that a neutron star will be only 10–15 km in diameter (Figure 11-1) and will have a mass of about 10^{14}g/cm^3. On Earth, a sugar-cube-sized lump of this material would weigh 100 million tons. This is roughly the density of an atomic nucleus, and to a certain extent, we can think of a neutron star as a gigantic atomic nucleus.

How massive can a neutron star be? That is a critical question, and a difficult one to answer because we don't know the strength of pure neutron material. Scientists can't make such matter in the laboratory, so its properties must be predicted theoretically. The most widely accepted calculations suggest that a neutron star cannot be more massive than 2–3 $M_\odot$. If a neutron star were more massive than that limit, the degenerate neutrons would not be able to support the weight, and the object would collapse (presumably into a black hole).

Simple physics predicts that neutron stars should be hot, spin rapidly, and have strong magnetic fields. We have seen that contraction heats a star, and the sudden collapse of a star would heat it to millions of degrees. Also, the small surface area of a neutron star would prevent the heat from escaping rapidly, so we should expect neutron stars to be very hot.

We should also expect them to spin rapidly. All stars rotate to some extent because they are formed from swirling clouds of interstellar matter. As such a star collapses, it must rotate faster, according to the law of conservation of angular momentum. Recall that we see this happen when ice skaters spin slowly with their arms extended and then speed up as they pull their arms closer to their bodies (see Figure 10-23). In the same way, a collapsing star spins faster as it pulls its matter closer to its axis of rotation. If the sun collapsed to a radius of 10 km, its period of rotation would decrease from 25 days to 0.001 second.

A collapsing star will also increase the strength of its magnetic field. Whatever magnetic field a star has is frozen into the star like blades of grass frozen

Figure 11-1 A tennis ball and a road map illustrate the relative size of a neutron star. Such an object, containing slightly more than the mass of the sun, would fit with room to spare inside the beltway around Washington, D.C. (Photo by author)

into a snowball. If the ball contracts, the grass is concentrated in a smaller volume. When a star collapses into a neutron star, it can concentrate its magnetic field and make it a billion times stronger. Since some stars have magnetic fields over a thousand times stronger than the sun's, a neutron star might have a field a billion to a trillion times stronger than the sun's. For comparison, that is about 10 million times stronger than any magnetic field ever produced in the laboratory.

Although the existence of neutron stars was predicted 50 years ago, there seemed to be no way to observe an object only 10 km in radius. Nevertheless, progress came not from theory but from observation.

The Discovery of Pulsars In November 1967, Jocelyn Bell, a graduate student at Cambridge University, England, found a peculiar pattern on the paper chart from a radio telescope. Unlike other radio signals from celestial bodies, this was a series of regular pulses (Figure 11-2). At first she and the leader of the project, Anthony Hewish, thought the signal was interference, but they found it day after day in the same place in the sky. Clearly, it was celestial in origin.

Another possibility, that it was from a distant civilization, led them to consider naming it LGM for Little Green Men. But within a few weeks the team found three more objects in other parts of the sky pulsing with different periods. The objects

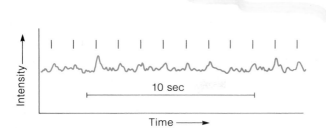

Figure 11-2 The first pulsar, CP 1919, was discovered November 28, 1967, when its regularly spaced pulses were noticed in the output from a radio telescope. The period is 1.33730119 seconds. Pulses are marked by tics.

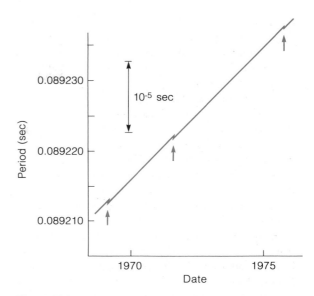

Figure 11-3 Glitches occur when a pulsar, gradually slowing down, suddenly speeds up by a small amount. Here three glitches in the pulsation of the Vela pulsar occurred over an 8-year interval.

were clearly natural, and the team dropped the name LGM in favor of **pulsar**—a contraction of *pulsing star*.

As more pulsars were found, astronomers argued over their nature. Periods ranged from 0.033 to 3.75 seconds and were nearly as exact as an atomic clock. Months of observation showed that many of the periods were slowly growing longer by a few billionths of a second per day, and that rare **glitches,** or sudden changes, could decrease the period of some pulsars (Figure 11-3). Something had to regulate the pulsation precisely, had to slow down gradually, but might be subject to occasional glitches.

Pulsars could not be stars. A normal star, even a small white dwarf, is too big to pulsate that fast. Nor could a star with a hot spot on its surface spin fast enough to produce the pulses. Even a small white dwarf would fly apart if it spun 30 times a second.

The pulses themselves lasted only about 0.001 second, and that was a clue. If a white dwarf blinked on and then off in that interval, we would not see a 0.001-second pulse. The near side of the white dwarf would be about 6000 km closer to us, and light from the near side would arrive 0.022 seconds before the light from the bulk of the white dwarf. Thus its short blink would be smeared out into a longer pulse. This is an important principle in astronomy—an object cannot change its brightness appreciably in an interval shorter than the time light takes to cross its diameter. If pulses from pulsars are no longer than 0.001 second, then the object cannot be larger than 300 km (190 miles) in diameter.

Only a neutron star is small enough to be a pulsar. In fact, a neutron star is so small, it can't

vibrate slow enough, but it can spin as fast as 1000 times a second without flying apart. The missing link between pulsars and neutron stars was found in late 1968, when astronomers discovered a pulsar at the heart of the Crab Nebula (Figure 10-13). The Crab Nebula is a supernova remnant, and theory predicts that some supernovae leave behind a neutron star. Since 1968, about 500 pulsars have been found. At least five are inside supernova remnants.

The fact that pulsars were found in supernova remnants and that the pulses were short hinted that pulsars are neutron stars. If we combine theory and observation, we can devise a model of a pulsar.

A Model Pulsar *Pulsar* is a misnomer. The periodic flashing of pulsars is linked to rotation, not pulsation. The spinning neutron star emits beams of radiation that sweep around the sky. When one of these beams sweeps over us, we detect a pulse, just as sailors see a pulse of light when the beam from a lighthouse sweeps over their ship. In fact, this model is called the **lighthouse theory.**

The lighthouse theory is generally accepted, and astronomers are becoming more confident of

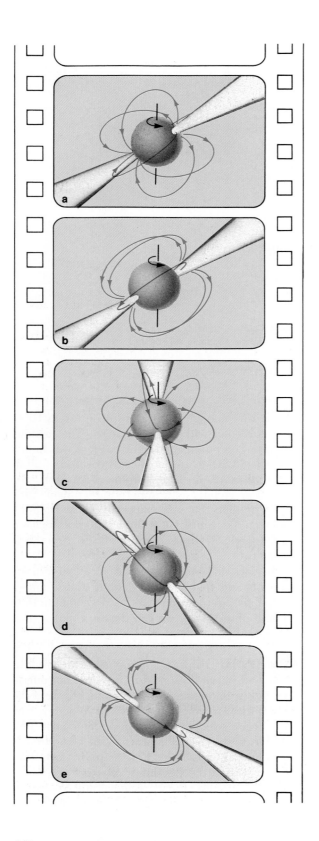

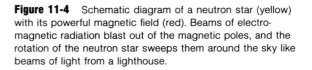

Figure 11-4 Schematic diagram of a neutron star (yellow) with its powerful magnetic field (red). Beams of electromagnetic radiation blast out of the magnetic poles, and the rotation of the neutron star sweeps them around the sky like beams of light from a lighthouse.

the mechanism that produces the beams. The theory suggests that the neutron star spins so fast and its magnetic field is so strong that it acts like a generator and creates an electric field around itself. This field is so intense that it rips charged particles, mostly electrons, out of the surface near the magnetic poles and accelerates them to high velocity. These accelerated electrons emit photons traveling in the same direction as the electrons. Thus the photons leave the neutron star in narrow beams emanating from the magnetic poles. If the magnetic axis is inclined with respect to the axis of rotation, as is the case with the earth and most of the planets in the solar system that have magnetic fields, then the neutron star will sweep the beams around the sky (Figure 11-4).

About 500 pulsars are known, but there are probably many more. Only when a pulsar's beams sweep over the earth do we detect its presence. In most cases, however, the beams of radio energy never point to the earth, so the pulsar remains invisible.

Two properties of pulsars support the lighthouse theory. First, many pulsars are slowing down—that is, their periods are increasing by a few billionths of a second each day. The amount of energy that a pulsar radiates into space per day, about 10^5 times more than the sun, is approximately equal to the amount of energy a spinning neutron star would lose by increasing its period by a few billionths of a second. This explains the source of the pulsar's energy. The spinning neutron star and its intense magnetic field convert energy of rotation into radiation (Figure 11-5).

The glitch is the second property of pulsars that supports the lighthouse theory. To see how a pulsar could produce glitches, we must consider the internal structure of a neutron star. Theoretical models of neutron stars are difficult to compute because no one knows exactly how pure neutron matter behaves. Most theorists agree, however, on the three principal layers in a neutron star (Figure 11-6). Near the surface the pressure is less, and

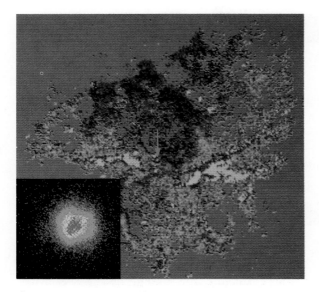

Figure 11-5 This false color image of the Crab Nebula shows that some areas (white) are over 90 percent helium. Presumably these atoms were made in the supernova that formed the neutron star (arrow). (Gordon M. MacAlpine and Alan K. Uomoto) Even 900 years after the explosion, the nebula is bright in X rays (inset), excited by the neutron star. (J. Trümper, Max-Planck Institute)

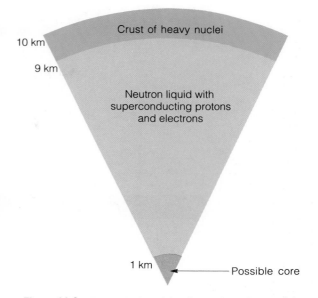

Figure 11-6 Theoretical models of a neutron star predict a crystalline crust about 1 km thick and 10,000 trillion times more rigid than steel. Deeper, the pressure produces a neutron liquid with some free protons and electrons. The magnetic field is rooted in this superconducting liquid. A solid core may exist in slightly more massive neutron stars.

atomic nuclei can exist as a rigid crystal layer about 1 km thick. The crust is roughly 10^{16} times more rigid than steel. Although this material is very strong, the tremendous gravity at the surface would prevent neutron star mountains from being more than a few millimeters high.

Below the crust the pressure is high enough to force the material into a liquid state made up mostly of neutrons with some free protons and electrons. Because the protons and electrons can move through the neutron liquid with almost no friction at all, the material is an almost perfect conductor of electricity—a **superconductor**. Apparently, the neutron star's magnetic field is anchored in this spinning mass of superconducting liquid.

Some theories of nuclear particle physics suggest that a neutron star might contain a solid core of massive particles. Whether such a core actually exists depends on details of quantum mechanics and on the behavior of gravity at very high strength. As these details are not understood yet, we cannot be sure that neutron stars have solid cores.

So far as glitches are concerned, the crust is the important layer in a neutron star. As a spinning neutron star slows, centrifugal force decreases, and gravity squeezes the neutron star tighter. Stresses build in the crust until it breaks in a "starquake," the neutron star equivalent of an earthquake. Once the crust is broken, the neutron star can shrink slightly, by about 1 mm, and rotation increases as the neutron star conserves angular momentum. This sudden speed-up increases the pulse rate, and we detect a glitch.

This could explain the glitches in the Crab Nebula pulsar, but those in the Vela pulsar seem too big and too common. A recent study suggests that, unlike the Crab Nebula pulsar, the Vela pulsar has a solid core and that circulation in the form of vortices in the liquid interior links the rotating core and the crust. A sudden change in these circulations could alter the rotation rate of the crust and thus cause a glitch. If this theory proves correct, astronomers may someday be able to use glitches to explore the interior of neutron stars, just as geologists use earthquakes to probe the interior of the earth.

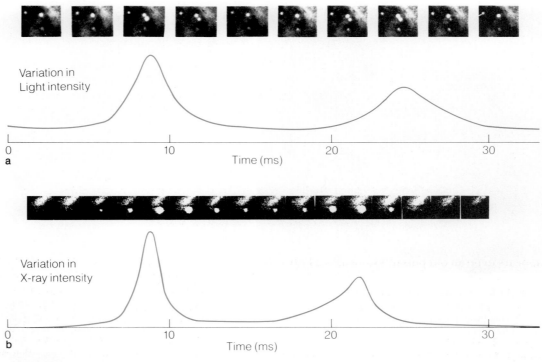

Figure 11-7 High-speed images of the Crab Nebula pulsar show it pulsing at visual wavelengths (a) and at X-ray wavelengths (b). The period of pulsation is 33 milliseconds (ms), and each cycle includes two pulses as its two beams of unequal intensity sweep over the earth. (a. © AURA, Inc., National Optical Astronomy Observatories, KPNO; b. Courtesy F. R. Harnden, Jr., from *The Astrophysical Journal,* published by the University of Chicago Press; © 1984 The American Astronomical Society)

The Evolution of Pulsars Theorists are beginning to understand how a pulsar ages. When it first forms, it is spinning very fast, perhaps nearly 100 times a second, and it contains a strong magnetic field. As it converts its energy of rotation into radiation, it gradually slows, and its magnetic field grows weaker. The average pulsar is apparently about 2×10^6 years old, and the oldest is about 10^7 years. Presumably, by the time a pulsar gets older than that, it is rotating too slowly to generate detectable beams.

If a pulsar contains a strong magnetic field and spins very fast, then it is capable of emitting very strong beams of radiation. In addition, it is capable of emitting shorter-wavelength photons than are older, slower pulsars. The Crab Nebula pulsar, the youngest known, emits pulses at radio, infrared, visible, X-ray, and gamma-ray wavelengths (Figure 11-7). In fact, the pulsar has been identified as a star at the center of the nebula long thought to be the remains of a supernova. No one knew it blinked on and off 30 times a second because the blinks blended together when viewed through a telescope or on photographic plates. Not until the star was observed electronically in 1969 were the blinks detected.

A total of four pulsars are known that produce visible pulses. All are fast-spinning neutron stars (Table 11-1), and three appear to be young neutron stars located inside supernova remnants. In addition to the Crab Nebula, the Vela pulsar produces optical pulses with a period of 0.089 second. It has an age of about 11,000 years—young for a pulsar. Another pulsar in the Large Magellanic Cloud (**LMC**) blinks 20 times a second and produces visible pulses. Its age is unknown, but it is located in a supernova remnant. A fourth visible light pulsar, **PSR 1937 + 21**, pulses very fast but

TABLE 11-1
Optical Pulsars

Location	Identification	Period (sec)	Age	In Supernova Remnant?
Crab Nebula	PSR 0531 + 21	0.033	900 years	Yes
Vela	PSR 0833 − 45	0.089	11,000 years	Yes
LMC	PSR 0540 − 69.3	0.050	Unknown	Yes
Vulpecula	PSR 1937 + 21	0.0016	Old	No

appears to be an old pulsar. (More about PSR 1937 + 21 follows.)

Evidently, we should expect to find the youngest pulsars inside supernova remnants, as in the case of the Crab Nebula pulsar. However, not every supernova remnant contains a pulsar, and not every pulsar is located inside a supernova remnant. Many supernova remnants probably contain pulsars whose beams never sweep over the earth. It will be difficult to detect such pulsars. Also, some pulsars have high proper motions, which suggests that a supernova explosion can occur slightly off center or can disrupt a binary system. Either would give a pulsar a high velocity, so it could leave its supernova remnant quickly. Of course, supernova remnants do not survive more than 50,000 years or so before they mix into the interstellar medium. Because the average pulsar is about 2×10^6 years old, its supernova remnant was lost long ago.

The explosion of supernova 1987A in February 1987 probably left behind a neutron star. The burst of neutrinos detected on Earth (see Chapter 10) is commonly accepted as evidence of the formation of a neutron star. We would not expect to see the neutron star at first because it would be hidden at the center of the expanding cloud of gas ejected into space. As that gas expanded and thinned, however, we would expect to detect X rays and gamma rays from the neutron star, and perhaps pulses if it emitted beams that swept over the earth. Yet five years after the explosion, no neutron star has been detected, and some astronomers are beginning to wonder if SN 1987A did give birth to a neutron star.

Although we would expect that a newborn pulsar would blink rapidly, the handful that pulse the fastest may be old. One of the fastest-known pulsars is catalogued as PSR 1937 + 21 in the constellation Vulpecula. It pulses 642 times a second and is slowing down only slightly. The energy stored in the rotation of a neutron star at this rate is equal to the total energy of a supernova explosion, so it seemed difficult at first to explain this pulsar. It now appears that PSR 1937 + 21 is an old neutron star that has gained mass and rotational energy from a companion in a binary system. Like water hitting a mill wheel, the matter falling on the neutron star has spun it up to 642 rotations per second. With its old, weak magnetic field, it slows down very slowly and will continue to spin for a very long time.

A number of other very fast pulsars, known as **millisecond pulsars** due to their short periods, have been found. At least two, PSR 1855 + 09 (186 pulses/sec) and PSR 1953 + 29 (163 pulses/sec), are members of binary systems. To understand such peculiar pulsars we must consider binary pulsars.

Binary Pulsars Of the hundreds of pulsars now known, a few are located in binary systems. These pulsars are of special interest because we can learn more about them by studying their orbital motion. In a few cases, mass transferred from their companion stars can generate X rays.

The first binary pulsar was discovered in 1974 when astronomers noticed that the pulse period of the pulsar PSR 1913 + 16, normally 0.059 second, was changing (Figure 11-8). When plotted

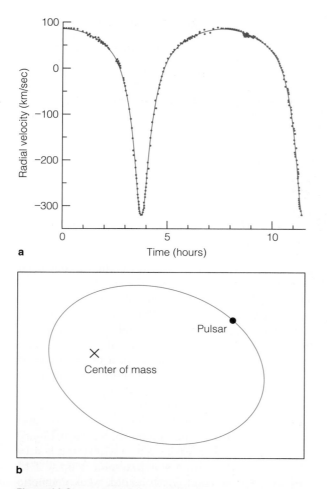

Figure 11-8 (a) The period of pulsation of pulsar PSR 1913 + 16 varies over an interval of 7.75 hours. If we treat this interval as the result of the Doppler effect, the resulting radial velocity curve resembles that of a spectroscopic binary. (b) Fitting an orbital solution (solid line) to the data points reveals that the pulsar revolves in an elliptical orbit around the center of mass of a binary star system.

against time, the period changes repeated every 7.75 hours.

Thinking of the Doppler shifts seen in the spectra of spectroscopic binaries, the radio astronomers realized that the pulsar had to be in a binary system with a period of 7.75 hours. When the orbital motion of the pulsar carries it away from the earth, we see the period slightly lengthened, just as the wavelength of light emitted by a receding source is lengthened. Then, when the pulsar rounds its orbit and approaches the earth, we see

the period slightly shortened. Reasonable assumptions about the size and mass of the companion star, combined with a detailed analysis of the orbital motion, suggest that the mass of the neutron star is about 2 $M_\odot$, in good agreement with theory.

The binary pulsar PSR 1913 + 16 is one of at least seven known binary pulsars. Such systems are now relatively quiet, but they may have once been violently active. The companion star was probably once a giant star losing mass to the neutron star, and that matter falling into the neutron star would have liberated tremendous energy. A single marshmallow dropped onto the surface of a neutron star from a distance of 1 AU would hit with an impact equivalent to a 3-megaton nuclear warhead. Even a small amount of matter transferred to a neutron star can generate high temperatures and release X rays and gamma rays.

Hercules X-1 is an example of such a system. Like radio signals from a pulsar, the X rays from this system arrive in pulses with a period of 1.2372253 seconds. Unlike radio pulsars, however, the pulses from Hercules X-1 vanish entirely for 5.8 hours every 1.7 days (Figure 11-9).

The origin of the pulses is still uncertain, but the most popular theory supposes that matter transferred from a companion star to a neutron star becomes entangled in its magnetic field and flows along the field to the magnetic poles, where its impact generates beams of X rays. The spinning neutron star sweeps the X-ray beams around the sky, producing the observed pulses.

Hercules X-1 contains a 2 $M_\odot$ star whose surface temperature is about 7000 K. The star is variable because the X rays from the neutron star heat one side of the star to about 20,000 K. As the system revolves, we see first the hot side and then the cool side of the star (Figure 11-10).

Doppler shifts in the star's spectrum show that the orbital period of the system is 1.7 days, which explains the disappearance of the X-ray pulses for 5.8 hours every 1.7 days. The system is an eclipsing binary, and the neutron star disappears behind the companion every 1.7 days. Analysis of the orbit tells us that the mass of the compact object is about 1.3–1 $M_\odot$, well below the limit for neutron stars.

Other complex variations appear to arise within the accretion disk. The X-ray activity varies with a period of 35 days, which theorists believe is the time it takes the disk to precess on its axis. Also, the

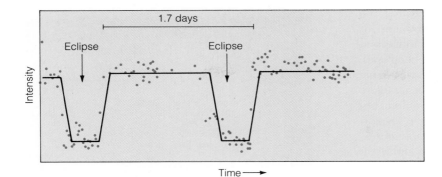

a

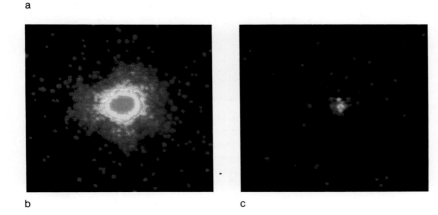

b   c

Figure 11-9 (a) X rays from Hercules X-1 disappear as the X-ray pulsar is eclipsed behind its stellar companion. (b and c) X-ray images show Hercules X-1 when it is "on" and "off." (Courtesy J. Trümper, Max-Planck Institute)

X rays and occasional bursts of gamma rays flicker with a period of 1.24 seconds, apparently the rotation period of the neutron star. Irregular gamma-ray bursts seem to occur because of irregularities in the flow of matter into and through the disk. Such variations in the accretion rate are common characteristics of these exotic X-ray binaries.

Just as normal stars in binary systems can evolve in peculiar ways when mass flows from one star to the other, neutron stars in binaries can alter the evolution of their companions. For example, the X-ray source 4U1820-30 is a binary system in which a neutron star orbits a white dwarf with a period of only 11 minutes. The separation between the stars is about a third the distance between the earth and moon. To explain how such a very close pair could originate, theorists suggest that a neutron star collided with a giant star and went into an orbit *inside* the star. The neutron star ate away the giant's envelope from the inside, leaving the white dwarf behind. Matter still flows from the white dwarf into an accretion disk around the neutron star.

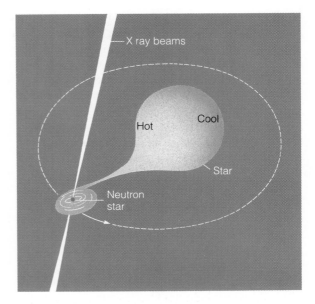

Figure 11-10 In Hercules X-1, X rays from the X-ray pulsar heat the near side of the star to 20,000 K. The rotating system shows us the hot side and the cool side of the star alternately, varying the visual brightness.

Another exotic millisecond pulsar, called the Black Widow, may solve a difficult problem for astronomers. If we are to believe that millisecond pulsars are old neutron stars that have been spun up in binary systems, we must explain why some millisecond pulsars have no binary companion. The Black Widow helps solve this problem because it is a fast pulsar (622 pulses per second) with a low-mass companion. Spectroscopic observations reveal that the blast of radiation and high-energy particles from the neutron star is boiling away the surface of the companion. The Black Widow pulsar is destroying its companion and will eventually be left as a solitary millisecond pulsar.

Modern astronomers have confirmed that neutron stars are real. Originally predicted theoretically, neutron stars have been firmly linked to observed pulsars. In the next section, we will try to repeat the process for the most massive compact objects. That is, we will try to link the theoretically predicted black holes with observed objects.

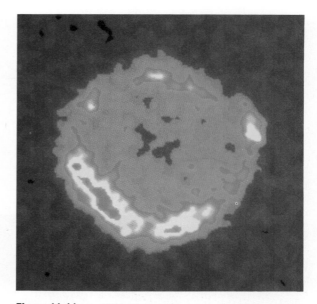

Figure 11-11 In 1572, Tycho Brahe saw a new star in the heavens, and the modern X-ray telescope EXOSAT reveals an expanding supernova remnant at the site of Tycho's star. But no star or pulsar remains. In some cases, supernova explosions may leave behind objects so massive they collapse into black holes rather than neutron stars. Whether that is true of Tycho's supernova, no one knows. (Courtesy J. Trümper, Max-Planck Institute)

11-2 BLACK HOLES

If, after a supernova explosion, the mass of the collapsing star is more than about $3\ M_\odot$, no known force in nature can stop it (Figure 11-11). The object reaches white-dwarf density, but the degenerate electrons cannot support the weight, and the collapse continues. When the object reaches neutron star density, the degenerate neutrons cannot support the weight, and the collapse goes on. The object quickly becomes smaller than an electron. No known force can stop gravity from squeezing the object to zero radius and infinite density.

As an object shrinks, its density and the strength of its gravity at its surface increases, and when an object shrinks to zero radius, its density and gravity become infinite. Mathematicians call such a point a **singularity.** In physical terms, we have difficulty thinking about infinite density and zero-radius objects, but even if such objects exist, they may not be visible to us. Theory predicts they will be hidden inside a region of space called a **black hole.**

Although black holes are difficult to discuss without drawing on the theory of general relativity and on sophisticated mathematics, we can use common sense and some simple physics to see why they form. Finding the velocity we need to escape from the gravity around a celestial body will help explain how black holes were first predicted theoretically and how they might be detected.

Escape Velocity Suppose we threw a baseball straight up. How fast must we throw it if it is not to come down? Of course, gravity would pull back on the ball, slowing it, but if the ball were traveling fast enough to start with, it would never come to a stop and fall back. Such a ball would escape from Earth. The **escape velocity** is the initial velocity an object needs to escape from a celestial body (Figure 11-12).

Whether we are discussing a baseball leaving Earth or a photon leaving a collapsing star, the escape velocity depends on two things: the mass of the celestial body and the distance from the center of mass to the escaping object. If the celestial body had a large mass, its gravity would be strong and we

Figure 11-12 Escape velocity, the velocity needed to escape from a celestial body, depends on mass. The escape velocity at the surface of a very small body would be so low we could jump into space. The escape velocity of the earth is much larger, about 11 km/sec (25,000 mph).

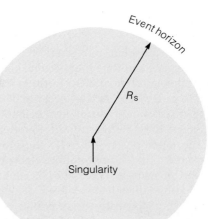

Figure 11-13 A black hole forms when an object collapses to a small size (perhaps to a singularity) and the escape velocity in its neighborhood is so great light cannot escape. The boundary of this region is called the event horizon because any event that occurs inside is invisible to outside observers. The radius of the region R_s is the Schwarzschild radius.

would need a high velocity to escape, but if we began our journey farther from the center of mass, the velocity needed would be less. For example, to escape from Earth, a spaceship would have to leave its surface at 11 km/sec (25,000 mph), but if we could launch spaceships from the top of a tower 1000 miles high, the escape velocity would be only 8.8 km/sec (20,000 mph). If we could make an object massive enough and/or small enough, its escape velocity could be greater than the speed of light. Such an object could never be seen because light could never leave it.

Rev. John Mitchell, a British amateur astronomer, was the first person to realize that Newton's laws of gravity and motion contained this implication. In 1783, he pointed out that an object 500 times the radius of the sun but of the same density would have an escape velocity greater than the speed of light. Then, "all light emitted from such a body would be made to return towards it." Mitchell had discovered the black hole.

Schwarzschild Black Holes If the core of a star collapses and contains more than 2–3 $M_\odot$, it will continue to collapse to a singularity and form a black hole. Some theorists believe that a singularity is impossible and that when we better understand the laws of physics, we will discover that the collapse halts before diameter zero.

It makes little difference to us, however. If the object becomes small enough, the escape velocity nearby is so large that no light can escape. We can receive no information about the object or about the volume of space near it, and we refer to this volume as a black hole. The boundary of this region is called the **event horizon,** because any event that takes place inside the surface is invisible to an outside observer (Figure 11-13). To see how such a region can exist, we must consider general relativity.

In 1916, Einstein published a mathematical theory of space and time that became known as the general theory of relativity. Einstein's equations showed that gravity could be described as a curvature of space–time, and almost immediately the astronomer Karl Schwarzschild found a way to solve the equations to describe the gravitational field around a single, nonrotating, electrically neutral lump of matter. That solution contained the first general relativistic description of a black hole; nonrotating, electrically neutral black holes are now known as Schwarzschild black holes. In recent decades, theorists such as Stephen W. Hawking have found a way to apply the mathematical equations of the general theory of relativity to charged, rotating black holes. For our discussion the differences are minor, and we may proceed as if all black holes were Schwarzschild black holes.

Schwarzschild's solution showed that if matter is packed into a small enough volume, then space–time is curved back on itself. Objects can still follow paths that lead into the region, but no path leads out, so nothing can escape, not even light. Thus the inside of the black hole is totally beyond the view of an outside observer. The event horizon is the boundary between the isolated volume of space–time and the rest of the universe, and the radius of the event horizon is called the **Schwarzschild radius R_s**—the radius within which an object must shrink to become a black hole (Figure 11-13).

Although Schwarzschild's work was highly mathematical, his conclusion is quite simple. The Schwarzschild radius (in meters) depends only on the mass of the object (in kilograms):

$$R_s = \frac{G\,M}{c^2}$$

In this simple formula, G is the gravitational constant, M is the mass, and c is the speed of light. A bit of arithmetic shows that a 1 $M_\odot$ black hole will have a Schwarzschild radius of 1.5 km, a 10 $M_\odot$ black hole will have a Schwarzschild radius of 15 km, and so on (Table 11-2). Even a very massive black hole would not be very large.

Any object could be a black hole if it were smaller than its Schwarzschild radius. For example, if we could squeeze Earth to a diameter of about

TABLE 11-2
The Schwarzschild Radius

	Mass ($M_\odot$)	R_s
Star	10	15 km
Star	3	4.5 km
Star	2	3 km
Sun	1	1.5 km
Earth	0.000003	0.45 cm

1 cm, its gravity would be so strong it would become a black hole. Fortunately, Earth will not collapse spontaneously into a black hole because its mass is less than the critical mass of 2–3 $M_\odot$. Only exhausted stellar cores more massive than this can form black holes under the sole influence of their own gravity. In this chapter, we are interested in black holes that might originate from the deaths of massive stars. These would have masses larger than 3 $M_\odot$. In the following chapters, we will encounter black holes whose masses might exceed 10^6 $M_\odot$.

Do not think of black holes as giant vacuum cleaners that will pull in everything in the universe. A black hole is just a gravitational field, and at a reasonably large distance its force is quite small. If the sun were replaced by a 1 $M_\odot$ black hole, the orbits of the planets would not change at all. The gravity of a black hole becomes extreme only when we approach close to it. The gravitational field of the earth is not dangerous because we can never come closer than 6400 km (4000 miles) to its center. But were the earth compressed into a black hole only 9 mm in diameter, we could come much closer to the center of the gravitational field, and then we would feel extreme gravitation. Thus we can imagine that the universe contains numerous black holes. So long as we and other objects stay a safe distance from the black holes, they will have no catastrophic effects.

A Leap into a Black Hole Because we cannot observe black holes up close, let us examine their peculiar properties by using our imaginations.

Figure 11-14 Leaping feet first into a black hole. A person of normal proportions (left) would be distorted by tidal forces (right) long before reaching the event horizon around a typical black hole of stellar mass. Tidal forces would stretch the body lengthwise while compressing it horizontally. Friction from this distortion would heat the matter to high temperatures.

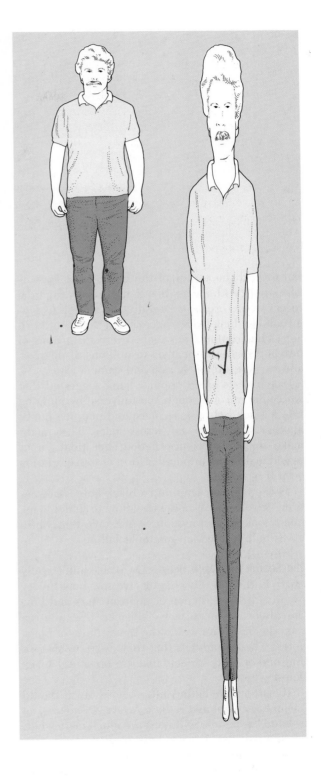

Let's imagine that we leap, feet first, into a Schwarzschild black hole.

If we were to leap into a black hole of a few solar masses from a distance of an astronomical unit, the gravitational pull would not be very large, and we would fall slowly at first. Of course, the longer we fell and the closer we came to the center, the faster we would travel. Our wristwatches would tell us that we fell for about 65 days before we reached the event horizon.

Our friends who stayed behind would see something different. They would see us falling more slowly as we came closer to the event horizon because, as explained by general relativity, clocks slow down in curved space–time. This is known as **time dilation.** In fact, our friends would never actually see us cross the event horizon. To them we would fall more and more slowly until we seemed hardly to move. Generations later, our descendants could focus their telescopes on us and see us still inching closer to the event horizon. We, however, would have sensed no slowdown and would conclude that we had crossed the event horizon after only about 65 days.

Another relativistic effect would make it difficult to see us with normal telescopes. As light travels out of a gravitational field, it loses energy, and its wavelength grows longer. This is known as the **gravitational red shift.** Although we would notice no effect as we fell toward the black hole, our friends would need to observe at longer and longer wavelengths in order to detect us.

While these relativistic effects seem merely peculiar, other effects would be quite unpleasant. Imagine again that we are falling feet first toward the event horizon of a black hole. We would feel our feet, which would be closer to the black hole, being pulled in more strongly than our heads. This is a tidal force, and at first it would be minor. But as we fell closer, the tidal force would become very large (Figure 11-14). Another tidal force would compress us as our left side and our right side both

TABLE 11-3
Four Black-Hole Candidates

Object	Location	Companion Star	Orbital Period	Mass of Compact Object
Cygnus X-1	Cygnus	O supergiant	5.6 days	6–11 $M_\odot$
LMC X-3	Dorado	B3 main-sequence	1.70 days	10 $M_\odot$
A0620-00	Monoceros	K main-sequence	7.752 hours	More than 3.18 $M_\odot$
V404 Cygni	Cygnus	G-K main-sequence	6.47 days	More than 6.26 $M_\odot$

fell toward the center of the black hole. For any black hole with a mass like that of a star, the tidal forces would crush us laterally and stretch us longitudinally long before we reached the event horizon. The friction from such severe distortions of our bodies would heat us to millions of degrees, and we would emit X rays and gamma rays.

Some years ago a popular book suggested that we could travel through the universe by jumping into a black hole in one place and popping out of another somewhere far across space. That might make for good science fiction, but tidal forces would make it an unpopular form of transportation even if it worked.

Our imaginary leap into a black hole is not entirely frivolous. We now know how to find a black hole. Look for a strong source of X rays. It may be a black hole into which matter is falling.

The Search for Black Holes Do black holes really exist? Earlier in this chapter, we saw how theory predicted the existence of neutron stars and how the discovery of pulsars confirmed that neutron stars do exist. Can we do the same thing for black holes? Theory predicts that they exist, but can we find one or more objects that are obviously black holes?

If matter were falling into a black hole, it should become very hot and radiate X rays. Of course, an isolated black hole will not have much matter falling inward, but a black hole in a binary system could gain mass from its companion, form a very hot accretion disk, and emit intense X rays. Thus, to find possible black holes, we must look at X-ray

binary systems. Some, like Hercules X-1, will contain neutron stars, but some may contain black holes.

We can tell the difference between a neutron star and a black hole in an X-ray binary if we can find the mass. Neutron stars must be less than 3 $M_\odot$. Any compact object in an X-ray binary with a mass of 3 $M_\odot$ or more must be a black hole.

So far astronomers have suggested four X-ray binaries as strong candidates for black holes (Table 11-3). In each case, the X rays are generated by matter flowing into a hot accretion disk around a compact object more massive than we would expect for a neutron star.

Cygnus X-1, the first X-ray object discovered in Cygnus, is apparently a hot supergiant O star and a compact object orbiting each other with a period of 5.6 days. The X rays are generated when matter in the supergiant's stellar wind flows into an accretion disk around the compact object (Figure 11-15). Although the compact object is invisible, we can estimate its mass from the Doppler shifts in the spectrum of the visible supergiant. If we assume that the supergiant has a mass of about 30 $M_\odot$, typical for a star of that temperature and luminosity, then the compact object must have a mass of 6–11 $M_\odot$, with 9 $M_\odot$ being most probable. This is well above the maximum mass for a neutron star.

Nevertheless, some observers have argued that the supergiant is not a normal star, that a third star may be present in the system, and so on. These possibilities mean that the compact object could, under certain peculiar circumstances, be a neutron

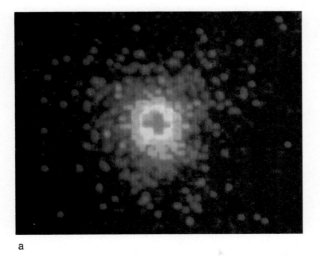

a

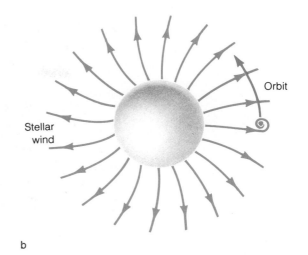

b

Figure 11-15 The X-ray source Cygnus X-1 (a) is generally believed to be a supergiant O star in a binary system with a black hole (b). Some of the gas from the O star's stellar wind flows into an accretion disk around the black hole, and the hot accretion disk emits the X rays we detect. (a. Courtesy J. Trümper, Max-Planck Institute)

star. Thus Cygnus X-1 is a good candidate for a black hole, but the evidence is not conclusive.

LMC X-3 is another black-hole candidate. (The letters LMC refer to the Large Magellanic Cloud, a small galaxy near our own Milky Way galaxy.) The compact object in LMC X-3 has a mass of at least 7 $M_{\odot}$ and probably 10 $M_{\odot}$, twice the mass of its companion. The system is a powerful source of X rays, and the visible light from the companion star is variable because the gravity of the compact object distorts the star into an egg shape. As the distorted star rotates, its total brightness varies. Yet the compact object is not detectable at visual wavelengths. Any reasonably normal star of that mass would be visible, so most astronomers now accept LMC X-3 as a black hole.

A third candidate is the X-ray system A0620-00 in the constellation Monoceros. It is only 1000 pc away and contains an ordinary main-sequence K star. The star and compact object orbit each other once every 7.75234 hours, and the light of the K star is variable because it is distorted into an egg shape by its companion. Analysis shows that the mass of the compact object must be greater than 3.18 $M_{\odot}$ and may be more than 7.3 $M_{\odot}$. Although this is slightly lower than the mass of the compact objects in Cygnus X-1 and LMC X-3, the K star seems normal in every way, and thus it is easier to

analyze the system and find the mass of the compact object. Consequently, A0620-00 seems a likely candidate for a black hole.

The most conclusive candidate known, V404 Cygni, drew astronomers' attention in 1989 when it emitted a powerful burst of X rays. By 1992, astronomers were able to prove that it consisted of a compact object orbiting a low mass companion star of spectral type G or K with a period of 6.47 days. Because the companion is such a low mass star, assumptions about its mass do not significantly affect the calculated mass of the compact object, which must be at least 6.26 $M_{\odot}$ and may be as large as 8–15 $M_{\odot}$—at least twice the mass limit for a neutron star.

Theory appears to be confirmed. Black holes do exist. In later chapters, we will discuss the possibility that black holes containing millions, perhaps billions, of solar masses lie in the cores of galaxies. For now, however, we will consider an object that is of interest not because it might contain a black hole, but because it shows how a compact object can produce tremendous energy when matter falls in at a high rate.

SS 433 What could be more exotic than a black hole devouring its stellar companion? One X-ray binary, SS 433, is so exotic it is almost grotesque. Its

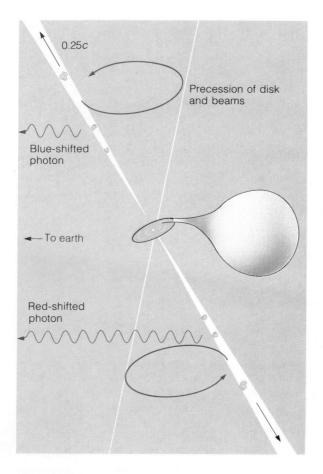

0.25c

Precession of disk
and beams

Blue-shifted
photon

To earth

Red-shifted
photon

Figure 11-16 The generally accepted model of SS 433 includes a compact object with a very hot accretion disk producing beams of radiation with embedded blobs of gas traveling at 25 percent the velocity of light. The precession of the disk swings the beams around a conical path every 164 days.

optical spectrum shows sets of spectral lines Doppler-shifted by about one-fourth the speed of light with one set shifted to the red and one set shifted to the blue. The object is both receding and approaching at fantastic speed.

Apparently, SS 433 is a binary system in which a massive compact object pulls matter from its companion star and forms an extremely hot accretion disk. The disk is so hot it is blasting radiation and matter away in beams aimed in opposite directions. The precession of the disk sweeps these beams around the sky once every 160 days, and we see both a red and a blue shift (Figure 11-16).

SS 433 is complex and probably contains a neutron star, but that is not why it is so important to modern astronomy. It is a prototype that illustrates how the gravitational field around a compact object such as a black hole can produce powerful beams of radiation and matter. We will see such phenomena 100 times more powerful when we begin to explore the galaxies—the topic of Unit 3.

SUMMARY

If the remains of a star collapse with a mass greater than the Chandrasekhar limit of 1.4 $M_\odot$, then the object cannot reach stability as a white dwarf. It must collapse to the neutron star stage, with a radius of 10–15 km and a density equal to that of an atomic nucleus. Such a neutron star can be supported by the pressure of its degenerate neutrons. But if the mass is greater than 2–3 $M_\odot$, then the degenerate neutrons cannot stop the collapse, and the object must become a black hole.

Theory predicts that a neutron star should rotate very fast, be very hot, and have a strong magnetic field. Such objects have been identified as pulsars, sources of pulsed radio energy. Pulsars are evidently spinning neutron stars, which emit beams of radiation from their magnetic poles. As they spin, they sweep the beams around the sky, and if the beams sweep over Earth, we detect pulses. This is known as the lighthouse theory. The spinning neutron star slows as it radiates energy into space, and glitches in its pulsation appear to be caused by starquakes in the crust of the neutron star.

A few pulsars have been found in binary systems, which allows astronomers to estimate the masses of the pulsars. Such masses are consistent with the predicted masses of neutron stars. In some binary systems such as Hercules X-1, mass flows into a hot accretion disk around the neutron star and causes the emission of X rays.

If a collapsing star has a mass greater than 2–3 $M_\odot$, then it must contract to a very small size— perhaps to a singularity, an object of zero radius. Near such an object, gravity is so strong not even light can escape, and we term the region a black hole. The surface of this region, called the event horizon, marks the boundary of the black hole. The Schwarzschild radius is the radius of this event horizon, amounting to only a few kilometers for black holes of stellar mass.

If we were to leap into a black hole, we would experience peculiar effects. Our friends who stayed behind would see two relativistic effects. Our clocks would slow down relative to our friends' clocks because of time dilation in the strong gravitational field. Also, the gravitational red shift would cause the light we emitted to be shifted to longer wavelengths. As we fell into the black hole, we would feel tidal forces that would deform and heat our mass until we grew hot enough to emit X rays. Any X rays emitted before we crossed the event horizon could escape.

To search for black holes, we must look for binary star systems in which mass flows into a compact object and emits X rays. If the mass of the compact object is greater than $2-3$ $M_\odot$, then the object is presumably a black hole. Four such objects have been located.

SS 433 is an X-ray binary that illustrates how rapid mass transfer can generate tremendous energy. Apparently, mass flows rapidly into an accretion disk around a compact object, perhaps a black hole, and the hot accretion disk ejects beams of radiation and gas in opposite directions at about 25 percent the speed of light.

NEW TERMS

neutron star

pulsar

glitch

lighthouse theory

superconductor

millisecond pulsar

singularity

black hole

escape velocity

event horizon

Schwarzschild radius R_s

time dilation

gravitational red shift

REVIEW QUESTIONS

1. Why is there an upper limit to the mass of neutron stars? Why is that upper limit not well known?

2. Explain in detail why we expect neutron stars to be hot, to spin fast, and to have strong magnetic fields.

3. Why does the short length of pulsar pulses eliminate normal stars as possible pulsars?

4. What observations support the lighthouse theory?

5. According to our model of a pulsar, if a neutron star formed with no magnetic field at all, could it be a pulsar? Why or why not?

6. Why did astronomers first assume that the millisecond pulsar was very young?

7. Why do we suspect that only very fast pulsars can emit visible pulses?

8. If the sun were replaced by a 1 $M_\odot$ black hole, how would the earth's orbit change?

9. What evidence do we have that black holes exist?

10. How do novae, Hercules X-1, and Cygnus X-1 resemble each other? How do they differ?

DISCUSSION QUESTIONS

1. Has the existence of neutron stars been sufficiently tested to be called a theory, or should it be called a hypothesis? What about the existence of black holes?

2. Why couldn't an accretion disk orbiting a giant star get as hot as an accretion disk orbiting a compact object?

PROBLEMS

1. If a neutron star has a radius of 10 km and rotates 642 times a second, what is the speed of the surface at the neutron star's equator in terms of the speed of light? (HINT: The circumference of a circle is $2\pi R$.)

2. A neutron star and a white dwarf have been found orbiting each other with a period of 11

minutes. If their masses are typical, what is the average distance between them? (HINT: See Box 8-5.)

3. If the earth's moon were replaced by a typical neutron star, what would the angular diameter of the neutron star be, as seen from Earth? (HINT: See Box 3-2.)

4. What is the Schwarzschild radius of Jupiter (mass = 2×10^{27} kg)? of a human adult (mass = 75 kg)? (HINT: See Appendix C for the values of G and c.)

5. If the inner accretion disk around a black hole has a temperature of 10^6 K, at what wavelength will it radiate the most energy? (HINT: See Box 6-2.)

6. What is the orbital period of a bit of matter in an accretion disk 2×10^5 km from a 10 $M_\odot$ black hole? (HINT: See Box 8-5.)

7. If SS 433 consists of a 20 $M_\odot$ star and a neutron star orbiting each other every 13.1 days, then what is the average distance between them? (HINT: See Box 8-5.)

RECOMMENDED READING

BACKER, DONALD, and SHRINIVAS R. KULKARNI. "A New Class of Pulsars." *Physics Today* 43 (March 1990), p. 26.

BAILYN, CHARLES. "Problems with Pulsars." *Mercury* 20 (March/April 1991), p. 55.

CLARK, DAVID H. *The Quest for SS 433*. New York: Viking Penguin, 1986.

DARLING, DAVID. "Space, Time, and Black Holes." *Astronomy* 8 (October 1980), p. 66.

———. "The Quest for Black Holes." *Astronomy* 11 (July 1983), p. 6.

GRAHAM-SMITH, SIR FRANCIS. "Pulsars Today." *Sky and Telescope* 80 (September 1990), p. 240.

GREENSTEIN, GEORGE. *Frozen Star*. New York: New American Library, 1984.

———. "Neutron Stars and the Discovery of Pulsars." *Mercury* 14 (March/April 1985), p. 34, and 14 (May/June 1985), p. 66.

HALL, DONALD E., "The Hazards of Encountering a Black Hole." *The Physics Teacher* 23 (December 1985), p. 540, and 24 (January 1986), p. 29.

HEWISH, A. "Pulsars After 20 Years." *Mercury* 18 (January/February 1989), p. 12.

MARGON, BRUCE. "Relativistic Jets in SS 433." *Science* 215 (15 January 1982), p. 247.

MARSCHALL, LAURENCE. *The Supernova Story*. New York: Plenum Press, 1988.

McCLINTOCK, JEFFREY. "Stalking the Black Hole in the Star Garden of the Unicorn." *Mercury* 14 (July/August 1987), p. 108.

———. "Do Black Holes Exist?" *Sky and Telescope* 75 (January 1988), p. 28.

PACZYNSKI, BOHDAN. "Binary Stars." *Science* 225 (20 July 1984), p. 275.

PARKER, BARRY. "In and Around Black Holes." *Astronomy* 14 (October 1986), p. 6.

RAWLEY, L. A., J. H. TAYLOR, M. M. DAVIS, and D. W. ALLAN. "Millisecond Pulsar PSR 1937 + 21: A Highly Stable Clock." *Science* 238 (6 November 1987), p. 761.

SCHORN, RONALD A. "SS 433—Enigma of the Century." *Sky and Telescope* 62 (August 1981), p. 100.

———. "Binary Pulsars: Back from the Grave." *Sky and Telescope* 72 (December 1986), p. 588.

SEEDS, M. A. "Stellar Evolution." *Astronomy* 7 (February 1979), p. 6. Reprinted in *Astronomy: Selected Readings*, ed. M. A. Seeds. Menlo Park, Calif.: Benjamin/Cummings, 1980, p. 95.

SEWARD, FREDERICK D. "Neutron Stars in Supernova Remnants." *Sky and Telescope* 71 (January 1986), p. 6.

SEXL, R., and H. SEXL. *White Dwarfs–Black Holes: An Introduction to Relativistic Astrophysics*. New York: Academic Press, 1979.

———. "Curved Space–Time near a Neutron Star." *Mercury* 9 (March/April 1980), p. 38.

TAARN, RONALD E. and BRUCE A. FRYXELL. "The Hydrodynamics of Accretion from Stellar Winds." *American Scientist* 77 (November/December 1989), p. 539.

VERSCHUUR, GERRIT. "On the Train of Exotic Pulsars." *Astronomy* 16 (December 1988), p. 22.

WHITE, NICOLAS E. "New Wave Pulsars." *Sky and Telescope* 73 (January 1987), p. 22.

SCIENCE FICTION

FORWARD, ROBERT L. *Starquake*. New York: Ballantine Books, 1985.

———. *Dragon's Egg*. New York: Ballantine Books, 1986.

NIVEN, LARRY. *Neutron Star*. New York: Ballantine Books, 1968.

UNIT 3

The Universe of
Galaxies

12

The Milky Way

We live inside one of the larger star systems in the universe. Our Milky Way galaxy is over 100,000 ly in diameter and contains over 100 billion stars. If Earth is the only inhabited planet in the galaxy, then the entire galaxy belongs to us, and, sharing equally, each person on Earth owns 50 stars plus assorted planets. Even if we must share with a few other inhabited planets, we are all vastly wealthy. Of course, we lack the transportation to visit our domains, and most people on this planet don't know how wealthy they are or even that they live inside a galaxy. In fact, it is only in the last century that the human race discovered the truth about the Milky Way.

The ancient Greeks named the faint band of light that stretches around the sky *galaxies kuklos*, the "milky circle." The Romans changed the name to *via lactea*, "milky road" or "milky way." Early in this century, astronomers

A hypothesis or theory is clear, decisive, and positive, but it is believed by no one but the man who created it. Experimental findings, on the other hand, are messy, inexact things, which are believed by everyone except the man who did that work.

Harlow Shapley,
Through Rugged Ways to the Stars

discovered that the faint band of the Milky Way is the glow of billions of stars in a great wheel-shaped star system, and that our system is only one of a great many such systems in the universe. Drawing on the Greek word for milk, we call them galaxies.

Almost every celestial object visible to our naked eyes is part of our Milky Way galaxy. Exceptions include the **Magellanic clouds,** small irregular galaxies located in the southern sky. They are not part of the Milky Way galaxy, but they do appear to be satellites of our galaxy. Another exception is the Andromeda galaxy, just visible to our unaided eyes as a faint patch of light in the constellation Andromeda.* Our galaxy probably looks much like the Andromeda galaxy (Figure 12-1a).

Consult the star charts at the end of this book to locate the Milky Way and the Andromeda galaxy in your night sky.

a

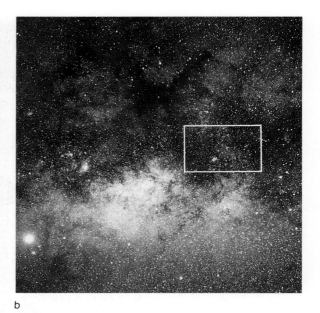

b

Figure 12-1 Our Milky Way galaxy. (a) If we could leave our galaxy and photograph it from a distance, it would look much like the Great Galaxy in Andromeda, a spiral galaxy about 2.3 million ly from us. (National Optical Astronomy Observatories) (b) Vast dust clouds block our view of the Milky Way. This photo looks toward the center of our galaxy (box) in the constellation Sagittarius. (Anglo-Australian Telescope Board)

Our own galaxy is difficult to study from our location inside it. At visual wavelengths, clouds of dust block our view like dark thunderclouds against a sunset (Figure 12-1b). Visual-wavelength photographs taken from Earth can explore only about 10 percent of our galaxy. But the dust clouds do not absorb strongly at the longer infrared or radio wavelengths. The **COBE** Satellite was able to see through the dust clouds and clearly photographed the thin disk of our galaxy (Figure 12-2).

As we explore our galaxy, we face three fundamental questions. First, we ask how our galaxy formed. We will discover that the chemical composition of the stars gives us an important clue to the history of the galaxy. Second, we wonder what causes the beautiful spiral arms that mark our galaxy and other galaxies like it. We will discover that star formation somehow creates the spiral arms.

And third, we will peer into the very center of our galaxy and ask what powerful object lies hidden there. Some astronomers suggest it is a black hole containing millions of solar masses.

As always when we try to answer scientific questions, we begin by gathering evidence for comparison with theories. In this case, our evidence includes the size, mass, and shape of the galaxy and the distribution of stars within it. Those observational facts point toward the answers to our three questions.

12-1 THE DISCOVERY OF THE GALAXY

Two hundred years ago, no one knew we lived in a galaxy. Astronomers speculated that stars were distributed more or less uniformly through space. In

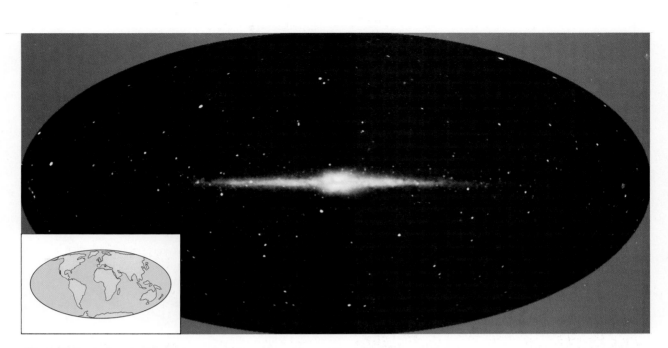

Figure 12-2 Just as the inset map projects the entire surface of the earth onto a single flat map, this infrared image from the COBE Satellite projects the entire sky onto a single oval. In the infrared, we can see through the dust in our galaxy, and thus we see the central bulge and thin disk. Although the Milky Way completely encircles the sky, the outer parts are not visible here. The center of the galaxy lies at the center of the image, and Orion lies at extreme right. (NASA)

the next 100 years, they concluded that all the stars we can see are part of a single star system only a few thousand parsecs in diameter. This system seemed to be vaguely wheel-shaped and roughly centered on the sun. Then, about one lifetime ago, a woman studying variable stars and a man studying star clusters unlocked one of nature's biggest secrets—that we live inside a galaxy.

Variable Stars Stars that change their brightness, so-called **variable stars,** are the keys to unlocking the secrets of our galaxy. Certain kinds of variable stars can reveal the distances to star clusters, which in turn can reveal the size and shape of our galaxy. Thus we begin with variable stars.

Certain stars can become variable after they leave the main sequence and swell into giants. Their evolution moves them back and forth in the giant region of the H–R diagram (see Chapter 10). If a star passes through a region called the **instability strip,** it can become unstable and start to pulsate as a variable star (Figure 12-3a).

Stars in the instability strip have the right temperature and luminosity to produce an energy-absorbing layer of gas in their outer envelopes. This layer stores and releases energy much as a spring does and makes the stars pulsate like beating hearts, expanding and contracting, growing slightly hotter and then slightly cooler.

The changes in surface area and temperature cause the brightness of these stars to vary (Figure 12-3b). We will refer to the time a star takes to go from bright to faint and back to bright as the **period of the variable star.** The periods range from a few hours to hundreds of days.

Though there are many types of variable stars, two kinds are especially important to our story of the discovery of the Milky Way galaxy. The **RR Lyrae variable stars,** named after variable star RR in the constellation Lyrae, have periods from 12 to 24 hours and are common in some star clusters. All RR Lyrae stars have approximately the same intrinsic brightness (absolute magnitudes of about +0.5).

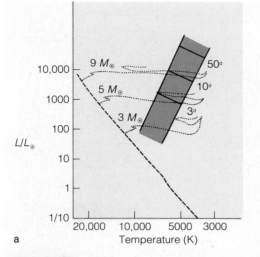

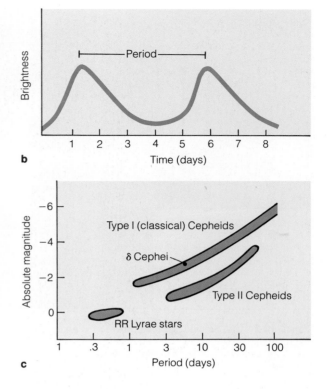

Figure 12-3 (a) When a star enters the instability strip (shaded), it becomes a variable star. Massive stars cross the strip at higher luminosities than lower-mass stars, and because of their higher mass and larger radius, they pulsate with longer periods. (b) A Cepheid variable can be identified by its brightness variation. (c) A period-luminosity diagram shows the different types of variables.

Another type of variable star is the **Cepheid,** named after δ Cephei, the first star of this type to be found. Cepheids have periods of from 1 to 60 days, but they do not all have the same absolute magnitude. More massive stars cross the instability strip higher in the diagram and are thus more luminous. Because of their greater mass and larger radius, they pulsate more slowly than the lower-mass stars. Consequently, there is a relation between the period and luminosity of Cepheid variables: the longer the period, the greater the luminosity.

First discovered by Harvard astronomer Henrietta Leavitt in 1912, this **period–luminosity relation** has become a powerful tool for the determination of distance. The period–luminosity diagram in Figure 12-3c shows the RR Lyrae stars and the two kinds of Cepheids. Type I Cepheids (also known as classical Cepheids) tend to be younger than Type II. An astronomer who identifies one of these stars in a star cluster or galaxy and determines the star's period of pulsation can find its absolute magnitude from the diagram. Comparing

this with the apparent magnitude yields the distance to the star cluster or galaxy (see Box 8-2).

The Size of the Galaxy In the years following World War I, a young astronomer named Harlow Shapley (1885–1972) began studying star clusters. He quickly combined his star clusters with Henrietta Leavitt's variable stars and discovered that the Milky Way system was much larger than anyone had supposed.

Shapley began by noticing that the two kinds of star clusters have different distributions in the sky. **Open star clusters** (Figure 12-4a) contain 100–1000 stars in a region 3–30 pc in diameter and are concentrated along the band of the Milky Way. The Pleiades (see Figure 9-4) is a well-known open cluster. **Globular star clusters** (Figure 12-4b) contain 100,000–1,000,000 stars crowded into a region about 25 pc in diameter and are not confined along the Milky Way. In fact, almost half of the known globular clusters are located in or near the constellation of Sagittarius (Figure 12-5). Shapley

a

b

Figure 12-4 (a) The "Jewel Box" is an open star cluster containing only a few hundred stars in a region about 8 pc in diameter. (b) The object known as 47 Tucanae is a large globular cluster containing over 1 million stars in a region about 50 pc in diameter—slightly larger than the typical globular cluster. (Both photos National Optical Astronomy Observatories)

assumed that this cloud of globular clusters was centered on the center of the galaxy. That assumption was, after all, quite logical. The motions of the globular clusters are ruled by the gravitation of the galaxy as a whole, and that gravitational field is, in effect, centered on the center of the galaxy. Thus Shapley's assumption was based on sound physical reasoning.

To estimate the size of the galaxy, Shapley had to find the distances to the globular clusters. They are clearly too far away to have measurable parallaxes, but they do contain variable stars. Henrietta Leavitt had studied such variable stars, but she knew only their apparent magnitudes. Cepheids are relatively rare, and there are none near enough to the sun to have measurable parallaxes. To find

the absolute magnitudes of the Cepheids, Shapley looked at their motions.

All stars are moving through space, and over periods of a few years these motions, known as **proper motions,** can be detected on photographs (Figure 12-6). The more distant stars have undetectably small proper motions, but the nearer stars have larger proper motions. Clearly, proper motions contain clues to distance. Although no Cepheids are close enough to have measurable parallaxes, Shapley found 11 with measurable proper motions. Through a statistical process, he was able to find their average distance and thus their average absolute magnitude. That allowed him to substitute absolute magnitudes for the apparent magnitudes on the period–luminosity diagram.

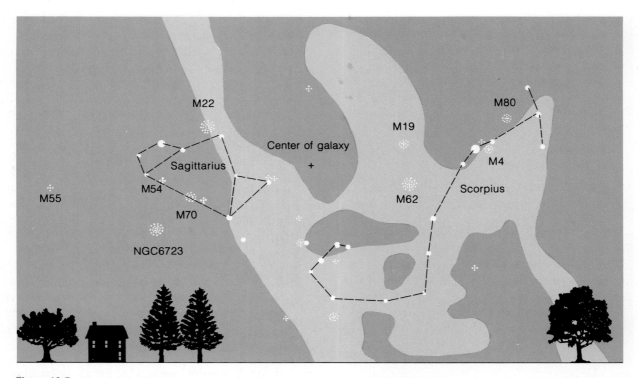

Figure 12-5 The globular clusters are concentrated toward the center of our galaxy. A few of the brighter globular clusters are shown here. The brightest are visible through binoculars on a dark night. The light blue region shows the extent of the visible Milky Way. Constellations are shown as they appear above the southern horizon on a summer night as seen from latitude 40°N, typical for most of the United States.

Figure 12-6 Photographs taken 10 years apart reveal the proper motion of one of the sun's nearest neighbors, Barnard's star, which is only 1.8 pc away. (Lick Observatory photograph)

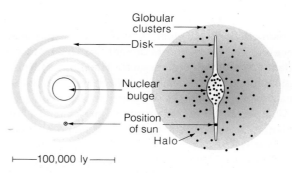

Figure 12-7 The Milky Way galaxy seen face on and edge on illustrates the shape and location of the disk, halo, and nucleus. Note the position of the sun and the distribution of the globular clusters.

In this way, Shapley **calibrated** the Cepheids for distance determinations. (The calibration of one parameter to find another is very common in astronomy.) From the diagram, he could determine the absolute magnitudes of the variable stars he saw in the globular clusters. The difference between the apparent and absolute magnitudes gave him the distance modulus and thus the distance to each cluster. (See also Box 8-2.)

When he plotted the direction of and distance to the globular clusters, he found that they were distributed in a swarm centered not on the sun, but on a point thousands of parsecs away in the direction of Sagittarius (Figure 12-7). He reasoned that since the motions of the clusters were dominated by the gravity of the entire galaxy, the center of the swarm should coincide with the center of the galaxy. Measuring the distance to this center revealed that the galaxy was much bigger than anyone had supposed. Earlier astronomers had underestimated the size of the galaxy because great clouds of dust and gas blocked their view.

Harlow Shapley was a brilliant and charming man. He may someday be as famous as Copernicus, for, like Copernicus, he changed the way we think of our place in nature. Before Shapley, we thought of our solar system as the center of the star system, perhaps the center of the universe. Shapley's finding placed us in the suburbs of a very large galaxy. As he said, "[It] is a rather nice idea because it means that man is not such a big chicken. He is incidental—my favorite term is 'peripheral.'"*

Soon other astronomers concluded that our Milky Way galaxy is only one of many such systems (which are discussed in Chapter 13). Thus we can think of our galaxy as representative of galaxies in general.

Components of the Galaxy Our galaxy, like many others, contains two components—a disk and a sphere (see Figure 12-7). Understanding the differences between these components will eventually suggest how our galaxy formed.

The **disk component** consists of all matter confined to the plane of rotation—that is, everything in the disk itself. This includes stars, star clusters, and nearly all of the galaxy's gas and dust.

We cannot quote a number for the thickness of the disk for two reasons. First, the disk does not have sharp boundaries. The stars become less crowded as we move away from the plane. Second, the thickness depends on the kind of star we study. The O stars lie within 300 ly of the galactic plane, but stars like the sun are much less confined. The disk defined by such stars is roughly 3000 ly thick.

The diameter of the disk and the position of the sun are also uncertain. Not only does the disk lack a sharp edge, but it is also heavily obscured by dust (see Figure 12-1b). Estimates place the diameter at 100,000 ly. Such distances are often expressed in **kiloparsecs (kpc).** Since 1 kpc equals 1000 pc (3260 ly), the diameter of our galaxy is about 30 kpc.

The position of the sun is uncertain because the dust clouds block our view of the center of the galaxy. Estimates of the distance to the center range from 10 kpc to 7 kpc or less. Recently, the International Astronomical Union urged astronomers to use a provisional distance of 8.5 kpc.

The most striking features of the disk component are the **spiral arms,** long spiral patterns of bright stars, star clusters, gas, and dust. Such spiral arms are easily visible in other galaxies, and we will see later that our own galaxy contains a spiral pattern.

The disk component contains two kinds of star clusters. The open clusters, described earlier in this chapter, are held together by their own gravitation, and although they do lose stars now and then, they can survive for a billion years or more. The disk also contains associations, groups of 10–100 stars so widely scattered in space that their mutual gravity cannot hold the association together. We find the stars moving together through space (Figure 12-8) because they were formed from a single gas cloud and have not yet wandered apart. Such associations are short-lived, and the stars eventually wander away. The youngest—O and B associations and T Tauri associations—are located along the spiral arms.

The second component of our galaxy is the **spherical component,** which includes all matter in our galaxy scattered in a roughly spherical distribution around the center. This includes the halo and the nuclear bulge.

The **halo** is a spherical cloud of thinly scattered stars and globular clusters. Because the halo con-

*Harlow Shapley, Through Rugged Ways to the Stars (New York: Scribner, 1969), p. 60.

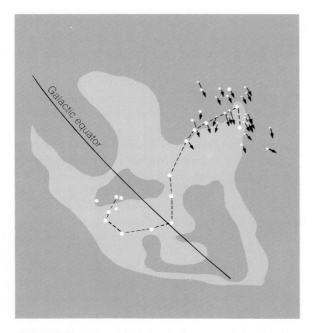

Figure 12-8 Many of the stars in the constellation Scorpius are members of an O and B association. They have formed recently from a single gas cloud and are moving together southwest along the Milky Way (light blue).

tains very little gas and dust, almost no star formation occurs there, and thus the halo contains very few young, bright stars. In fact, the vast majority of halo stars are old, cool, lower-main-sequence stars and giants much like the stars in globular clusters.

Because globular clusters contain so many stars in such small regions, the clusters are very stable and have survived for billions of years. From their H–R diagrams, we can estimate their ages at 10–14 billion years, making them the oldest clusters in our galaxy. Recent work suggests that some globular clusters may be as old as 18 billion years.

The **nuclear bulge** is the dense swarm of stars with a radius of about 2 kpc surrounding the center of the galaxy (see Figure 12-7). Although it is slightly flattened, we include it in the spherical component. Its stars are similar to halo stars, and it tends to lack gas and dust. We cannot observe it easily because of thick dust in the disk, which blocks our view, but observations at radio, infrared, and X-ray wavelengths can penetrate this dust. Studies over the last 10 years have revealed regions of star formation deep inside the nuclear bulge.

The disk, halo, and nuclear bulge contain vast numbers of stars. A basic property of any object is its mass, and so we must ask, What is the total mass of our galaxy?

The Mass of the Galaxy   When we needed to find the masses of stars, we studied the orbital motions of the pairs of stars in binary systems. To find the mass of the galaxy, we must look at the orbital motions of the stars within the galaxy. Every star in our galaxy follows an orbit around the center of mass of the galaxy, and thus we say the galaxy rotates. That rotation can reveal the total mass of the galaxy. Consequently, our discussion of the mass of the galaxy is also a discussion of the rotation of the galaxy and the orbits of the stars within the galaxy.

We can find the orbits of stars by finding how they move. Of course, we could use the Doppler effect to find a star's radial velocity (see Box 6-4). In addition, if we could measure the distance to and proper motion of a star, we could find the velocity of the star perpendicular to the radial direction. Combining all of this information, we could then find the shape of the star's orbit.

Disk stars move in nearly circular orbits that lie in the plane of the galaxy (Figure 12-9). The sun, for example, is a disk star and moves about 220 km/sec in the direction of Cygnus, carrying Earth and the other planets along with it. Since its orbit is a circle with a radius of 8.5 kpc, it takes about 240 million years to complete one orbit.

We can use the orbital motion of the sun to find the mass of the galaxy if we think of the sun and the center of the galaxy as two objects in orbit around each other. Of course, the distances are so vast, we notice no effect of this motion during our lifetimes or even over centuries. Just as in a binary star system, the period and the radius of the orbit can tell us the mass. In this case, we know the period is about 240 million years, and we know the radius is about 8.5 kpc, so the mass of the Milky Way galaxy must be about 100 billion solar masses.

This estimation is uncertain for a number of reasons. First, we don't know the radius of the sun's orbit with much certainty. We estimate it at 8.5 kpc, but it could be as low as 7 kpc or as high as 10 kpc. Second, this estimate includes only the mass inside the sun's orbit. Mass spread uniformly outside the sun's orbit will not affect its orbital

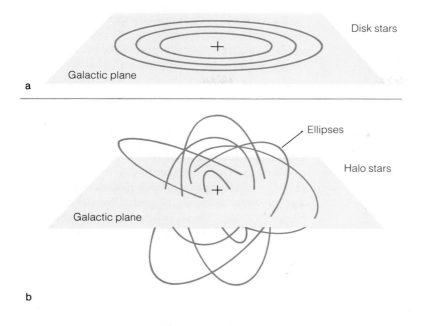

Disk stars

Galactic plane

a

Ellipses

Halo stars

Galactic plane

b

motion. Thus 100 billion solar masses is a lower limit for the mass of the galaxy, but we don't know how much to increase our estimate to include the rest of the galaxy.

This estimate is also uncertain because it depends on our knowledge of the rotation of the galaxy, and that rotation is very complex. The motion of the stars near the sun shows that the disk does not rotate as a solid body. Each star follows its own orbit, and stars in some regions have shorter or longer orbital periods than the sun. This is called **differential rotation** (Figure 12-10) and is different from the rotation of a solid body—a record on a turntable, for instance. Whereas three spots lined up on a record would stay together as the record turned, three stars lined up in the galaxy will draw apart because some have shorter orbital periods.

A graph of the orbital velocity of stars at various orbital radii in the galaxy is called a **rotation curve** (Figure 12-11). If all of the mass were concentrated at the center, then orbital velocity would be high near the center and would decline as we moved outward. This has been called *Keplerian motion* because it is described by Kepler's third law of planetary motion. Recall that the third law says that the square of the orbital period of a planet orbiting the sun is proportional to the cube of the radius of its orbit. If all of the mass in the galaxy were concentrated near its center, we would ex-

Galactic center

Sun

Figure 12-10   The differential rotation of the galaxy means that stars at different distances from the center have different orbital periods. In this example, the star just inside the sun's orbit has a shorter period and gains on the sun, while the star outside falls behind.

pect the rotation curve to obey the third law. That is, the orbital period squared should increase in proportion to the cube of the orbital radius. We would see this in the rotation curve as a decrease in the orbital velocity with increasing radius (Figure 12-11).

But the motion would not be Keplerian if the mass of the galaxy were not mostly concentrated near the center of the galaxy. If the mass were distributed uniformly through the galaxy, for example, then stars near the center would not enclose much mass inside their orbits and their velocities would be lower than expected. Stars farther

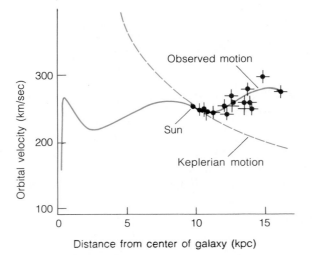

Figure 12-11 If most of the galaxy's mass were inside the sun's orbit, we would expect the rotation curve to obey Kepler's third law outside the sun's orbit. But the observed orbital velocities in the outer disk are higher than expected. This suggests that the galaxy contains significant amounts of mass outside the orbit of the sun. Points with error bars represent the orbital velocities of dense gas clouds in the disk of the galaxy. (Adapted from data by Leo Blitz)

from the center would enclose more mass inside their orbits and thus would have higher orbital velocities. We might see this non-Keplerian motion as a rotation curve that rises with increasing radius.

The observed rotation curve is quite complex (Figure 12-11), but it implies that much of the mass of our galaxy is invisible. If the visible disk and halo included all of the matter in our galaxy, then the rotation curve should rise to a peak of about 170 km/sec at 8 kpc and then decline. Observations clearly show that the rotation curve rises higher than predicted and does not decline at large distances. Thus the galaxy must include much more mass than what we can see.

The rising rotation curve seems to require that the galaxy be surrounded by an extended distribution of mass called the **galactic corona.** Estimates suggest this corona extends to 7 times the visible radius of the galaxy and contains 10–100 times the mass of the visible galaxy. We will see in the next chapter that other galaxies seem to have galactic coronas.

Such a corona is not just an extension of the halo stars, dust, or gas because we do not detect it at optical, infrared, or radio wavelengths. It must be composed of some "dark matter" that we cannot detect with existing telescopes. Some theorists now speculate that the corona is composed of massive, exotic subatomic particles, though such particles have not yet been detected either in the galaxy itself or in the laboratory. Thus the mass of the galaxy and the extent of the halo are uncertain.

Motion in the halo is quite different from that in the disk. Each halo star and globular cluster follows its own randomly tipped elliptical orbit (see Figure 12-9b). These orbits carry the stars and clusters far out into the spherical halo, where they move slowly, but when they fall back into the inner part of the galaxy, their velocities increase. Thus motions in the halo do not resemble a general rotation but are more like the random motions of swarming bees.

As halo stars pass through the disk at steep angles, they move across the orbits of disk stars. Because they do not move in the same direction as the sun, they seem to have unusually high velocities. Thus we may find halo stars in the disk (see Figure 12-7), but they are only passing through and can be recognized by their high velocities with respect to the sun.

The differences between the disk component and the spherical component illuminate our galaxy's past. We have already seen that the two components differ in number of stars, amount of gas and dust, and orbital shape and orientation. In the next section, we will discover that they also differ in chemical composition. That clue, combined with our knowledge of stellar evolution, will lead us to consider the formation of our galaxy.

12-2 THE ORIGIN OF THE MILKY WAY

Just as paleontologists reconstruct the history of life on Earth from the fossil record, astronomers try to reconstruct our galaxy's past from the fossil it left behind as it evolved. That fossil is the spherical component of the galaxy. The stars we see now in the halo formed when the galaxy was young. By studying those stars, their chemical composition,

TABLE 12-1
Stellar Populations

| | Population I | | Population II | |
	Extreme	Intermediate	Intermediate	Extreme
Location	Spiral arms	Disk	Nuclear bulge	Halo
Metals (%)	3	1.6	0.8	Less than 0.8
Shape of orbit	Circular	Slightly elliptical	Moderately elliptical	Highly elliptical
Average age (yr)	100 million and younger	0.2–10 billion	2–10 billion	10–14 billion

and their distribution in the galaxy, we can try to understand how the galaxy formed.

Stellar Populations   Near the end of World War II, astronomers realized that there were two types of stars in the galaxy. The type they were accustomed to studying was that located in the disk, like those stars near the sun. These they called **population I** stars. The second type, called **population II** stars, are usually found in the halo, in globular clusters, or in the central bulge. In other words, the two stellar populations are associated with the two components of the galaxy.

The stars of the two populations are very similar. They burn nuclear fuels and evolve in nearly identical ways. They differ only in their abundance of atoms heavier than helium, atoms that astronomers refer to collectively as **metals**. (Note that this is not the way the word *metal* is commonly used by nonastronomers.) Population I stars are metal-rich, containing 2–3 percent metals, while population II stars are metal-poor, containing only about 0.1 percent metals. The metal content of the star defines its population.

Population I stars belong to the disk component of the galaxy and are sometimes called disk population stars. They have circular orbits in the plane of the galaxy and are relatively young stars that formed within the last few billion years. The sun is a population I star, as are the Type I Cepheids discussed previously.

Population II stars belong to the spherical component of the galaxy and are sometimes called the halo population stars. These stars have randomly tipped, elliptical orbits, and are old stars. The metal-poor globular clusters are part of the halo population, as are the RR Lyrae and Type II Cepheids.

Since the discovery of stellar populations, astronomers have realized that there is a gradation between populations (Table 12-1). Extreme population I stars, like the stars in Orion, are found only in the spiral arms. Slightly less metal-rich population I stars, called intermediate population I stars, are located throughout the disk. The sun is such a star. Stars even less metal-rich, such as stars in the nuclear bulge, belong to the intermediate population II. The most metal-poor stars are those in the halo and in globular clusters. These are extreme population II stars.

Why do the disk and halo stars have different metal abundances? Those chemical differences tell astronomers that the stars have different ages and thus must have formed at different stages in the life of the galaxy. To use this clue to tell the story of our galaxy, we must discuss the cycle of element building.

The Element-Building Cycle   We saw in Chapter 10 how elements heavier than helium are built up by the nuclear reactions inside evolving stars. Theory suggests that atoms less massive than iron are built before the supernova explosion, and a small number of atoms heavier than iron are made by nuclear reactions during the high-density, high-temperature phase of the supernova explosion. A chart of the abundance of the elements in the universe illustrates the low abundance of atoms heavier than iron (Figure 12-12).

When the galaxy first formed, there should have been no metals because stars had not yet manufactured any. The gas from which the galaxy condensed must have been almost pure hydrogen (80 percent) and helium (20 percent). (Where the hydrogen and helium came from is a mystery we will save till Chapter 15.)

The first stars to form from this gas were metal-poor, and now, 10–14 billion years later, their spectra still show few metal lines. Of course, they may have manufactured some atoms heavier than helium, but because the stars' interiors are not mixed, those heavy atoms stay trapped at the centers of the stars where they were produced and do not affect the spectrum (Figure 12-13). The population II stars that we see today in the halo are the survivors of an earlier generation of stars to form in the galaxy.

Most of the first stars evolved and died, and the most massive became supernovae and enriched the interstellar gas with metals. Succeeding generations of stars formed from gas clouds that were more enriched, and each generation added to the enrichment through the supernovae of massive stars. By the time the sun formed, 5 billion years ago, the element-building process had added about 1.6 percent metals. Since then the metal abundance has increased further, and stars forming now, in the Orion Nebula for example, incorporate 2–3 percent metals and represent the metal-rich population I stars. Thus metal abundance varies between populations because of the production of heavy atoms in successive generations of stars.

The lack of metals in the spherical component tells us it is very old, a fossil left behind by the

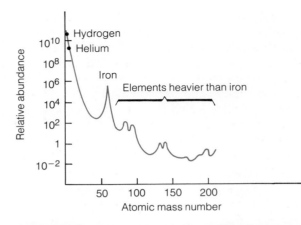

Figure 12-12 The abundance of the elements. Because they are made only during supernova explosions, the elements heavier than iron are rare.

galaxy when it was young and drastically different from its present disk shape. Thus the study of element building and stellar populations leads us to consider the first of our three questions about the galaxy: How did it form?

The Formation of the Galaxy   In the late 1950s and early 1960s, as astronomers came to understand populations and element building, they developed a theory that describes the formation of the galaxy. We will begin by discussing that theory, and then we will see how new observations are forcing astronomers to consider new, alternative theories.

The traditional theory says that early in the history of the universe, before stars made many metal atoms, the galaxy began as a roughly spherical cloud of hydrogen and helium gas contracting under the influence of its own gravity. As it contracted, its density grew until the gas began to fragment

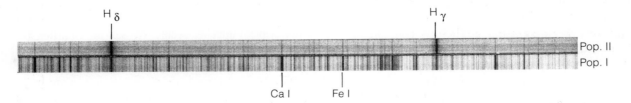

Figure 12-13   Spectra of population II and population I stars of similar spectral type show hydrogen lines of equal strength, but the lines of heavier atoms are conspicuously weak in the population II star's spectrum. Labeled lines are calcium (Ca) and iron (Fe). (Lick Observatory photograph)

into individual clouds. Because the original gas was turbulent, the clouds formed with random velocities. When the gas density grew high enough, some of the gas began to form clusters of metal-poor stars. Because the galaxy was approximately spherical at this time, these first star clusters formed in a spherical distribution that we see today as the halo.

These star clusters may not have looked like globular clusters. Many probably contained too few stars and were too scattered through space to hold themselves together. They gradually dissociated, freeing stars that wandered through the spherical cloud. Clusters that had more stars packed into a smaller volume survived, although they, too, lost members occasionally. Today, over 10 billion years since they formed, the surviving clusters are the highly stable globular clusters.

Although the galaxy began as a spherical distribution of gas, it immediately began to collapse into a rotating disk. Randomly moving eddies in the cloud collided with one another and the turbulent motions canceled out, leaving the galaxy with a uniform rotation. A low-density cloud of gas rotating uniformly around an axis cannot maintain a spherical shape. A star can remain spherical because it has high internal pressure to balance the weight of the gas, but in a cloud, where the density is low, this pressure is not effective. Like a blob of pizza dough spun into the air, the cloud must flatten into a rotating disk (Figure 12-14).

According to the traditional theory, this collapse of the gas into a disk took about 1 billion years and ended almost all star formation in the halo. Thus all halo stars and clusters should be about the same age. The collapse into a disk did not, however, alter the orbits of the stars that had formed when the galaxy was spherical. The halo population stars were left behind as a fossil of the early galaxy. Subsequent generations of stars formed in flatter distributions. The intermediate population I stars, for instance, are scattered hundreds of light-years above and below the plane of the disk. The gas distribution in the galaxy has now

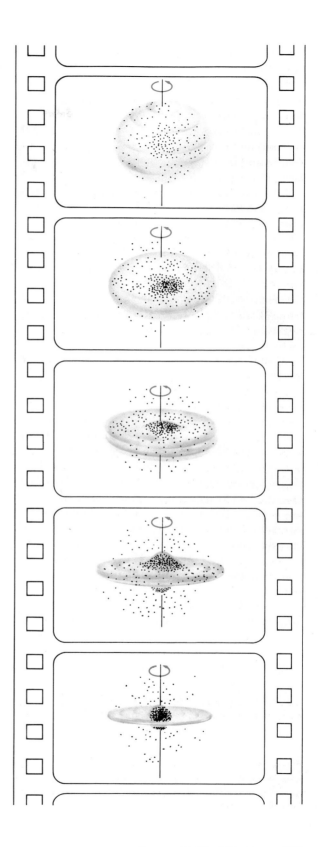

Figure 12-14 According to the traditional theory, the galaxy began as a spherical cloud of gas (shaded) in which stars and star clusters (dots) formed. As the rotating gas cloud collapsed into a disk, the halo stars were left behind as a fossil of the early galaxy.

become so flat that the newest stars, the extreme population I stars, are confined to a disk only 300 ly thick.

In addition to a change in distribution, the shapes of stellar orbits also changed as the galaxy flattened. When the galaxy was young, the turbulent gas moved at random and the stars that formed then took up orbits with random orientations and random shapes. As the galaxy collapsed into a disk, the random gas motion canceled out, and the gas took up more circular orbits, as did the forming stars. This explains why the oldest stars in the galaxy have the most elliptical orbits.

Simultaneously with these changes in distribution and orbital shape, the metal abundance in the stars grew with every generation, producing the stellar populations summarized in Table 12-1. The populations and their properties provide a record of our galaxy's past.

Even though this traditional theory has been very successful, new observations are forcing astronomers to consider alternative theories. For example, searches for extremely metal-poor stars have turned up a few that contain so few metals they seem to have formed before the globular clusters. One star contains 30,000 times less metals than the sun. Other research has shown that globular clusters have ages spanning a range of 3 billion years or more. The traditional theory says the globular clusters formed before the collapse of the gas cloud into a disk, a period of only 1 billion years. Apparently, the formation of the galaxy was a more complex process than traditionally thought.

New, so-called renegade theories are being developed. One, for example, proposes that the gas in the early universe formed small clouds, which made a few stars and fell together to build the larger, spherical, turbulent cloud that gave birth to the galaxy. Another theory suggests that our galaxy was born by the merger of two or more young galaxies. We will see in the next chapter that galaxies can merge with each other.

Although today's astronomers can tell the story of stellar evolution with some confidence, they cannot describe the evolution of our galaxy in equal detail. Even if we did know how the Milky Way galaxy formed and evolved, there would still be mysteries hidden in the star clouds of the galaxy. In the following sections, we will consider our two remaining questions—the cause of the spiral arms and the nature of the galactic core.

12-3 SPIRAL ARMS AND STAR FORMATION

The most striking feature of galaxies like the Milky Way is the system of spiral arms that wind outward through the disk. These arms contain swarms of hot, blue stars, clouds of dust and gas, and young star clusters. These young objects hint that the spiral arms involve star formation. As we try to understand the spiral arms, we face two problems. First, how can we be sure our galaxy has spiral arms when our view is obscured by dense clouds of dust? Second, why doesn't the differential rotation of the galaxy destroy the arms? The solution to both problems involves star formation.

Tracing the Spiral Arms Studies of other galaxies show us that spiral arms contain hot, blue stars. Thus one way to study the spiral arms of our own galaxy is to locate these stars. Fortunately, this is not difficult, since O and B stars are often found in associations and, being very bright, are easy to detect across great distances. Unfortunately, at these great distances their parallax is too small to measure, so their distances must be found by other means, usually by spectroscopic parallax (see Box 8-4).

O and B associations near the sun are not located randomly (Figure 12-15). They form three bands, indicating that there are three segments of spiral arms near the sun. If we could penetrate the dust clouds, we could locate other O and B associations and trace the spiral arms farther, but like a traveler in a fog, we see only the region near us.

Objects used to map spiral arms are called **spiral tracers.** O and B associations are good spiral tracers because they are bright and easy to see at great distances. Other tracers include young open clusters, clouds of hydrogen ionized by hot stars (emission nebulae), and certain kinds of variable stars.

Notice that all spiral tracers are young objects. O stars, for example, live for only a few million years. If their orbital velocity is about 250 km/sec, they cannot have moved more than about 500 pc since they formed. This is less than the width of a spiral arm. Since they don't live long enough to move away from the spiral arms, they must have formed there.

The youth of spiral tracers gives us an important clue to the nature of the arms. Somehow the arms

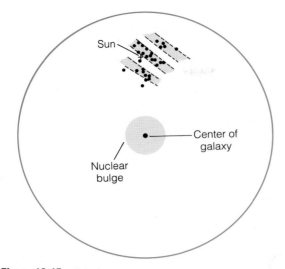

Figure 12-15 The O and B associations near the sun lie along three bands (shaded). These are segments of spiral arms.

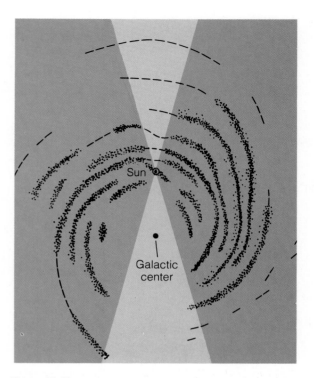

Figure 12-16 A map of our galaxy at 21-cm radiation shows traces of a spiral pattern in the distribution of cool hydrogen. (Courtesy Gerrit Verschuur)

are associated with star formation. Before we can follow this clue, however, we must extend our map of spiral arms to show the entire galaxy.

Radio Maps of Spiral Arms The dust clouds that block our view at visual wavelengths are transparent at radio wavelengths because radio waves are much longer than the diameter of the dust particles. When we point a radio telescope at a section of the Milky Way, we receive 21-cm radio signals coming from cool hydrogen in a number of spiral arms at various distances across the galaxy. Fortunately, the signals can be unscrambled by measuring the Doppler shifts of the 21-cm radiation, except in the directions toward and away from the nucleus. There the orbital motions of gas clouds are perpendicular to the line of sight, and all of the radial velocities are zero. Thus the radio map shown in Figure 12-16 reveals spiral arms throughout the disk of the galaxy, except in the wedge-shaped regions toward and away from the center.

Radio astronomers can also use the strong radio emission from carbon monoxide (CO) to map the location of giant molecular clouds in the plane of the galaxy. Recall from Chapter 9 that these clouds are sites of active star formation. Maps constructed from such observations reveal that the giant molecular clouds are located in large structures that resemble segments of spiral arms (Figure 12-17).

Radio maps combined with optical data reveal the spiral pattern of our galaxy (Figure 12-18). The segments we see near the sun are part of a spiral pattern that continues throughout the disk. But the spiral arms are rather irregular and are interrupted by bends, spurs, and gaps. The stars we see in Orion, for example, appear to be a detached segment of a spiral arm. There are significant sources of error in the radio mapping method, but many of the irregularities along the arms seem real, and photographs of nearby spiral galaxies show similar features. One study compared all of the available data on our galaxy's spiral pattern with the patterns seen in other galaxies and constructed a model of what our galaxy probably looks like from a distance (Figure 12-19).

The most important feature in the radio maps is easy to overlook—spiral arms are regions of higher gas density. Spiral tracers told us that the arms contain young objects, and we suspected active star formation. Now radio maps confirm our suspicion by telling us that the material needed to make stars is abundant in spiral arms.

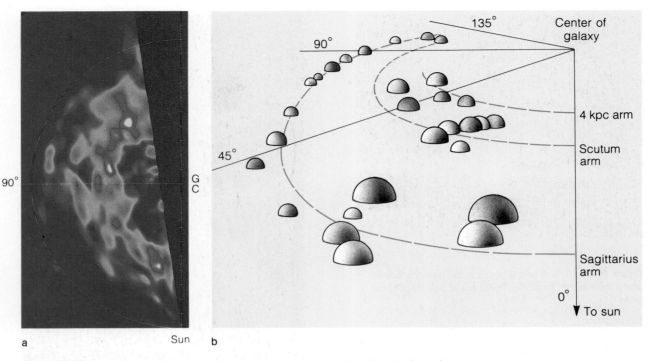

Figure 12-17 (a) The CO distribution in the galactic plane as seen from directly above the galactic center. (Dan P. Clemens) (b) The location of giant molecular clouds in our galactic plane is shown here as seen from a location 2 kpc directly above the sun. Angles in these diagrams are galactic longitudes measured clockwise from the sun. (Adapted from a diagram by T. M. Dame, B. G. Elmegreen, R. S. Cohen, and P. Thaddeus)

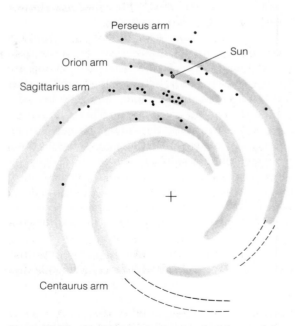

Figure 12-18 Combined optical and radio data reveal the overall spiral pattern of the Milky Way.

The Density Wave Theory Having mapped the spiral pattern, we can ask, What are spiral arms? We can be sure that they are not physically connected structures such as bands of a magnetic field holding the gas in place—a theory that was considered about three decades ago. Like a kite string caught on a spinning hubcap, such arms would be wound up and pulled apart by differential rotation within a few tens of millions of years. Yet spiral arms are common in disk-shaped galaxies and must be reasonably permanent features.

The most popular theory since the 1950s is called the **density wave theory.** It proposes that spiral arms are waves of compression rather like sound waves, which move around the galaxy, triggering star formation. Because these waves move slowly, orbiting gas clouds overtake the spiral arms from behind and create a moving traffic jam within the arms.

We can imagine the density wave as a traffic jam behind a truck moving slowly along a highway.

a

Figure 12-19 Our galaxy in perspective. (a) An observer a few million light-years away might have this view of our galaxy. The cross marks the position of the sun. (Painting by M. A. Seeds based on a study by G. De Vaucouleurs and W. D. Pence) (b) Like our own Milky Way, the galaxy M83 appears to have a strong two-armed spiral pattern with branches and spurs. (Anglo-Australian Telescope Board)

b

Seen from an airplane overhead, the jam seems a permanent though slow-moving feature. But individual cars overtake the truck, move up through the jam, await their opportunity, and pass the truck. So too do clouds of gas overtake the spiral density wave, become compressed in the "traffic jam," and eventually move out in front of the arm, leaving the slower-moving density wave behind.

Mathematical models of this process have been very successful at generating spiral patterns that look like our own and other spiral galaxies. In each case, the density wave takes on a regular two-armed spiral pattern winding outward from the nuclear bulge to the edge of the disk. In addition, the theory predicts that the spiral arms will be stable over a period of a billion years. The differential rotation does not wind them up because they are not physically connected structures.

Of course, star formation will occur where the gas clouds are compressed. Stars pass through the spiral arms unaffected, like bullets passing through a wisp of fog, but large clouds of gas slam into the spiral density wave from behind and are suddenly compressed (Figure 12-20). We saw in Chapter 9

that sudden compression could trigger the formation of stars in a gas cloud. Thus new star clusters should form along the spiral arms.

The brightest stars, the O and B stars, live such short lives that they never travel far from their birthplace and are found only along the arms. Their presence is what makes the spiral arms glow so brightly. Lower-mass stars, like the sun, live longer and have time to move out of the arms and continue their journey around the galaxy. The sun may have formed in a star cluster about 5 billion years ago when a gas cloud smashed into a spiral arm. Since that time, the sun has escaped from its cluster and made about 20 trips around the galaxy, passing through spiral arms many times.*

The density wave theory is very successful in explaining the properties of spiral galaxies, but it has two problems. First, what stimulates the formation of the spiral pattern? Of course, some pro-

*Some scientists have suggested that the periodic passage of the sun through the denser gas in spiral arms could affect the amount of sunlight reaching Earth and thus cause ice ages. This is an interesting but highly speculative idea.

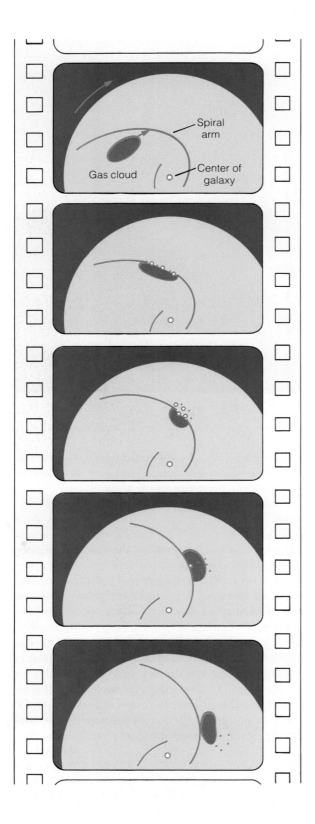

In the figure: Spiral arm, Gas cloud, Center of galaxy

cess must generate the spiral arms in the first place, but the pattern must be restimulated or it would die away over a period of a billion years or so. One possibility is that the galaxy is naturally unstable to certain disturbances, just as a guitar string is unstable to certain vibrations. Any sudden disturbance —the rumble of a passing truck, for example— can set the string vibrating at its natural frequencies. Similarly, minor fluctuations in the galaxy's disk or gravitational interactions with passing galaxies might generate a density wave.

The second problem with the density wave theory involves the spurs and branches in the arms of our own and other galaxies. Computer models of density waves produce regular, two-armed spiral patterns. Some galaxies, called grand-design galaxies, do indeed have symmetric two-armed patterns, but others do not. Other galaxies have a great many short spiral segments, giving them a fluffy appearance. These galaxies have been termed **flocculent,** meaning "woolly." Our galaxy is probably an intermediate galaxy. How can we explain these variations if the density wave theory always produces two-armed, grand-design, spiral arms? Perhaps the solution lies in an alternative theory that involves the compressional triggering of star formation.

Self-Sustaining Star Formation The formation of massive stars along a spiral arm could lead to the formation of more stars in a self-sustaining process that maintains the arms as semipermanent features of the galaxy. Recent computer simulations of this theory have been successful in producing flocculent spiral patterns.

Self-sustaining star formation may occur if the birth of a massive star compresses neighboring clouds of gas (Figure 12-21). This compression could be caused by the outward rush of heated gas as the massive star becomes luminous, or by the expansion of a supernova remnant produced by the death of a massive star. Since massive stars live such short lives, their birth and the production of a

Figure 12-20 According to the density wave theory, gas clouds overtake the spiral arm from behind and smash into the density wave. The compression triggers the formation of stars. The massive stars (open circles) are so short-lived that they die before they can leave the spiral arm. The less massive stars (dots) emerge from the front of the arm with the remains of the gas cloud.

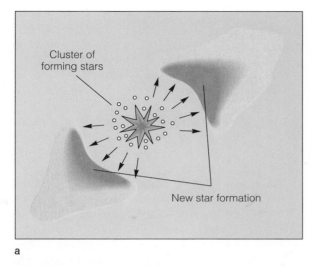

a

b

Figure 12-21 (a) Self-sustaining star formation may occur when a cluster of forming stars develops a massive member. The energetic birth of a massive star or the sudden expansion of a supernova remnant can compress nearby gas clouds and trigger new star formation. (b) The Eta Carinae Nebula is a region of active star formation in which massive stars near the center of the nebula are driving gas and dust outward. Such regions are likely sites of self-sustaining star formation. (National Optical Astronomy Observatories)

supernova are separated by a mere instant in the evolution of a galaxy. Whichever process compresses the neighboring gas clouds, star formation could begin in those clouds with sufficient compression. If star formation created additional massive stars, the process could be self-sustaining.

If self-sustaining star formation works—and we do see instances where it happens—then we should expect star formation to take place in patches throughout the disk of the galaxy. Differential rotation would pull these patches into segments of spiral arms because the inner edges of the patches would orbit about the center of the galaxy faster than the outer edges.

Computer simulations of self-sustaining star formation do produce patterns that have spiral features (Figure 12-22), but they do not have the symmetrical grand-design patterns we see in some galaxies. It seems likely that some galaxies are dominated by density waves and develop into beautiful two-armed spiral galaxies. Other galaxies may be dominated by self-sustaining star formation and thus lack grand-design patterns. Instead, those galaxies may be flocculent. The difference may lie

in the trigger for the spiral density wave. A galaxy that does not interact with other galaxies may not be able to generate a density wave and may consequently be flocculent.

Most galaxies, however, may be influenced by both mechanisms. Figure 12-23, for example, shows a model of spiral structure in a galaxy where a spiral density wave establishes the overall design and where star formation is triggered both by the collision of interstellar gas clouds with the spiral arm and by self-sustaining star formation. Our Milky Way galaxy may be similar, with a density wave producing a grand-design pattern and self-sustaining star formation creating the spurs and branches (see Figure 12-19a).

Spiral arms in the galaxy's disk are not entirely understood, but astronomers clearly feel they are on the right track. The very center of our galaxy, however, hides a much more difficult problem.

Figure 12-22 Comparison of a galaxy and a model. This ultraviolet image of spiral galaxy M 81 shows regions of active star formation along its spiral arms. A computer model based on self-sustaining star formation has been tipped to fit the galaxy, and red dots show regions of star formation in the model. The agreement between the galaxy and the model suggests that self-sustaining star formation can account for some spiral patterns. (NASA Astro image; model adapted from work by Humberto Gerola)

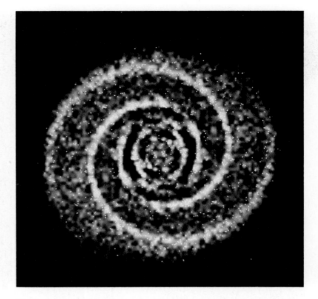

Figure 12-23 A computer model of a galaxy in which both the spiral density wave and self-sustaining star formation operate. Gray and white areas represent gas clouds. White dots represent young stars. (Reprinted courtesy William W. Roberts and *The Astrophysical Journal,* published by the University of Chicago Press; © 1984 The American Astronomical Society)

12-4 THE NUCLEUS OF THE GALAXY

The most mysterious region of our galaxy is its very center, the nucleus. Hidden behind thick clouds of gas and dust, it is totally invisible at optical wavelengths. But radio, infrared, X-ray, and gamma-ray signals penetrate the clouds and give us a picture of stars crowded together, a disk of gas spinning at the center, and clouds of gas rushing outward. The last of our three questions about the Milky Way galaxy is deceptively simple: What lies at the center? To attack that question we must once again compare observations with theories.

Observations If we examine the Milky Way on a dark night, we might notice a slight thickening in the direction of the constellation Sagittarius, but nothing specifically identifies that as the direction toward the heart of the galaxy (Figure 12-24). Even Shapley's study of globular clusters identified

the center of the galaxy only approximately. When radio astronomers turned their telescopes toward Sagittarius, they found a collection of radio sources, the most powerful of which, **Sagittarius A,** lies at the expected location of the galactic core (Figure 12-25).

Radio observations suggest that the central regions of the galaxy are now expanding. The Doppler shifts in the 21-cm radiation show that much of the neutral hydrogen is in a disk hundreds of parsecs in diameter that is rotating and expanding outward.

Observations at 21-cm wavelength show us where low-density gas is located, but radio observations of emission from the CO molecule show us the location of much denser gas clouds. Such observations reveal a ring of molecular clouds only 250 pc from the center of the galaxy, which are moving outward at 140 km/sec.

Infrared observations tell us that the stars are tremendously crowded in the central parsecs of

Figure 12-24 The Milky Way stretches from horizon to horizon in this photo of the entire sky. Cassiopeia is near the horizon at lower left and Sagittarius at upper right. Note the dark clouds of dust that obscure the shape of the Milky Way, and the slight thickening in Sagittarius, the direction toward the center of the galaxy. (Steward Observatory)

the galaxy. Infrared radiation at wavelengths shorter than two-millionths of a meter (2000 nm) comes almost entirely from cool stars. Observations at these wavelengths show that the stars are only about 1000 AU apart. Near the sun, stars are about 330,000 AU (1.6 pc) apart. Infrared photons with wavelengths longer than four-millionths of a meter (4000 nm) come almost entirely from interstellar dust warmed by the crowded stars.

Radio observations show that the central region contains an inclined clumpy disk of gas with a central cavity about 10 ly in diameter. Much of the gas is ionized by radiation from hot stars. Because the clumps should smooth out within about 100,000 years, astronomers suspect a central eruption drove most of the central gas outward, forming the cavity, within the last 100,000 years.

Some observations imply that a large amount of mass lies at the very center of the galaxy. For example, infrared emission from ionized neon at 12.8 microns (12,800 nm) shows that gases near the center are orbiting very rapidly. Only 0.3 pc (62,000 AU) from the center the orbital velocity is 200 km/sec. These gas clouds have orbital periods

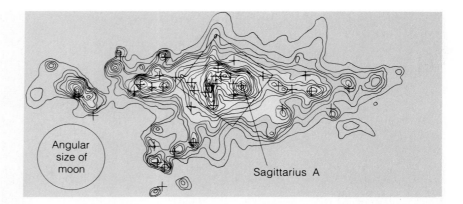

Figure 12-25 A radio map of the Sagittarius region of the Milky Way (see box in Figure 12-1) reveals an intense radio source, Sagittarius A, at the expected location of the center. Crosses mark far-infrared sources associated with star formation. Box marks boundaries of Figure 12-26a. Angular size of the moon shown for comparison. (Adapted from observations by W. L. Altenhoff, D. Downes, T. Pauls, and J. Schraml; and by S. F. Odenwalk and G. G. Fazio)

of about 9000 years, and Kepler's third law (see Box 8-5) tells us that the orbits must enclose almost 3 million solar masses. Seemingly, such a large amount of mass could not exist as stars and escape detection. But what could it be? If the mass is a single object, it must be very small. Radio interferometry shows that the central object must be no more than 20 AU in diameter.

What could be so massive and so small? Since the mid-1970s, astronomers have been asking themselves that question. One exciting theory involves a gigantic black hole.

The Massive Black Hole Theory To account for the nature of the center of our galaxy, astronomers must explain how a large amount of mass can occupy a small space, and a massive black hole fits the bill quite well. A black hole of 3 million solar masses would be only about 13 times larger than the sun—quite small in terms of the galaxy. Also, a massive black hole could produce energy and might explain some of the other phenomena seen in the central regions. Before we can decide, however, we must examine the theory more carefully, compare it with observations, and consider alternatives.

We do observe large amounts of energy emanating from the center of the galaxy, but it is not certain that the energy comes from a central object. The central regions contain dense swarms of K and M stars, and there may be some very hot stars there as well. Also, we do see evidence that eruptions have occurred in the central regions, so a central body that produced bursts of energy might be appropriate.

A massive black hole could supply almost unlimited energy. There is no known limit to the mass of a black hole, and the more massive it is, the more energy a bit of matter that falls in can release. And other energy sources are limited. A supernova, for example, converts very little of its mass into energy before blowing itself apart, and there is a limit to the number of massive stars capable of producing supernovae that can occupy a small region.

Such a massive black hole could not have formed from the collapse of a single star but would have gradually developed as gas clouds fell toward the center of the galaxy. We will see later that other galaxies may contain massive black holes.

A radio map made at a wavelength of 190 cm has revealed what seems to be a low-energy jet extending about 30 pc away from the center of the galaxy along the southern axis of rotation. We saw when we discussed SS 433 that jets of matter can be ejected from black holes swallowing matter. No such jet is seen extending to the north, but extremely high-resolution radio maps show continuous arcs of hot gas filaments following what appears to be a magnetic field (Figure 12-26a). This suggests to theorists that the core of the galaxy contains a rapidly rotating body of hot gas capable of generating such a magnetic field, perhaps an accretion disk around a black hole.

At the highest resolution, the VLA radio telescope can detect a spiral swirl of matter in the central 3 pc of the galaxy, the hole in the disk of gas and dust that whirls around the center. This swirl is not related to the spiral arms but appears to be matter spiraling inward toward the central object (Figure 12-26b and 12-26c). This inflow ap-

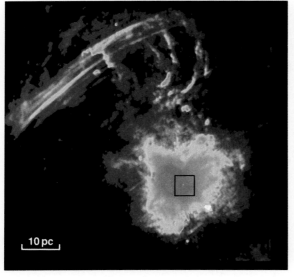

a

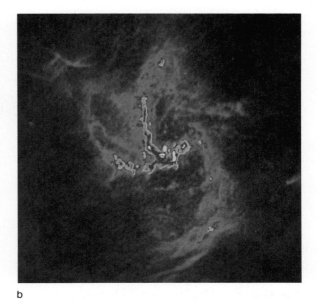

b

Figure 12-26 Sagittarius A. (a) This VLA map shows filaments of excited gas 50 pc long arching out of the center of our galaxy as if constrained by a magnetic field. Compare with boxed region in Figure 12-25. (National Radio Astronomy Observatory) (b) A detailed map of the center (shown roughly by the box in a.) reveals the excited inner edge of a disk of gas and dust with gas flowing into a central object. (N. Killeen and Kwok-Yung Lo) (c) The central spiral appears to lie in the open center of the larger disk of neutral gas. It is believed to show matter flowing into a central object.

c

pears to add a few tenths of a solar mass per year to the central mass, and accretion at that rate onto a massive black hole could easily provide the observed energy.

Many astronomers believe that galaxies commonly contain massive black holes at their centers, and in Chapter 14 we will discuss peculiar galaxies in which eruptions are commonly attributed to massive black holes. Nevertheless, while a massive black hole may exist at the center of our galaxy, observations do not require that it do so. One alternative theory proposes that in-falling gas forms stars that, through gravitational encounters, are ejected from the central region. The most massive of the stars near the center could heat the gas and dust.

Despite all the observations and theories, we cannot yet be sure what lies at the center of the Milky Way galaxy. One way to extend our research is to compare our galaxy with others. We begin that strategy in the next chapter.

SUMMARY

The Milky Way galaxy is a star system roughly 100,000 ly in diameter and containing about 200 billion stars. Its true dimensions were not known until early in this century, when Harlow Shapley began studying globular clusters. In the clusters, he found a type of variable star that had been studied some years before by Henrietta Leavitt. She had found that the period of variation of these Cepheid variables was related to their luminosity. Shapley used these variable stars to find the distance to the globular clusters and thus discover that the Milky Way galaxy was much larger than anyone had previously supposed.

The Milky Way galaxy contains two components, the disk component and the spherical component. Disk stars are metal-rich population I stars moving in circular orbits that lie in the plane of the disk. Because stars that are farther from the center move slower in their orbits, the disk rotates differentially.

The spherical component consists of a nuclear bulge at the center and a halo of thinly scattered stars and globular clusters that completely envelops the disk. Halo stars are metal-poor population II stars moving in random, elliptical orbits.

Studies of the motions of stars and star clusters in the outer parts of the galaxy show that it is rotating more rapidly than was expected. This seems to mean that the galaxy is surrounded by a galactic corona extending roughly seven times farther than was originally thought. This also means our galaxy and presumably other galaxies are much more massive than was previously thought.

The distribution of populations through the galaxy suggests a way the galaxy could have formed from a spherical cloud of gas that gradually flattened into a disk. The younger the stars, the more metal-rich they are and the more circular and flat their orbits are.

The very youngest objects lie along spiral arms within the disk. They live such short lives they don't have time to move from their place of birth in the spiral arms. Maps of these spiral tracers and cool hydrogen clouds reveal the spiral pattern of our galaxy.

The spiral density wave theory suggests that the spiral arms are regions of compression that move through the disk. When an orbiting gas cloud smashes into the compression wave, the gas cloud forms stars. Another process, self-sustaining star formation, may act to modify the arms as the birth of massive stars triggers the formation of more stars by compressing neighboring clouds.

The nucleus of the galaxy is invisible at visual wavelengths, but radio, infrared, X-ray, and gamma-ray radiation can penetrate the dust clouds. These wavelengths reveal crowded central stars and heated clouds of dust. Some of the central features are expanding outward.

The very center of the Milky Way is marked by a radio source, Sagittarius A, that is also a source of infrared radiation, X rays, and gamma rays. The core must be less than 20 AU in diameter and must contain about $3 \times 10^6\ M_\odot$. Many astronomers believe that this central object is a black hole.

NEW TERMS

Magellanic clouds

variable stars

instability strip

period of a variable star

RR variable Lyrae star

Cepheid

period–luminosity relation

open star cluster

globular star cluster

proper motion

calibration

disk component

kiloparsec (kpc)

spiral arm

spherical component

halo

nuclear bulge

differential rotation

rotation curve

galactic corona

population I and II

metal

spiral tracer

density wave theory

flocculent galaxy

self-sustaining star formation

Sagittarius A

REVIEW QUESTIONS

1. Why is it difficult to specify the dimensions of the disk and halo?

2. Why didn't astronomers before Shapley realize how large the galaxy is?

3. How did Shapley calibrate the period–luminosity relation for use in finding distances?

4. Why do we conclude that metal-poor stars are older than metal-rich stars?

5. Contrast the orbital motion of disk and halo stars. How is their motion related to the origin of the galaxy?

6. Why are all spiral tracers young?

7. Why couldn't spiral arms be physically connected structures? What would happen to them?

8. Why does self-sustaining star formation produce clouds of stars that look like segments of spiral arms?

9. What evidence do we have that the center of our galaxy is occupied by a massive black hole?

DISCUSSION QUESTIONS

1. Why do some star clusters lose stars more slowly than others?

2. How would this chapter be different if interstellar dust clouds did not absorb starlight?

3. The Hubble Space Telescope eventually will be able to see objects much fainter and smaller than those seen through the largest telescopes on Earth. Will it help us better understand Sagittarius A? Why or why not?

PROBLEMS

1. Make a scale sketch of our galaxy in cross section. Include the disk, sun, nucleus, halo, and some globular clusters. Try to draw the globular clusters to scale size.

2. Because of dust clouds, we can see only about 5 kpc into the disk of the galaxy. What percentage of the galactic disk can we see? (HINT: Consider the area of the entire disk and the area we can see.)

3. If the fastest passenger aircraft can fly 1600 km/hr (1000 mph), how long would it take to reach the sun? the galactic center? (HINT: 1 pc = 3 × 10^{13} km.)

4. If the RR Lyrae stars in a globular cluster have apparent magnitudes of 14, how far away is the cluster? (HINT: See Box 8-2.)

5. If interstellar dust makes an RR Lyrae star look 1 magnitude fainter than it should, by how much will we overestimate its distance? (HINT: See Box 8-2.)

6. If a globular cluster is 10 minutes of arc in diameter and 8.5 kpc away, what is its diameter? (HINT: Use the small-angle formula from Box 3-2.)

7. If we assume that a globular cluster 4 minutes of arc in diameter is actually 25 pc in diameter, how far away is it? (HINT: Use the small-angle formula from Box 3-2.)

8. If the sun is 5 billion years old, how many times has it orbited the galaxy?

9. If the true distance to the center of our galaxy is found to be 7 kpc and the orbital velocity of the sun is 220 km/sec, what is the minimum mass of the galaxy? (HINTS: Find the orbital period of the sun and then see Box 8-5.)

10. Infrared radiation from the center of our galaxy with a wavelength of about 2 × 10^{-6} m (2000 nm) comes mainly from cool stars. Use this wavelength as λ_{max} and find the temperature of the stars. (HINT: See Box 6-2.)

OBSERVATIONAL ACTIVITY: A NAKED-EYE VARIABLE STAR

Of the few thousand variable stars known, most are much too faint to see with the naked eye. However, a few very interesting stars are quite bright, and we can see their brightness vary without the aid of telescopes. In fact, the star we will discuss, δ Cephei, was among the first variable stars discovered.

δ Cephei is easy to locate in the northern sky. Use the star charts at the end of this book to find the constellation Cepheus and then use Figure 12-27 to locate δ Cephei.

The light variations of δ Cephei are easily visible to the naked eye. Estimate the brightness of the star by comparing it with other stars in the constellation. Keep careful records of your obser-

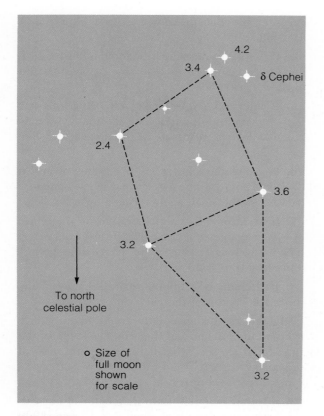

Figure 12-27 Cepheid variable stars are named after star δ in Cepheus. It varies from 3.6 to 4.3 magnitude. The magnitudes of nearby stars are given here. The constellation is shown as it appears on October evenings.

vations, including the date, time, brightness of δ Cephei, and condition of the sky. After a number of nights, you can graph these observations to produce a light curve. The period of δ Cephei is 5.36634 days.

RECOMMENDED READING

BERENDZEN, RICHARD, RICHARD HART, and DANIEL SEELEY. *Man Discovers the Galaxy.* New York: Science History Publications, 1976.

BLITZ, LEO. "Giant Molecular-Cloud Complexes in the Galaxy." *Scientific American* 246 (April 1982), p. 84.

BURNHAM, ROBERT. "Strange Doings at the Milky Way's Core." *Astronomy* 18 (October 1990), p. 39.

CATCHPOLE, ROBIN M. "A Window on Our Galaxy's Core." *Sky and Telescope* 75 (February 1988), p. 154.

CLARK, GAIL O. "Stellar Populations: Key to the Clusters." *Astronomy* 14 (October 1986), p. 106.

COMMINS, NEIL, and LAURENCE MARSHALL. "How Do Spiral Galaxies Spiral?" *Astronomy* 15 (December 1987), p. 6.

DAME, THOMAS M. "The Molecular Milky Way." *Sky and Telescope* 76 (July 1988), p. 22.

EICHER, DAVID. "The Case for Density Waves." *Astronomy* 18 (July 1990), p. 28.

FRIEL, EILEEN. "A Symposium on Stellar Populations." *Mercury* 13 (November/December 1984), p. 165.

HIRSCHFELD, ALAN. "How Far Is the Galactic Center?" *Sky and Telescope* 68 (December 1984), p. 498.

KILLIAN, ANITA. "Galactic Center Update." *Sky and Telescope* 71 (March 1986), p. 255.

PALMER, SAMUEL. "Unveiling the Hidden Milky Way." *Astronomy* 17 (November 1989), p. 32.

SCOVILLE, NICK, and JUDITH S. YOUNG. "Molecular Clouds, Star Formation and Galactic Structure." *Scientific American* 250 (April 1984), p. 42.

SMITH, DAVID H. "A Sideways Look at Galactic History." *Sky and Telescope* 72 (August 1986), p. 111.

SMITH, DAVID H., and LEIF J. ROBINSON. "Dissecting the Hub of Our Galaxy." *Sky and Telescope* 68 (December 1984), p. 494.

TOWNES, CHARLES H., and REINHARD GENZEL. "What Is Happening at the Center of Our Galaxy?" *Scientific American* 262 (April 1990), p. 46.

VERSCHUUR, GERRIT. *Interstellar Matters.* New York: Springer-Verlag, 1989.

———. "The Magnetic Milky Way." *Astronomy* 18 (June 1990), p. 32.

WHITNEY, C. A. *The Discovery of Our Galaxy.* New York: Knopf, 1971.

13

Distance
Hubble (law)

Galaxies

Previous chapters took us to the stars, star clusters, spiral arms, and finally the core of the Milky Way galaxy. Yet we have hardly begun our exploration of the universe. If our journey were compared to a trip around the world, we would not yet have traveled 1000 feet. In this chapter, we leave behind the familiar Milky Way and penetrate deep into space, out among the galaxies.

Beyond the edge of our own galaxy, we can detect billions of other galaxies. Figure 13-1 shows the location of 34,729 of the brighter galaxies, and in it we see two important effects. First, we see almost no galaxies along the path of the Milky Way. This **zone of avoidance** is caused by the dust clouds in our galaxy, which block our view of distant galaxies. Second, we notice that even when we look away from the plane of our galaxy, we see that the other galaxies are not uniformly scat-

The ability to theorize is highly personal; it involves art, imagination, logic, and something more.

Edwin Hubble,
The Realm of the Nebulae

tered through space. Galaxies are clumped together in clusters, and the clusters are grouped into even larger structures.

In this chapter, we will try to understand how galaxies form and evolve, and we will discover that the amount of dust and gas in a galaxy is critically important. We will also discover that interactions between galaxies in clusters can influence their structure and evolution.

Before we can begin building theories, however, we must gather some basic data concerning galaxies. We must examine the different kinds of galaxies and note the differences in their shapes. We must also determine the basic properties of galaxies— diameter, luminosity, and mass. As was the case with stars, our first task is to determine distances. Once we know what galaxies are like, we will be ready to try to explain their origin and evolution.

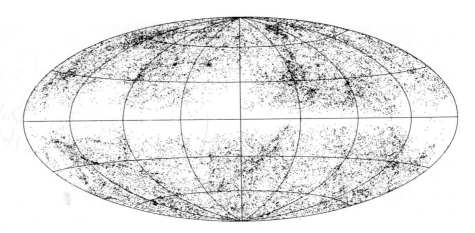

Figure 13-1 A total of 34,729 galaxies are plotted in this map of the entire sky. The plane of the Milky Way runs along the center from right to left. The dust clouds in our galaxy block our view of other galaxies, producing the region of missing galaxies known as the zone of avoidance. Note the clumpy distribution of galaxies. (Courtesy of Nigel A. Sharp)

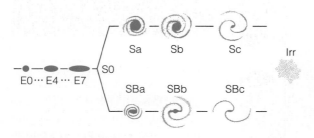

Figure 13-2 The tuning-fork diagram organizes the galaxy types—elliptical (E), spiral (S), barred spiral (SB), and irregular (Irr)—according to star formation history. It is not an evolutionary diagram.

13-1 THREE TYPES OF GALAXIES

Although billions of galaxies are visible from Earth, we may classify them into three basic types according to their shapes. Just as the shapes of organisms may tell a biologist how a species evolved, the shapes of galaxies may give us clues to their evolution.

The three basic classes of galaxies are elliptical, spiral, and irregular. We can subdivide these classes to account for small variations in form among similar galaxies. To organize these classes into an easy-to-remember system, astronomers usually arrange them in a **tuning-fork diagram,** as first devised by Edwin Hubble (Figure 13-2).

Elliptical Galaxies Galaxies that appear to be round or elliptical in shape, have almost no visible gas or dust, lack hot, bright stars, and have no spiral pattern are called **elliptical galaxies** (Figure 13-3a). The stars are most crowded toward the center, and the outer parts of some large ellipticals are peppered with hundreds of globular clusters (Figure 13-3b).

Elliptical galaxies are classified according to shape. If we measure on a photograph distances a and b, the largest and smallest diameters through the center of an elliptical galaxy, then the equation

$$10(a - b)/a$$

gives an index of its shape. This index, rounded to the nearest integer, ranges from 0, for circular outlines, to 7, for the most elliptical outlines observed. Thus elliptical galaxies are classified into subgroups ranging from E0 to E7.

From their outlines, we might expect elliptical galaxies to be spheres flattened by rotation about their shortest diameter—that is, to be shaped like thick hamburger buns. Detailed observations, however, suggest that at least some are shaped like slightly flattened footballs, longer than they are wide and wider than they are thick. Radial velocities show that these galaxies are not rotating about their shortest diameter but are rotating in a complex tumbling motion.

The spectra of elliptical galaxies indicate that they are limited to a single generation of stars. A few studies have detected small dust and gas clouds where a few new stars are forming, but for the most part all of their stars are old. Lack of extensive gas and dust means that few new stars

a

b

c

Figure 13-3 Elliptical galaxies contain little gas and dust and few bright stars. Thus they are nearly featureless clouds of stars. (a) Here the central object is a giant elliptical galaxy attended by smaller ellipticals. (b) M 87 is a giant elliptical galaxy surrounded by a swarm of over 500 globular clusters. Objects to the lower right are smaller galaxies. (c) Dwarf ellipicals are very small. Leo I, an E4 dwarf elliptical, contains only about 4 times as much mass as a large globular cluster. (a. and c. Courtesy Rudolph E. Schild; b. National Optical Astronomy Observatories)

can form, so the only stars we see are old, lower-mass stars like those in our galaxy's spherical component. The massive stars would have died long ago.

Spiral Galaxies Galaxies that contain an obvious disk component, gas, dust, and hot, bright stars are called **spiral galaxies.** The dust is especially obvious in those galaxies that we view edge-on (Figure 13-4).

Among spiral galaxies we identify three distinct types: S0 galaxies, normal spirals, and barred spirals. Unlike our galaxy, the S0 galaxies show no obvious spiral arms, have very little gas and dust,

and contain very few hot, bright stars (Figure 13-5). However, they have an obvious disk component with a large nuclear bulge at the center. They appear to be intermediate between elliptical and spiral galaxies.

Normal spiral galaxies can be further subclassified into three groups according to the size of their nuclear bulge and the degree to which their arms are wound up (Figure 13-6). Spirals that have little gas and dust, larger nuclear bulges, and tightly wound arms are classified Sa. Sc galaxies have large clouds of gas and dust, small nuclear bulges, and very loosely wound arms. The Sb galaxies are intermediate between Sa and Sc. Since there is more

Figure 13-4 As seen from Earth, the spiral galaxy NGC 4565 lies nearly edge-on, making the nuclear bulge and the thinness of the disk easily apparent. Dust in the disk is clearly evident as a dark line crossing in front of the central bulge. Compare with Figure 12-2. (Courtesy Rudolph E. Schild)

gas and dust in the Sb and Sc galaxies, we find more young, hot, bright stars along their arms. The Andromeda galaxy (see Figure 12-1a) and our own Milky Way galaxy are Sb galaxies.

A minority of spirals have an elongated nuclear bulge with spiral arms springing from the ends of the bar. These **barred spiral galaxies** are classified SBa, SBb, or SBc, according to the same criteria listed for normal spirals (Figure 13-7). The elongated shape of the nuclear bulge is not well understood, but some astronomers working with computer models have succeeded in imitating the rotating bar structure. It appears to occur when an instability develops in the stellar distribution within the rotating galaxy, altering the orbits of the inner stars and generating a stable, elongated nuclear bulge. Thus, except for the peculiar rotation in the nuclear bulge, barred spirals are similar to normal spiral galaxies.

Irregular Galaxies Certain galaxies that have a chaotic appearance, with large clouds of gas and dust mixed with both young and old stars, are known as

a

b

Figure 13-5 S0 galaxies are clearly disk galaxies, but they contain very little gas and dust and thus are making few new stars. (a) NGC 4526 is a typical S0 galaxy with a prominent nuclear bulge and extended disk but little dust. (b) The S0 galaxy NGC 4111 is seen edge on, but no dust is visible. Compare the dust in this galaxy with that in the edge-on spiral in Figure 13-4. (a. National Optical Astronomy Observatories; b. Rudolph E. Schild)

Figure 13-6 (a) NGC 3898 is a typical Sa galaxy with a large nuclear bulge and little gas and dust to outline its tightly wound spiral arms. (b) NGC 2841 is an Sb galaxy, and (c) NGC 628 is an Sc galaxy where abundant gas and dust clouds outline the spiral arms with bright knots of star formation. (d) The Sc galaxy NGC 2997 clearly shows dust lanes and luminous clouds of newborn stars along the spiral arms. (a., b., and c. Courtesy Rudolph E. Schild; d. Anglo-Australian Telescope Board)

irregular galaxies. They have no obvious spiral arms or nuclear bulge. The Large and Small Magellanic clouds are the best-studied examples of irregular galaxies (Figure 13-8).

It is difficult to decide how common irregular galaxies are because they are faint. In catalogs of galaxies, about 70 percent are spiral, but that is the result of a selection effect. Spiral galaxies contain hot, bright stars and are thus very luminous and

easy to see. Many ellipticals are faint, so in general ellipticals are probably more common than spirals, and irregulars may make up only about 25 percent of all galaxies.

Photographs suggest that irregular galaxies are small, but are they smaller than spirals? Although we have classified the galaxies, we will know little about them until we are able to measure their basic parameters.

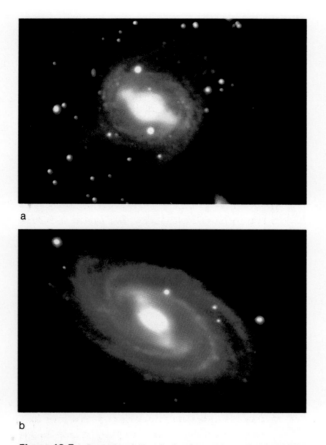

a

b

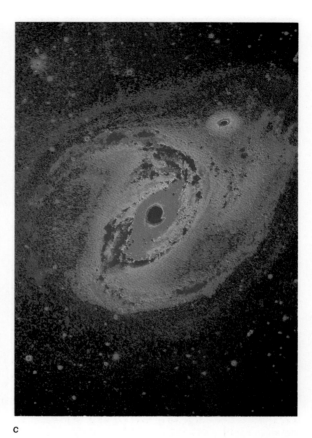

c

Figure 13-7 Barred spiral galaxies have elongated nuclei with spiral arms springing from the ends of the central bar. Here NGC 4650 (a) is an SBa galaxy, and NGC 3992 (b) is an SBb. The image of NGC 1097 (c) has been photographically enhanced to exaggerate the color differences between the red bar and the blue spiral arms. (a. National Optical Astronomy Observatories; b. Smithsonian Astrophysical Observatory; c. Halton Arp)

13-2 MEASURING THE PROPERTIES OF GALAXIES

Unfortunately, the diameter, luminosity, and mass of a galaxy are not obvious in photographs, but to understand galaxies, astronomers must measure these properties. Just as in our study of stellar characteristics (Chapter 8), the first step in our study of galaxies is to find out how far away they are. Once we know a galaxy's distance from us, its size and luminosity are relatively easy to find. Later in this section, we will see that finding mass is more difficult.

Distance The distances to galaxies are so large it is not convenient to measure them in light-years, par-

secs, or even kiloparsecs. Instead, we will use the unit **megaparsec (Mpc),** or 1 million pc. One Mpc equals 3.26 million ly, or approximately 2×10^{19} miles.

To find the distance to a galaxy, we must search among its stars for a familiar object whose luminosity or diameter we know. Such objects are called **distance indicators.**

Because their period is related to their luminosity (see Figure 12-3c), Cepheid variable stars are reliable distance indicators. If we know the period of the star's variation, we can use the period–luminosity diagram to learn its absolute magnitude. By comparing its absolute and apparent magnitudes, we can find its distance.

Figure 13-8 The Large Magellanic Cloud is an irregular galaxy that is only about 50 kpc away. It and the Small Magellanic Cloud are apparently in orbit around our Milky Way galaxy. The brilliant pink emission is the Tarantula Nebula. See Figure 10-14. (Copyright R. J. Dufour, Rice University)

For example, suppose we observed that the longest-period Cepheid variables in a certain galaxy had apparent magnitudes of 19. The period–luminosity relation tells us that the longest-period Cepheids have absolute magnitudes of −6, so these stars appear 25 magnitudes fainter than their absolute magnitudes. That is, their distance modulus is 25, which corresponds to a distance of 1 Mpc. (See Box 8-2.)

Cepheids can tell us the distance to any galaxy in which individual Cepheids are visible. Unfortunately, only about 30 galaxies are close enough to have clearly visible Cepheids. Generally, galaxies beyond about 6 Mpc are too distant for us to see Cepheids, but in 1986 astronomers using the latest CCD imaging techniques succeeded in finding two Cepheids in the galaxy M 101. This galaxy lies between 6.3 and 7.9 Mpc away, so we can expect new advances in imaging to extend our ability to find Cepheids in nearby galaxies.

Studies of the few dozen galaxies whose distances are known from Cepheids reveal large numbers of bright giants and supergiants, large globular clusters, and occasional novae at maximum brightness. The brightest supergiants and novae have absolute magnitudes of about −9, and globular clusters about −10, respectively, so they, too, can be used as distance indicators. If we found bright supergiants in a galaxy, we could measure their apparent magnitude and find the distance to the galaxy. The same is true for novae and large globular clusters.

Planetary nebulae have also been used as accurate distance indicators. A team of astronomers found that planetary nebulae have predictable absolute magnitudes when observed at the wavelength of 500.7 nm, the wavelength of an emission line of twice-ionized oxygen. They were able to compute distances to galaxies by observing the apparent brightness of planetary nebulae within those galaxies.

Another distance indicator is the cloud of ionized hydrogen called an **H II region,** which forms around a very hot star. The Great Nebula in Orion is an H II region. Studies of the nearest galaxies, whose distances are known from other distance indicators, show that these H II regions have predictable diameters. If we detect H II regions in a distant galaxy, we can measure their angular diameter and assume they have the same diameter as the H II regions in other galaxies. Then we can find the distance by using the small-angle formula (see Box 3-2). The H II regions are important distance indicators because they are easy to detect and because they could, in principle, give us distances beyond the limit for most other distance indicators. Unfortunately, the diameter of an H II region depends on the kind of star at its center and the density of its gas. Thus different kinds of galaxies may have different-sized H II regions. This limits the dependability of the method.

Occasionally a supernova flares in a distant galaxy. At maximum, supernovae have absolute magnitudes of −16 to −20, so they could give us estimates of the distances to their host galaxies.

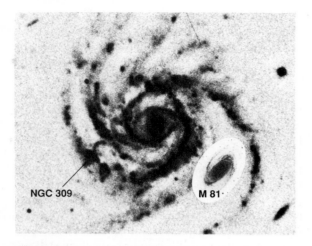

Figure 13-9 Once we know the distances to galaxies, we can compare them. Here NGC 309 and M 81 (inset) are reproduced to the same scale. Clearly NGC 309 is much larger. M 81 also appears in Figure 12-22. (Courtesy David L. Block and B. Dumoulin; NGC 309 photo by T. Kinman; M 81 image from Palomar Sky Survey)

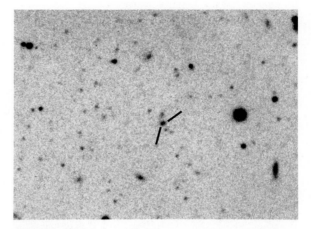

Figure 13-10 This photographic plate of the visible galaxy associated with the radio source 3C 13 is reproduced as a negative to enhance detail. The galaxy is about 3.7 Mpc (12 billion ly) away, and even the best plates show little more than a blur. (Hyron Spinrad)

However, supernovae themselves are not well understood, so we can't be sure of the true absolute magnitude at maximum. Also, they are rare—we might wait decades for a supernova to occur in any particular galaxy.

Notice how distance indicators for more distant galaxies are calibrated using nearby galaxies. (We discussed this calibration technique in Chapter 12.) For example, the brightest stars were calibrated using distances found from Cepheid variables, and H II regions might be calibrated using distances found from the brightest stars. Astronomers speak of the distance scale as if it were a pyramid in which each layer rests on the layer below. Of course, the foundation stones of the distance scale are the parallaxes of stars in our own galaxy and the Cepheid variables in nearby galaxies.

To measure distances to the farthest galaxies, we must calibrate the galaxies themselves as distance indicators. Studies of nearby galaxies indicate that an average galaxy like the Milky Way has a luminosity of about 16×10^9 solar luminosities, corresponding to an absolute magnitude of -20.5. If we see a similar galaxy and measure its apparent magnitude, we can find its distance (see Box 8-2). Other types of galaxies have different luminosities, so astronomers must know what kind of galaxy they are observing (Figure 13-9). Unfortunately, there

is considerable error in this method because of differences in luminosity among galaxies of the same type. Averaging the brightest galaxies in a cluster reduces some of the error and determines distances to the most distant galaxies.

In spite of the uncertainties in distance measurements, it is clear that galaxies are far apart and are scattered through the universe out to tremendous distances. The nearest large galaxy, the Andromeda galaxy, is 0.66 Mpc (2.15 million ly) away. The most distant objects identifiable on photographs as galaxies are about 3000 Mpc (10 billion ly) away from us (Figure 13-10). Radio telescopes can detect faint sources of radio signals that are evidently distant galaxies much more luminous at radio wavelengths than at visible-light wavelengths. Thus the radio telescopes can detect them even though they are beyond the limit of the largest optical telescopes.

The tremendous distance between the galaxies produces an effect akin to time travel. When we look at a galaxy millions of light-years away, we do not see it as it is now, but as it was millions of years ago when its light began the journey toward Earth. Thus, when we look at a distant galaxy, we look into the past by an amount called the **look-back time,** a time in years equal to the distance to the galaxy in light-years.

The look-back time to nearby objects is usually not significant. The look-back time to the moon is 1.3 seconds, to the sun only 8 minutes, and to the nearest star about 4 years. The Andromeda galaxy has a look-back time of about 2 million years, a mere eye blink in the lifetime of a galaxy. But when we look at more distant galaxies, the look-back time becomes an appreciable part of the age of the universe. We will see evidence in Chapter 15 that the universe began no more than 20 billion years ago. Thus, when we look at the most distant visible galaxies, we are looking back 10 billion years to a time when the universe may have been significantly different. This effect will be important in this and the next chapter.

The Hubble Law   Although astronomers find it difficult to measure the distance to a galaxy, they often estimate such distances using a relation that was first noticed early in this century.

In the years after World War I, astronomers began studying the spectra of many galaxies, and they soon noticed that almost all of the spectra had spectral lines shifted toward longer wavelengths. (Astronomers commonly refer to this as a red shift.) Interpreted as consequences of the Doppler effect, these red shifts implied that the galaxies were receding.

In 1929, the American astronomer Edwin Hubble announced a general law of red shifts now known as the **Hubble law.** This law says that a galaxy's velocity of recession equals a constant times its distance. Thus the more distant a galaxy is, the faster it recedes from us. The constant H, now known as the **Hubble constant,** is very difficult to determine.

A number of modern studies of the recession of the galaxies suggest that H equals about 50 km/sec/Mpc.* Other studies, however, yield values as high as 100 km/sec/Mpc. The uncertainty arises from the difficulty of determining the distances to galaxies. Different groups of astronomers have found distances in different ways and have arrived at different values for H. In Chapter 15, we will see that this has an important effect on our understanding of the history of the universe, but here it

*H has the units of a velocity divided by a distance. These are usually written as km/sec/Mpc, meaning km/sec per Mpc.

The Hubble law is important because it is commonly interpreted to show that the universe is expanding. In Chapter 15, we will discuss the implications of this expansion, but here we will use the Hubble law as a practical way to estimate the distance to a galaxy. Simply stated, a galaxy's velocity of recession divided by the Hubble constant equals its distance. (See Box 13-1.) This makes it relatively easy to find galactic distances, since large telescopes can photograph the spectrum of a distant galaxy and reveal its red shift even though distance indicators such as variable stars are totally invisible.

We cannot abandon distance indicators and use the Hubble law exclusively, however. In Chapter 14, we will discuss peculiar galaxies that, according to some astronomers, may not obey the Hubble law. To detect such departures from the law, we must measure distances with distance indicators and use the red shifts only for estimates.

Diameter and Luminosity   The distance to a galaxy is the key to finding its diameter and its luminosity. We can easily photograph a galaxy and measure its angular diameter in seconds of arc. If we know the distance, we can use the small-angle formula (see Box 3-2) to find its linear diameter. Also, if we measure the apparent magnitude of a galaxy, we can use the distance to find its absolute magnitude and thus its luminosity (see Box 8-2).

The results of such observations are shown in Table 13-1. Irregular galaxies tend to be small objects of low luminosity. Although they are common, they are easy to overlook. Our Milky Way galaxy is large and luminous compared with most spiral galaxies, though we know of a few spiral galaxies that are even larger and more luminous— at least 50 percent larger in diameter and about 10 times more luminous. Elliptical galaxies cover a wide range of diameters and luminosities. The largest, called giant ellipticals, are five times the size of our Milky Way. But many elliptical galaxies are very small, dwarf ellipticals, only 1 percent the diameter of our galaxy.

Clearly, the diameter and luminosity of a galaxy do not determine its type. Some small galaxies are irregular and some are elliptical. Some large galaxies are spiral and some are elliptical. We need more

The Hubble law is a simple relationship between the radial velocity of a galaxy and its distance. A galaxy's radial velocity V_r in kilometers per second equals a constant H times the distance d to the galaxy in megaparsecs:

$$V_r = Hd$$

The constant H, known as the Hubble constant, is believed to be about 70 km/sec/Mpc. That is, for every megaparsec between two galaxies, they will recede from each other by an additional 70 km/sec. In general, the distance to a galaxy in megaparsecs equals its radial velocity divided by the Hubble constant.

Example: If a galaxy has a radial velocity of 700 km/sec, how far away is it? *Solution:* If we assume the Hubble constant is 70 km/sec/Mpc, then the distance in megaparsecs is 700 divided by 70, or 10 Mpc.

The precise value of the Hubble constant is not known and may be anywhere between 50 and 100 km/sec/Mpc. We will discuss the importance of this uncertainty in Chapter 15. For now it means simply that distances derived from the Hubble law are only estimates.

TABLE 13-1
Normal Galaxies Compared to the Milky Way*

	Elliptical	Spiral	Irregular
Diameter	0.01–5	0.2–1.5	0.05–0.25
Luminosity	0.00005–5	0.005–10	0.00005–0.1
Mass	0.000001–50	0.005–2	0.0005–0.15

*In units of diameter, luminosity, and mass of the Milky Way galaxy.

data before we can build a theory for the origin and evolution of galaxies.

Of the three basic parameters that describe a galaxy, we have found two—diameter and luminosity. The third, as was the case for stars, is more difficult to discover.

Mass Although the mass of a galaxy is difficult to determine, it is an important quantity. It tells us how much matter the galaxy contains, which gives us clues to the galaxy's origin and evolution. In this section, we will examine two fundamental ways to find the masses of galaxies.

One way is called the **rotation curve method.** We begin by photographing the galaxy's spectrum at different points along its diameter and plotting the Doppler shift velocities in a rotation curve like that in Figure 13-11a. This tells us how fast the galaxy is rotating. The sizes of the orbits the stars follow around the galaxy's center are related to the

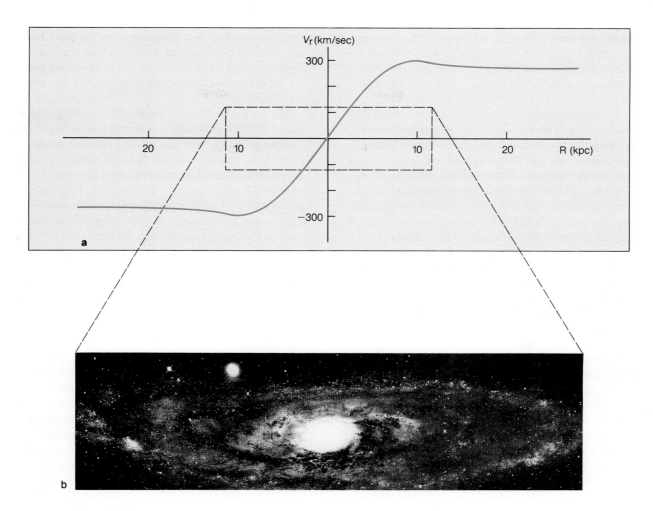

Figure 13-11 The rotation curve of the Andromeda galaxy (a) shows that orbital velocity in the outer regions does not decline. This implies that the galaxy is surrounded by a corona of dark matter. Compare with Figure 12-11. The photo (b) shows only the brighter part of the galaxy, which has a diameter of about 23 kpc. If the entire galaxy were visible in the night sky, it would span over 5°—ten times the diameter of the full moon. (a. Diagram adapted from 21-cm observations by Morton S. Roberts and Robert N. Whitehurst; b. Lick Observatory)

size of the galaxy, easily found from its angular diameter and its distance. If we divide the circumference of the orbit by the orbital velocity of the stars there, we find their orbital period. We can then find the mass of the system by using the radius of the orbit and the orbital period, just as we did for binary stars (see Box 8-5).

The rotation curve is the best method for finding the masses of galaxies, but it only works for the nearer galaxies, whose rotation curves can be ob-

served. More distant galaxies appear so small we cannot measure the radial velocity at different points along the galaxy. Also, recent studies of our own galaxy and others show that the outer parts of the rotation curve do not decline to lower velocities (Figure 13-13a). This shows that the outermost visible parts of some galaxies do not travel slower and tells us that the galaxies contain large amounts of mass outside this radius, perhaps in extended galactic coronas like the one that seems to surround our own galaxy (see Chapter 12). Because the rotation curve method can be applied only to nearby galaxies and because it cannot determine the masses of galactic coronas, we must look at the second way to find the masses of galaxies.

The **cluster method** of finding galactic mass depends on the motions of galaxies within a cluster. If we measure the radial velocities of many galaxies in a cluster, we find that some velocities are larger than others because of the orbital motions of the individual galaxies in the cluster. Given the range of velocities and the size of the cluster, we can ask how massive a cluster of this size must be to hold itself together with this range of velocities. Dividing the total mass of the cluster by the number of galaxies in the cluster yields the average mass of the galaxies. This method contains the built-in assumption that the cluster is not flying apart. If it is, our result is too large. Since it seems likely that most clusters are held together by their own gravity, the method is probably valid.

A related way of measuring a galaxy's mass is called the **velocity dispersion method.** It is really a version of the cluster method. Instead of observing the motions of galaxies in a cluster, we observe the motions of matter within a galaxy. In the spectra of some galaxies, broad lines indicate that stars and gas are moving at high velocities. If we assume the galaxy is bound by its own gravity, we can ask how massive it must be to hold this moving matter within the galaxy. This method, like the one before, assumes that the system is not coming apart.

The structure and evolution of a star are determined by its mass, but this is clearly not so for galaxies. There is a wide range of masses among the elliptical, spiral, and irregular galaxies. The smallest contain about 10^{-6} as much mass as the Milky Way galaxy, and the largest contain as much as 50 times more (see Table 13-1).

Dark Matter in Galaxies When astronomers began measuring the masses of galaxies, they were surprised because their answers seemed much too large. Evidently, galaxies contain much more mass than what we can see. Since this mass is not visible, it has been called **dark matter.**

Large clusters of galaxies, for example, do not appear to contain enough mass to hold themselves together. The velocities of the galaxies are so large the clusters would fly apart if they contained only the mass we can see. Because these clusters have survived for billions of years, astronomers conclude they must contain large amounts of dark matter whose gravitational attraction holds the clusters together.

We have seen that the rotation curves of galaxies imply that they contain much more mass than the visible portions, but X-ray observations provide even more dramatic proof. X-ray studies of E and S0 galaxies show that many of them are filled with very hot, low-density gas. The gas is so hot, the gas atoms travel at high velocity and would easily escape from the galaxies if their masses consisted only of the matter we can see. To hold that hot gas inside, the galaxies must contain vast amounts of dark matter.

This is not a small effect. Galaxies seem to contain 10–100 times more dark matter than visible matter. The universe we see—the luminous matter—has been compared to the foam on an invisible ocean.

Astronomers have proposed a number of theories to explain why the dark matter is so difficult to detect. One suggestion is that it is made up of low-luminosity stars and brown dwarfs, but such objects should be detectable by their infrared radiation. Other proposals suggest that dark matter is made up of neutrinos with mass or exotic subatomic particles such as **WIMPs** (see Box 9-1). So far no laboratory experiments have been able to measure a mass for the neutrino, and no WIMPs have ever been detected, so dark matter remains one of the fundamental unresolved problems of modern astronomy. We will see in Chapter 15 that the presence of dark matter could affect the origin and evolution of the universe as a whole.

For the moment, we must put aside the mysteries of dark matter and ask a simple question: How did the galaxies form and evolve? The answer will lead us to consider collisions between galaxies.

13-3 THE ORIGIN AND EVOLUTION OF GALAXIES

Our goal in this chapter has been to build a theory to explain the origin and evolution of galaxies. In the previous chapter, we developed a theory that described the origin of our own Milky Way galaxy, and, presumably, other galaxies formed similarly

Figure 13-12   The Coma cluster of galaxies contains at least 1000 galaxies and is especially rich in E and S0 galaxies. The two giant galaxies near its center are an E and an S0. Although the Coma cluster is over 100 Mpc distant, it is so large that, if we could see it in the evening sky, it would appear to be eight times the diameter of the full moon. (Courtesy L. Thompson, © National Optical Astronomy Observatories; Tucson)

from the contraction of large clouds of gas. But why did some galaxies become spiral, some elliptical, and some irregular? Clues to that mystery lie in the clustering of galaxies.

Clusters of Galaxies   As we saw in Figure 13-1, the distribution of galaxies over the sky is not uniform. Most galaxies lie in clusters. Our own Milky Way is a member of a cluster containing over 20 galaxies, and over 2700 other clusters have been catalogued within 4 billion ly.

For our discussion, we can sort clusters of galaxies into two groups: rich and poor. **Rich galaxy clusters** contain over a thousand galaxies, mostly elliptical, scattered through a spherical volume about 3 Mpc (10^7 ly) in diameter. Such a cluster is generally very crowded, with the galaxies more concentrated toward the center. Rich clusters often contain one or more giant elliptical galaxies at their centers.

The Coma cluster (located in the constellation Coma Berenices) is an example of a rich cluster (Figure 13-12). It lies over 100 Mpc from us and contains at least 1000 galaxies, mostly Es and S0s. Its galaxies are highly crowded around a central giant elliptical and a large S0. The distance to the Coma cluster is not accurately known because it is far beyond the distance to which Cepheid variables are visible.

One of the nearest clusters to us, the Virgo cluster, contains over 2500 galaxies and is, by our definition, a rich cluster. It does contain a giant elliptical galaxy, M 87, near its center, but it is not very crowded and contains mostly spiral galaxies. Although the Virgo cluster is exceptional as a rich cluster, it is critically important because it is nearby. At a distance of about 15 Mpc (50×10^6 ly), no Cepheid variable stars are visible, so the distance must be found from the brightest stars, globular clusters, planetary nebulae, and so on. The calibra-

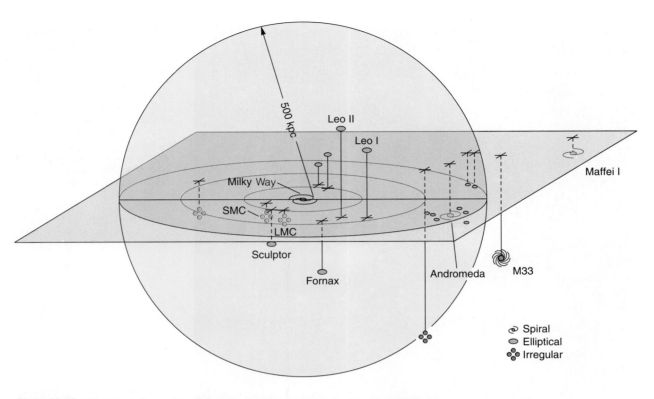

Figure 13-13 The Local Group. Our galaxy is a member of a poor galaxy cluster called the Local Group. In this three-dimensional map, the galaxies are shown with respect to the plane of our galaxy. Most of the galaxies are small ellipticals. Other member galaxies may not have been seen because they lie near the plane of the Milky Way and thus are obscured by dust clouds within our galaxy.

tions of many distance indicators depend on the distance to the Virgo cluster, so it is subject to intense study.

Poor galaxy clusters contain fewer than 1000 galaxies and are irregularly shaped. Instead of being crowded toward the center, poor clusters tend to have subcondensations, small groupings within the clusters.

Our own Local Group, which contains the Milky Way, is a good example of a poor cluster (Figure 13-13). It contains slightly more than 20 members scattered irregularly through a volume about 1 Mpc in diameter. Of the brighter galaxies, 14 are elliptical, 3 are spiral, and 4 are irregular.

The total number of galaxies in the Local Group is uncertain because a few galaxies lie in the plane of the Milky Way and are hidden by the obscuring dust clouds. For instance, the Italian astronomer Paolo Maffei, studying infrared photographs, discovered two galaxies hidden behind the dust

clouds of the Milky Way in the constellation Cassiopeia. Maffei 1 is a giant elliptical that would be visible to the naked eye if there were no dust in the way, and Maffei 2 is a spiral galaxy that probably contains many bright stars. Neither object is detectable at visual wavelengths because of the dust clouds in the Milky Way. Although Maffei 2 lies about 5 Mpc distant and cannot be a member of the Local Group, Maffei 1 is only about 1 Mpc distant, and thus lies on the outer fringes of our small cluster of galaxies.

The clumpiness of the Local Group illustrates the subclustering in poor galaxy clusters. The two largest galaxies, the Milky Way and Andromeda, are the centers of two subclusters. The Milky Way galaxy is accompanied by the Magellanic clouds and about three other dwarf galaxies. The Andromeda galaxy is attended by seven dwarf elliptical galaxies and a smaller spiral, M 33.

In general, rich clusters tend to have a larger

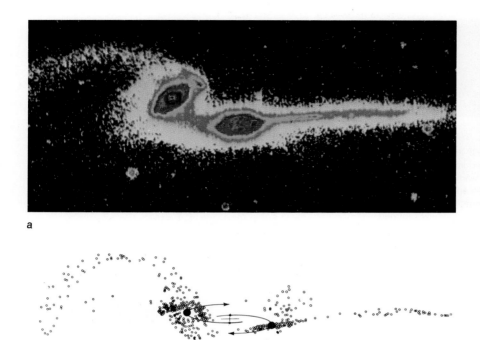

Figure 13-14 (a) The Mice are a pair of interacting galaxies with peculiar tails (shown here in a false-color image). (National Optical Astronomy Observatories) (b) A computer simulation of a close encounter between normal galaxies produces similar tails. (From "Violent Tides Between Galaxies" by A. Toomre and J. Toomre. Copyright © 1973 by Scientific American, Inc. All rights reserved.)

a

b

percentage of E and S0 galaxies than poor clusters do. This suggests that a galaxy's environment is important in determining its structure, and it has led astronomers to suspect that collisions between galaxies, which should occur much more often in the crowded rich clusters, can affect their evolution.

Colliding Galaxies The average separation between galaxies is only about 20 times their diameter. Thus, like two elephants blundering about under a circus tent, galaxies should bump into each other once in a while. Stars, on the other hand, almost never collide with each other. In the region of the galaxy near the sun, the average separation between stars is about 10^7 times their diameter. Thus a collision between two stars is about as likely as a collision between two gnats flitting about in a domed stadium.

Large telescopes reveal hundreds of galaxies that appear to be colliding. One of the most famous pairs of colliding galaxies, NGC 4676A and NGC 4676B, is called the Mice because of the tail-like deformities (Figure 13-14). In addition to tails,

some interacting galaxies seem to be connected by bridges of gas and stars.

When two galaxies collide, they pass through each other, and there is no direct contact between the stars of one galaxy and the stars of the other. The interstellar gas and dust in the galaxies does interact as gas clouds collide, and the sudden compression can trigger rapid star formation (Figure 13-15), or the gas in one galaxy can be stripped away by the other galaxy.

It is mainly the galaxies' gravitational fields that twist and deform the galaxies, producing tails and bridges. It is not even necessary for the galaxies to penetrate each other—a close encounter is enough to produce strong tidal forces that distort the galaxies.

Because these collisions can last hundreds of millions of years, we cannot see the galaxies change, but computer simulations of collisions can reproduce much of the structure that we see in real galaxies (Figure 13-16). Even certain peculiar galaxies such as the **ring galaxies** (Figure 13-17) are now believed to be produced by occasional head-on collisions.

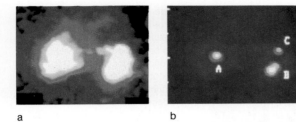

a   b

Figure 13-15 The galaxies NGC 3690 and IC 694 are interacting with each other. An infrared image of the pair (a) shows that they are bright with dust warmed by newly formed stars and that they are connected by a bridge of matter. The brightest parts of the image (b) show the nuclei of the galaxies A and B and a huge cloud of newly formed stars at C and another near the nucleus B. (Courtesy Ian McLean, Joint Astronomy Center, Hawaii)

If two galaxies collide at low initial velocity, it is possible for them to merge. A small galaxy merging with a large galaxy will be gradually pulled apart by tidal forces, and its stars will spread through the larger galaxy in a process called **galactic cannibalism.** Computer simulations of such mergers show that the galaxies will throw off shells of stars as the nuclei of the two galaxies spiral into each other, and such shells of stars have been observed around real galaxies (Figure 13-18).

Apparently, our own Milky Way is interacting with the Large and Small Magellanic clouds. Those two galaxies appear to have passed through the outer disk of our galaxy about 200 million years ago, and a long, curving bridge of gas now connects them with our galaxy. The Small Magellanic Cloud has been disrupted into at least two clouds separated by 20,000 ly (Figure 13-19). Our own galaxy appears to be distorted by the interaction, with one edge of the disk bent down and the other bent up like the brim of a hat.

Galactic Evolution Before we try to build a theory of galactic evolution, we must pause long enough to eliminate a tempting but discredited idea. When the tuning-fork diagram first appeared, astrono-

Figure 13-16 A computer simulation of a collision between galaxies produces a pattern very similar to M 51, the Whirlpool galaxy (bottom) and its small companion. (From "Violent Tides Between Galaxies" by A. Toomre and J. Toomre. Copyright © 1973 by Scientific American, Inc. All rights reserved. Lick Observatory photograph.)

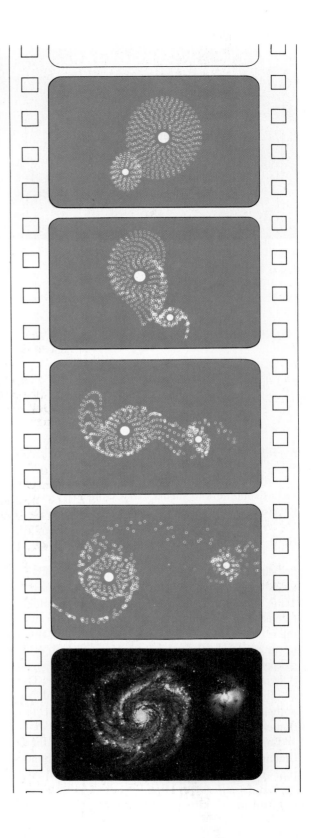

Figure 13-17 Ring galaxies are rare and sometimes lack nuclei. In this case, however, the nucleus is visible as the bright object within the ring. Such systems are believed to be the product of collisions between galaxies. (National Optical Astronomy Observatories)

mers theorized that it was an evolutionary diagram showing how elliptical galaxies evolved into spiral and then into irregular galaxies. Unfortunately, this simple theory doesn't work. Studies of elliptical galaxies show that they contain little or no gas and are not making many new stars. Thus an elliptical can never become a spiral.

In addition, evolution can't go from irregular to spiral to elliptical. Only four irregular galaxies, those in the Local Group, are close enough to study in detail, but all four contain both young and old stars. Thus they cannot be young in comparison with spiral and elliptical galaxies.

The tuning-fork diagram is not an evolutionary diagram. Galaxies do not evolve from elliptical to spiral or vice versa, any more than cats evolve into dogs. The tuning fork is a star formation diagram showing that different types of galaxies have had different histories of star making.

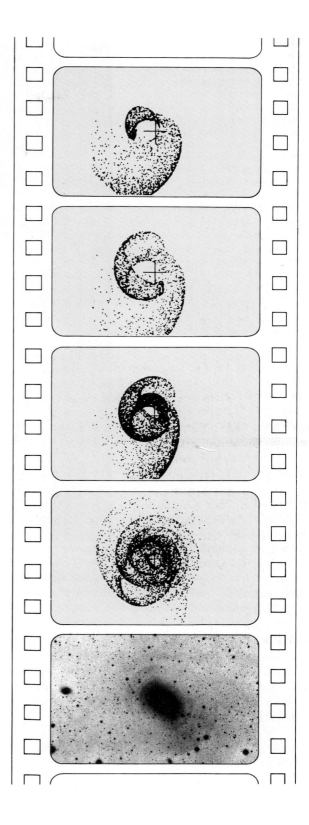

Figure 13-18 Computer simulations of merging galaxies suggest that they can produce shells of stars as the two galaxies whirl around their common center of mass. Only one galaxy is shown in this sequence. Such shells are seen around galaxies such as NGC 3923 (bottom), shown here as a negative image specially enhanced to reveal low-contrast features. (Model courtesy Francois Schweizer and Alar Toomre; photo courtesy David Malin, Anglo-Australian Telescope Board)

Figure 13-19 The Small Magellanic Cloud (SMC) is an irregular galaxy. Radio and optical studies show that the SMC is actually two clouds (inset). One cloud of stars (white) is about 20,000 ly closer than the other cloud (shaded). (Photo courtesy National Optical Astronomy Observatories; diagram adapted from data by Don S. Mathewson and Vincent L. Ford)

Clearly, a galaxy's past history of star formation is critical to understanding its evolution. Elliptical galaxies have used up their star-forming gas and dust. They cannot make many new stars, and those that remain are cool, lower-main-sequence stars and red giants. The Infrared Astronomy Satellite found elliptical galaxies very faint in the infrared because they lack dust. But spirals have conserved their gas and dust, and can still make new stars. Our theory must explain this difference in star formation history.

For a number of decades, astronomers have suspected that the rotation of a galaxy is an important factor. A galaxy that forms from a gas cloud with little angular momentum might contract rapidly, compress its gas to form stars quickly, and thus consume its gas and dust, making a single generation of stars. We might see such a galaxy today as an elliptical. But a galaxy that formed from a gas cloud with more angular momentum would contract more slowly, making stars generation after generation as it flattened into a disk. Such a galaxy would develop into a spiral.

Angular momentum may be a factor, but the excess of E and S0 galaxies in rich clusters suggests that collisions are more important. A galaxy in a rich cluster should suffer a higher-than-usual number of collisions with other galaxies. Collisions could trigger star formation and drive the galaxy to consume its gas and dust rapidly. Also, some collisions may strip gas and dust out of a galaxy.

This suggests that elliptical galaxies are the

products of collisions, and some astronomers now suspect that the average elliptical galaxy has been built up by the merger of a number of smaller galaxies. The giant elliptical galaxies in rich clusters may have cannibalized a number of smaller galaxies, or they may have gained mass lost by other colliding galaxies in the cluster. If this is true, then spiral galaxies are systems that have experienced very few collisions. More common in poor clusters where collisions are rare, they have been able to form stars more slowly and still have star-forming gas and dust left.

Our theory does not account for the irregular galaxies very well. They may have been formed from turbulent gas clouds, or they may be galaxies that have been disrupted by encounters with other galaxies, as have the Large and Small Magellanic clouds.

Our theory of galactic evolution is incomplete, but it is very new. Astronomers are still gathering data. A theory begins as speculation, passes through a stage we might call a working hypothesis, and finally graduates to become a well-formed, widely tested theory. Our story of galactic evolution is presently no more than a working hypothesis, but the new giant telescopes now being built and the Hubble Space Telescope will provide new data on the structure of galaxies that will allow astronomers to refine the theory. Two decades ago, astronomers hesitated to discuss the evolution of galaxies. Two decades from now, we may tell the story of the galaxies with the same confidence we now have when we discuss the evolution of stars.

The Distribution of Galaxies in Space Although we can see some hints of how galaxies evolve, the actual origin of galaxies from clouds of gas is not well understood. We will see in Chapter 15 that the universe is believed to have begun suddenly, with the production of great clouds of hydrogen and helium. We must assume that the galaxies condensed from those gas clouds, but the structures we see in the distribution of galaxies (see Figure 13-1) show that that process was quite complex.

Clusters of galaxies are associated in groups called **superclusters.** The Local Group is part of the Local Supercluster, a roughly disk-shaped swarm of galaxy clusters 50–75 Mpc in diameter. Superclusters are linked into even larger structures variously described as filaments, pancakes, or walls. The voids between these larger structures seem to be almost empty of galaxies, and some astronomers have likened the distribution of superclusters to the foamy distribution of water within soap suds.

One dramatic example of these large structures was found in 1989. By mapping the location of roughly 11,000 galaxies lying in four strips across the sky, astronomers found a great sheet of galaxies 15 million ly thick and at least 500 million ly long (Figure 13-20). Dubbed the Great Wall, it is the largest structure known.

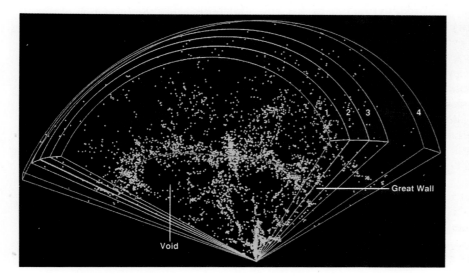

Figure 13-20 The Great Wall (dashed line), a concentration of galaxies 500 million ly long, is revealed in this survey of four slices of the sky. Our galaxy is located at the vertex of the slices, which cover only 1/100,000th of the visible universe. Voids such as the one at left appear almost empty of galaxies, and other filamentary structures and voids are visible in other parts of the map. (See also p. 16.) (Courtesy M. J. Geller and J. P. Huchra, Center for Astrophysics)

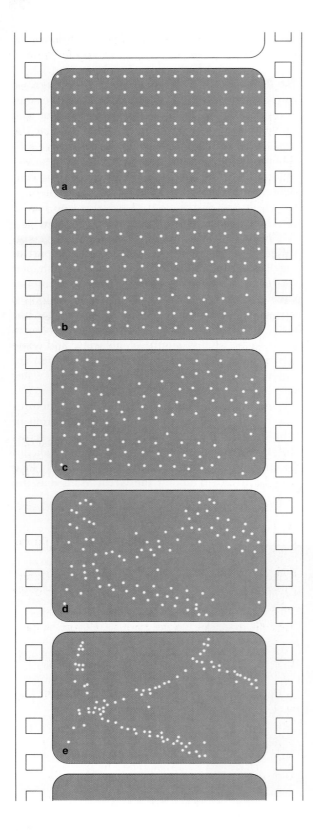

Figure 13-21 The formation of filaments, walls, and voids is shown here beginning with a young universe uniformly filled with identical masses (a). Computer simulations show that small irregularities can grow into the observed distribution of galaxies (e). This animation is adapted from models that include the gravitational attraction of neutrinos with nonzero masses.

This distribution of the galaxies in walls and filaments appears to be related to the motions of the galaxies. In 1989, astronomers discovered that our Milky Way and neighboring galaxies are falling at 400–1000 km/sec toward a region 50 Mpc away in the direction of the constellation Centaurus. Presumably, this motion is caused by the gravitational attraction of a region containing roughly 1 million times the mass of our galaxy. Although there is a concentration of galaxies in that region, the object, apparently a large supercluster of galaxies, has not been identified. Astronomers refer to it as the Great Attractor.

It is not clear how the filamentary distribution of the galaxies formed. When the universe was young, the gas may have collected into filaments and walls and then formed galaxies, or the galaxies may have formed first and then drawn themselves together to form superclusters, filaments, and walls. Theoretical models can produce both such structures, but the models require some form of dark matter (Figure 13-21). We may not understand the origin of the filaments and walls until we solve the mystery of dark matter.

Astronomers understand the origin of stars much better than they understand the origin of galaxies, perhaps because the formation of stars is still going on while galaxies appear to have formed long ago when the universe was young. We will continue our discussion of the origin of the universe in Chapter 15, but first, in the next chapter, we must examine certain peculiar galaxies that appear to be suffering titanic eruptions in their cores.

SUMMARY

We can divide galaxies into three classes—elliptical, spiral, and irregular—with subclasses specifying the galaxy's shape. The elliptical galaxies contain little gas and dust and few bright, young stars.

Spiral and irregular galaxies have large amounts of gas and dust and are actively making new stars.

To measure the properties of galaxies, we must first find out how far away they are. For the nearer galaxies, we can judge distances using distance indicators, objects whose luminosity or diameter is known. The most accurate distance indicators are Cepheid variable stars. Other distance indicators are bright giants and supergiants, globular clusters, planetary nebulae, and novae. Another type of distance indicator, the H II regions, have known diameters. To use them, we compare their angular diameter with their known linear diameter. In addition, we can estimate the distance to the farthest galaxy clusters using the average luminosity of the brightest galaxies.

The Hubble law shows that the radial velocity of a galaxy is proportional to its distance. Thus we can use the Hubble law to estimate distances. The galaxy's radial velocity divided by the Hubble constant equals its distance in megaparsecs.

The masses of galaxies can be measured in two basic ways—the rotation curve method and the velocity dispersion method. The rotation curve method is more accurate but can be applied only to nearby galaxies. Both methods suggest that galaxies contain 10–100 times more dark matter than visible matter.

Galaxies occur in clusters. Our own galaxy is a member of the Local Group, a small cluster. A galaxy in a rich cluster may collide with other galaxies more often than a galaxy in a poor cluster, and such collisions can force a galaxy to form new stars and use up its gas and dust. Collisions can also strip gas out of a galaxy. This may explain why elliptical and S0 galaxies are more common in rich clusters than in poor clusters. Spiral galaxies may be star systems that have not experienced many collisions.

Clusters of galaxies are organized into superclusters, and superclusters are linked together in a network of filaments and walls. The regions between seem to be great voids where there are few if any galaxies. How these filaments, walls, and voids developed is not yet understood.

NEW TERMS

zone of avoidance

tuning-fork diagram

elliptical galaxy

spiral galaxy

barred spiral galaxy

irregular galaxy

megaparsec (Mpc)

distance indicator

H II region

look-back time

Hubble law

Hubble constant (H)

rotation curve method

cluster method

velocity dispersion method

dark matter

rich galaxy cluster

poor galaxy cluster

ring galaxy

galactic cannibalism

supercluster

REVIEW QUESTIONS

1. Why is there a zone of avoidance? If we lived in an elliptical galaxy, would we see a zone of avoidance?

2. Draw and label a tuning-fork diagram. Why can't galaxies evolve from elliptical to spiral? from spiral to elliptical?

3. What is the difference between an Sa and an Sb galaxy? between an S0 and an Sa galaxy? between an Sb and an SBb galaxy? between an E7 and an S0 galaxy?

4. Why wouldn't white dwarfs make good distance indicators?

5. Why isn't the look-back time important among nearby galaxies?

6. How is the velocity dispersion method of finding the masses of galaxies related to the cluster method?

7. What evidence do we have that there is dark matter in galaxies?

8. Why should collisions be more common in rich clusters than in poor clusters?

9. How can collisions explain the excess of E and S0 galaxies seen in rich clusters?

DISCUSSION QUESTIONS

1. Should the Infrared Astronomy Satellite have found irregular galaxies bright or faint in the infrared? Why?

2. Why do we believe that galaxy collisions are likely, but star collisions are not?

PROBLEMS

1. If a galaxy contains a Type I (classical) Cepheid with a period of 30 days and an apparent magnitude of 20, what is the distance to the galaxy?

2. If you find a galaxy that contains globular clusters that are 2 seconds of arc in diameter, how far away is the galaxy? (HINTS: Assume a globular cluster is 25 pc in diameter, and see Box 3-2.)

3. If a galaxy contains a supernova that at its brightest has an apparent magnitude of 17, how far away is the galaxy? (HINTS: Assume the absolute magnitude of the supernova is -19, and see Box 8-2.)

4. If a galaxy has a radial velocity of 2000 km/sec and the Hubble constant is 70 km/sec/Mpc, how far away is the galaxy? (HINT: See Box 13-1.)

5. If you find a galaxy that is 20 minutes of arc in diameter and you measure its distance to be 1 Mpc, what is its diameter? (HINT: See Box 3-2.)

6. We have found a galaxy in which the outer stars have orbital velocities of 150 km/sec. If the radius of the galaxy is 4 kpc, what is the orbital period of the outer stars? (HINTS: 1 pc = 3.08×10^{13} km, and 1 yr = 3.15×10^7 sec.)

7. A galaxy has been found that is 5 kpc in radius and whose outer stars orbit the center with a period of 200 million years. What is the mass of the galaxy? On what assumptions does this result depend? (HINT: See Box 8-5.)

RECOMMENDED READING

BARNES, JOSHUA, LARS HERNQUIST, and FRANCOIS SCHWEIZER. "Colliding Galaxies." *Scientific American* 265 (August 1991), p. 40.

BURNS, JACK O. "Dark Matter in the Universe." *Sky and Telescope* 68 (November 1984), p. 396.

DRESSLER, ALAN. "Observing Galaxies Through Time." *Sky and Telescope* 82 (August 1991), p. 126.

GELLER, MARGARET J., and JOHN P. HUCHRA. "Mapping the Universe." *Sky and Telescope* 82 (August 1991), p. 134.

GINGERICH, OWEN, and BARBARA WELTHER. "Harlow Shapley and the Cepheids." *Sky and Telescope* 70 (December 1985), p. 540.

GREGORY, S., and N. MORISON. "The Formation of Galaxies and Clusters." *Mercury* 14 (May/June 1985), p. 85.

HARRIS, WILLIAM E. "Globular Clusters in Distant Galaxies." *Sky and Telescope* 81 (February 1991), p. 148.

HARTLEY, KAREN. "Elliptical Galaxies Forged by Collisions." *Astronomy* 17 (May 1989), p. 42.

HODGE, PAUL W. "The Local Group: Our Galactic Neighborhood." *Mercury* 16 (January/February 1987), p. 2.

KAUFMAN, MICHELE. "Tracing M 81's Spiral Arms." *Sky and Telescope* 73 (February 1987), p. 135.

KEEL, WILLIAM C. "Crashing Galaxies: Cosmic Fireworks." *Sky and Telescope* 77 (January 1989), p. 18.

KRAUSS, LAWRENCE M. "Dark Matter in the Universe." *Scientific American* 255 (December 1986), p. 58.

MACROBERT, ALAN. "No Missing Mass?" *Sky and Telescope* 70 (July 1985), p. 22.

MARSCHALL, LAURENCE. "Superclusters: Giants of the Cosmos." *Astronomy* 12 (April 1984), p. 6.

PERATT, ANTHONY. "Simulating Spiral Galaxies." *Sky and Telescope* 68 (August 1984), p. 118.

RUBIN, VERA C. "Dark Matter in Spiral Galaxies." *Scientific American* 248 (June 1983), p. 96.

SCHRAMM, DAVID N. "Origin of Cosmic Structure." *Sky and Telescope* 82 (August 1991), p. 140.

SCHWEIZER, FRANCOIS. "Colliding and Merging Galaxies." *Scientific American* 231 (17 January 1986), p. 227.

SILK, JOSEPH. "Formation of Galaxies." *Sky and Telescope* 72 (December 1986), p. 582.

SMITH, DAVID H. "Spirals from Order and Chaos." *Sky and Telescope* 74 (August 1987), p. 136.

TOOMRE, A., and J. TOOMRE. "Violent Tides Between Galaxies." *Scientific American* 233 (December 1973), p. 38.

VERSCHUUR, GERRIT. "Is the Milky Way an Interacting Galaxy?" *Astronomy* 16 (January 1988), p. 26.

14

Galaxies with Active Nuclei

Of the millions of galaxies visible through Earth-based telescopes, a surprising number do not fit the standard categories of elliptical, spiral, and irregular. Some of these peculiar galaxies appear to result from interactions between galaxies. Other peculiar galaxies have brilliant nuclei filled with high-temperature gas churning at high velocities, and a few show direct evidence of the ejection of matter in long jets racing out of the nucleus. This suggests that events of titanic violence are occurring in the centers of some galaxies.

Radio astronomers find the sky speckled with faint radio sources, most of which are distant galaxies radiating so much radio energy that they are detectable by radio telescopes even if they are much too far away to be recorded on photographs. Astronomers now believe that these galaxies are undergoing tremendous eruptions at

Virtually all scientists have the bad habit of displaying feats of virtuosity in problems in which they can make some progress and leave until the end the really difficult central problems.

M. S. Longair,
High Energy Astrophysics

their centers. That is, they have active nuclei.

Over the past 20 years, astronomers have come to understand the nature of these active nuclei. In this chapter, we will trace the growth of that understanding by considering the observations of these active galaxies and the models that astronomers have built to explain those observations.

A surprising number of astronomers object to the quotation that opens this chapter, perhaps because it seems to belittle scientists. But it is a realistic description of how scientists attack a problem. Big problems are made of little problems, and scientists begin by solving the little problems, learning from those solutions, and moving ever closer to the center of the big problem. Perhaps no better example can be found than modern astronomy's attack on "the really difficult central problem" in the nuclei of active galaxies.

14-1 ACTIVE GALACTIC NUCLEI

In the 1950s, radio astronomers began to notice that some galaxies were powerful sources of radio energy. Of course, all galaxies, including our own Milky Way, emit some radio energy, but some emit over 10 million times more radio energy than a normal galaxy. Such objects were first known as **radio galaxies.** But newer observations reveal that many of these galaxies also radiate at infrared, ultraviolet, and X-ray wavelengths, so these objects have become known as **active galaxies.** In many cases, the source of the activity lies in the nucleus of the galaxy, and such objects are known as **active galactic nuclei (AGN).**

Double-Lobed Radio Galaxies A double-lobed radio galaxy emits radio energy from two regions, called lobes, located on opposite sides of the galaxy. In some cases, the galaxy itself is silent at radio wavelengths.

Radio lobes are generally much larger than the galaxy they accompany (Figure 14-1). Many are as large as 60 kpc (200,000 ly) in diameter, twice the size of the Milky Way galaxy. From tip to tip, the radio lobes span hundreds of kiloparsecs. One of the largest-known radio galaxies is 3C 236 (the 236th object in the *Third Cambridge Catalogue of Radio Sources*). In fact, it is one of the largest objects in the universe, with radio lobes that reach across 5.8 Mpc (19×10^6 ly). Another radio galaxy, 3C 345, has radio lobes that span 24 Mpc.

Radio lobes have two properties that hint at their origin: they radiate synchrotron radiation and they are often most intense on the side away from the central galaxy. Recall that synchrotron radiation is produced when high-speed electrons spiral through a magnetic field (see Chapter 10). Though the field is at least 1000 times weaker than Earth's, it fills a tremendous volume and represents vast stored energy. In addition, the high-speed electrons also contain energy. The total energy in a radio lobe equals about 10^{53} J, approximately the energy equivalent of converting 10^6 $M_\odot$ directly into energy. Clearly, the process that creates radio lobes is powerful.

The second hint to the nature of the radio lobes is that many radio lobes emit more intensely from the side away from the galaxy (Figure 14-2). Not

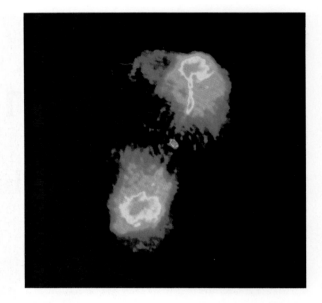

Figure 14-1 3C 388 is a typical double-lobed radio galaxy. The visible galaxy lies at the center and appears to supply energy to the radio lobes on either side of it. Such radio lobes can span millions of parsecs. (National Radio Astronomy Observatory, operated by Associated Universities, Inc. under contract with the National Science Foundation; observers J. O. Burns and W. A. Christiansen)

all radio lobes have these **hot spots,** but many do.

Synchrotron radiation and hots spots suggest an explanation for double-lobed radio galaxies now known as the **double-exhaust model.** According to this model, if an eruption in the core of the galaxy drove jets of high-energy gas outward in opposite directions, they could blow up a pair of radio lobes like bubbles in the intergalactic medium. The ejected gas would be ionized and would drag along part of the galaxy's magnetic field, and the high-speed electrons in the gas would produce synchrotron radiation. The beams could produce hot spots where they pushed against the far side of each radio lobe (Figure 14-3).

A prime example of a double-lobed radio galaxy is Cygnus A, the second-brightest radio source in the sky (see Figure 14-2). It radiates about 10 million times more radio energy than the Milky Way galaxy, all from two lobes containing tangled magnetic fields and bright leading edges. At least one jet is visible in radio maps. On optical photographs, the central galaxy is oddly distorted, but

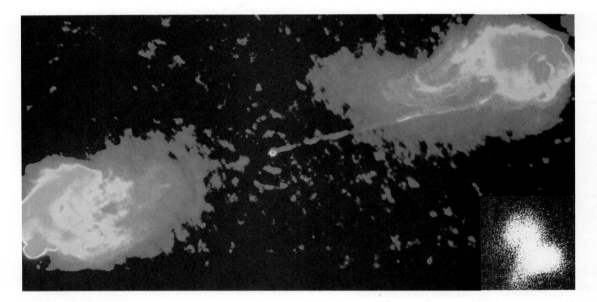

Figure 14-2 Cygnus A is a powerful, double-lobed radio galaxy. This radio map produced by the VLA radio telescope shows hot spots as brightenings on the outer edges of the lobes. Note the long, thin jet of matter flowing from the galaxy to the right-hand lobe. The jet is at least 50 kpc long—about 50 percent longer than the diameter of our galaxy. A fainter counterjet is also visible leading to the left lobe. In a visible-light photograph, the deformed central galaxy (inset) is roughly the size of the region between the radio lobes. (National Radio Astronomy Observatory; Palomar Observatory photograph)

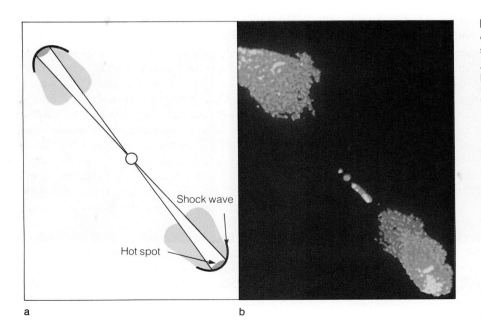

Shock wave

Hot spot

a   b

Figure 14-3 (a) In the double-exhaust model, the hot spots are produced by beams, or jets, of high-speed particles pushing into the intergalactic gas. (b) In this false-color radio map of the galaxy 3C 219, an active nucleus (red) ejects a narrow jet pointing toward one of the two radio lobes containing hot spots (yellow). (National Radio Astronomy Observatory/ Associated Universities, Inc.; observers A. H. Bridle and R. A. Perley)

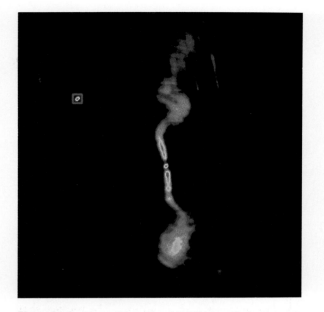

Figure 14-4 The active galaxy 3C 449 has spiral patterns in its radio lobes. They may be produced by a source of jets that orbits the nucleus at a distance of about 1 kpc. An alternate model suggests that the jets (each about 20 kpc long) precess much like the beams produced by the much smaller system SS 433. At visual wavelengths, the galaxy appears to be an elliptical galaxy about ten times the size of the central spot in this false-color image. (Courtesy Richard A. Perley)

Cygnus A is so far away, about 2000 Mpc, that the best photos show no detail.

In some cases, such as Cygnus A (see Figure 14-2), we can see jets of high-energy gas flowing from the nucleus of the galaxy out into the radio lobes. In galaxies such as 3C 449 (Figure 14-4), the jets are twisted by the motion of the galaxy.

Another fascinating radio galaxy is Centaurus A (Figure 14-5), a double-lobed source so close it covers 10° in the sky—as large as the bowl of the Big Dipper. Located between the radio lobes is NGC 5128, a peculiar giant elliptical galaxy encircled by a ring of dust and newly forming stars. Long-exposure photographs reveal a faint jet extending 40 kpc (1.3×10^5 ly) toward the northern radio lobe—as if matter were still being ejected from the galaxy.

More detailed studies of this galaxy reveal additional evidence of a central eruption. Higher-resolution radio maps provided by the VLA radio telescope show smaller lobes only 10 kpc from the center of the galaxy. These maps also show a jet of excited gas extending from the core to the northwest. X-ray maps from the Einstein satellite also show this jet. By linking together radio telescopes in Australia and South Africa, a team of radio astronomers has found a radio jet only about 1 pc

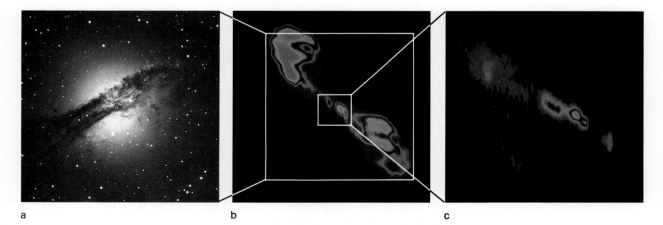

a   b   c

Figure 14-5 The radio source Centaurus A consists of a large pair of radio lobes at the center of which can be seen the peculiar galaxy NGC 5128 (a). Within this galaxy is a smaller set of radio lobes shown here in a false-color radio map (b). Roughly ten times smaller still, a jet of high-energy gas (c) extends out of the core of the galaxy (red). The small jet is evidently responsible for the much larger radio lobes. (a. National Optical Astronomy Observatories; b. and c. National Radio Astronomy Observatory, operated by Associated Universities, Inc. under contract with the National Science Foundation)

long at the very center of the galaxy. This jet also points toward the northwest radio lobe. Clearly, the center of the galaxy has erupted repeatedly over a long period of time to produce the lobes we detect.

Doppler shifts show that the bright, spherical part of the galaxy is rotating about an axis that lies in the plane of the dust ring, but the dust ring is rotating about an axis perpendicular to the plane of the ring. Thus the system has been interpreted as a merger in progress between a giant elliptical galaxy and a smaller spiral. We will see later how such a merger might trigger an eruption in a galaxy's core.

If our double-exhaust model is correct, the radio lobes are very tenuous. They do not contain sufficient gravity or strong enough magnetic fields to hold themselves together. Apparently, the lobes are held together by their collision with the intergalactic gas.

The presence of an intergalactic gas is clearly shown by the **head–tail radio galaxies.** These objects appear in a radio map as a bright head trailing a long tail. Apparently, these galaxies are ejecting material in beams shooting in two opposite directions, as in the case of normal double-lobed radio galaxies. The motion of the intergalactic medium past the galaxy, however, sweeps the ejected material away in a long tail like the plume from a twin smokestack (Figure 14-6a). This interpretation is supported by the observation that head–tail galaxies occur in clusters of galaxies where the intergalactic medium is relatively dense.

Some head–tail galaxies have complex structures such as separate blobs or spirals along the two streams. The blobs suggest that the beams may not remain "on" all the time, and the spirals suggest that the axis along which the beams shine may be precessing and spewing relativistic particles out of the galaxy in swirling streams (Figure 14-6b). Thus the head–tail radio galaxies appear to be caused by eruptions in the cores of the galaxies.

The double-exhaust model helps us understand the overall geometry of double-lobed radio galaxies, but it does not help us understand the central problem: What mechanism at the center of a galaxy can produce such eruptions? To pursue that part of the problem, we must consider active galaxies that lack radio lobes.

Galaxies with Erupting Nuclei Some galaxies have active nuclei but lack radio lobes. **Seyfert galaxies,** for example, are spiral galaxies with unusually bright, tiny cores that fluctuate in brightness (Fig-

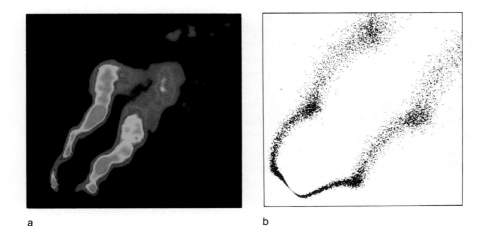

a
b

Figure 14-6 (a) VLA radio image of 3C 129 showing the spiral form of the streams produced by this head–tail galaxy. (b) Computer-generated stream pattern from a model using a precessing-beam source. (Map courtesy of Laurence Rudnick, University of Minnesota, and Jack Burns, University of New Mexico; model courtesy of Vincent Icke, University of Minnesota, and *The Astrophysical Journal,* published by the University of Chicago Press, © 1981 the American Astronomical Society)

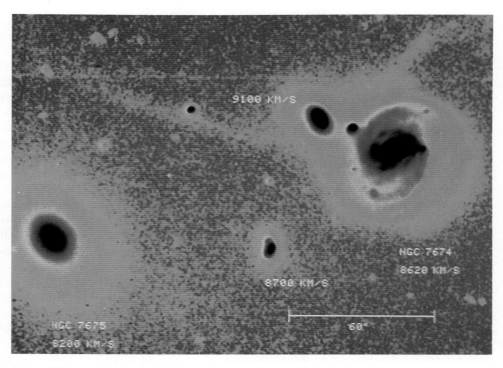

Figure 14-7 Seyfert galaxy NGC 7674 (upper right) has a small, bright core. This false-color image reveals tails of material to the top and to the upper left, which suggest that NGC 7674 is interacting with nearby galaxies. Such interactions may trigger Seyfert outbursts. (John W. Mackenty, Institute for Astronomy, University of Hawaii)

ure 14-7). Most Seyfert galaxies are powerful sources of infrared radiation, and some are radio sources. Of the roughly 120 known, 25 emit X rays and one is known to emit gamma rays.

Spectra of the nuclei reveal emission lines, evidence of very highly excited gas, but further study of the spectra divides Seyfert galaxies into two classes. Type 1 Seyferts have broad spectral lines typical of gas velocities over 1000 km/sec, whereas Type 2 Seyferts have narrow spectral lines, so the gas velocities must be significantly lower.

Since 2 percent of all spiral galaxies are Seyferts, we can conclude one of two things: either active cores occur in 2 percent of all spiral galaxies, or all spiral galaxies erupt 2 percent of the time. The heart of our own galaxy suggests that all spirals erupt now and then. We may be living in a galaxy that was a Seyfert only a few hundred million years ago.

Whereas Seyferts are spiral galaxies, **BL Lac objects** (also called *blasers*) appear to be elliptical galaxies. These objects are named after the first one discovered—object BL in the constellation Lacerta (the Lizard). They have no absorption or emission lines in their spectra, but when the light of the bright core is blocked, we see the faint spectrum of an elliptical galaxy. Evidently, BL Lac objects are the active cores of elliptical galaxies.

Some galaxies with active nuclei and no radio lobes emit jets. M 87, for example, is a giant elliptical galaxy located in the nearby Virgo cluster. Short-exposure photographs reveal a bright central nucleus and a jet of ionized matter 1800 pc long squirting out of the nucleus (Figure 14-8). Radio observations with the VLA radio telescope reveal that the jet emits synchrotron radiation, proving that it contains a magnetic field and high-speed electrons. Knots of gas along the jet may be instabilities in the flowing gas. By comparing VLA maps made years apart, astronomers have been able to see the motion of the knots as they flow outward at about 48 percent the speed of light.

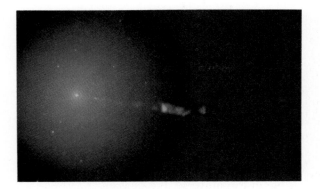

Figure 14-8 A jet 1800 pc long flows to the right out of the core of the giant elliptical galaxy M87 in this near infrared image from the Hubble Space Telescope. Note the narrow width of the jet and the small knots. The stars in the galaxy are strongly concentrated toward the center, further evidence of a massive black hole. Some theorists propose that a 3-billion solar mass black hole lies at the center of the galaxy. Faint starlike objects in the galaxy are globular clusters. (NASA)

What could power such eruptions? A few years ago, M 87 gave astronomers a clue. Observations of the core showed that the stars are tremendously crowded (Figure 14-8) and that the spectral lines are broad. The broad lines show that the stars move at high velocity. If the stars are held in the core by gravity, the core must be very massive—about $5 \times 10^9 M_\odot$. When this was first found, many astronomers suspected the matter was a massive black hole. It isn't certain that M 87 contains a massive black hole, but the concept of black holes in the cores of galaxies is popular. Similar studies suggest that the Seyfert galaxy NGC 4151 contains a central body of $10^8 M_\odot$ with matter flowing inward. Also, the nearby Andromeda galaxy has 10^7–$10^8 M_\odot$ at its core. If these central masses are black holes, matter falling in could power the active cores.

If we combine the double-exhaust model with the proposal that active galactic nuclei harbor massive black holes, we can create a comprehensive model that helps us understand not only what happens inside active galaxies but also what causes galaxies to become active.

A Unified Model of Active Galactic Nuclei If at least some galaxies contain massive black holes at their centers, then matter flowing into the black hole could release tremendous energy. As we saw in Chapter 11, the matter would form an accretion disk around the black hole. As friction slowed the matter enough to fall into the black hole, the inner part of the accretion disk would become very hot.

The hot inner parts of an accretion disk can produce jets of hot gas and radiation streaming out of the nucleus along the axis of rotation of the disk. If the disk were thick, the jets would be focused into the narrow beams commonly observed. Jets are believed to be common in active galactic nuclei.

Our unified model can explain why we see two types of Seyfert galaxies. Although we cannot resolve the core of a Seyfert galaxy and see the accretion disk, we can imagine that the two types of galaxies arise because of different inclinations of the accretion disk. Type 1 Seyfert galaxies seem to be inclined so that light reaches us from deep inside the hole in the accretion disk and the jet is pointed nearly toward us. Thus we see broad spectral lines because the gas is very hot and moving rapidly and the Doppler shift makes the lines broad. Type 2 Seyfert galaxies are inclined so that the disk is nearly edge on, and thus we don't receive light directly from the central hole and jet. The light we see comes from slower-moving gas farther from the black hole, and thus we see narrow spectral lines.

In the same way, we see a BL Lac object when the thick accretion disk at the center of an elliptical galaxy is tipped so that we look directly into the central opening and the jet is pointed nearly toward us. The intense light from the central region drowns out the rest of the galaxy.

The unified model also explains what makes galaxies become active. If all or at least some galaxies contain a massive black hole, then a galaxy can become active only when matter is flowing into the black hole. So long as no matter flows inward, the black hole is quiescent and the galaxy remains inactive. But a collision with another galaxy could disrupt the orbital motion of matter in the galaxy and feed the black hole, triggering an eruption.

Recent observations suggest that interactions between galaxies are important in triggering eruptions. A high-quality image of the peculiar galaxy NGC 5128 (see Figure 14-5) shows that it is surrounded by faint shells of scattered stars (see Figure 13-18). Calculations show that shells occur when galaxies collide. Another recent study found that Seyfert galaxies are three times more common among interacting pairs of galaxies and

a

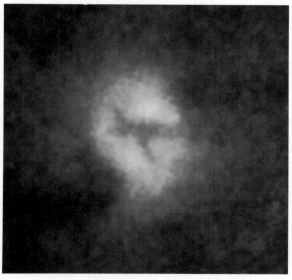

b

Figure 14-9 (a) The core of peculiar galaxy Arp 220 in a true color image from the Hubble Space Telescope. Known as a "starburst" galaxy, Arp 220 appears to be the result of a collision between two galaxies that created active nuclei and stimulated the formation of star clusters up to 10 times the mass of globular clusters. Such collisions are the suspected trigger of most active galactic nuclei. (b) The core of otherwise normal galaxy M51 shows a thick dust lane believed to be the edge-on disk around a central black hole. (See Figure 13-16.) (NASA)

that roughly 25 percent have distorted shapes suggesting tidal interactions with other galaxies. Perhaps collisions are necessary to trigger episodes of Seyfert behavior. Models show that collisions could direct enough matter into a central black hole to power an active nucleus (Figure 14-9).

We can't be sure that massive black holes lie at the centers of active galaxies, but the evidence supporting the unified model continues to mount. In fact, the unified model of active galactic nuclei can help us understand the most distant objects visible in the universe, the quasars.

14-2 QUASARS

Quasars (also called *quasi-stellar objects*, or *QSOs*) are small, powerful, but extremely remote sources of radiation. Astronomers are now using the unified model of active galactic nuclei to understand quasars as very distant, very powerful active galaxies. Their significance, however, goes far beyond this interpretation. They are so distant that their look-back times are tremendous—as much as 90 percent of the age of the universe. Thus the discovery of quasars in the 1960s gave astronomers an intriguing new view of the most distant parts of universe.

The Discovery of Quasars In the early 1960s, radio interferometers (see Chapter 5) showed that a number of radio sources are much smaller than normal radio galaxies. Photographs of the location of these radio sources did not reveal a central galaxy, not even a faint wisp, but rather a single star-like point of light. The first of these objects so identified was 3C 48, and later the source 3C 273 was added. Though these objects emitted radio signals like those from radio galaxies, they were obviously not normal radio galaxies. Even the most distant photographable galaxies look fuzzy, but these seemed like stars. Their spectra, however, were totally unlike stellar spectra, so the objects were called quasi-stellar objects (Figure 14-10).

For a few years, the spectra of quasars were a mystery. A few unidentifiable emission lines were superimposed on a continuous spectrum. But in 1963, Maarten Schmidt at Hale Observatories tried red-shifting the hydrogen Balmer lines to see if they could be made to agree with the lines in 3C 273's spectrum. At a red shift of 15.8 percent, three lines clicked into place (Figure 14-11). Other quasar spectra quickly yielded to this approach, revealing even larger red shifts.

The red shift Z is the change in wavelength $\Delta\lambda$ divided by the unshifted wavelength λ_0:

$$\text{red shift} = Z = \frac{\Delta\lambda}{\lambda_0}$$

Astronomers commonly use the Doppler formula to convert these red shifts into radial velocities. Thus the large red shifts of the quasars imply large radial velocities.

To understand the significance of these large radial velocities, we must draw on the Hubble law (see Box 13-1). Recall that the Hubble law states that galaxies have radial velocities proportional to their distances, and thus the distance to a quasar is equal to its radial velocity divided by the Hubble constant. The large red shifts of the quasars imply that they must be at great distances.

The red shift of 3C 273 is 0.158, and the red shift of 3C 48 is 0.37. These are large red shifts, but not as large as the largest then known for galaxies, about 1.0. Soon, however, quasars were found with red shifts much larger than that of any known

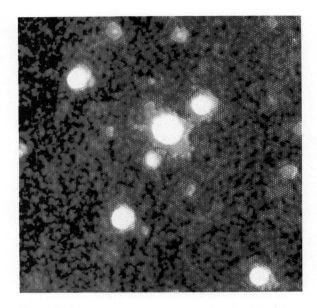

Figure 14-10 Quasars look starlike in photographs and are not obviously galaxies, thus the name ''quasi-stellar objects.'' In this CCD image of the quasar 3C 275.1 (bright image near center) made with the 4-m telescope atop Kitt Peak, a small amount of nebulosity is visible around the quasar. (National Optical Astronomy Observatories)

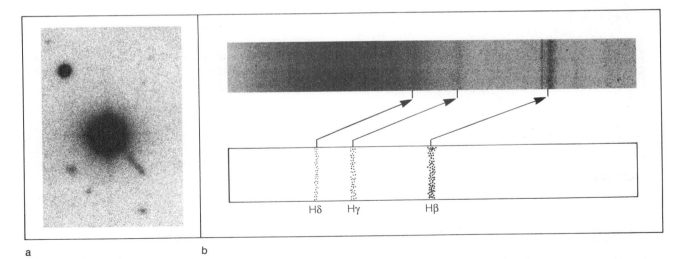

a   b

Figure 14-11 (a) This negative image of 3C 273 shows the bright quasar at the center surrounded by faint fuzz. Note the jet protruding to lower right. (b) The spectrum of 3C 273 (top) contains three hydrogen Balmer lines red-shifted by 15.8 percent. The drawing shows the unshifted positions of the lines. (Courtesy Maarten Schmidt)

The Relativistic Red Shift

Many quasars have red shifts greater than 1, and if we put such large red shifts into the Doppler formula from Box 6-4, we get velocities greater than the speed of light. This should not worry us, however, for two reasons. First, the formula given in Box 6-4 is only an approximation good for small velocities. It is correctly called the classical Doppler formula. For very high velocities, we should use the **relativistic red shift** formula. Second, the red shifts of the galaxies are not true Doppler shifts. As we will see in Chapter 15, they are caused by the expansion of the universe as a whole. Although the true formula for the radial velocity of a galaxy or quasar depends on details of the expansion of the universe not yet known, it does not differ dramatically from the relativistic Doppler formula. Thus we can use the relativistic Doppler formula to show how relativity limits the velocities of objects even when their red shifts exceed 1.

The theory of relativity gives the following formula for the Doppler effect, where V_r is the radial velocity, c is the speed of light, and Z is the red shift:

$$\frac{V_r}{c} = \frac{(Z + 1)^2 - 1}{(Z + 1)^2 + 1} \text{ where } Z = \frac{\Delta\lambda}{\lambda_0}$$

Example: If a quasar has a red shift of 2, what is its radial velocity? *Solution:* First, Z + 1 equals 3, and thus $(Z + 1)^2$ equals 9. Then the velocity in terms of the speed of light is:

$$\frac{V_r}{c} = \frac{9 - 1}{9 + 1} = \frac{8}{10} = 0.8$$

That is, the quasar has a radial velocity equal to 80 percent the speed of light.

Figure 14-12 illustrates how V_r/c depends on Z. For small red shifts, the velocity is the same in both the classical and the relativistic case, but as Z gets larger, the difference between the classical approximation and the true velocity increases. No matter how large Z becomes, the velocity can never quite equal the speed of light.

Figure 14-12 At high velocities, the relativistic red shift must be used in place of the classical approximation. Note that no matter how large Z gets, the speed can never equal that of light.

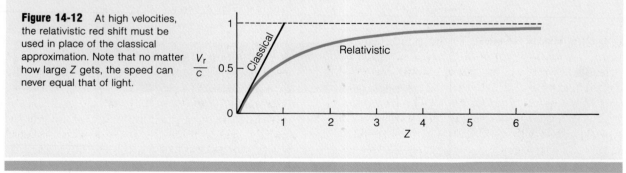

galaxy. Some quasars are evidently so far away that galaxies at those distances are very difficult to detect. Yet the quasars are easily photographed. This leads us to the conclusion that quasars must have 10–1000 times the luminosity of a large galaxy. Quasars must be superluminous.

Soon after quasars were discovered, astronomers detected fluctuations in brightness over times as short as a few days. Recall from our discussion of pulsars (see Chapter 11) that an object cannot change its brightness appreciably in less time than it takes light to cross its diameter. The rapid fluctuations in quasars showed that they are small objects, not more than a few light-days or light-weeks in diameter.

Thus by the late 1960s astronomers faced a

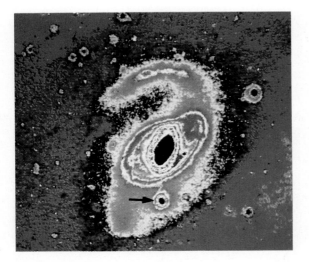

Figure 14-13 Supporters of the local hypothesis argue that quasar Markarian 205 (arrow) is connected to the galaxy NGC 4319 and thus must be no more distant than the galaxy. Therefore the large red shift of the quasar is not a measure of the distance to the galaxy. Most astronomers, however, conclude that the quasar and galaxy are at different distances and only appear to be connected. (Courtesy Peter Wehinger)

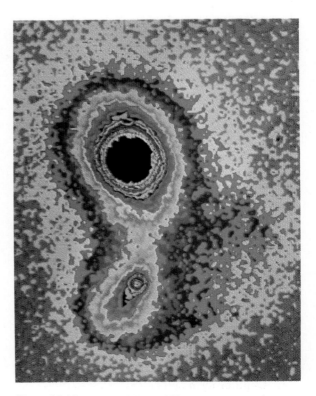

Figure 14-14 Quasar 0351 + 026 (top) appears to be interacting with a faint galaxy (below). The two objects have similar red shifts and are thus presumed to be at the same distance. This image has been computer-enhanced and reproduced in false color to bring out details of interaction. (National Optical Astronomy Observatories)

problem: How could quasars be superluminous but also be very small? Since then, evidence has accumulated that quasars are the active cores of very distant galaxies. Thus our unified model of active galactic nuclei can help us understand them.

Quasars in Distant Galaxies To this point, astronomers have found about 5000 quasars, of which 97 percent are radio-quiet. Most quasars have red shifts larger than 1, which implies very large velocities of recession. (See Box 14-1.) The largest red shifts exceed 4, indicating that those quasars are receding at over 90 percent the speed of light. Nearly all astronomers assume these large red shifts imply the quasars are very far away.

One theory proposed during the 1960s, known as the **local hypothesis,** suggested that the quasars are local, not distant, objects and thus not superluminous (Figure 14-13). However, this theory failed to explain what caused the large red shifts of the quasars. Although a few astronomers still argue for it, the local hypothesis has now been largely abandoned in light of a growing mass of evidence showing that quasars are indeed located in distant galaxies.

For instance, long-exposure photographs of some quasars reveal that they are surrounded by extremely faint images of galaxies (Figure 14-14). The red shifts of the galaxies are the same as the red shift of the quasar, implying that the cluster and the quasar lie at the same distance. Related observations show that quasars are often grouped together, just as clusters and superclusters of galaxies are grouped. Presumably, this shows that quasars are located in some of the galaxies in these clusters.

Another important piece of evidence is the structure that is visible in the images of some quasars. For example, 3C 273 shows a jet of material (see Figure 14-11a) much like the jet in M 87. Many quasars, including 3C 273, show a slight fuzziness, and some have surrounding wisps. Spectra of this quasar fuzz are identical with spectra of

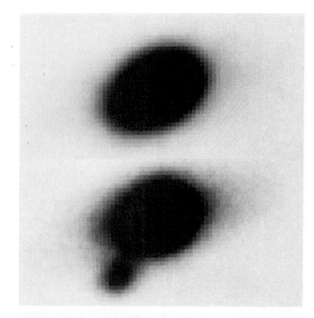

Figure 14-15 Two negative images of the quasar 1059 + 730. The lower image shows an extra point of light— apparently a supernova that exploded in the galaxy associated with the quasar. (Courtesy Bruce Campbell)

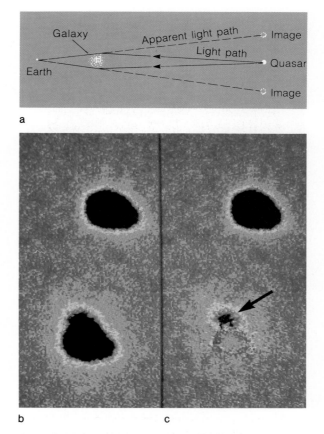

Figure 14-16 The gravitational lens effect. (a) A distant quasar can appear to us as multiple images if its light is deflected and focused by the mass of an intervening galaxy. The galaxy, being much fainter than the quasar, is not easily visible from the earth. (b) The two images of quasar 0957 + 561 appear as black blobs in this computer-enhanced false-color image. (c) When the upper image is subtracted from the lower image, we are able to detect the faint image of the giant elliptical galaxy (arrow) whose gravity produces the gravitational lens effect. (Courtesy Alan Stockton, Institute for Astronomy, University of Hawaii)

distant galaxies with the same red shift as that of the quasar. Quasar 3C 273, for example, is in a giant elliptical galaxy.

If quasars are the cores of active galaxies, then the unified model of active galactic cores would explain the jets visible near some quasars. In addition, radio maps of some quasars reveal double radio lobes just like those that flank some active galaxies.

Still more evidence appeared when astronomers noticed a new point of light near quasar QSO 1059 + 730. The point of light was a supernova exploding in the galaxy that hosts the quasar (Figure 14-15). The appearance of the supernova confirms that the quasar is located in a distant galaxy.

A discovery made in 1979 has gone even further to prove that quasars lie at great distances. The object 0957 + 561 lies just a few degrees west of the bowl of the Big Dipper and consists of two quasars separated by only 6 seconds of arc. The spectra of these quasars are identical and even have the same red shift (1.4136). Quasar spectra are as different as fingerprints, so when two quasars so close together proved to have the same spectra

and the same red shift, astronomers concluded that they were separate images of the same quasar.

The two images are formed by a **gravitational lens,** an effect first predicted by Einstein in 1936 but never seen before. A galaxy between us and the quasar bends the light from the quasar and focuses it into multiple images (Figure 14-16a). The galaxy itself is much too far away to be easily visible against the glare of the quasar, but Alan Stockton has used computer image processing to subtract one image of the quasar from the other. This re-

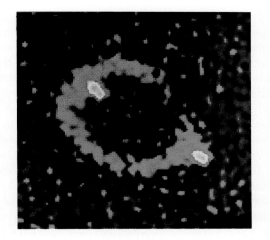

Figure 14-17 Radio source MG 1131 + 0456 appears as a nearly complete ring in this VLA image. In this example of a gravitational lens, a massive galaxy too faint to be visible bends radio waves from a very distant active galaxy to form the ring. Such gravitational lenses are known as Einstein rings. (Courtesy National Radio Astronomy Observatory/ Associated Universities, Inc., J. N. Hewitt, and E. L. Turner)

veals the nucleus of a giant elliptical galaxy (Figure 14-16b and 14-16c), probably the brightest member of a cluster. The red shift of the galaxy is only 0.36, much less than that of the quasar (1.4136).

The gravitational lens effect producing the multiple images of quasar 0957 + 561 is clear evidence that the quasar is very distant. It has to be far beyond the giant elliptical galaxy, and that galaxy is so distant it is hardly visible.

To date, at least six examples of gravitational lenses have been found. The VLA image in Figure 14-17 is a radio-wavelength gravitational lens of the sort called an **Einstein ring.** Radio energy from a distant active galaxy is imaged by a massive galaxy between us and the active galaxy. The massive galaxy is itself too faint to be seen, but it bends the radio waves from the distant active galaxy into a nearly perfect ring. Such Einstein rings can occur only when the lensing body is nearly in line with the distant object.

A Model Quasar Drawing on our experience with active galaxies, it is now possible to expand our model of active galactic nuclei to include quasars. In fact, we see a continuum of activity, from the modest energy released by the nuclei of galaxies like our own, to that released by Seyferts, BL Lac objects, double-lobed radio galaxies, and so on up to the powerful eruptions of quasars.

Consequently, our model of a quasar consists of a galaxy containing a massive black hole into which large amounts of matter are flowing. An influx of 1 $M_\odot$ per year into a $10^7 - 10^9$ $M_\odot$ black hole could power a typical quasar.

The energy generated near the black hole is thought to drive jets of ionized gas and radiation that blast outward along the axis of rotation of the disk. In this respect, the best-guess model of quasars resembles a massive version of SS 433, which appears to be a normal neutron star or black hole ejecting jets of gas and radiation. The radio energy we receive from quasars may come from the gas in these jets, and the continuous spectrum we see may form from the hot gas in the inner disk. Because this is a very small region, rapid fluctuations can occur.

The best evidence is that quasar emission lines do not fluctuate rapidly, so they need not arise in the small core. Rather, they could originate in hot gas surrounding the core in a region many light-years in diameter. This gas is kept hot by the synchrotron radiation streaming out of the core, and the width of the emission lines indicates that the gas is turbulent, apparently because of the intense radiation.

The orientation of the accretion disk is clearly important. If the accretion disk of a quasar is tipped so that a jet is pointed directly toward us, we may see one kind of quasar. But if the disk is tipped slightly so the jet is not directed exactly toward us, we may see a slightly different kind of quasar. Astronomers are now using this unified model to sort out the different kinds of quasars and active galaxies so we can understand how they are related. For example, using infrared radiation to penetrate dust, astronomers observed the core of the double-lobed radio galaxy Cygnus A (see Figure 14-2) and found an object much like a quasar. Astronomers have begun to refer to such objects as buried quasars.

As in the case of active galaxies, many quasars are found in distorted galaxies that are interacting with nearby companions. Such a tidal interaction could throw matter into a central black hole and trigger a quasar outburst. Figure 14-18 shows three quasars that appear to illustrate different stages in such an encounter.

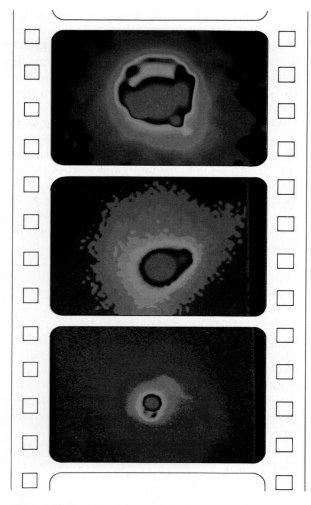

Figure 14-18 These three quasars appear to illustrate three stages in the triggering of a quasar by the tidal interaction between galaxies. The quasar IRAS 00275-2859 (top) is located in a galaxy interacting with a relatively distant companion (not visible in this image) and thus represents an early stage. In the galaxy containing quasar PG 1613 (middle), the galaxies are closer together. The nucleus of the companion is visible at the 2 o'clock position, and the outline of the interacting galaxies is distorted. Markarian 231 (bottom) is located in a severely distorted galaxy containing radial streaks and a tail to the lower right. Such features are common when galaxies merge, suggesting that Markarian 231 represents a late stage in the merger of two galaxies. (National Optical Astronomy Observatories)

It is important to remember that the look-back time to quasars is very large. We see them as they were when the universe was young, and so quasars may represent an early stage in the history of the universe. The collapse of great gas clouds to form galaxies would have released tremendous energy and could have led to the formation of massive black holes at the centers of galaxies. In addition, when the universe was younger, it had not expanded to the present extent, a subject we will discuss in the next chapter. Thus, when galaxies were young, they tended to be closer together and must have suffered tidal distortions more often. We have seen that such encounters can trigger eruptions in the centers of active galaxies and may be related to the formation of quasars.

Our study of peculiar galaxies has led us far out in space and back in time to quasars. The light now arriving from the most distant quasars left when the universe was only about one-fourth its present age. The quasars stand at the very threshold of the study of the universe itself. In Chapter 15, we take up that study.

SUMMARY

Some galaxies appear to be suffering eruptions in their cores. Such active core galaxies are powerful sources of energy and appear to be related to the in-fall of matter into central, massive black holes.

Double-lobed radio galaxies emit radio energy from lobes on either side of the galaxy. In some cases, we can see jets leading from the core of the galaxy out into the lobes. The double-exhaust model proposes that the core of the galaxy ejects relativistic particles in two jets that punch into the intergalactic medium and blow up the radio lobes like balloons. The impact of the beams on the intergalactic medium produces the hot spots detected on the outer edges of many radio lobes. We see many active galaxies producing jets, and we can see instances where the jets are blown back by the intergalactic medium to produce head–tail galaxies.

Some galaxies, such as Seyfert galaxies and BL Lac objects, have small, brilliant cores. Giant elliptical galaxies such as M 87 have bright cores with jets.

The energy source in active galactic nuclei appears to be mass flowing into massive black holes. Active galaxies are often distorted by interactions with other galaxies, and such tidal interactions could throw matter into the central black holes and thus trigger eruptions.

The quasars appear to be related objects. Their spectra show emission lines with very large red shifts. The large red shifts are interpreted to mean the quasars are at great distance and thus must be superluminous. Yet, because they fluctuate in brightness over intervals as short as a week or less, quasars must also be very small. The detection of quasars in clusters of galaxies, gravitational lenses, and the galaxylike spectra of quasar fuzz show that quasars are indeed located inside galaxies at great distances.

Growing evidence suggests that quasars are the active cores of very distant galaxies. As in the case of active galaxies, quasars are often found in galaxies that are distorted or interacting with other galaxies. The rapid flow of matter into a central, massive black hole could produce so much energy that the quasars would be visible at great distances.

Because they lie at great distances, we see quasars as they were long ago. The look-back time to the most distant quasar is more than 15 billion years. Thus they may represent an early stage in the formation of galaxies. The first collapse of large gas clouds may form massive black holes and release tremendous energy in a small region. Hot gas around the region could produce the observed emission lines.

NEW TERMS

radio galaxy	head–tail radio galaxy
active galaxy	Seyfert galaxy
active galactic nuclei (AGN)	BL Lac object
	quasar
double-lobed radio galaxy	local hypothesis
	relativistic red shift
hot spot	gravitational lens
double-exhaust model	Einstein ring

REVIEW QUESTIONS

1. What is the difference between the terms *radio galaxy* and *active galaxy?*

2. What evidence do we have that eruptions in the cores of galaxies are common?

3. What evidence do we have that radio lobes are inflated by jets from active galactic nuclei?

4. What evidence do we have that M 87 contains a black hole?

5. How does our unified model of active galactic nuclei explain the difference between the two types of Seyfert galaxies?

6. Why do we conclude that quasars must be very small but superluminous?

7. What evidence do we have that quasars are located inside distant galaxies?

8. How does our model of active galactic nuclei relate quasars to active galaxies?

DISCUSSION QUESTIONS

1. How does the geometry of a quasar resemble the geometry of an active galactic nuclei? of SS 433? of a protostar?

2. Are active galaxies fundamentally different from normal galaxies, or are they normal galaxies in an active phase?

PROBLEMS

1. The total energy stored in a radio lobe is about 10^{53} J. How many solar masses would have to be converted to energy to produce this energy? (HINTS: Use $E = mc^2$. One solar mass equals 2×10^{30} kg.)

2. If the jet in NGC 5128 is traveling at 5000 km/sec and is 40 kpc long, how long will it take for gas to travel from the core of the galaxy to the end of the jet? (HINT: 1 pc equals 3×10^{13} km.)

3. Cygnus A is roughly 225 Mpc away, and its jet is about 50 seconds of arc long. What is the length of the jet in parsecs? (HINT: See Box 3-2.)

4. If the average giant elliptical galaxy is 50 kpc in diameter and the visible galaxy associated with 3C

449 (see Figure 14-4) is 1 minute of arc in diameter, calculate an estimate of the distance to the galaxy. (HINT: See Box 3-2.)

5. The active core of a galaxy 3.25 Mpc from us is observed to have an angular diameter of 0.0015 seconds of arc. What is its linear diameter? (HINT: See Box 3-2.)

6. If a quasar is 1000 times more luminous than an entire galaxy, what is the absolute magnitude of such a quasar? (HINTS: The absolute magnitude of a bright galaxy is about −21. See Box 8-2.)

7. If the quasar in Problem 6 were located at the center of our galaxy, what would its apparent magnitude be? (HINTS: See Box 8-2 and ignore dimming by dust clouds.)

8. What is the radial velocity of 3C 48 if its red shift is 0.37? (HINT: See Box 14-1.)

9. If the Hubble constant is 70 km/sec/Mpc, how far away is the quasar in Problem 8? (HINT: See Box 13-1.)

10. The hydrogen Balmer line Hβ has a wavelength of 486.1 nm. It is shifted to 563.9 nm in the spectrum of 3C 273. What is the red shift? (HINT: What is $\Delta\lambda$?)

RECOMMENDED READING

ARP, HALTON. *Quasars, Redshifts, and Controversies.* Berkeley, Calif.: Interstellar Media, 1987.

BALICK, BRUCE. "Quasars with Fuzz." *Mercury* 12 (May/June 1983), p. 81.

BARNES, JOSHUA, LARS HERNQUIST, and FRANCOIS SCHWEIZER. "Colliding Galaxies." *Scientific American* 265 (August 1991), p. 40.

BLANDFORD, ROGER D., MITCHELL C. BEGELMAN, and MARTIN J. REES. "Cosmic Jets." *Scientific American* 246 (May 1982), p. 124.

BURNS, JACK O., and R. MARCUS PRICE. "Centaurus A: The Nearest Active Galaxy." *Scientific American* 249 (November 1983), p. 56.

———. "Chasing the Monster's Tail." *Astronomy* 18 (August 1990), p. 28.

CAPRIOTTI, EUGENE R. "Seyfert Galaxies." *Natural History* 90 (February 1981), p. 82.

COURVOISIER, THIERRY J.-L., and E. IAN ROBSON. "The Quasar 3C 273." *Scientific American* 264 (June 1991), p. 50.

DOWNS, ANN. "Radio Galaxies." *Mercury* 15 (March/April 1986), p. 34.

FEIGELSON, ERIC D., and ETHAN J. SCHREIER. "The X-Ray Jets in Centaurus A and M 87." *Sky and Telescope* 65 (January 1983), p. 6.

FIENBERG, RICHARD TRESCH. "IRAS and the Quasars." *Sky and Telescope* 73 (January 1987), p. 13.

GORENSTEIN, MARC V. "Charting Paths Through Gravity's Lens." *Sky and Telescope* 66 (November 1983), p. 390.

GOTT, J. RICHARD, III. "Gravitational Lenses." *American Scientist* 71 (March/April 1983), p. 150.

HOFF, DARREL B. "Laboratory Exercises in Astronomy —Quasars." *Sky and Telescope* 63 (January 1982), p. 20.

KANIPE, JEFF. "M-87: Describing the Indescribable." *Astronomy* 15 (May 1987), p. 6.

———. "Quest for the Most Distant Objects in the Universe." *Astronomy* 16 (June 1988), p. 20.

———. "Anatomy of a Cosmic Jet." *Astronomy* 16 (July 1988), p. 30.

———. "The Curious Shapes of Cosmic Jets." *Astronomy* 17 (March 1989), p. 40.

LEA, SUSAN. "The Astrophysical Zoo: M 87." *Mercury* 12 (January/February 1983), p. 25.

MARAN, STEPHEN P. "Cosmic Collisions." *Natural History* 102 (August 1987), p. 22.

MORRISON, N., and S. GREGORY. "Centaurus A: The Nearest Active Galaxy." *Mercury* 13 (May/June 1984), p. 75.

———. "A Remarkable Image of Cygnus A." *Mercury* 14 (March/April 1985), p. 55.

PERATT, ANTHONY L. "Are Black Holes Necessary?" *Sky and Telescope* 66 (July 1983), p. 19.

REES, MARTIN J. "Black Holes in Galactic Centers." *Scientific American* 263 (November 1990), p. 56.

RODRIGUEZ, LUIS. "Cosmic Jets: Bipolar Outflows in the Universe." *Astronomy* 12 (June 1984), p. 66.

SCHWARZSCHILD, BERTRAM. "Probing the Early Universe with Quasar Light." *Physics Today* (November 1987), p. 17.

SHAWCROSS, WILLIAM E. "Multiple Quasars and Gravitational Lenses." *Sky and Telescope* 60 (December 1980), p. 486.

SMITH, DAVID H. "Mysteries of Cosmic Jets." *Sky and Telescope* 69 (March 1985), p. 213.

SULENTIC, JACK W. "Are Quasars Far Away?" *Astronomy* 12 (October 1985), p. 66.

TRIMBLE, VIRGINIA, and LODEWIJK WOLTJER. "Quasars at 25." *Science* 234 (10 October 1986), p. 155.

TURNER, EDWIN. "Quasars and Gravitational Lenses." *Science* 223 (23 March 1984), p. 1255.

WRIGHT, A., and H. WRIGHT. *At the Edge of the Universe.* New York: Wiley (Horwood), p. 1989.

Cosmology

The Universe, as has been observed before, is an unsettlingly big place, a fact which for the sake of a quiet life most people tend to ignore.

Douglas Adams,
The Restaurant at the End of the Universe

What is the biggest number? A billion? How about a billion billion? That is only 10^{18}. How about 10^{100}? That very large number is called a googol, a number that is at least a billion billion times larger than the total number of atoms in all of the galaxies we can observe. Can you name a number bigger than a googol? Try a billion googols, or better yet, a googol of googols. You can go even bigger than that, 10 raised to a googol. (That's called a googolplex.) No matter how large a number you name, we can name a bigger number. That is what infinity means—big without limit.

If the universe is infinite in size, then you can name any distance you want, and the universe is bigger than that. Try a googol to the googol light-years. If the universe is infinite, there is more universe beyond that distance.

Is the universe infinite or finite? In this chapter, we will try to answer that question and oth-

ers, not by playing games with big numbers, but through **cosmology**, the study of the universe as a whole.

If the universe is not infinite, then we face the edge–center problem. Suppose that the universe is not infinite, and you journey out to the edge of the universe. What do you see: a wall of cardboard? a great empty space? nothing, not even space? What happens if you try to stick your head out beyond the edge? These are almost nonsensical questions, but they illustrate the problem we have when we try to think about an edge to the universe. Modern cosmologists believe the universe cannot have an edge. In this chapter, we will see how the universe might be finite but have no edge.

If the universe has no edge, then it can have no center. We find the centers of things—globular clusters, sheets of paper, bowling balls, and pizzas—by referring to their edges.

If the universe is not eternal (infinite in age), then it had a beginning, and we must think about how and when the universe began. Of course, we must also think about the possibility that the universe may eventually end.

15-1 THE STRUCTURE OF THE UNIVERSE

We begin our study of cosmology by considering the most basic property of the universe—its geometry in space and time. To help us, we will discuss the two fundamental observations of cosmology. The first of these observations is easy to make: it gets dark at night. That observation can tell us that the universe had to have a beginning.

Olbers' Paradox and the Necessity of a Beginning   We have all noticed that the night sky is dark. However, reasonable assumptions about the geometry of the universe can lead us to the conclusion that the night sky should glow as brightly as a star's surface. This conflict between observation and theory is called **Olbers' paradox** after Heinrich Olbers, a Viennese physician and astronomer who discussed the problem in 1826.

However, Olbers' paradox is not Olbers'. The problem of the dark night sky was first discussed by Thomas Digges in 1576 and was further analyzed by such astronomers as Johannes Kepler in 1610 and Edmond Halley in 1721. Olbers gets the credit through an accident of scholarship on the part of modern cosmologists who did not know of previous discussions. What's more, Olbers' paradox is not a paradox. We will be able to understand why the night sky is dark by revising our assumptions about the nature of the universe.

To begin, let's state the so-called paradox. Suppose the universe is static, infinite, eternal, and uniformly filled with stars. (The aggregation of stars into galaxies makes no difference to our argument.) If we look in any direction, our line of sight must eventually reach the surface of a star (Figure 15-1). Consequently, every point on the surface of the sky should be as bright as the surface of a star, and it should not get dark at night.

Of course, the most distant stars would be much fainter than the nearer stars, but there would be a greater number of distant stars than nearer stars. The intensity of the light from a star decreases according to the inverse square law, so distant stars would not contribute much light. However, the farther we look in space, the larger the volume we survey. Thus the number of stars we see at any given distance increases as the square of the distance. The two effects cancel out, and the stars at any given distance contribute as much total light as the stars at any other distance. Then, given our assumptions, every spot on the sky must be occupied by the surface of a star, and it should not get dark at night.

Imagine the entire sky glowing with the brightness of the surface of the sun. The glare would be overpowering. In fact, the radiation would rapidly heat the earth and all other celestial objects to the average temperature of the surface of the stars, 1000 K at least. Thus we can pose Olbers' paradox in another way: Why is the universe so cold?

Olbers assumed that the sky was dark because clouds of matter in space absorb the radiation from distant stars. But this interstellar medium would gradually heat up to the average surface temperature of the stars, and the gas and dust clouds would glow as brightly as the stars.

Today, cosmologists believe they understand why the sky is dark. Olbers' paradox makes the incorrect prediction that the sky should be bright because it is based on two incorrect assumptions: the universe is neither static nor infinitely old.

In Chapter 13, we saw that the galaxies are receding from us. The distant stars in these galaxies are receding from the earth at high velocity, and their light is Doppler-shifted to long wavelengths. We can't see the light from these stars because their light is red-shifted, and the energy of the photons is reduced to levels we cannot detect. Expressed in another way, the universe is cold because the photons from very distant stars arrive with such low energy they cannot warm objects by any significant amount. Although this explains part of the problem, the red shifts of the distant galaxies are not enough to make the night sky as dark as it appears. The red shifts explain no more than 10 percent of the problem.

The second part of the explanation was first stated by Edgar Allan Poe in 1848. He proposed that the night sky was dark because the universe was not infinitely old but had been created at some time in the past. The more distant stars are so far away that light from them has not reached us yet. That is, if we look far enough, the look-back time is

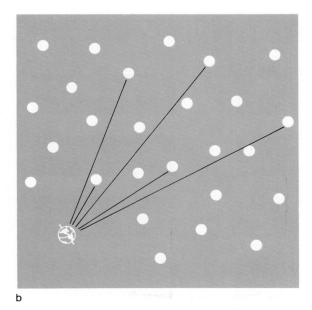

a

b

Figure 15-1 (a) Every direction we look in a forest eventually reaches a tree trunk, and we cannot see out of the forest. (b) If the universe is infinite and uniformly filled with stars, then any line from Earth should eventually reach the surface of a star. This predicts that the night sky should glow as brightly as the surface of the average star, a puzzle commonly referred to as Olbers' paradox. (Photo courtesy Janet Seeds)

greater than the age of the universe, and we look back to a time before stars began to shine. Thus the night sky is dark because the universe is not infinitely old.

This is a powerful idea because it clearly illustrates the difference between the universe and the observable universe. The universe is everything that exists, and it could be infinite. But the observable universe is the part that we can see. We will learn later that the universe is 10–20 billion years old. In that case, the observable universe has a radius of 10–20 billion light-years. Do not confuse the observable universe, which is finite, with the universe as a whole, which could be infinite.

The assumptions that we made when we described Olbers' paradox were at least partially in error. This illustrates the importance of assumptions in cosmology and serves as a warning that our commonsense expectations are not dependable. All of astronomy is reasonably unreasonable—that is, reasonable assumptions often lead to unreasonable results. That is especially true in cosmology, so we must always examine our assumptions with particular care.

Basic Assumptions in Cosmology Although we could make many assumptions, three are basic—homogeneity, isotropy, and universality.

Homogeneity is the assumption that matter is uniformly spread throughout space. Obviously, this is not true on the small scale, because we can see matter concentrated in planets, stars, and galaxies. Homogeneity refers to the large-scale distribution. If the universe is homogeneous, we should be able to ignore individual galaxies and think of matter as an evenly spread gas in which each particle is a cluster of galaxies. Recent observations of voids and filaments suggest that the universe is homogeneous only on the largest scale.

Isotropy is the assumption that the universe looks the same in every direction, that it is isotropic. On the small scale this is not true, but if we ignore local variations like galaxies and clusters of galaxies, the universe should look the same in any direction. For example, we should see roughly the same number of galaxies in every direction. Again, observations suggest that the universe is isotropic only on the largest scale.

The most easily overlooked assumption is uni-

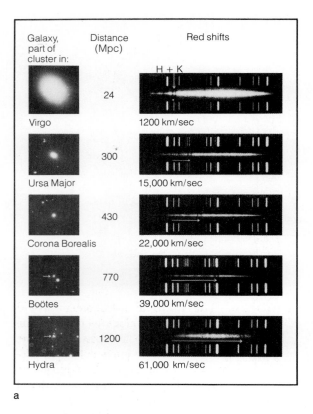

Galaxy, part of cluster in:	Distance (Mpc)	Red shifts
Virgo	24	1200 km/sec
Ursa Major	300	15,000 km/sec
Corona Borealis	430	22,000 km/sec
Boötes	770	39,000 km/sec
Hydra	1200	61,000 km/sec

a

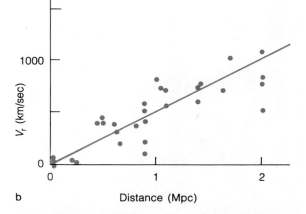

b

Figure 15-2 (a) The red shifts in these galaxy spectra appear as displacements of the H and K lines of calcium toward the red (arrows). The red shifts are interpreted as showing that the galaxies are receding and the universe is expanding. (Palomar Observatory photograph) (b) The graph that changed the universe. Hubble's first plot of the velocities of and distances to galaxies did not probe very far into space, but it clearly showed that more distant galaxies recede at higher velocity and thus that the universe expands. Because of errors in his calculation of distances, Hubble's first estimate of the Hubble constant was too large.

versality, which holds that the physical laws we know on Earth, such as the law of gravity, apply everywhere in the universe. Although this may seem obvious at first, some astronomers challenge universality by pointing out that when we look out in space we look back in time. If the laws of physics change with time, we may see peculiar effects when we look at distant galaxies. For now we will assume that the physical laws observed on Earth apply everywhere in the universe.

The assumptions of homogeneity and isotropy lead to an assumption so fundamental it is called the **cosmological principle.** According to this principle, any observer in any galaxy sees the same general features of the universe. For example, all observers should see the same kinds of galaxies. As in previous assumptions, we ignore local variations and consider only the overall appearance of the universe, so the fact that some observers live in galaxies in clusters and some live in isolated galaxies is only a minor irregularity.

Evolutionary changes are not included in the cosmological principle. If the universe is expand-ing and the galaxies are evolving, observers living at different times may see galaxies at different stages. The cosmological principle says that once observers correct for evolutionary changes, they should see the same general features.

The cosmological principle is actually an extension of the Copernican principle. Copernicus said that Earth is not in a special place; it is just one of a number of planets orbiting the sun. The cosmological principle says that there are no special places in the universe. Local irregularities aside, one place is just like another. Our location in the universe is typical of all other locations.

If we accept the cosmological principle, then we may not imagine that the universe has an edge or a center. Such locations would not be the same as all other locations.

The first fundamental observation of cosmology, that the night sky is dark, has led us to deep insights about the nature of the universe. The second fundamental observation will be equally revealing.

The Expansion of the Universe The second fundamental observation of cosmology is that the spectra of galaxies contain red shifts that are proportional

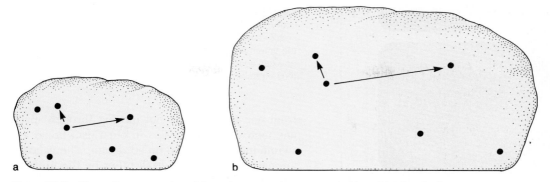

Figure 15-3 The uniform expansion of the universe can be represented by raisins in a loaf of raisin bread (a). As the dough rises (b), raisins originally near each other move apart more slowly, and raisins originally farther apart move away from each other more rapidly. A colony of bacteria living on any raisin would find that the velocities of recession of the other raisins were proportional to their distances.

to their distances (Figure 15-2a). These red shifts are commonly interpreted as Doppler shifts due to the recession of the galaxies, which is why we say the universe is expanding.

Edwin P. Hubble discovered the velocity–distance relationship in 1929 using the spectra of only 46 galaxies (Figure 15-2b). Since then, the Hubble law, as it's known, has been confirmed for hundreds of galaxies out to great distances. This law (see Box 13-1) is clear evidence that the universe is expanding uniformly and has no center.

To see how the Hubble law implies uniform, centerless expansion, imagine that we make a loaf of raisin bread (Figure 15-3). As the dough rises, the expansion pushes the raisins away from one another. Two raisins that were originally separated by only 1 cm move apart rather slowly, but two raisins that were originally separated by 4 cm of dough are pushed apart more rapidly. The uniform expansion of the dough causes the raisins to move away from each other at velocities proportional to their distances. According to the Hubble law, the larger the distance between two galaxies, the faster they recede from each other. This is exactly the result we expect from uniform expansion.

The raisin bread also shows that the expansion has no identifiable center. Bacteria living on one of the raisins would see themselves surrounded by a universe of receding raisins. (Here we must assume that they cannot see to the edge of the loaf—that is, they must not be able to see to the edge of their universe.) It does not matter on which raisin the

bacteria live. The bacterial astronomers would measure the distances and velocities of recession and derive a bacterial Hubble law showing that the velocities of recession are proportional to the distances. So long as they cannot see the edge of the loaf, they cannot identify any raisin as the center of the expansion.

Similarly, we see galaxies receding from us, but no galaxy or point in space is the center of the expansion. Any observer in any galaxy would see the same expansion that we see.

Although astronomers and cosmologists commonly refer to these red shifts as Doppler shifts and often speak of the recession of the galaxies, relativity provides a more elegant explanation. Einstein's theory of general relativity explains the expansion of the universe as an expansion of space–time itself. A photon traveling through this space–time is stretched as space–time expands, and the photon arrives with a longer wavelength than it had when it left (Figure 15-4). Photons from distant galaxies travel for a longer time and are stretched more than photons from nearby galaxies. Thus the expansion of the universe is an expansion of the geometry and not just a simple recession of the galaxies.

This explains, by the way, why the equation for the relativistic red shift given in Box 14-1 is not quite correct for galaxies. That equation comes from the special theory of relativity and expresses the Doppler shift caused by the motion of an object through space. But the galaxies, in receding

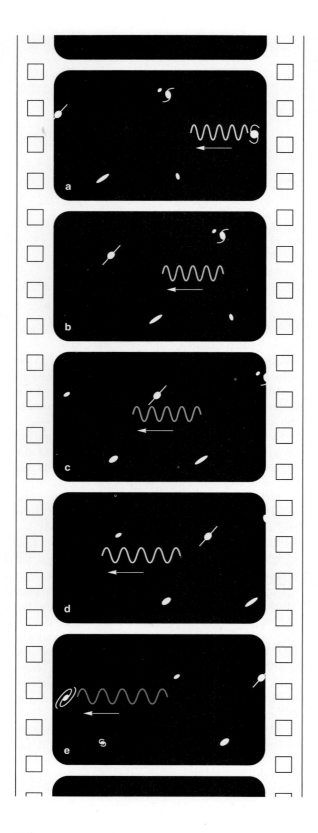

Figure 15-4 General relativity explains that the red shift of light from distant galaxies is produced by the stretching of space–time between the time a photon is emitted (a) and the time it arrives at the earth (e). Photons from distant galaxies spend more time traveling through space and are thus stretched more than photons from nearby galaxies.

from one another, are not moving through space any more than the raisins in our raisin bread are swimming through the bread dough. The galaxies are carried away from each other by the uniform expansion of space itself.

We need not worry about the actual form of the red shift equation; it is different for different assumptions about the nature of space and time. Rather, we must focus on the central concept of modern cosmology, the curvature of space–time.

The Geometry of Space–Time How can the universe expand if it does not have any extra space to expand into? How can it be finite if it doesn't have an edge? The properties we have ascribed to the universe seem to violate common sense, but we can understand some of these if we apply a few ideas from general relativity.

In Chapter 11, we saw that the presence of mass can curve space–time, and we sense that curvature as a gravitational field. Einstein's theory also predicts that the universe as a whole is curved. (See Box 15-1.)

If the universe is infinite, then it has no edge, and we say it is unbounded. If it has no boundary, then it can have no center because we define the centers of things by reference to their edges. In terms of curvature, such possible universes are said to be open or flat.

If the universe is finite, however, we say it is closed. Only a finite number of cubic centimeters exist in a closed universe. Yet such a universe need not have a boundary. To see why, think of an ant living on an orange (Figure 15-6). The ant can walk all over the orange, and if it leaves dirty footprints, it might eventually say, "I have visited every square centimeter in my two-dimensional universe, so it must be finite. But I can't find any edge. My universe is finite but unbounded." Our universe could similarly be of finite volume if it is curved such that it is closed.

We must not try to visualize the expansion of the universe as an outer edge moving into previ-

The Curvature of Space

Most modern cosmological theories are based on Einstein's general theory of relativity and its main feature, curved space–time. This curvature of space–time is important in cosmology because the mass of the universe can produce a general curvature in space–time that determines the motion of the universe.

To illustrate curvature, we can use a two-dimensional analogy of our three-dimensional universe. Suppose an ant was confined to a two-dimensional surface. Not only must the ant stay on the surface, but the ant's light cannot travel perpendicular to the surface. This surface could be flat (zero curvature), spherical (positive curvature), or saddle-shaped (negative curvature) (Figure 15-5). Because the ant cannot leave the surface, it might be unaware of the true curvature of its universe.

One way the ant could detect the curvature would be to draw circles and measure their areas. On the flat surface a circle would always have an area of exactly πr^2, but on the positively curved surface its area would be less than πr^2, and on the negatively curved surface it would be more. Drawing small circles would not suffice because the area of small circles would not differ noticeably from πr^2, but if the ant could draw big enough circles, it could actually measure the curvature of its two-dimensional universe.

We are three-dimensional creatures, but our universe might still be curved, and we could measure that curvature by measuring the volume of spheres. If space–time is flat, then no matter how big a sphere we measure, its volume will always be $\frac{4}{3}\pi r^3$. But in positively curved space–time, spheres would have less volume than this; in negatively curved space–time, they would have more. The spheres must be many megaparsecs in radius for the difference to be detectable.

We could measure the volume of such spheres by counting the number of galaxies within a cer-

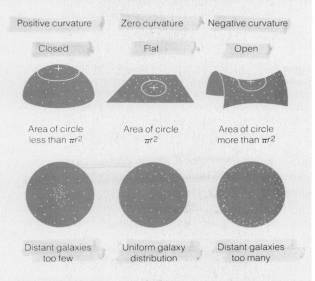

Figure 15-5 In two-dimensional space, curvature distorts the area of a circle; in three-dimensional space, it distorts the volume of a sphere. We could measure distortion and detect curvature by counting galaxies.

tain distance r from Earth. If galaxies are homogeneously scattered through space, then the number within distance r should be proportional to the volume of the sphere. If, as we count to greater and greater distances, we find the number of galaxies increasing proportional to $\frac{4}{3}\pi r^3$, space–time is flat. However, if we find an excess of distant galaxies, space–time is negatively curved, and if we find a deficiency, space–time is positively curved.

By thinking of the two-dimensional analogy shown in Figure 15-5, astronomers usually refer to a universe with positive curvature as a **closed universe.** A zero-curvature universe is said to be a **flat universe,** and a universe with negative curvature is an **open universe.** Notice that flat and open universes are infinite, but closed universes are finite.

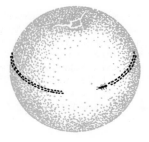

Figure 15-6 An ant confined to the two-dimensional surface of an orange could explore the entire surface without coming to an edge. Were it to leave dirty footprints, it might realize that its two-dimensional universe was finite but unbounded.

ously unoccupied space. Open, flat, or closed, the universe has no edge, so it does not need additional room to expand. The universe contains all of the volume that exists, and the expansion is a change in the nature of space–time that causes that volume to increase.

According to all the evidence and consistent with general relativity, we live in an expanding universe. If the universe were a videotape, we could run it backward and see the universe contracting. Gradually, the galaxies would approach one another until they began to merge. The total volume of the universe would decrease until all the matter and energy that exists was trapped in a very high-temperature, high-density state—a state called the big bang.

15-2 THE BIG BANG

Our two fundamental observations, the darkness of the night sky and the red shifts of the galaxies, tell us the universe had a beginning and is expanding. Those ideas draw our imaginations back in time to the moment when the universe began. The **big bang theory** proposes that the universe began in a high-temperature, high-density state. In this section, we will discuss observations of the big bang, and we will see how the big bang gave birth to matter. Finally, we will see how the fate of the universe depends on its average density.

The Big Bang as a Universe-Wide Event Although our imaginations try to visualize the big bang as a lo-

calized event, we must keep firmly in mind that the big bang did not occur at a single place but filled the entire volume of the universe. The matter of which we are made was part of that big bang, so we are inside the remains of that event, and the universe continues to expand all around us. We cannot point in any particular direction and say, ''The big bang occurred over there.'' The big bang occurred everywhere (Figure 15-7).

Although the big bang occurred long ago, the theory predicts that we can detect it because of the finite speed of light. When we look at a distant galaxy, we do not see it as it is now, but as it was long ago. We see the most distant galaxies as they were soon after they formed from the chaos of the big bang. It doesn't matter what direction we look because we are inside and galaxies surround us on all sides.

Suppose that we look even further, past the most distant objects, backward in time to the fiery clouds of matter in the big bang (Figure 15-7c). From these great distances, we receive light that was emitted by the hot gas soon after the universe began. Again, it does not matter what direction we look in space because we are inside.

The radiation that comes to us from this distance has a tremendous red shift. The farthest visible galaxies have red shifts of about 1.0, and the farthest quasars almost 5, but the radiation from the big bang has a red shift of about 1000. Thus the visible light emitted by the big bang arrives at Earth as infrared radiation and short radio waves.

The Primordial Background Radiation If radiation is now arriving from the big bang, then it should be detectable. The story of that discovery begins in the mid-1960s when two Bell Laboratories physicists, Arno Penzias and Robert Wilson, were using a horn antenna to measure the radio brightness of the sky (Figure 15-8). Their measurements showed a peculiar noise in the system, which they at first attributed to pigeons living inside the antenna. After relocating the birds and cleaning out their droppings, the scientists still detected the noise. Perhaps they would have enjoyed cleaning the antenna more if they had known they would win the 1978 Nobel Prize for physics for the discovery they were about to make.

The explanation for the noise goes back to 1948, when George Gamow predicted that the

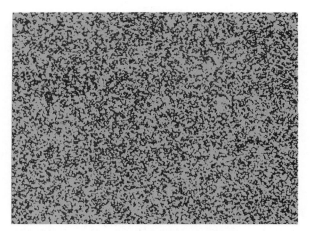

a A region of the universe during the big bang

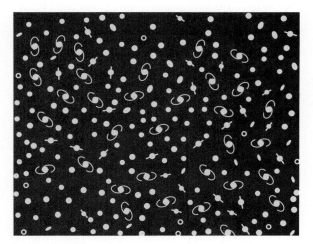

b A region of the universe now

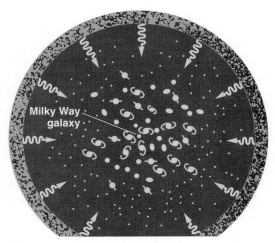

c The present universe as it appears from our galaxy

Figure 15-7 Three views of a small region of the universe centered on our galaxy. (a) During the big bang explosion, the region is filled with hot gas and radiation. (b) Later, the gas forms galaxies, but we can't see the universe this way because the look-back time distorts what we see. (c) Near us we see galaxies, but farther away we see young galaxies (dots) and at a great distance we see radiation (arrows) coming from the hot clouds of the big bang. Note that (c) shows the *observable* universe, and thus we are at the center.

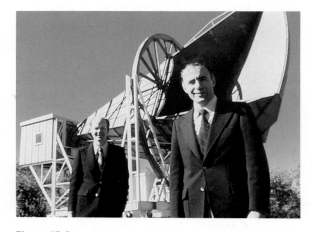

Figure 15-8 Robert Wilson and Arno Penzias (right) pose before the horn antenna with which they discovered the primordial background radiation. (Bell Laboratories)

early big bang would be very hot and would emit copious black-body radiation. A year later, physicists Ralph Alpher and Robert Herman pointed out that the large red shift of the big bang gas clouds would lengthen the wavelengths of the radiation and make the hot gas clouds seem very cold. There was no way to detect this radiation until the mid-1960s, when Robert Dicke at Princeton concluded the radiation should be just strong enough to detect with newly developed techniques. Dicke and his team began building a receiver.

When Penzias and Wilson heard of Dicke's work, they recognized the noise they had detected as radiation from the big bang, the **primordial background radiation.**

The background radiation was measured at many wavelengths as astronomers tried to confirm that it did indeed follow a black-body curve. Some measurements could be made from the ground, but balloon and rocket observations were critical and

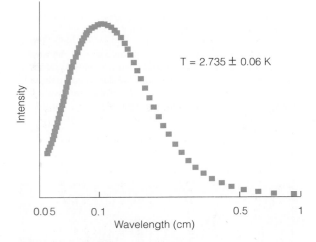

Figure 15-9 Primordial background radiation from the big bang is observed in the infrared and radio parts of the spectrum. These COBE results show the radiation fits a black-body curve with a temperature of 2.735 K.

difficult. A few disturbing observations suggested that the radiation departed inexplicably from a black-body curve and thus might not come from the big bang.

In November 1989, the orbiting Cosmic Background Explorer (**COBE**) began taking the most accurate measurements of the background radiation ever made. Within 2 months, the verdict was in (Figure 15-9). To an accuracy of better than 1 percent, the radiation follows a black-body curve with a temperature of 2.735 ± 0.06 K—in good agreement with accepted theory.

It may seem strange that the hot gas of the big bang seems to have a temperature of only 2.7 K, but recall the tremendous red shift. We see light that has a red shift of about 1000—that is, the wavelengths of the photons are about 1000 times longer than when they were emitted. The gas clouds that emitted the photons had a temperature of about 3000 K, and they emitted black-body radiation with a λ_{max} of about 1000 nm (see Box 6-2). Although this is in the near infrared, the gas would also have emitted enough visible light to glow orange-red. But the red shift has made the wavelengths about 1000 times longer, so λ_{max} is about 1 million nm (equivalent to 1 mm). Thus the hot gas of the big bang seems to be 1000 times cooler, about 2.7 K.

Although the radiation is almost perfectly isotropic, precise observations reveal small departures from complete isotropy. The background radiation has slightly shorter wavelengths—it seems hotter—in one direction and has slightly longer wavelengths—it seems cooler—in the opposite direction. This difference is caused by the motion of our Milky Way galaxy along with the Local Group at about 540 km/sec toward the massive Virgo cluster. This motion causes a slight blue shift in the direction of motion and that makes the background radiation appear slightly hotter. In the opposite direction there is a red shift and the radiation looks cooler. (Smaller irregularities in the background radiation will be discussed in the last section of this chapter.)

The greatest importance of the primordial background radiation lies in our interpretation of it as radiation from the big bang. If that interpretation is correct, the radiation is evidence that a big bang did occur. That evidence is so strong that the **steady state theory,** which held that the universe was eternal and unchanging, has been abandoned. (See Box 15-2.) Nearly all astronomers now accept that the universe evolves and probably began with a big bang.

The Origin of Matter Modern cosmologists have been able to reconstruct the history of the early universe to reveal how energy and matter interacted during the big bang to produce the matter of which we are made. As we recount that history, remember that the big bang did not occur in a specific place. The big bang filled the entire volume of the universe from the first moment.

We cannot begin our history at time zero, because we do not understand the physics of matter and energy under such extreme conditions, but we can come close. At a time earlier than 0.0001 second after the beginning, the universe was filled with high-energy photons having a temperature greater than 10^{12} K and a density greater than 5×10^{13} g/cm^3. When we say the photons had a given temperature, we mean that the photons were the same as black-body radiation emitted by an object of that temperature. Thus the photons in the early universe were gamma rays of very short wavelength and therefore very high energy. When we say that the radiation had a certain density, we refer to

Through the 1950s and most of the 1960s, astronomers had an alternate to the big bang theory. The steady state theory proposed that the universe did not evolve, that it always had the same general properties. Stars and galaxies might age and die, but new stars and new galaxies would be born to take their place and thus preserve the general properties of the universe.

One of the most distinctive features of the steady state theory was caused by the expansion of the universe. As the universe expands, its average density should decrease. The steady state theory, however, held that the universe did not change, so it proposed that matter was created continuously to maintain the density of the universe. This newly created matter, presumably in the form of hydrogen atoms, collected in great clouds between the receding galaxies and eventually gave birth to new galaxies to take the place of aging galaxies no longer able to make new stars.

The steady state theory remained a controversial although exciting idea in cosmology until the late 1960s. By then, the primordial background radiation had been widely accepted as evidence that a big bang had occurred, and the steady state theory was gradually abandoned.

Einstein's equation $E = m_0 c^2$. We can express a given amount of energy in the form of radiation as if it were matter of a given density. In the early universe, the radiation had a density roughly equal to the density of an atomic nucleus.

If two photons have enough energy, they can interact and convert their energy into a pair of particles—a particle of normal matter and a particle of antimatter. In the early universe, photons had sufficient energy to produce proton–antiproton pairs and neutron–antineutron pairs. However, when a particle collides with its antiparticle, the particles are annihilated, and the mass is converted into a pair of gamma-ray photons. Thus the early universe was a soup of energy continuously switching from photons to particles and back again.

While all of this went on, the universe was expanding, and the wavelengths of the photons were lengthened by the expansion. This lowered the energy of the gamma rays, and the universe cooled. By the time the universe was 0.0001 second old, its temperature had fallen to 10^{12} K. By this time the average energy of the gamma rays had fallen below the energy equivalent to the mass of a proton or a neutron, so the gamma rays could no longer produce such heavy particles. The particles combined with their antiparticles and quickly converted most of the mass into photons.

It would seem from this that all the protons and neutrons should have been annihilated by their antiparticles, but for quantum mechanical reasons a small excess of normal particles existed. For every billion protons annihilated by antiprotons, one survived with no antiparticle to destroy it. Thus we live in a world of normal matter, and antimatter is very rare.

Although the gamma rays did not have enough energy to produce protons and neutrons, they could produce electron–positron pairs, which are about 1800 times less massive than protons and neutrons. This continued until the universe was about 4 seconds old, at which time the expansion had cooled the gamma rays to the point where they could no longer create electron–positron pairs. Thus the protons, neutrons, and electrons of which our universe is made were produced during the first 4 seconds in the history of the universe.

This soup of hot gas and radiation continued to cool and eventually began to form atomic nuclei. High-energy gamma rays can break up a nucleus, so the formation of such nuclei could not occur until the universe had cooled somewhat. By the time the universe was 3 minutes old, protons and neutrons could link to form deuterium, the nucleus of a heavy hydrogen atom, and by the end of the next minute, further reactions began converting deuterium into helium. But no heavier atoms could be built because no stable nuclei exist with atomic weights of 5 or 8 (in units of the hydrogen atom). Cosmic element building during the big bang had to proceed rapidly, step by step, like someone hopping up a flight of stairs (Figure 15-10). The lack of stable nuclei at atomic weights of 5 and 8 meant there were missing steps in the stairway, and the step-by-step reactions could not jump over these gaps.

By the time the universe was 30 minutes old, it had cooled sufficiently that nuclear reactions had stopped. About 25 percent of the mass was helium nuclei, and the rest was in the form of protons—hydrogen nuclei. This is the cosmic abundance we see today in the oldest stars. (The heavier elements, remember, were built by nucleosynthesis inside many generations of massive stars.) The cosmic abundance of helium was fixed during the first minutes of the universe.

Although hydrogen and helium nuclei existed after the first 4 minutes, the universe was still too hot to permit the nuclei to capture electrons and become neutral atoms. The photons were still energetic and could ionize any atom that momentarily formed as a nucleus captured an electron. Free electrons deflect photons very easily, and thus the radiation could not travel through the gas easily. The gas was opaque, ionized matter, and it was locked together with the radiation. During this time, the universe was dominated by the radiation (Figure 15-11), but as the universe continued to expand, the gas cooled and the photons were stretched to longer wavelengths and thus to lower energy.

By the time the universe was nearly 1 million years old, the expansion had stretched the photons to such long wavelengths they no longer had the energy to ionize atoms. The atomic nuclei captured free electrons, the gas became transparent, and the radiation was free to travel through the universe. The temperature at the time of this **recombination** was about 3000 K. We see these photons arriving now as primordial background radiation.

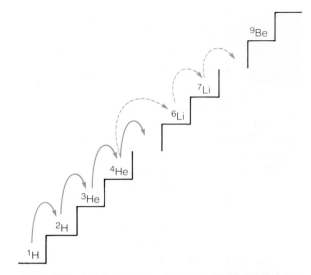

Figure 15-10 Cosmic element building. During the first few minutes of the big bang, temperatures and densities were high, and nuclear reactions built heavier elements. Because there are no stable nuclei with atomic weights of 5 or 8, the process built very few atoms heavier than helium.

After recombination, the universe was no longer dominated by radiation. Instead, matter was free to move under the influence of gravity, so we say that the universe after recombination is dominated by matter (Figure 15-11). How the matter cooled and collected into clouds and eventually gave birth to galaxies in clusters and superclusters is not well understood. We will return to this problem at the end of this chapter.

For now, we must turn our attention away from the first moments of the big bang and instead consider its final moments—the ultimate fate of the universe.

The End of the Universe: A Question of Density Will the universe ever end? Will it go on expanding forever with stars burning out, galaxies exhausting their star-forming gas and becoming cold, and dark systems expanding forever through an endless, dark universe? How else could the universe end?

Earlier, we decided that the universe could be either open, flat, or closed. The geometry of the universe determines how it will end, so we must consider these three possible universes.

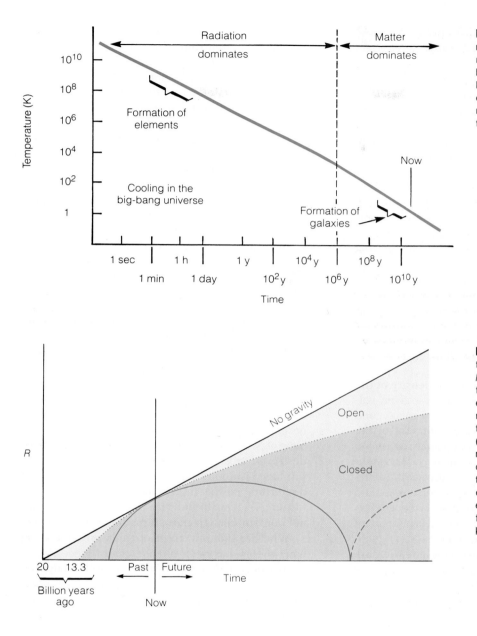

Figure 15-11 During the first few minutes of the big bang, some matter became deuterium and helium and a few heavier atoms. Later, when radiation was no longer dominant, galaxies formed, and nuclear reactions inside stars made the rest of the chemical elements.

Figure 15-12 The expansion of the universe as a function of time. *R* is some measure of the size of the universe. Open-universe models expand without end (shaded region), but closed models pass through repeated big bangs (curved solid line). Dotted line marks the dividing line between open and closed models. Note that the estimated age of the universe depends on the rate at which the expansion is slowing down. (This figure assumes *H* = 50 km/sec/Mpc.)

The general curvature of the entire universe is determined by its density. If the average density of the universe is equal to the **critical density**, or 4×10^{-30} g/cm³, space–time will be flat. If the average density of the universe is less than the critical density, the universe is negatively curved and open. If the average density is greater than the critical density, the universe is positively curved and closed (see Box 15-1).

We decided earlier that an open universe and a flat universe are both infinite. Both of these universes will expand forever. The gravitational field of the material in the universe, present as a curvature of space–time, will cause the expansion to slow; but, if the universe is open, it will never come to a stop (Figure 15-12). If the universe is flat, it will barely slow to a stop after an infinite time. Thus if the universe is open or flat, it will expand forever, and the galaxies will eventually become black, cold, solitary islands in a universe of darkness.

If the universe is closed, however, its fate will be quite different. In a closed universe, the gravitational field, present as curved space–time, is sufficient to slow the expansion to a stop and make the universe contract. Eventually, the contraction will compress all matter and energy back into the high-energy, high-density state from which the universe began. This end to the universe, a big bang in reverse, has been called the big crunch. Nothing in the universe could avoid being destroyed by this "crunch."

Some theorists have suggested that the big crunch will spring back to produce a new big bang and a new expanding universe. This theory, called the **oscillating universe theory,** predicts that the universe undergoes alternate stages of expansion and contraction. Recent theoretical work, however, suggests that successive bounces of an oscillating universe would be smaller until the oscillation ran down.

It is quite difficult to measure the density of the universe. We could count galaxies in a given volume, multiply by the average mass of a galaxy, and divide by the volume, but we are not sure of the average mass of a galaxy, and many galaxies are too small to see even when they are nearby. We would also have to include the mass equivalent to the energy present in the universe, because mass and energy are related. The best attempts yield a density of about 5×10^{-31} g/cm^3, about 10 percent of the mass needed to close the universe.

This figure does not, however, include the dark matter, and as we saw in Chapter 13, many observations suggest that much of the mass in the universe has not yet been detected. Galaxies have invisible, massive coronas, and clusters of galaxies contain more mass than the sum of the visible galaxies. Apparently, 90–99 percent of the mass of the universe is dark matter.

Theorists have suggested a wide range of subatomic particles that could make up the dark matter. Axions, photinos, and WIMPs are only a few, but none of these theoretical particles has been detected. Another theory is that neutrinos have a small mass. Because there are about 10^8 neutrinos in the universe for every normal particle, even a tiny mass for the neutrino could add up to significant dark matter. As yet, however, no studies have detected any neutrinos with mass.

The dark matter is one of the great mysteries of our age. If dark matter does exist, it could easily make the true density of the universe equal to or greater than the critical mass. Many astronomers believe that the observed density will eventually be found to equal the critical density, and the universe will be found to be exactly flat. As yet, however, the dark matter has not been detected.

If dark matter is not sufficient to account for a closed universe, then the universe must be open. In fact, isotopes such as deuterium and lithium-7 suggest that the universe is not closed. Deuterium is an isotope of hydrogen in which the nucleus contains a proton and a neutron. This element is so easily converted into helium that none can be produced in stars. In fact, stars destroy what deuterium they have by converting it into helium. But ultraviolet observations of the interstellar medium made with the Hubble Space Telescope reveal that about 15 out of every million hydrogen atoms in space are actually deuterium. That deuterium, plus the deuterium that has been destroyed in stars, must have been made during the first minutes of the big bang. If the universe contained enough normal matter to be closed, then during the big bang it would have been so dense it would have destroyed most of the deuterium and would have produced isotopes such as lithium-7. That we have deuterium in the interstellar medium and little lithium-7 seems to indicate that the universe cannot contain enough normal matter to be closed.

Whether the universe is open or closed, the big bang theory has been phenomenally successful in explaining the origin and evolution of the universe. At present, no other theory is a serious contender. But there are important questions yet to be resolved concerning the big bang.

15-3 REFINING THE BIG BANG: THREE QUESTIONS

Many questions continue to puzzle cosmologists, but the three we examine here will illustrate some important trends in current research. First, we must see how today's cosmologists are applying quantum mechanics to the study of the early universe. Second, we must discuss the controversy over the age of the universe. And finally, we must try to understand how the big bang made the vast clusters of galaxies that we see today.

The Quantum Universe   A newly developed theory, combining general relativity and quantum mechanics, may be able to tell us how the big bang began and why the universe is in the particular state that we observe. Some theorists claim that the new theory will explain why there was a big bang in the first place. To introduce the new theory, we can consider two unsolved problems in the current big bang theory.

One of the problems is called the **flatness problem.** The universe seems to be balanced between an open and a closed universe—that is, it seems nearly flat. Given the vast range of possibilities, from zero to infinite, it seems peculiar that the density of the universe is so close to the critical density that would make it flat. If dark matter is as common as it seems, the density may be even closer to that level than presently estimated.

Even a small departure from critical density when the universe was young would be magnified by subsequent expansion. Had the universe been only a tiny bit less dense during the first instants of the big bang, it would now have too low a density to form stars and galaxies. To be so near critical density now, the density of the universe during its first moments must have been within 1 part in 10^{49} of the critical density at that moment. So the flatness problem is: Why is the density of the universe so nearly equal to the critical density? That is, why is the universe so nearly flat?

Another problem with the big bang theory is the isotropy of the primordial background radiation. Once we correct for the motion of our galaxy, we see the same background radiation in all directions to at least 1 part in 1000. Yet when we look at the background radiation coming from two points in the sky separated by more than a degree, we look at two parts of the big bang that were not causally connected when the radiation was emitted. That is, when recombination occurred and the gas of the big bang became transparent to radiation, the universe was not old enough for any signal to have traveled from one of these regions to the other. Thus the two spots we look at did not have time to exchange heat and even out their tempera-

tures. Then how did every part of the entire big bang universe get to be so precisely the same temperature by the time of recombination? This is called the **horizon problem** because the two spots are said to lie beyond their respective light-travel horizons.

The key to these two problems and to others involving subatomic physics may lie with a new theory. It has been called the **inflationary universe** because it predicts a sudden expansion when the universe was very young, an expansion even more extreme than that predicted by the big bang theory.

To understand the inflationary universe, we must consider recent attempts to unify the four forces of nature. Recall from Chapter 9 that physicists know of only four forces—gravity, the electromagnetic force, the strong nuclear force, and the weak nuclear force.

For many years, theorists have tried to unify these forces—that is, they have tried to describe the forces with a single mathematical law (Figure 15-13). A century ago, James Clerk Maxwell showed that the electric force and the magnetic force were really the same effect, and we now count them as a single electromagnetic force. About 15 years ago, theorists succeeded in unifying the electromagnetic force and the weak force in what they called the electroweak force, effective only for processes at very high energy. At lower energies, the electromagnetic force and the weak force behave differently. Now theorists have found ways of unifying the electroweak force and the strong force at even higher energies. These new theories are called **Grand Unified Theories** or **GUTs.**

Studies of GUTs suggest that when the universe began, all of the forces were unified as a single force. About 10^{-35} second after the big bang, the expanding universe cooled to the point at which the forces of nature began to separate from one another. This released tremendous amounts of energy, which suddenly inflated the universe by a factor between 10^{20} and 10^{30}. At that time the part of the universe that we can see now, the entire observable universe, was no larger than the volume of an atom, but it suddenly inflated to the volume of a cherry pit and then continued its slower expansion to its present dimensions. (Remember that we are speaking here of the observable universe,

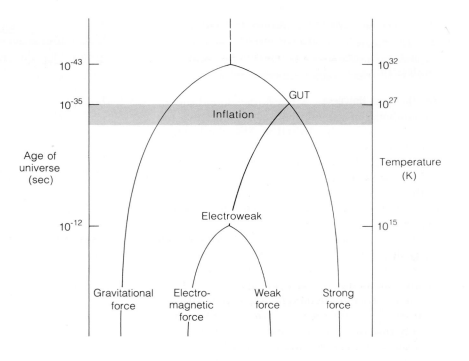

Figure 15-13 When the universe was very young and hot (top), the four forces of nature were indistinguishable. As the universe began to expand and cool, the forces separated and triggered a sudden inflation in the size of the universe.

which is finite, and not the entire universe, which may be infinite.)

This theory can solve the flatness problem and the horizon problem. The sudden inflation of the universe would have forced whatever curvature it had toward zero, just as inflating a balloon makes a small spot on its surface flatter. Thus we now see the universe as nearly flat because of that sudden inflation long ago. In addition, because the part of the universe we can see was once no larger in volume than an atom, it had plenty of time to equalize its temperature before the inflation occurred. Now we see the same temperature for the background radiation in all directions.

The inflationary universe is based, in part, on quantum mechanics, and a slightly different aspect of quantum mechanics may explain why there was a big bang at all. Theorists believe that a universe totally empty of matter could be unstable and decay spontaneously by creating pairs of particles until it was filled with the hot, dense state we call the big bang. This theoretical discovery has led some cosmologists to believe that the universe could have been created by a chance fluctuation in space–time. In the words of physicist Frank Wilczyk,

"The reason there is something instead of nothing, is that 'nothing' is unstable."

The inflationary universe and other theoretical phenomena predicted by quantum mechanics have become an important and exciting part of modern cosmology. But another area of excitement and controversy is centered around an aspect of traditional observational cosmology that began with Edwin Hubble's discovery of the expanding universe.

The Age of the Universe In 1929, Edwin Hubble announced that the universe was expanding and reported the rate of that expansion, the number now known as the Hubble constant (H) (see Box 13-1). Within a few years, astronomers had proposed the idea that the universe began in a big bang and had used H to extrapolate backward to find the age of the universe. But when they used the constant that Hubble had determined, 530 km/sec/Mpc, they found that the universe was only about half as old as the earth. Something was wrong.

If we know H, we can estimate the age of the universe. The Hubble constant tells us how fast the

BOX 15-3
The Hubble Constant and the Age of the Universe

The Hubble constant H is a measure of the rate of expansion of the universe. It contains enough information to estimate the age of the universe.

To discover how long ago the universe began expanding, divide the distance D to a galaxy by the velocity V_r with which that galaxy recedes. The result, known as the Hubble time, is the time the galaxy took to travel the distance, assuming that it has maintained constant velocity.

Because D is measured in megaparsecs and V_r in kilometers per second, we must convert D into kilometers by multiplying by 3.085×10^{19}, the number of kilometers in 1 Mpc. Then dividing by V_r yields the age of the universe in seconds. To convert to years, we must divide by 3.15×10^7, the number of seconds in a year. Thus the age of the universe in years is:

$$T = \frac{D}{V_r} \times \frac{3.085 \times 10^{19}}{3.15 \times 10^7} \approx \frac{D}{V_r} \times 10^{12} \text{ years}$$

But D/V_r is just $1/H$, so we can simplify our formula for the age of the universe:

$$T \approx \frac{1}{H} \times 10^{12} \text{ years}$$

If H is 50 km/sec/Mpc, the universe is about 20×10^9 years old, assuming there is no gravity to slow the expansion. To find the true age of the universe, we must know the extent to which gravity has slowed the expansion, and that depends on the average density of the universe.

---

galaxies are receding from one another, and, assuming that they have always traveled at their present rate, we simply ask how long it took for the galaxies to reach their present separations. (See Box 15-3.) Hubble's original value for H gave an age of about 2 billion years, but we now know that his distances were too small. The modern, corrected value of H gives an age of 10–20 billion years. For comparison, the earth is no older than 4.6 billion years.

When we use H to make these simple calculations of the age of the universe, we are assuming that the universe has always expanded at its present rate. But we saw earlier that gravity slows the expansion of the universe, so the galaxies traveled faster in the past, and the estimate for the age of the universe—known as the **Hubble time**—must be an upper limit. That is, it would be the age of the universe if there were no gravity slowing the expansion (solid line in Figure 15-12). To find the true age, we must know the extent to which gravity has slowed the expansion, and that depends on the average density of the universe. As we saw earlier in this chapter, the average density is highly uncertain.

As an example, let us assume that H is 50 km/sec/Mpc. Then, if gravity has not slowed the expansion of the universe, it is 20 billion years old. But we know the universe contains some matter. If the average density of the universe is equal to the critical density, the density needed to close the universe (dotted line in Figure 15-12), the universe must be two-thirds as old as the estimate just presented—about 13.3×10^9 years. If the density is less than that, the universe is open, and its age is between 13.3×10^9 and 20×10^9 years. If the density is more than the critical value, the universe is younger than 13.3×10^9 years.

Other evidence suggests that the universe is older than this. Measured ages for globular clusters based on the turn-off points in their H–R diagrams range from 14.8 to 20.4 billion years, with uncertainties as large as ±3.9 billion years. Another study suggests that the universe must be between

16.8 and 23.8 billion years old to allow time for nucleosynthesis in massive stars to build the present amount of heavy elements. Such estimates of the age of the universe suggest that H cannot be much larger than 70 km/sec/Mpc.

Measuring the value of H is difficult for two reasons. First, we must find the distances to galaxies that are so far away we cannot see such dependable distance indicators as the Cepheid variables. Hubble's distances were much too small, and his value of H was 530 km/sec/Mpc, much larger than modern estimates, which range from 50 to 100 km/sec/Mpc.

A second reason it is difficult to measure H is related to the motions of the galaxies themselves. Our Milky Way galaxy and the Local Group of galaxies in which we live seem to be falling toward the massive Virgo and Hydra–Centaurus superclusters. Different teams of astronomers correct for this motion differently when they calculate the radial velocities of other galaxies. Also, recent studies have begun to suggest that streaming motions occur among the galaxies. Our Local Group, the Virgo supercluster, and the Hydra–Centaurus supercluster may all be streaming in the same direction, perhaps due to the forces that produced the voids and filaments in the distribution of galaxies. If these streaming motions do occur, then the velocities of distant galaxies may be affected, thus distorting estimates of H.

The current controversy over the value of H and the age of the universe is of fundamental importance. Some astronomers believe H to be about 50 km/sec/Mpc, and others believe it to be as large as 100 km/sec/Mpc. Still other studies yield numbers between these extremes. A recent study using distances derived from the rate of expansion of supernovae remnants in other galaxies gives a value of 65 km/sec/Mpc.

To learn the age of the universe, we must make two very difficult measurements. We must find the true value of H, which means we must understand the streaming motions of the galaxies. In addition, we must know the true density of the universe, and

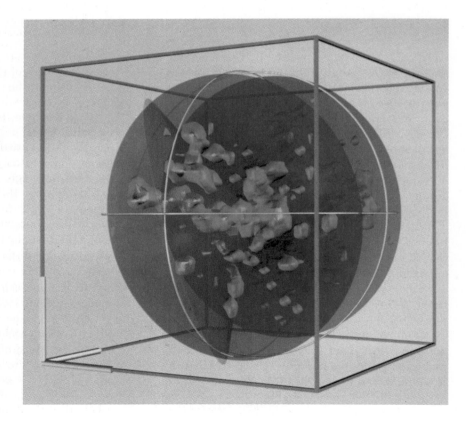

Figure 15-14 The Pisces–Cetus supercluster is shown as green and yellow regions in this plot of the location of 382 galaxy clusters within 340 Mpc of Earth. The supercluster, extending from the center to the near left face of the cube, is one of the largest known structures in the universe. Growing evidence shows that clusters of galaxies are organized into such vast structures. (Courtesy R. Brent Tully)

Figure 15-15 A cluster of galaxies slightly more than 4 billion ly away shows little detail even in the largest telescopes. Beyond these distances, galaxies are almost undetectable unless they have active nuclei, which we see as quasars. (Palomar Observatory)

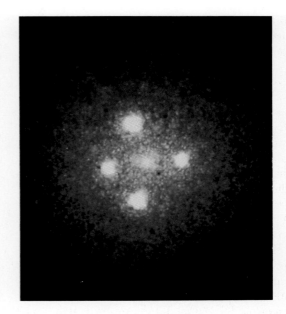

Figure 15-16 We can easily compare a quasar and a galaxy in this image of the gravitational lens G2237 + 305. A galaxy only 400 million ly distant produces four images of the quasar, which lies 8 billion ly away. Of the galaxy we see only its bright core in the center of the four images of the quasar. Astronomers believe the distant quasars are the active nuclei of galaxies too distant to see, so the existence of quasars at large look-back times assures us that galaxies formed early in the history of the universe. (NASA/European Space Agency)

that means we must solve the problem of the dark matter. These are difficult problems, but an even more fundamental problem lies hidden in the primordial background radiation. How did the galaxies form?

The Origin of Structure Modern cosmologists are working to understand how the structure of the universe could have formed from the gas of the big bang. By *structure* we mean not only individual galaxies, but also the clusters and superclusters of galaxies, plus the vast walls such as the Great Wall. There is growing evidence that such immense structures are common in the universe (Figure 15-14).

This structure is a problem for cosmologists because the primordial background radiation is extremely uniform and thus the gas of the big bang must have been extremely uniform at the time of recombination. If the universe was so uniform when it was young, how did it get so lumpy now?

The problem of the origin of structure has developed slowly since the 1960s when the background radiation was first discovered. Then the most distant galaxies had red shifts no greater than

1. These galaxies had look-back times no greater than 4 billion years (Figure 15-15), so it was not difficult to imagine how the uniform gas of the big bang could eventually form galaxies. That is, there seemed to be plenty of time between recombination and the origin of galaxies.

Through the 1970s and 1980s, however, that age of mystery narrowed. Astronomers recognized quasars as the active nuclei of galaxies much farther away than 4 billion ly. Galaxies at such great distances are almost beyond detection, but the brilliant quasars assure us that such galaxies exist (Figure 15-16). These galaxies must have formed early in the history of the universe. At present, the most distant quasars have look-back times that are 93 percent the age of the universe. Thus the age of mystery when the galaxies formed must be less than 7 percent the age of the universe.

The discovery of the Great Wall and other immense structures makes the problem more puzzling. If it is difficult to explain how galaxies

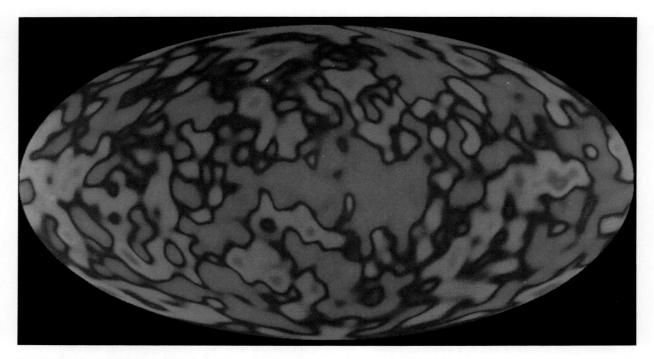

Figure 15-17 Over 70 million measurements by the COBE satellite went into this all-sky map of the primordial background radiation. Patchy areas are irregularities in the background radiation that suggest the gas of the big bang was not perfectly uniform. The gravitational influence of dark matter could have forced these denser regions to grow into the clouds of galaxies we see today. (NASA/Goddard Space Flight Center)

formed quickly from the uniform gas of the big bang, it is even more difficult to explain vast walls.

How did all of this structure originate? The universe had to be very uniform during recombination and then clump quickly to form the structure we see. Cosmologists now work on a two-pronged solution: Dark matter provides the gravitation, and galaxy seeds provide nuclei around which structure grows.

We have seen clear evidence that dark matter must exist, and cosmologists are now trying to build models that show how the gravitation of dark matter can cause the matter of the early universe to clump quickly into large structures. One kind of dark matter is called hot dark matter. If it exists it would consist of particles that travel at near the speed of light, such as neutrinos with nonzero masses. Cold dark matter, on the other hand, would consist of slow-moving particles such as WIMPs, axions, and similar as yet undetected particles. Cold dark matter, because its particles trav-

el slower, seems better able to clump together quickly.

Galaxy seeds around which this clumping occurs may be defects in the fabric of space–time. As the forces of nature separate from one another in what cosmologists call a phase transition, it is possible that defects occur between regions with different properties. We might compare such defects with the flaws we see in ice cubes. When the water freezes beginning at the outer edges, different parts of the ice cube have different crystal orientations, and as the water freezes inward toward the center, these different regions eventually meet and form defects that we see inside the ice cube as lines and sheets. Defects in space–time would possess strong gravitational fields and could act as galaxy seeds around which dark matter clumps to create galaxies, clusters, walls—all of the structure we see.

Most cosmologists believe that these phase transitions occurred very early (as in the inflationary universe theory), so any resulting structure should be visible as irregularities in the background radiation. Until 1992, the best observations could detect no irregularities in the background radiation, and that made astronomers wonder where the structure in the universe could have come from. But in the spring of 1992, George Smoot and a team of scientists announced that data from the COBE satellite reveal irregularities in the background radiation of about 0.0006 percent (Figure 15-17). This discovery is of tremendous importance because it tells us that the seeds of cosmic structure were present as irregularities in the big bang at the time of recombination. The great clouds of galaxies we see today may have had their origin when a moment of inflation created defects in space–time that grew by the gravitational influence of dark matter.

The three questions we have reviewed here— the inflationary universe, the age of the universe, and the origin of structure—in no way invalidate the big bang theory. They are refinements. One cosmologist likened the big bang theory to the theory that the world is round. Geologists cannot predict earthquakes yet, but they do not reject that the earth is round. Cosmologists continue to revise their theories, but the big bang seems a cornerstone of modern cosmology.

Although we have now traced the origin of the universe, the origin of the elements, and the birth and death of stars, we have left out one important class of objects—planets. How does the earth fit into this grand scheme of origins? That is the subject of the next unit.

SUMMARY

The fact that the night sky is dark shows that the universe is not infinitely old. If it were infinite in extent and age, then every spot on the sky would glow as brightly as the surface of a star. This problem, known incorrectly as Olbers' paradox, illustrates how important assumptions are in cosmology.

The basic assumptions of cosmology are homogeneity, isotropy, and universality. Homogeneity says that matter is spread uniformly through the universe. Isotropy says the universe looks the same in any direction. Both deal only with general features. Universality assumes that the laws of physics known on Earth apply everywhere. In addition, the cosmological principle asserts that the universe looks the same from any location.

The Hubble law implies that the universe is expanding. Tracing this expansion backward in time, we come to an initial high-density state commonly called the big bang. From the Hubble constant, we can conclude that the expansion began 10–20 billion years ago.

The universe can be infinite or finite, but it cannot have an edge or a center. Such regions would violate the cosmological principle. Instead, we assume that the universe occupies curved space–time and thus could be finite but unbounded. Depending on the average density of the universe, it could be open, flat, or closed. Measurements of the density of the universe are uncertain because of the presence of dark matter, which we cannot detect easily.

During the first few minutes of the big bang, about 25 percent of the matter became helium, and the rest remained hydrogen. Very few heavy atoms were made. As the matter expanded, instabilities caused the formation of clusters of galaxies, which are still receding from each other.

Whether the universe expands forever or slows to a stop and begins to contract depends on the amount of matter in the universe. If the average density is greater than the critical density of 4×10^{-30} g/cm^3, it will provide enough gravity to slow the expansion and force the universe to collapse. The collapse will smash all matter back to a high density from which a new big bang may emerge. Such a universe is termed closed because the curvature of space is positive. If the average density is less than 4×10^{-30} g/cm^3, gravity will be unable to stop the expansion, and it will continue forever. Such a universe is termed open because the space curvature is negative.

The inflationary universe theory combines quantum mechanics, general relativity, and cosmology. It predicts that the universe underwent a sudden inflation when it was only 10^{-35} seconds old. This seems to explain why the universe is so flat and why the primordial background radiation is so isotropic.

We can estimate an upper limit to the age of the universe from the Hubble constant, but we need to know the average density of the universe to find a true age. Also, the value of the Hubble constant is not well known. It is believed to be between 50 and 100 km/sec/Mpc. If it is more than about 70 km/sec/Mpc, the age of the universe will be less than the ages determined for some star clusters.

Although the primordial background radiation is highly isotropic, it is not perfectly uniform. Data from the COBE satellite reveal small irregularities in the background radiation which imply that the gas of the big bang was not perfectly uniform. These irregularities may have originated during a moment of inflation when phase transitions produced defects in space–time. Theorists believe that such defects, aided by the gravitational influence of dark matter, could act as galaxy seeds and grow quickly into the galaxies, clusters, superclusters, and walls of galaxies we see today.

NEW TERMS

cosmology	steady state theory
Olbers' paradox	recombination
homogeneity	critical density
isotropy	oscillating universe theory
universality	flatness problem
cosmological principle	horizon problem
closed universe	inflationary universe
flat universe	grand unified theories (GUTs)
open universe	Hubble time
big bang theory	galaxy seeds
primordial background radiation	

REVIEW QUESTIONS

1. Would the night sky be dark if the universe were only 5 billion years old and not expanding? Explain your answer.

2. How can we be located at the center of the observable universe if we accept the Copernican principle?

3. Why can't an open universe have a center? Why can't a closed universe have a center?

4. What evidence do we have that the universe is expanding? that it began with a big bang?

5. Why couldn't atomic nuclei exist when the universe was younger than 3 minutes old?

6. Why is it difficult to determine the present density of the universe?

7. How does the inflationary universe theory resolve the flatness problem? the horizon problem?

8. If the Hubble constant is really 100 km/sec/Mpc, then much of what we understand about the evolution of stars and star clusters must be wrong. Explain why.

9. Why do we conclude that the universe must have been very uniform during its first million years?

10. What is the difference between hot dark matter and cold dark matter? What difference does this make to cosmology?

DISCUSSION QUESTION

1. Do you think Copernicus would have accepted the cosmological principle? Why or why not?

PROBLEMS

1. Use the data in Figure 15-2a to plot a velocity–distance diagram, find H, and determine the approximate age of the universe.

2. Figure 15-2b shows Hubble's first velocity distance diagram. Use it to determine the value of H and the Hubble time that he derived. If his distances were too small by a factor of 7, what answer should he have obtained?

3. If a galaxy is 8 Mpc away from us and recedes at 456 km/sec, how old is the universe, assuming that gravity is not slowing the expansion? How old is the universe if it is flat?

4. If the temperature of the big bang had been 10^6 K at the time of recombination, what wavelength of maximum would the primordial background radiation have as seen from the earth?

5. If the average distance between galaxies is 2 Mpc and the average mass of a galaxy is $10^{11}M_\odot$, what is the average density of the universe? (**HINTS:** The volume of a sphere is $\frac{4}{3}\pi r^3$. The mass of the sun is 2×10^{33} g.)

6. If the value of the Hubble constant were found to be 60 km/sec/Mpc, how old would the universe be if it was not slowed by gravity? if it were flat?

7. Figure 15-12 is based on an assumed Hubble constant of 50 km/sec/Mpc. How would you change the diagram to fit a Hubble constant of 70 km/sec/Mpc?

RECOMMENDED READING

ANDERSEN, PER H. "Large-scale Structure, Streaming and Galaxy Formation." *Physics Today* 40 (October 1987), p. 19.

CLOSE, FRANK. *The Cosmic Onion*. New York: American Institute of Physics, 1986.

CORNELL, J. *Bubbles, Voids, and Bumps in Time*. New York: Cambridge University Press, 1989.

DAVIES, PAUL. "Everyone's Guide to Cosmology." *Sky and Telescope* 81 (March 1991), p. 250.

DISNEY, M. *The Hidden Universe*. New York: Macmillan, 1984.

DRESSLER, ALAN. "The Large-Scale Streaming of Galaxies." *Scientific American* 257 (September 1987), p. 46.

FERRIS, T. "The Radio Sky and the Echo of Creation." *Mercury* 13 (January/February 1984), p. 2.

FLAM, FAYE. "In Search of a New Cosmic Blueprint." *Science* 254 (22 November 1991), p. 254.

GELLER, MARGARET J., and JOHN P. HUCHRA. "Mapping the Universe." *Sky and Telescope* 82 (August 1991), p. 134.

GREGORY, STEPHEN A. "The Structure of the Visible Universe." *Astronomy* 16 (April 1988), p. 42.

GUTH, ALAN H., and PAUL J. STEINHARDT. "The Inflationary Universe." *Scientific American* 250 (May 1984), p. 116.

HALLIWELL, JONATHAN J. "Quantum Cosmology and the Creation of the Universe." *Scientific American* 265 (December 1991), p. 76.

HARRISON, EDWARD R. "The Dark Night Sky Riddle: A 'Paradox' That Resisted Solution." *Science* 226 (23 November 1984), p. 941.

————. *Masks of the Universe*. New York: Macmillan, 1985.

JONES, BRIAN. "The Legacy of Edwin Hubble." *Astronomy* 17 (December 1989), p. 38.

KIPPENHAHN, RUDOLF. "Light from the Depths of Time." *Sky and Telescope* 73 (February 1987), p. 140.

LINDE, ANDREI. "The Universe: Inflation Out of Chaos." *New Scientist* 105 (7 March 1985), p. 14.

PERATT, ANTHONY L. "Plasma Cosmology." *Sky and Telescope* 83 (February 1992), p. 136.

SCHRAMM, DAVID N. "The Origin of Cosmic Structure." *Sky and Telescope* 82 (August 1991), p. 140.

SHU, FRANK H. "The Expanding Universe and the Large-Scale Geometry of Space-Time." *Mercury* 12 (November/December 1983), p. 162.

SILK, JOSEPH. "Probing the Primeval Fireball." *Sky and Telescope* 79 (June 1990), p. 600.

SMITH, DAVID H., and ALAN MacROBERT. "Looking into the Big Bang." *Sky and Telescope* 77 (June 1988), p. 593.

TUCKER, W., and K. TUCKER. *The Dark Matter*. New York: Morrow, 1988.

WEINBERG, STEVEN. *The First Three Minutes*. New York: Basic Books, 1977.

————. "Origins." *Science* 230 (4 October 1985), p. 15.

WESSON, PAUL S. "Olbers' Paradox Solved at Last." *Sky and Telescope* 77 (June 1988), p. 594.

Planets

in Perspective

C H A P T E R
16

The Origin of the Solar System

Microscopic creatures live in the roots of your eyelashes. Don't worry. Everyone has them, and they are harmless. (*Demodex folliculorum* has been found in 97 percent of individuals with healthy skin.) They hatch, fight for survival, mate, lay eggs, and die in the tiny spaces around the roots of our eyelashes without doing us any harm. Some live in renowned places—the eyelashes of the Queen of England, of a glamorous movie star, of a famous athlete—but the tiny beasts are not self-aware; they never stop to say, "Where are we?" Humans are much more intelligent; we can wonder where we are in the universe and how we came to be here.

In this chapter, we begin exploring our solar system: its sun, nine planets, and some scattered gas, dust, and debris. Our planetary system occupies no more than 10^{-30} of the volume of the observable universe. If the words

**What place is this?
Where are we now?**

Carl Sandburg,
Grass

in this book represent the volume of the observable universe, the solar system would be less than a trillionth of a trillionth of one letter. However, the significance of the solar system—for us at least—far exceeds its relative size.

There are many reasons for studying the solar system. Not the least of these reasons is that understanding how planets form and evolve teaches us more about our own planet. Our planet now has so many people on it that we are altering our environment in dramatic ways, and it is possible that we could make Earth uninhabitable. Understanding the origins of the barren wastes of Mars and the sulfuric acid fogs of Venus may teach us how to preserve our own more comfortable climate.

Astronomers need no justification for studying the solar system, for it is the only collection of nonluminous bodies in the universe that we can study in detail.

Other stars may have planets, but such planets are not visible from Earth. Yet there may be more planets in the universe than stars, so, if only to complete our study of celestial objects, we must examine the solar system.

An intriguing reason for studying the planets of our solar system is the search for life beyond Earth. As if enchanted, some matter on Earth's surface lives and is aware of its existence. To search for other living beings, we must examine the surfaces of planets—stars are too hot and space is too cold for the evolution of life forms. So far as we know, life seems to require the moderate conditions found only on the surface of some planets. To focus our search for life, we must understand how planets form and evolve.

Above all, we study the solar system because it is our home in the universe. Because we are an intelligent species, we have the right—and even the responsibility—to wonder what we are and where we are. Our kind have inhabited this solar system for at least a million years, but only within a single lifetime have we begun to understand what a solar system is. Like sleeping passengers on a train, we waken, look out at the passing scenery, and mutter to ourselves, "What place is this? Where are we now?"

16-1 THE GREAT CHAIN OF ORIGINS

We are linked through a great chain of beginnings that leads backward through time to the first instant of the big bang. The gradual discovery of the links in that chain is one of the most exciting adventures of the human intellect. In earlier chapters, we discussed some of that story: the origin of the universe, the galaxies, the stars, and the elements. Here we will explore further to consider the formation of planets.

A Review of the Origin of Matter Your body is composed of mass that came into existence within minutes of the big bang explosion. The protons, neutrons, and electrons were created by the energy present as gamma rays; by the time the universe was only 3 minutes old, these subatomic particles had combined to form atomic nuclei.

About 25 percent of the mass became helium, an inert chemical element that is not present in your body. Most of the matter in the big bang, however, remained in the form of single protons, the nuclei of hydrogen atoms. Today roughly 75 percent of the mass in the universe is still hydrogen, including the hydrogen atoms in the water molecules in your body. These hydrogen nuclei have survived unchanged since the big bang.

Within a billion years or so after the big bang, matter began to collect into galaxies and form stars. Nuclear fusion reactions inside the stars arranged the protons and neutrons into new partnerships and created heavier elements. The carbon, nitrogen, and oxygen so important throughout your body, the calcium in your bones, and all of the most common heavy elements up to and including the iron in your blood have been cooked up by nuclear fusion inside many generations of stars.

The elements heavier than iron were created by rapid nuclear fusion, which occurred during supernovae explosions. These elements are much less common than elements lighter than iron, but some of them, such as the iodine in your thyroid gland, play important roles in your body chemistry. The supernovae explosions also spread these newly created elements back into the interstellar medium, where they could be incorporated in the formation of new stars. Indeed, the expanding supernovae remnants sometimes trigger the formation of new stars.

The sun did not form until two or three generations of stars in our Milky Way galaxy had lived, made heavy elements, and died. When the sun did form, about 5 billion years ago, it incorporated gas enriched with heavy elements. How a tiny fraction of that matter came together to form the earth, the oceans, and your body is a story we will explore in the rest of this book. As we explore we must remember the great chain of origins that produced the chemical elements. As geologist Preston Cloud remarked, "Stars have died that we might live."

The Origin of Planets Over roughly the last two centuries, astronomers have proposed two kinds of theories for the origin of the planets. Catastrophic theories proposed that the planets formed from some improbable cataclysm such as the collision of the sun and another star. Gradualistic theories proposed that the planets formed naturally with the sun. Since about the time of World War II, evidence has accumulated to support the gradualistic theories. In fact, nearly all astronomers now believe that planets form naturally as a by-product of star formation.

In earlier chapters, we saw how stars can form from the gravitational contraction of gas and dust clouds. In at least some cases, that contraction must be triggered by compression of the gas cloud, perhaps by a nearby supernova explosion.

As stars form in these contracting clouds, they remain surrounded by cocoons of dust and gas, and the rotation of the cloud causes that dust and gas to form a spinning disk around the protostar. When the center of the star grows hot enough to ignite nuclear reactions, its surface quickly heats up, becomes more luminous, and blows away the gas and dust cocoon.

Modern theories suppose that planets form in the rotating disks of gas and dust around young stars (Figure 16-1). Infrared observations of T Tauri stars, for instance, show that some are surrounded by gas clouds rich in dust, and spectra show that these stars are blowing away their nebulae at speeds up to 200 km/sec. The presence of bipolar flows from young stars confirms that such systems contain gas and dust distributed in a disk-shaped cloud (see Chapter 9).

Our own planetary system probably formed in such a disk-shaped cloud around the sun. When the sun became luminous enough, the remaining gas and dust were blown away into space, leaving the planets orbiting the sun. This is known as the **solar nebula theory** because the planets form from the nebula around the protosun (Figure 16-1). If the theory is right, then planets do form as a by-product of star formation, and thus most stars should have planetary systems.

Unfortunately, we cannot check our theory by looking for planets orbiting other stars. Even the nearest stars are so distant that present Earth-based telescopes cannot detect the faint light reflected from planets that might be orbiting the stars. The Hubble Space Telescope will not be able to detect such planets because of the glare of the nearby stars. The telescope's mirror would need to be about 200 times smoother to reduce scattered light.

But observational evidence does suggest that nearby stars have planets. The Infrared Astronomy

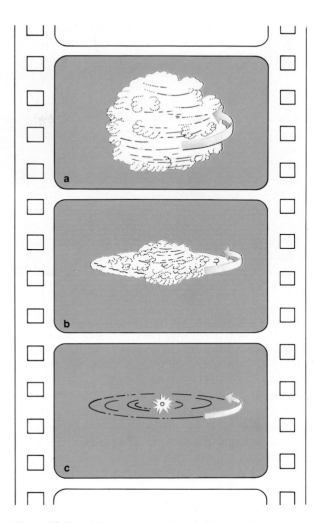

Figure 16-1 (a) Because the solar nebula was rotating, (b) it contracted into a disk, and (c) the planets formed with orbits lying in nearly the same plane.

Satellite detected shells of warm dust around more than a dozen nearby stars, including the bright star Vega. Earth-based observations have detected disks of gas and dust around at least three stars. For example, by blocking the bright radiation from the star itself, astronomers have been able to record the fainter emissions of a dusty disk orbiting the star β Pictoris (Figure 16-2). Other disks have been detected around HL Tauri and T Tauri. These appear to be the kind of circumstellar disks in which planets form.

If our theory is right, the planets formed from the nebula around the young sun. Can we find any

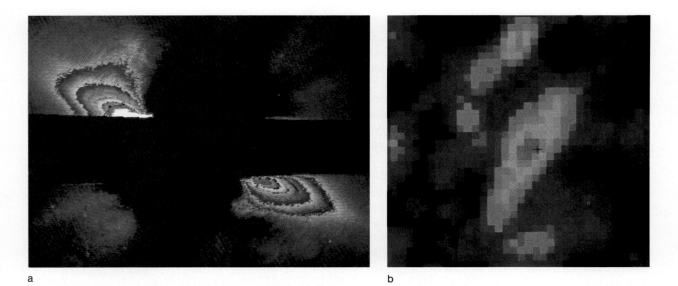

Figure 16-2 (a) This false-color image shows a nearly edge-on disk of dusty material orbiting the nearby star β Pictoris. The star was hidden behind the horizontal bar across the image. (Francesco Paresce and Christopher Burrows, Space Telescope Science Institute and European Space Agency) (b) A false-color radio map shows the distribution of the molecule CO in what appears to be a disk around the star HL Tauri. (Anneila I. Sargent and Steven Beckwith)

supporting evidence in the present solar system? To answer this question, we must survey the solar system and identify its most significant characteristics, for if our theory is correct, these characteristics were determined by the nature of the solar nebula.

16-2 A SURVEY OF THE SOLAR SYSTEM

In this section, we will make a rapid survey of our solar system noting its main features. In the next three chapters, we will discuss the bodies of our solar system in detail, but here we will simply try to note the basic characteristics of our solar system. Those characteristics are the raw data on which we will build a theory of its origin.

The solar system is composed almost entirely of empty space (Figure 16-3). Imagine that we reduce the solar system until the earth is the size of a grain of table salt, about 0.3 mm (0.01 inch) in diameter. The moon is a speck of pepper about 1

cm (0.4 inch) away, and the sun is the size of a small plum 4 m (13 ft) from Earth. Mercury, Venus, and Mars are grains of salt. Jupiter is an apple seed 20 m (66 ft) from the sun, and Saturn is a smaller seed more than 36 m (120 ft) away. Uranus and Neptune are slightly larger than average salt grains, and Pluto is a speck of pepper more than 150 m (500 ft) from the central plum. These tiny specks of matter scattered around the sun are all that remains of the solar nebula.

Rotation and Orbital Motion The planets revolve around the sun in orbits that lie close to a common plane. The orbit of Mercury, the closest planet to the sun, is tipped 7° to Earth's, and Pluto's orbit is tipped 17.2°. The rest of the planets' orbital planes are inclined by no more than 3.4°. Thus, the solar system is basically disk-shaped.

The rotation of the sun and planets on their axes also seems related to this disk shape. The sun rotates with its equator inclined only 7.25° to the earth's orbit, and most of the other planets' equators are tipped less than 30°. The rotations of Venus and Uranus are peculiar, however. Venus ro-

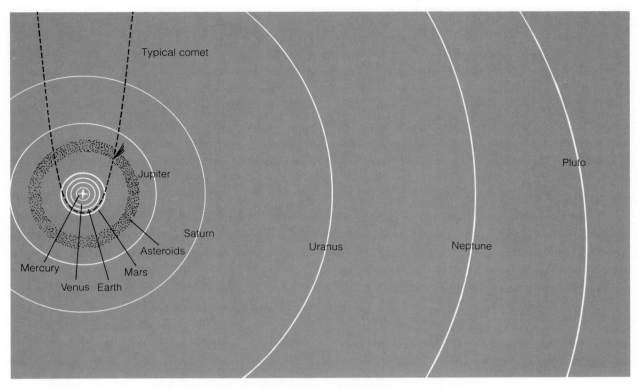

Figure 16-3 The solar system. The radii of the orbits are drawn to scale here. Note the eccentricity of Pluto's orbit. At this scale only the sun would be visible. The planets are too small to be seen.

tates backward compared with the other planets, and Uranus rotates on its side (with its equator almost perpendicular to its orbit).

Apparently, the preferred direction of motion in the solar system—counterclockwise as seen from the north—is also related to its disk shape. All the planets revolve counterclockwise around the sun, and with the exception of Venus and Uranus, they rotate counterclockwise on their axes.[*]

Thus, with only a few exceptions, most of which are understood, revolution and rotation in the solar system follow a disk theme. But sharp eyes might detect another significant pattern in the planetary orbits: each planet is a little less than twice as far from the sun as its inward neighbor (see Figure 16-3). In 1766, Johann Titius found a simple se-

quence of numbers that reproduces these distances, and because it was first reported by Johann Bode in 1772, the sequence is now known as the **Titius–Bode rule.** (See Box 16-1.)

No satisfactory theory explains why the orbits of the planets are spaced as they are, but we can interpret the Titius–Bode rule to mean that the planets were formed in some systematic way. That is, they could not have been formed by some random process. Whatever theory we propose to explain the origin of the planets, that theory must build the planets in an orderly way.

Two Kinds of Planets Perhaps the most important clue we have to the origin of the solar system is the division of the planets into two categories: terrestrial, or Earthlike, and Jovian, or Jupiterlike. The **terrestrial planets** are small, dense, rocky worlds with less atmosphere than the Jovian planets. The terrestrial planets—Mercury, Venus, Earth, and

Recall from Chapter 3 that astronomers distinguish between the words revolve *and* rotate. *A planet revolves around the sun but rotates on its axis.*

The Titius–Bode Rule

The Titius–Bode rule specifies a simple series of steps that produce a list of numbers matching the sizes of the planetary orbits. To construct this list, we write down the number 0 and below that 3. Continuing downward, we make each number twice the preceding number—0, 3, 6, 12, 24, 48, and so forth. After adding 4 to each number, we divide each by 10. The resulting numbers approximate the radii of the planetary orbits in astronomical units (see Table 16-1).

The rule describes the inner planets well, even including the asteroid belt, but the match for the outer solar system is not as good.

If we leave out Neptune, Pluto fits well, but the omission of Neptune is not justified by any physical evidence.

Some astronomers contend that the agreement is mere chance. There is no physical reason for any of the steps in the Titius–Bode rule. In fact, no matter what the sizes of the orbits, a mathematician could find a sequence of steps that would generate matching numbers. However, the simplicity of the Titius–Bode rule leads many astronomers to view it as significant. It may be telling us that there was nothing random about the formation of the planets.

TABLE 16-1
Average Planetary Distance from the Sun

Planet	Titus–Bode Prediction (AU)	Observed (AU)
Mercury	(0 + 4)/10 = 0.4	0.387
Venus	(3 + 4)/10 = 0.7	0.723
Earth	(6 + 4)/10 = 1.0	1
Mars	(12 + 4)/10 = 1.6	1.524
Asteroids	(24 + 4)/10 = 2.8	2.77 ave.
Jupiter	(48 + 4)/10 = 5.2	5.203
Saturn	(96 + 4)/10 = 10.0	9.539
Uranus	(192 + 4)/10 = 19.6	19.18
Neptune		30.06
Pluto	(384 + 4)/10 = 38.8	39.44

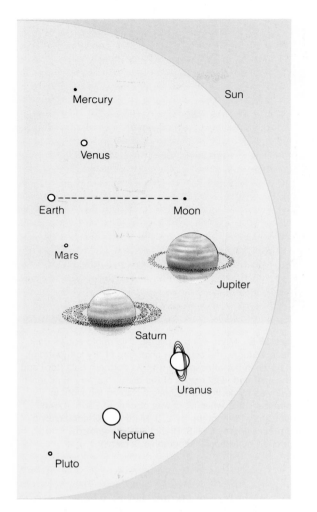

Figure 16-4 The relative sizes of the planets compared to the solar disk.

Imagine a normal brick and another brick made of styrofoam. The density of the normal brick is much higher than that of the styrofoam brick. In the case of planets, we will discover that the Jovian planets are much less dense than the terrestrial planets.

All the terrestrial planets are scarred by craters. Mercury's surface looks much like the moon's, with thousands of overlapping craters (Figure 16-5). Because Mercury is very hot and only 40 percent larger than the moon, it has held no atmosphere. Venus is nearly as large as Earth and has a thick atmosphere. In fact, its surface is perpetually hidden below a dense layer of clouds, so only by mapping its surface with radar have astronomers found craters on Venus. Earth, too, has some craters, although erosion rapidly wears away such features. Mars is about half Earth's diameter and has a much thinner atmosphere, which permits us to see its surface easily. Space probe photos show that some of the markings visible from Earth are in fact craters (Figure 16-6). In addition, planetary probes have found craters on the satellites of Mars, Jupiter, Saturn, and Uranus. This suggests that craters are a characteristic of every object in the solar system with a surface capable of retaining such features.

The terrestrial planets have densities ranging from 3.3 to 5.5 g/cm³. Typically, they contain cores of iron and nickel surrounded by a mantle of dense rock. In contrast, the Jovian planets are rich in hydrogen and helium, so their average density is low—less than 1.75 g/cm³. Saturn's very low density, 0.7 g/cm³, is less than the density of water, so the planet would float if we could find a bathtub big enough. Nevertheless, all the Jovian planets are massive. Jupiter is more than 318 times as massive as Earth, and Saturn is about 95 Earth masses. Uranus and Neptune are smaller—15 and 17 Earth masses, respectively.

Although photographs of Jovian planets reveal swirling cloud patterns (Figure 16-7), their atmospheres are not very deep compared to their radii. In fact, if these so-called gas giants were shrunk to a few centimeters in diameter, their gaseous hydrogen and helium atmospheres would be no deeper than the fuzz on a badly worn tennis ball.

Mathematical models predict that the interiors of the Jovian planets are mostly hydrogen in two different states (Figure 16-8). Beneath the cloud

Mars—lie in the inner solar system. By contrast, the **Jovian planets** are large, gaseous, low-density worlds. The Jovian planets—Jupiter, Saturn, Uranus, and Neptune—lie in the outer solar system beyond the asteroids (Figure 16-4). Note that Pluto does not fit into either category very well. It is small like the terrestrial planets but lies far from the sun and has a low density like the Jovian planets.

Density is a very important concept in planetary astronomy, so it is worth reviewing here. Density is a measure of the amount of matter in a given volume and tells us how concentrated that matter is.

a

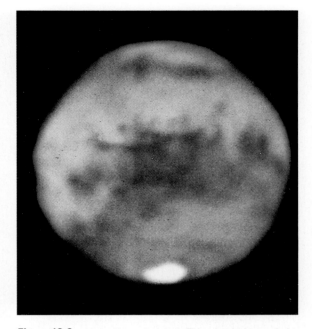

Figure 16-6 Mars, the red planet. This photo shows dark surface features, the southern polar cap of dry ice (bottom), and hazes in the atmosphere over its northern pole (top) and at the east and west edges of the disk. This very high-quality image was made from atop Mauna Kea on the night in September 1988 when Mars was closest to Earth. (Lowell Observatory Photograph by L. J. Martin; color compositing and processing by USGS, Branch of Astrogeology, supported by National Geographic Society.)

b

Figure 16-5 Both Mercury (a) and the moon (b) are heavily cratered. These images are reproduced to the same scale to show the relative sizes of Mercury and the moon. A scale image of Earth would be four times the diameter of the lunar image. (a. NASA/JPL, b. Lick Observatory photograph)

belts of these planets are layers where pressure forces the hydrogen into a liquid state. In the larger Jovian planets, Jupiter and Saturn, the pressure at deeper layers is so high that the hydrogen atoms can no longer hold onto their electrons. With the electrons free to move about, the material (called **liquid metallic hydrogen**) is an excellent electrical conductor. The centers of the Jovian planets are believed to be hot, Earth-sized cores of heavy elements often called "rocky cores," although the heat and pressure would not allow rock to exist as it does on Earth.

Another characteristic of the Jovian planets is their large satellite systems. Jupiter has at least 16 known moons; 4 of them, the **Galilean satellites**—named after Galileo, who discovered them—are visible through a small telescope or even binoculars (Figure 16-9). Saturn, Uranus, and Neptune have systems of satellites. A few of these moons are

a

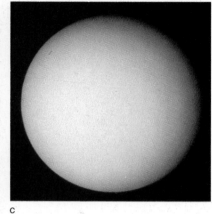

c

d

b

Figure 16-7 (a) Jupiter and its cloud belts and Great Red Spot photographed by Voyager 1. (NASA/JPL) (b) Saturn and its belts and rings photographed by the Hubble Space Telescope. (NASA) (c) Uranus photographed by the Voyager 2 spacecraft. Cloud belts are not visible because of methane ice-crystal haze high in the planet's atmosphere. (NASA/JPL) (d) Neptune imaged by the Voyager 2 spacecraft. (NASA)

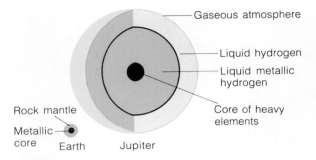

Figure 16-8 Cross sections of Earth and Jupiter illustrate the differences between the interiors of terrestrial and Jovian planets. Earth contains a metallic core with a rock mantle. Jupiter, 11.18 times larger in diameter, is composed mostly of liquid hydrogen, with a core composed of heavy elements such as iron, nickel, and silicon.

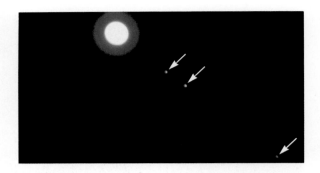

Figure 16-9 Jupiter has at least 17 satellites, but only the 4 Galilean satellites are visible in small telescopes. In this photo, the disk of Jupiter is overexposed to reveal 3 of the fainter Galilean satellites at lower right. The 4th satellite was behind the planet when this photo was taken. (Grundy Observatory photograph)

Figure 16-10 The icy rings of Saturn are so reflective that they are easily visible through a small telescope. This photo shows what we might expect to see through a 6-inch telescope. For more detailed photos of the rings, see Chapter 18. (Grundy Observatory photograph)

Figure 16-11 The asteroid Gaspra as photographed by the Galileo spacecraft. The asteroid is 20 by 12 by 11 km and rotates once every 7.04 hours. Its irregular shape and numerous craters testify that it has suffered many impacts. The color here is exaggerated to reveal subtle color variation. Gaspra would look gray to the human eye. (NASA/Galileo Imaging Team)

larger than our own moon, but most are small, icy worlds not visible in modest size telescopes.

All the Jovian planets have rings. Saturn has a bright ring of icy particles visible through even a small telescope (Figure 16-10). Jupiter, Uranus, and Neptune have rings made of dark, rocky, difficult-to-detect material. All four ring systems show evidence of gravitational interactions between the rings and the satellites.

We will discuss the satellites and rings of the Jovian planets in later chapters. Here it is sufficient to note that the Jovian planets have large numbers of satellites and ring systems, whereas the terrestrial planets have few satellites and no ring systems at

a

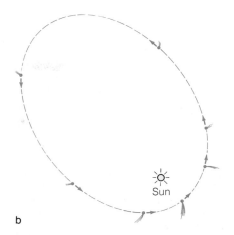

b

Figure 16-12 (a) A comet may remain visible in the evening or morning sky for weeks as it moves through the inner solar system. Comet West was in the sky during March 1976. (Lick Observatory photograph) (b) A comet in a long elliptical orbit becomes visible when the sun's heat vaporizes its ices and pushes the gas and dust away in a tail.

all. These facts will help us understand how planets were formed in the solar nebula.

Space Debris The sun and planets are not the only remains of the solar nebula. The solar system is littered with three kinds of space debris: asteroids, comets, and meteoroids. Although this material represents a tiny fraction of the mass of the system, it is a rich source of information about the origin of the planets.

The **asteroids** (also called minor planets) are small, rocky worlds, most of which orbit the sun between the orbits of Mars and Jupiter. A few asteroids follow orbits that bring them into the inner solar system, and several occasionally pass within a few tens of millions of miles of Earth. Some are located in Jupiter's orbit, and some have been found as far away as the orbit of Saturn.

About 200 of these objects are more than 100 km (60 miles) in diameter, and more than 2000 are more than 10 km (6 miles) in diameter. There are probably half a million greater than 1 km (0.6 mile) and billions that are smaller. Because even the largest are only a few hundred kilometers in diameter, telescopes reveal no surface features. For years, however, astronomers have noted that many asteroids vary in brightness as if they were irregular in shape and rotated as they orbit the sun. Only the largest seem to be spherical.

Expectations were confirmed in 1991 when the Galileo spacecraft on its way to Jupiter passed near the asteroid Gaspra and photographed it (Figure 16-11). The asteroid is highly irregular in shape and is pitted by craters. It has evidently suffered many collisions in its history.

Apparently, the asteroids are composed of material that failed to build a planet at a distance of 2.8 AU from the sun, perhaps due to the influence of massive Jupiter just outside the asteroid belt.

In contrast to asteroids, the brightest **comets** are impressively beautiful objects (Figure 16-12a). Most comets are faint, however, and are difficult to locate even at their brightest. A comet may take months to sweep through the solar system, during which time it appears as a glowing head with an extended tail of gas and dust.

According to the **dirty snowball theory** of comets, the tail, which can be more than 1 AU in length for a bright comet, is produced by the nucleus, a ball of dirty ices (mainly water and carbon dioxide) only a few dozen kilometers in diameter. When the nucleus enters the inner solar system, the sun's radiation begins to vaporize the ices, releasing gas and dust. The radiation pressure from sunlight and the solar wind pushes the gas and dust away in the form of a long tail that always points away from the sun (Figure 16-12b).

The nuclei of comets are icy bodies left over from the origin of the planets. Thus we must conclude that at least some parts of the solar nebula were rich in ices. We will see later in this chapter how important ices were in the formation of the Jovian planets, and we will discuss comets in more detail in Chapter 19.

Unlike the stately comets, **meteors** flash across the sky in momentary streaks of light (Figure 16-13). They are commonly called "shooting stars." Of course, they are not stars but small bits of rock and metal falling into Earth's atmosphere and bursting into incandescent vapor about 80 km (50 miles) above the ground because of friction with the air. This vapor condenses to form dust, which settles slowly to the ground adding about 10,000 tons per year to Earth's mass.

Technically, the word *meteor* refers to the streak of light in the sky. In space, before its fiery plunge, the object is called a **meteoroid,** and if any part of it survives its fiery passage to Earth's surface, it is called a **meteorite.** Most meteoroids are specks of dust, grains of sand, or tiny pebbles. Almost all the meteors we see in the sky are produced by meteoroids that have masses less than 1 g. Only rarely is one massive enough and strong enough to survive its plunge and reach Earth's surface.

A few thousand meteorites have been found, and we will discuss their particular forms in Chapter 19. We mention meteorites here for one specific clue they give us concerning the solar nebula: meteorites tell us the age of the solar system.

The Age of the Solar System If the solar nebula theory is correct, the planets should be about the same age as the sun. The most accurate way to find the age of a celestial body is to bring a sample into a laboratory and determine its age by radiometric dating—that is, by measuring the amount of radioactive elements remaining in the sample. So far the

Figure 16-13 A meteor is the streak of glowing gases produced by a bit of material falling into the earth's atmosphere. Friction with the air vaporizes the material about 80 km (50 miles) above the earth's surface. (Lick Observatory photograph)

only bodies that have been dated in this way are Earth, the moon, and meteorites.

The oldest Earth rocks so far discovered and dated are about 3.9 billion years old. That does not mean that Earth formed 3.9 billion years ago. The surface of Earth is active, and the crust is continually destroyed and re-formed from material welling up from beneath the crust (see Chapter 17). Thus the age of these oldest rocks tells us only that Earth is *at least* 3.9 billion years old.

One of the most exciting results of the Apollo lunar landings was the lunar rocks brought back to Earth's laboratories, where they could be dated. Because the moon's surface is not being recycled like Earth's, some parts of it might have survived unaltered since early in the history of the solar system. Dating the rocks showed the oldest to be 4.48 billion years old. Thus the solar system must be *at least* 4.48 billion years old.

Another important source for determining the age of the solar system is meteorites. Radiometric dating of meteorites yields a range of ages, with the oldest about 4.6 billion years old. This figure is widely accepted as the age of the solar system.

Characteristic Properties of the Solar System

1. Disk shape of the solar system
   Orbits in nearly the same plane
   Common direction of rotation and revolution

2. Titius–Bode rule

3. Two planetary types:
   Terrestrial—inner planets: high-density
   Jovian—outer planets: low-density

4. Planetary ring systems for Jupiter, Saturn, Uranus, and Neptune

5. Space debris—asteriods, comets, and meteors
   Composition
   Orbits

6. Common ages of about 4.6 billion years for Earth, the moon, meteorites, and the sun

One last celestial body deserves mention: the sun. Astronomers estimate the age of the sun as about 5 billion years, but this is not a radiometric date because they cannot obtain a sample of solar material. Instead, they estimate the sun's age from the radiometric ages of Earth, the moon, and meteorites. Computer models of the sun give only approximate ages, but they generally agree with the age of the earth.

Apparently, all the bodies of the solar system formed at about the same time some 4.6 billion years ago. This is the last item we add to our list of significant properties (see Table 16-2).

16-3 THE STORY OF PLANET BUILDING

The challenge of modern planetary astronomy is to compare the characteristics of the solar system to the solar nebula theory and tell the story of how the planets formed. The chemical composition of the nebula determined the chemical composition of the planets, but before the planets could form, the gases of the solar nebula had to condense into solid bits. Those solid bits then collected to form

larger and larger bodies, which eventually grew into the planets. Finally, the remaining gas of the solar nebula was dispersed, and planet building stopped. Thus our story begins with the nebula that gave birth to the sun—the solar nebula.

The Chemical Composition of the Solar Nebula Everything we know about the nature of the solar system and the nature of star formation suggests that the solar nebula was a fragment of an interstellar gas cloud. Such a cloud would have been mostly hydrogen with some helium and tiny traces of the heavier elements.

This is precisely what we see in the composition of the sun (see Table 6-2). Analysis of the solar spectrum shows that the sun is mostly hydrogen, with 25 percent of its mass being helium and only about 2 percent being heavier elements. Of course, nuclear reactions have fused some hydrogen into helium, but this happens in the sun's core and has not affected its surface composition. Thus the composition revealed in its spectrum is essentially the composition of the gases from which it formed.

This must have been the composition of the solar nebula. We can see that in the Jovian planets, where hydrogen and helium are abundant and heavier elements are rare. Yet the Jovian planets do not have as large a percentage of hydrogen and helium as the sun does. These gases are lightweight and escape into space more easily than the heavier elements. Thus we can conclude that the Jovian planets have lost a small amount of the lighter elements originally present in the solar nebula.

The terrestrial planets, in contrast, contain very little hydrogen and helium. These small planets have such low masses they were unable to keep the hydrogen and helium from leaking into space. Indeed, because of their small masses they were probably unable to capture very much of these light-weight gases from the solar nebula. The terrestrial planets are dense worlds because they are composed of the heavier elements from the solar nebula.

Although we can see how the chemical composition of the solar nebula is reflected in the present composition of the sun and planets, a very important question remains: How did gas and dust in the solar nebula come together to form the solid matter of the planets? We must answer this question in two stages. First, we must understand how gas and dust formed billions of small solid particles, and then we must explain how these particles built the planets.

The Condensation of Solids from the Gas   The key to understanding the process that converted the nebular gas into solid matter is the variation in density among solar system objects. We have already noted in Table 16-2 that the four inner planets are high-density terrestrial bodies, whereas the outermost planets are low-density giant planets (except for Plato, which is small and low-density). This division is due to the different ways gases condensed into solids in the inner and outer regions of the solar nebula.

Even among the four terrestrial planets, we find a pattern of subtle differences in density. Merely listing the observed densities of the terrestrial planets does not reveal the pattern because Earth and Venus, being more massive, have stronger gravity and have squeezed their interiors to higher densities. We must look at the **uncompressed densities**—the densities the planets would have if their gravity did not compress them. These densities (Table 16-3) show that the closer a planet is to the sun, the higher its uncompressed density.

This density variation probably originated when the solar system first formed solid grains. The kind of matter that condensed in a particular region would have depended on the temperature of the gas there. In the inner regions, the temperature may have been 1500 K or so. The only materials that could have formed grains at this temperature were compounds with high melting points, such as metal oxides and pure metals, which are very dense. Farther out in the nebula it was cooler, and silicates (rocky material) could have condensed. These are less dense than metal oxides and metals. In the cold outer regions, ices of water, methane, and ammonia could have condensed. These are low-density materials.

The sequence in which the different materials condense from the gas as we move away from the sun is called the **condensation sequence** (Table 16-4). It suggests that the planets, forming at different distances from the sun, accumulated from different kinds of materials. Thus the inner planets formed from high-density metal oxides and metals, and the outer planets formed from low-density ices.

TABLE 16-3

Observed and Uncompressed Densities

Planet	Observed Density (g/cm³)	Uncompressed Density (g/cm³)
Mercury	5.44	5.4
Venus	5.24	4.2
Earth	5.52	4.2
Mars	3.93	3.3
(Moon)	3.36	3.35

We must also remember that the solar nebula did not remain the same temperature throughout the formation of the planets, but may have grown progressively cooler. Thus a particular region of the nebula may have begun by producing solid particles of metals and metal oxides but, after cooling, began producing particles of silicates. Allowing for the cooling of the nebula makes our theory much more complex, but it also makes the processes by which the planets formed much more understandable.

The Formation of Planetesimals   In the development of a planet, three groups of processes operate. First, grains of solid matter grow larger, eventually reaching diameters ranging from a few centimeters to several kilometers. These objects, called **planetesimals,** are believed to be the bodies that the second group of processes collects into planets. Finally, a third set of processes clears away the solar nebula. The study of planet building is the study of these three groups of processes.

According to the solar nebula theory, planetary development in the solar nebula began with the growth of dust grains. These specks of matter, whatever their composition, grew from microscopic size by two processes: condensation and accretion.

A particle grows by **condensation** when it adds matter one atom at a time from a surrounding gas. Thus snowflakes grow by condensation in Earth's

TABLE 16-4
The Condensation Sequence

Temperature (K)	Condensate	Planet (Estimated Temperature of Formation) (K)
1500	Metal oxides	Mercury (1400)
1300	Metallic iron and nickel	
1200	Silicates	
1000	Feldspars	Venus (900)
680	Troilite (FeS)	Earth (600) Mars (450)
175	H_2O ice	Jovian (175)
150	Ammonia–water ice	
120	Methane–water ice	
65	Argon–neon ice	Pluto (65)

atmosphere. In the solar nebula, dust grains were continuously bombarded by atoms of gas, and some of these stuck to the grains. A microscopic grain capturing a gas atom increases its mass by a much larger fraction than a gigantic boulder capturing a single atom. Thus condensation can increase the mass of a small grain rapidly, but as the grain grows larger, condensation becomes less effective.

The second process is **accretion,** the sticking together of solid particles. In building a snowman, we roll a ball of snow across the snowy ground so that it grows by accretion. In the solar nebula, the dust grains were, on the average, no more than a few centimeters apart, so they collided with one another frequently. Their mutual gravitation was too small to hold them to one another, but other effects may have helped. Static electricity generated by their passage through the gas could have held them together, as could compounds of carbon that might have formed a sticky surface on the grains. Ice grains might have stuck together better than some other types. Of course, some collisions might break up clumps of grains; on the whole, however, accretion must have increased grain size. If it had not, the planets would not have formed.

There is no clear distinction between a very large grain and a very small planetesimal, but we can consider an object a planetesimal when its diameter becomes a centimeter or so. Objects this size and larger were subject to new processes that tended to concentrate them. One important effect may have been that the growing planetesimals collapsed into the plane of the solar nebula. Dust grains could not fall into the plane because the turbulent motions of the gas kept them stirred up, but the larger objects had more mass, and the gas motions could not have prevented them from settling into the plane of the spinning nebula. This would have concentrated the solid particles into a thin plane about 0.01 AU thick and would have made further planetary growth more rapid.

This collapse of the planetesimals into the plane is analogous to the flattening of a forming galaxy. However, an entirely new process may have become important once the plane of planetesimals formed. Computer models show that the rotating disk of particles should have been gravitationally unstable and would have broken up into small clouds (Figure 16-14). This would further concentrate the planetesimals and help them coalesce

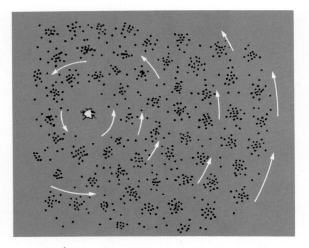

Figure 16-14 Gravitational instabilities in the rotating disks of planetesimals may have forced them to collect in clumps, accelerating their growth.

into objects up to 100 km (60 miles) in diameter. Thus the theory predicts that the nebula became filled with trillions of planetesimals ranging in size from pebbles to tiny planets. As the largest began to exceed 100 km in diameter, new processes began to alter them, and a new stage in planet building began, the growth of protoplanets.

The Growth of Protoplanets The coalescing of planetesimals eventually formed **protoplanets,** massive objects destined to become planets. We must discuss not only the processes that built the planets but also the processes that modified them. According to one theory, the planets formed as solid bodies and later melted; according to another, they formed in a molten state. We will see that this makes a significant difference to theories of the origin of the planetary atmospheres.

If planetesimals collided with each other at orbital velocities, it is unlikely that they would have stuck together. The average orbital velocity in the solar system is about 30 km/sec (67,000 mph). Head-on collisions at this velocity would have pulverized the material. However, the planetesimals moved in the same direction in the nebular plane and thus avoided head-on collisions. Instead, they merely rubbed shoulders at low relative velocities. Such collisions would more likely fuse them together than shatter them.

In addition, some adhesive effects probably helped. Sticky coatings and electrostatic charges on the surfaces of the smaller planetesimals probably aided formation of larger bodies. Collisions would have fragmented some of the surface rock, but if the planetesimals were large enough, their gravity would have held onto some fragments, forming a layer of soil composed entirely of crushed rock. Such a layer on the larger planetesimals may have been effective in trapping smaller bodies.

The largest planetesimals would grow the fastest because they had the strongest gravitational field. Not only could they hold onto a cushioning layer to trap fragments, but their stronger gravity could attract additional material. These planetesimals grew to protoplanetary dimensions, sweeping up more and more material. When massive enough, they trapped some of the original nebular gas to form primitive atmospheres. At some point, they crossed the boundary between planetesimals and protoplanets.

The Jovian planets have no surfaces so their formation seems simpler than that of the terrestrial planets with their dense interiors and silicate crusts. In fact, planetary astronomers currently debate two different theories about the formation of the terrestrial planets.

According to the **homogeneous accretion** theory, the protoplanets, including Earth, grew by the accretion of planetesimals that had the same general composition. These accumulated to form a planet-sized ball of material that was of homogeneous composition throughout (Figure 16-15). Had the planet accumulated rapidly, the energy released by the in-falling planetesimals, the **heat of formation,** would have melted the planet; its denser material would have sunk to the center, and the lighter material would have floated to the surface. Because the homogeneous accretion theory holds that the planet remained homogeneous during formation, it assumes that the planet formed slowly to give the heat time to radiate into space.

According to this theory, once the planet formed, heat began to accumulate in its interior from the decay of short-lived radioactive elements such as ^{26}Al and others. This heat eventually melted the planet and allowed it to differentiate. **Differentiation** is the separation of material according to density. When the homogeneous planet melted, the heavy metals such as iron and nickel settled to the core, while the lighter silicates floated to the surface to form a low-density crust.

According to the homogeneous accretion theory, Earth's present atmosphere was not its first. The first atmosphere consisted of gases trapped from the solar nebula—mostly hydrogen and helium. They were driven off by the heat, aided perhaps by outbursts from the infant sun, and new gases baked from the rocks to form a secondary atmosphere. This creation of a planetary atmosphere from the planet's interior is called **outgassing.**

The homogeneous accretion theory is being challenged by a modified version called the **heterogeneous accretion** theory. This theory assumes that the solar nebula cooled during the formation of the planets so that they did not accumulate from planetesimals of common composition. As planet building began, the first particles to condense in the inner solar system were metals and metal oxides, so the protoplanets began by accreting metallic cores. Later, as the nebula cooled, more silicates could form, and the protoplanets added silicate mantles and crusts. In addition, the planets probably grew rapidly enough so that the heat released by the in-falling material melted the planets, and they further differentiated as they formed.

The heterogeneous accretion theory suggests that our present atmosphere was not produced entirely by outgassing. Much of our atmosphere and our rich supply of water may have accumulated late in the formation of the planets as the earth swept up volatile-rich planetesimals formed in the cooling solar nebula. Such icy planetesimals may have formed in the outer parts of the solar nebula and were then scattered by encounters with the Jovian planets.

Clearly, the heterogeneous accretion theory is much more sophisticated than the homogeneous accretion theory. Probably neither theory is entirely correct, but they illustrate the processes that created the planets in our solar system. So far we have applied these processes primarily to Earth,

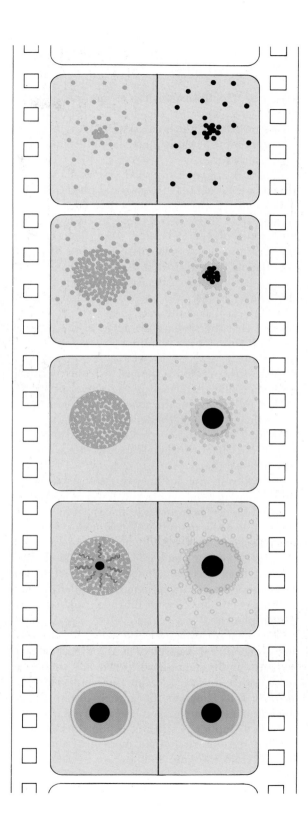

Figure 16-15 Two theories of planetary origins. Left: A planet grows by homogeneous accretion, is heated by radioactive decay, and differentiates. Right: A planet grows by heterogeneous accretion, accumulating first a metallic core, then a mantle, and then a crust.

but we can now ask how well they explain the entries in Table 16-2.

Explaining the Characteristics of the Solar System
Table 16-2 contains the list of distinguishing characteristics of the solar system. Any theory of the origin of the solar system should explain these characteristics.

The disk shape of the solar system is inherited from the solar nebula. The sun and planets revolve and rotate in the same direction because they formed from the same rotating gas cloud. The orbits of the planets lie in the same plane because the rotating solar nebula collapsed into a disk, and the planets formed in that disk.

The Titius–Bode rule (see Box 16-1) is more difficult to explain. We don't understand why the planets are spaced the way they are, but we can suggest that the spacing is related to the way the growing protoplanets dominated regions of the solar nebula.

The condensation sequence explains how the terrestrial planets formed in the inner part of the solar system. Most of the solar nebula consisted of gases that could not condense at the higher temperatures near the sun, so the earthlike planets grew from what little metals and silicates were able to condense there.

In contrast, the Jovian planets formed in the outer solar nebula, where the lower temperature allowed the gas to form large amounts of ices, perhaps three times more ices than silicates. Thus the Jovian planets grew rapidly and became massive. Some theoretical models indicate that Jupiter formed from this abundance of material in less than a thousand years, although most theories predict many millions of years for the formation of the Jovian planets.

The heat of formation (the energy released by in-falling matter) was tremendous for these massive planets. Jupiter must have grown hot enough to glow with a luminosity about 1 percent that of the present sun. However, because it never got hot enough to generate nuclear energy as a star would, it cooled. Jupiter is still hot inside. In fact, both Jupiter and Saturn radiate more heat than they absorb from the sun, so they are evidently still cooling.

When the Jovian planets grew sufficiently large, they attracted vast amounts of nebular gas and thus grew rich in light gases such as hydrogen and helium. The terrestrial planets could not do this because they never became massive enough and because the gas in the inner nebula was hotter and more difficult to trap.

The large satellite systems of the Jovian planets may be made up, in part, of captured planetesimals. The inner planets have only three satellites (Earth has one and Mars two), apparently because the terrestrial planets are less massive and less able to capture satellites.

The ring systems of the Jovian planets are a true mystery. Saturn's ring system may have originated with the planet, or it may have formed later, but the rings around Jupiter, Uranus, and Neptune are too delicate to have survived for 4.6 billion years. We will discuss the origin of the ring systems when we examine these planets in detail in Chapter 18.

Many astronomers see the formation of the Jovian planets and their satellites as miniature versions of the origin of the solar system from the solar nebula. A rotating gas cloud flattens into a disk with a massive body growing in the center. Some ring systems may be fossils of these flattened gas clouds, and the condensation of the Jovian satellites may have been similar to the condensation of the planets in the solar nebula.

Our general understanding of the origin of the solar system also explains the origin of asteroids, meteors, and comets. They appear to be the last of the debris left behind by the solar nebula. These objects are such important sources of information about the history of our solar system that we will discuss them in detail in Chapter 19. But for now, what happened to the solar nebula?

The End of the Solar Nebula
The planets appear to have grown in the solar nebula at the same time that the sun was forming, about 4.6 billion years ago. Indeed, much of the planetary growth may have taken place in darkness because the sun may not have become luminous yet. But once the sun became hot and luminous, it began to clear the nebula of gas and dust and brought planet building to a halt.

Four effects helped to clear the nebula. The most important was **radiation pressure.** When the sun became a luminous object, light streaming from its surface pushed against the particles of the solar nebula. Large bits of matter like planetesimals and planets were not affected, but low-mass specks of dust and individual gas atoms were driven outward and eventually pushed from the system. This is not a sudden process, and it may not have occurred at the same time everywhere in the nebula. Before sunlight could begin clearing the outer nebula, it first had to push its way through the inner nebula.

The second effect that helped clear the nebula was the solar wind, the flow of ionized hydrogen and other atoms away from the sun's upper atmosphere. This flow is a steady breeze that rushes past the earth at about 400 km/sec (250 miles/sec). When the sun was young, it may have had an even stronger solar wind, and irregular fluctuations in its luminosity, like those observed in T Tauri stars, may have produced surges in the wind that helped push dust and gas out of the nebula.

The third effect was the sweeping up of space debris by the planets (the final accretion of the planets). Some astronomers believe that this was very efficient; others suspect that it was less important than the first two processes. Certainly it helped. The extensive cratering of the terrestrial planets and the satellites of Earth, Mars, and Jovian planets is evidence that most of the remaining solar nebula was gobbled up by the planets about 4 billion years ago (Figure 16-16). Since then the solar system has been relatively clear of solid debris.

The fourth effect was the ejection of material from the solar system by close encounters with planets. If a small object such as a planetesimal passes close to a planet, it can gain energy from the planet's gravitational field and be thrown out of the solar system. Ejection is more probable for encounters with more massive planets, so the Jovian planets were probably very efficient at ejecting the icy planetesimals that formed in their region of the nebula.

Together these four effects cleared away the solar nebula. This may have taken considerable time, but eventually the solar system was relatively clear, the planets could no longer gain mass, and planet building ended.

We have made a quick survey of our solar system and used its general characteristics to develop a theory for its origin. In the next three chapters, we must reexamine the solar system in detail, testing the solar nebula theory further and trying to understand the evolution of planetary bodies.

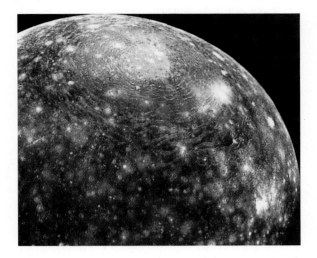

Figure 16-16 The heavily cratered surface of Jupiter's satellite Callisto is shown in this composite of images made by the Voyager 2 spacecraft. Nearly all solid surfaces in the solar system are blasted with impact scars, showing that the solar system once contained much more debris than at present. The smallest features visible in this photo are about 7 km in diameter. (NASA)

SUMMARY

We can reconstruct the process by which the solar system formed by studying the characteristic properties of the system. These properties suggest that the sun and planets formed at about the same time from the same cloud of gas and dust—the solar nebula.

One of the most striking properties of the solar system is its disk shape. The orbits of the planets lie in nearly the same plane, and they all revolve around the sun in the same direction, counterclockwise as seen from the north. This is also true of many of the satellites of the planets. With only two exceptions, the planets rotate counterclockwise around axes roughly perpendicular to the plane of the solar system. This disk shape and the motion of the planets appear to have originated in the solar nebula.

Another striking feature of the solar system is the division of the planets into two families. The terrestrial planets, which are small and dense, lie in the inner part of the system. The Jovian planets are large, low-density worlds in the outer part of the system. In general, the closer a planet lies to the sun, the higher its uncompressed density.

The solar system is now filled with smaller bodies such as asteroids, comets, and meteoroids. The asteroids are small, rocky worlds, most of which orbit the sun between Jupiter and Mars. They appear to be material left over from the formation of the solar system.

Another important characteristic of the solar system bodies is their similar ages. Radiometric dating tells us that the earth, moon, and meteorites are no more than about 4.6 billion years old. Thus it seems our solar system took shape about 4.6 billion years ago.

The solar nebula theory proposes that the solar system began as a contracting cloud of gas and dust that flattened into a rotating disk. The center of this cloud became the sun, and the planets eventually formed in the disk of the nebula. Because the nebula was disk-shaped, the planetary orbits lie in nearly the same plane, and most rotation in the solar system has the same direction—the direction in which the nebula was originally rotating.

According to the condensation sequence, the inner part of the nebula was so hot only high-density minerals could form solid grains. The outer regions, being cooler, condensed to form icy material of lower density. The planets grew from these solid materials, with the denser planets forming in the inner part of the nebula and the lower-density Jovian planets forming farther from the sun.

Planet building probably began as dust grains, which grew by condensation and accretion into planetesimals ranging from a few centimeters to a few kilometers in diameter. These planetesimals settled into a thin plane around the sun and accumulated into larger bodies, the largest of which grew the fastest and eventually they became protoplanets.

According to the homogeneous accretion theory, the terrestrial planets grew slowly at low temperature from the accumulation of planetesimals of homogeneous composition. Later the planets grew hot from the decay of radioactive elements, and their interiors melted and differentiated into core, mantle, and crust. The heterogeneous accretion theory proposes that the planets formed rapidly, grew hot from their heat of formation, and differentiated as they formed. The first planetesimals were metals and metal oxides, but as the nebula cooled, more silicates condensed and built mantles and crusts.

The Jovian planets probably grew rapidly from icy material and became massive enough to attract and hold vast amounts of nebular gas. Heat of formation raised their temperatures very high when they were young, and Jupiter and Saturn still radiate more heat than they absorb from the sun.

Once the sun became a luminous object, it cleared the nebula as its radiation and solar wind pushed material out of the system. The planets helped by absorbing some planetesimals and ejecting others from the system. Once the solar system was clear of debris, planet building ended.

NEW TERMS

solar nebula theory

Titius–Bode rule

terrestrial planets

Jovian planets

liquid metallic
   hydrogen

Galilean satellites

asteroids

comet

dirty snowball theory

meteor

meteoroid

meteorite

uncompressed density

condensation sequence

planetesimal

condensation

accretion

protoplanet

homogeneous accretion

heat of formation

differentiation

outgassing

heterogeneous accretion

radiation pressure

REVIEW QUESTIONS

1. What produced the helium now present in the sun's atmosphere? in Jupiter's atmosphere? in the sun's core?

2. What evidence do we have the stars commonly form with disks of gas and dust surrounding them?

3. If a careful search of all the nearest stars found no planetary systems, the solar nebula theory would be in doubt. Why?

4. According to the solar nebula theory, why is the sun's equator nearly in the plane of Earth's orbit?

5. If you visited another planetary system circling a G star like our sun, would you expect the outer planets to be Jovian, or might this be peculiar to our solar system?

6. Why is almost every solid surface in the solar system marked with impact craters?

7. If you visited the planetary system described in Question 5 and determined its age by radioactive dating, would you expect to get 4.6 billion years, or might this be peculiar to our solar system?

8. What is the difference between condensation and accretion? between homogeneous accretion and heterogeneous accretion?

9. Why are the terrestrial planets so much less massive than the Jovian planets?

10. Why did planet building in the solar nebula come to a stop?

DISCUSSION QUESTIONS

1. In your opinion, if a planetary system originated with one of the first stars to form in our galaxy, how might the process of planetary development differ from the formation of our solar system? (**HINT**: To what population would the star belong?)

2. If the solar nebula theory is correct, then there are probably more planets in our galaxy than stars. Why?

PROBLEMS

1. If you observed the solar system from the nearest star (1.3 pc away), what would the maximum angular separation be between Jupiter and the sun? (**HINT**: See Box 3-2.)

2. The brightest planet in our sky is Venus, which is sometimes as bright as apparent magnitude −4 when it is at a distance of about 1 AU. How many times fainter would it look from a distance of 1 pc (206,265 AU)? What would its apparent magnitude be? (**HINTS**: Remember the inverse square law; see Box 2-2.)

3. If a planet existed beyond Pluto, where would it lie according to the Titius–Bode rule? (Searches for such a planet have found nothing.)

4. What is the smallest diameter of a crater you can identify in Figure 16-5a? (**HINT**: See Appendix D to find the diameter of Mercury in kilometers.)

5. In Table 16-3, which object's observed density differs least from its uncompressed density? Why?

6. What composition might we expect for a planet that formed in a region of the solar nebula where the temperature was about 100 K?

7. Suppose that Earth grew to its present size in 1 million years through the accretion of planetesimals averaging 100 g each. On the average, how many planetesimals did Earth capture per second? (**HINT**: See Appendix D to find Earth's mass.)

8. If you stood on Earth during its formation as described in Problem 7 and watched a region covering 100 m², how many impacts would you expect to see in an hour? (**HINTS**: Assume that Earth had its present radius. The surface area of a sphere is $4\pi r^2$.)

9. The velocity of the solar wind is roughly 400 km/sec. How long does it take to travel from the sun to Pluto?

RECOMMENDED READING

BARNS-SVARNEY, PATRICIA. "The Chronology of Planetary Bombardments." *Astronomy* 16 (July 1988), p. 20.

BEATTY, J. KELLY, BRIAN O'LEARY, and ANDREW CHAIKIN. *The New Solar System*, 3rd ed. New York: Cambridge University Press, 1990.

BECKWITH, STEVEN, and ANNEILA SARGENT. "HL Tauri: A Site for Planet Formation?" *Mercury* 16 (November/December 1988), p. 178.

CAMERON, A. G. W. "The Origin and Evolution of the Solar System." *Scientific American* 233 (September 1975), p. 32.

CHAIKIN, ANDREW. "Pieces of the Sky." *Sky and Telescope* 63 (April 1982), p. 344.

COUPER, HEATHER. "In Search of Solar Systems." *New Scientist* 106 (13 November 1986), p. 34.

FALK, S. W., and D. N. SCHRAMM. "Did the Solar System Start with a Bang?" *Sky and Telescope* 58 (July 1979), p. 18.

GILLETT, STEPHEN. "The Rise and Fall of the Early Reducing Atmosphere." *Astronomy* 13 (July 1985), p. 66.

HARTMANN, W. K. "In the Beginning." *Astronomy* 4 (June 1976), p. 6.

———. "Cratering in the Solar System." *Scientific American* 236 (January 1977), p. 84.

———. *Moons and Planets*, 3rd ed. Belmont, Calif.: Wadsworth, 1993.

JAKI, S. L. "The Titius–Bode Law: A Strange Bicentenary." *Sky and Telescope* 43 (May 1972), p. 280.

McBRIDE, KEN. "Looking for Extra Solar Planets." *Astronomy* 12 (October 1984), p. 6.

McSWEEN, HARRY Y. "Chondritic Meteorites and the Formation of Planets." *American Scientist* 77 (March/April 1989), p. 144.

MARAN, STEPHEN P. "Where Do Comets Come From?" *Natural History* 91 (May 1982), p. 80.

MARVIN, URSULA B. "Search for Antarctic Meteorites." *Sky and Telescope* 62 (November 1981), p. 423.

MURRAY, BRUCE, MICHAEL C. MALIN, and RONALD GREELEY. *Earthlike Planets.* San Francisco: W. H. Freeman, 1981.

PONNAMPERUMA, CYRIL, ed. *Comparative Planetology.* New York: Academic Press, 1978.

REEVES, H. "The Origin of the Solar System." *Mercury* 6 (March/April 1977), p. 80.

RINGWOOD, A. E. *Origin of the Earth and Moon.* New York: Springer-Verlag, 1979.

WALDROP, M. MITCHELL. "Glimpses of Solar Systems in the Making." *Science* 235 (27 February 1987), p. 971.

WETHERILL, G. W. "The Formation of the Earth from Planetesimals." *Scientific American* 244 (June 1981), p. 163.

———. "Dating Very Old Objects." *Natural History* 91 (September 1982), p. 14.

———. "Occurrence of Earth-Like Bodies in Planetary Systems." *Science* 253 (2 August 1991), p. 535.

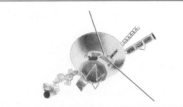

FURTHER EXPLORATION

VOYAGER, the Interactive Desktop Planetarium, can be used for further exploration of the topics in this chapter. This chapter corresponds with the following projects in *Voyages Through Space and Time:* Project 5, The Inner Planets; Project 6, The Outer Planets.

C H A P T E R

17

The Earthlike Planets

The first two people to visit the moon responded in dramatically different ways. Armstrong responded to the significance of the first human step on the surface of another world, but Aldrin responded to the moon itself. It *is* desolate and it *is* magnificent. But it is not unusual. Most planets in the universe probably look like the moon, and astronauts may someday walk on such worlds and compare them with our moon.

The comparison of one planet with another is called **comparative planetology,** and it is one of the best ways to analyze the worlds in our solar system. We will learn much more by comparing planets than we would by studying them individually.

The basis for this comparison is Earth. We know a great deal about Earth and it is our home, so it is natural for us to compare other worlds with our own. But in addition, our planet is a planet of

That's one small step for a man . . . one giant leap for mankind.

Neil Armstrong,
on the moon

Beautiful, beautiful. Magnificent desolation.

Edwin E. Aldrin, Jr.,
on the moon

extremes. Its interior is molten and generates a magnetic field. Its crust is active, with moving sections that push against each other and trigger earthquakes, volcanoes, and mountain building. Almost 75 percent of Earth's surface is covered by a layer of liquid water with an average depth of 3 km. Even Earth's atmosphere is extreme. Processes have altered Earth's air from its natural composition to a highly unusual, oxygen-rich sea of gas. Once we understand Earth's complex properties, we should find it easier to understand other planets that resemble Earth.

17-1 PLANET EARTH

If Earth is to be the benchmark against which we compare other Earthlike planets, then we must examine our planet in detail. We

Figure 17-1 The four stages of planetary development. First: Differentiation into core and crust. Second: Cratering. Third: Flooding of lowlands by lava or water or both. Fourth: Slow surface evolution.

will consider its history, interior, crust, and atmosphere.

A Four-Stage History Earth passed through a four-stage history (Figure 17-1) as it evolved from a forming protoplanet to its present state.

The first stage, *differentiation*, occurred during or soon after the formation of the planet, depending on whether the planet formed by heterogeneous or homogeneous accretion (see Chapter 16). In any case, Earth appears to have developed a molten metallic core as the heavy iron and nickel drained to the center. The lighter silicates floated to the top and eventually solidified to form a thin, brittle crust.

The second stage, *cratering*, began as the crust solidified. Because the solar nebula was filled with rocky debris, the young Earth was heavily battered by meteorites that pulverized the newly formed crust and created a moonlike landscape. The last of the large meteorites blasted out crater basins hundreds of kilometers in diameter. As the solar nebula cleared, the amount of debris decreased, and the level of cratering fell rapidly to its present slow rate.

The third stage, *flooding* of the basins, began as the decay of radioactive elements heated Earth's interior. Lava welled up through fissures in the crust and flooded the deeper basins. Later, as the atmosphere cooled, water condensed and fell as rain, filling the basins and forming the first oceans.

The fourth stage, *slow surface evolution*, has continued for the last 3.5 billion years or more. Earth's surface is in constant motion as sections of crust slide over each other, build mountains, and shift continents. In addition, moving water and air erode the surface and wear away geological features.

All terrestrial planets are believed to have passed through these four stages, but some planets, because of their mass or temperature, have emphasized some stages over others. Our goal in this section is to study Earth's interior, crust, and atmosphere in order to establish a base for comparison with other planets. Only by understanding Earth in detail can we understand planets in general.

Earth's Interior The theory of the origin of planets from the solar nebula predicts that Earth should have melted and differentiated into a dense metallic core with a low-density silicate crust. But what evidence do we have that it did differentiate?

Earth's mass divided by its volume tells us its average density, 5.52 gm/cm³. (See Data File 2.) But the density of Earth's rocky crust is only about 2.8 gm/cm³. Clearly, a large part of Earth's interior must be made of material denser than rock.

High temperatures and tremendous pressure in Earth's interior make any direct exploration impossible. Even the deepest oil wells extend only a few kilometers down and don't reach through the crust. It is quite impossible to drill far enough to sample Earth's core. Yet Earth scientists have studied the interior and found clear proof that Earth did differentiate. This analysis of Earth's interior is possible because earthquakes produce seismic waves that travel through the interior and eventually register on seismographs all over the world (see Data File 2, figure a).

Seismic studies show that the interior consists of three parts: a metallic core; a dense, rocky mantle; and a thin, low-density crust. The core is composed of iron and nickel at a density of 14 g/cm³, denser than lead. The radius of the core is about 55 percent of Earth's radius. At a temperature of about 7000 K, it is hotter than the surface of the sun, but it is not a gas. The tremendous weight pressing inward on the core keeps it solid at its center, although the outer parts of the core, being under lower pressure, are liquid. The central solid core measures only 22 percent of Earth's radius and is difficult to detect. We will see later that the liquid part of the core creates Earth's magnetic field.

The **mantle** is the layer of dense rock and metal oxides that lies between the molten core and the surface. The paths of seismic waves in the mantle show that it is not molten, but it is not precisely solid, either. Mantle material behaves as a **plastic,** a material with the properties of a solid but capable of flowing under pressure. The asphalt used in paving roads is a common example of a plastic. It shatters if struck with a sledgehammer, but it bends under the weight of a heavy truck. Just below Earth's crust, where the pressure is less than at greater depths, the mantle is most plastic.

Earth's rocky crust is quite thin, about 35 km (22 miles) under the continents and only 5 km (3 miles) under the oceans. Owing to its lower density (2.5–3.5 gm/cm³), this material floats on the denser mantle (3.5–5.8 gm/cm³) (Figure 17-2). Unlike the mantle, the crust is brittle and breaks much more easily than the plastic mantle. We will see that this is a characteristic property of many planetary crusts.

Apparently, Earth's magnetic field is a direct result of its rapid rotation and its molten core. The origin of planetary magnetic fields is not yet well understood, but the best current theory is the **dynamo effect.** It supposes that the liquid core is stirred by convection. The rotation of Earth couples this motion into a circulation that generates electric currents throughout the core. Since the highly dense, molten iron–nickel alloy is a better electrical conductor than copper, the material commonly used for electrical wiring, the currents can flow freely and generate a magnetic field.

Though this description of Earth's magnetism seems adequate, many mysteries remain. For example, rocks retain traces of the magnetic field in which they solidify, and some contain fields that point backward. That is, they imply that Earth's magnetic field was reversed at the time they solidified. Careful analysis of such rocks indicates that Earth's field has reversed itself every million years or so, with the north magnetic pole becoming the south magnetic pole and vice versa. These reversals are poorly understood but may be related to changes in the core convection.

Convection in Earth's core is important because it generates the magnetic field. As we will see in the next section, convection in the mantle constantly remakes Earth's surface.

Plate Tectonics Earth's surface is active. It is constantly destroyed and renewed as large sections of crust move about. Geologists refer to this phenomenon as **plate tectonics.** *Plate* refers to the sections of moving crust, and *tectonics* comes from the Greek word for "builder." Plate tectonics is the builder of Earth's surface, because interactions between plates destroy old terrain, push up mountains, and create new crust.

The energy that moves the plates comes from the hot interior. Convection currents of hot mantle material rise from deep layers and spread out under the crust, while other, cooler regions sink. This circulation in the mantle apparently drags along large crustal sections at speeds of a few centimeters per year (see Figure 17-2).

Earth

Average distance from the sun	1.00 AU (1.495979 × 10⁸ km)
Eccentricity of orbit	0.0167
Maximum distance from the sun	1.0167 AU (1.5210 × 10⁸ km)
Minimum distance from the sun	0.9833 AU (1.4710 × 10⁸ km)
Inclination of orbit to ecliptic	0°
Average orbital velocity	29.79 km/sec
Orbital period	1.00 y (365.26 days)
Period of rotation (with respect to the sun)	24ʰ00ᵐ00ˢ
Inclination of equator to orbit	23°27′
Equatorial diameter	12,756 km
Mass	5.976 × 10²⁴ kg
Average density	5.52 g/cm³
Surface gravity	1.0 Earth gravities
Escape velocity	11.2 km/sec
Surface temperature	−50° to 50°C (−60° to 120°F)
Average albedo	0.39
Oblateness	0.0034

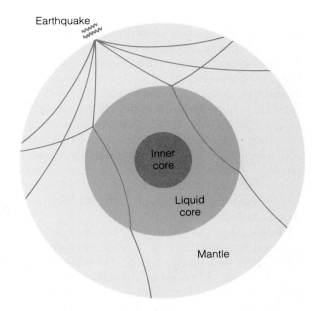

a. Earth's interior can be explored by timing seismic waves as they travel through the interior and reflect off of boundaries between layers.

b. A typical view of Earth. Liquid water oceans cover 75 percent of the surface.

c. Earth photographed from space. Weather patterns and oceans dominate the planet.
(NASA)

Figure 17-2 The solid crust of Earth is about 35 km thick under the continents but only 5 km thick under the oceans. It floats on the plastic mantle, and convection currents in the mantle spread the seafloor open and move the continents.

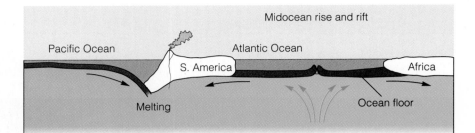

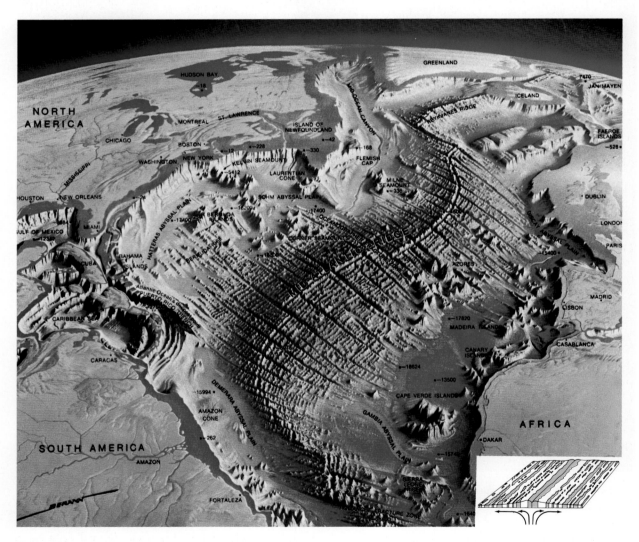

Figure 17-3 If we could drain the Atlantic Ocean, we would see the midocean rift snaking along the crest of the midocean rise from Iceland into the South Atlantic. The ocean floor spreads apart along the rift, pushing the continents apart. As magma rises into the rift and solidifies, it records the alternation of Earth's magnetic field (inset). These alternating bands of normal (dark) and reversed (white) magnetism in the ocean floor are clear evidence of seafloor spreading. (Courtesy Alcoa)

Figure 17-4 Deep trenches in the ocean floor mark places where plates descend (for example, near Japan and Chile), and midocean rises show where the ocean floor is spreading apart (for example, mid-Atlantic). The Hawaiian–Emperor chain of extinct undersea volcanos extends northwest from Hawaii. This map was constructed from gravitational perturbations on the Seasat earth satellite. (NASA)

Direct evidence of plate tectonics lies in the ocean floor. The crust there is thin and composed primarily of dense **basalt,** rock characteristic of solidified lava. These seafloors are not at all flat. Running across the ocean beds are undersea mountain ranges called **midocean rises,** and splitting these rises are chasms called **midocean rifts** (see Figure 17-2). The rocks of the midocean rises are young, but that only hints at the explanation. The real evidence appeared when scientists measured the residual magnetism of the seafloor and found that some sections retained fields that were the reverse of Earth's present field. Evidently, these regions solidified from molten rock during Earth's periodic magnetic reversals. That these regions show alternating, symmetric bands of magnetism running parallel to the midocean rifts indicates that Earth is creating new crust along the rifts while the seafloors spread outward.

The Atlantic Ocean is an especially good example (Figure 17-3). Molten material wells up along the midocean rift and solidifies to form young basalt. As the crustal plates spread apart at about 2–4 cm per year, the midocean rift opens and new material rises to form more crust. Thus midocean rises are composed of young rock. Parallel magnetic bands form because Earth's field reverses every

million years or so, and the solidifying rock records these changes in alternating strips (Figure 17-3), just as the moving tape in a tape recorder makes a record of the changing magnetic field in the recording head.

If Earth is adding crust in one region, it must be destroying it in another. Crust vanishes in trenches along the coasts of some continents where the spreading seafloor slides downward (Figure 17-4; see also Figure 17-2). The collision of moving plates often crumples the crust and pushes up mountains. In addition, the descending seafloor melts along with its accumulated sediment and releases low-density molten rock that rises to the surface and produces volcanism. The Andes Mountains along the west coast of South America and the associated volcanism are due to the descent of the Pacific Ocean floor beneath the continent. The same process produces the volcanoes (including Mount St. Helens) and earthquakes of Washington state, western Canada, Alaska, and Japan. The deep trenches where the ocean floor descends are visible in Figure 17-4.

The seafloor does not always slip below a continent. The floor of the Atlantic Ocean is locked to North and South America and is pushing the continents westward. Tracing this continental drift

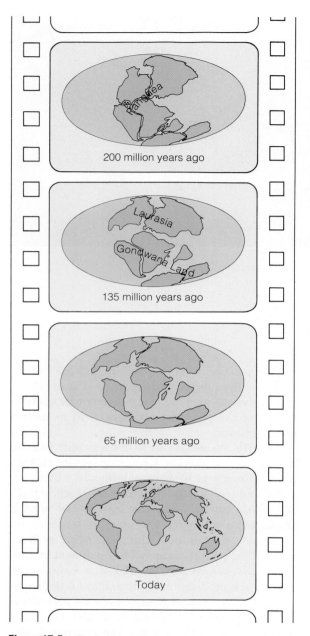

Figure 17-5 Continental drift has broken the original land mass Pangaea first into Laurasia and Gondwanaland and eventually into the continents we know. (Adapted from R. S. Dietz and J. C. Holden, *Scientific American* 223, April 1970)

backward in time, Earth scientists find that North and South America were in contact with Europe and Africa only 200 million years ago.

Understanding seafloor spreading and continental drift permits us to construct a history of Earth's surface. The early motions of Earth's crust are unknown, but about 250 million years ago all of the land masses were joined in one great continent, Pangaea (Figure 17-5). About 200 million years ago, currents in the mantle broke Pangaea into two land masses, Laurasia in the north and Gondwanaland in the south. These land masses further fragmented and drifted into their present configuration.

Today, the plates continue to drift at the urging of the currents in the mantle. They split apart to form new seas, come together to crumple the crust into mountains, and slip against each other to generate earthquakes. The first sign of a continent's splitting is a long, straight, deep depression called a **rift valley**. Africa has recently split from Arabia, opening a rift valley now filled by the Red Sea. Other rift valleys extend southward from the Red Sea across eastern Africa. In contrast, the collision of Africa with the east coast of North America roughly 250 million years ago folded the crust into the Appalachian Mountains, just as the collision of India with southern Asia is now building the Himalaya Mountains. In addition, sections of crust in direct contact may slip past each other, as in the case of the Pacific plate carrying part of southern California northward along the San Andreas fault and generating frequent earthquakes.

Plotting the location of earthquakes and volcanism on a world map reveals the edges of the moving plates (Figure 17-6). The Pacific plate, bounded by a ring of earthquake zones, descends under the Eurasian plate along the North Pacific and Japanese islands. This active zone of circum-Pacific earthquakes and volcanoes is often called the ring of fire. Earthquakes and volcanoes also occur along the midocean rises where new crust forms and the seafloor spreads.

Because Earth's crust is active, all geological features gradually change. The oldest existing portions of the crust, the Canadian shield and portions of South Africa and Australia, are only about 3.9 billion years old. The constant churning of Earth's surface has wiped away all record of older crust.

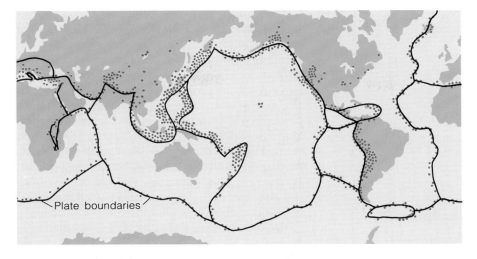

Figure 17-6 Earthquakes (shown as dots) tend to outline the crustal plates. Volcanism follows a similar pattern, and much of the Pacific is ringed by volcanic peaks that extend from Chile through Central America, Alaska, and south through Japan—the "ring of fire." The volcanism and associated earthquakes in Hawaii are produced where a hot spot in the mantle penetrates upward through a plate. A few earthquakes, such as those near St. Louis and the American east coast, are caused by old, secondary faults produced by the tension and compression of plate motions.

Earth's Atmosphere In discussing Earth's atmosphere, we face three related problems: How did it originate? How has it evolved? How is human activity altering it? These problems are important not only in the context of our discussion of other planets but also in terms of our lives on Earth.

Our planet's first atmosphere, its **primeval atmosphere,** was once thought to have contained hydrogen, helium, methane, ammonia, and other gases drawn from the solar nebula. Modern studies of the solar system's formation, however, hint that the planets formed by heterogeneous accretion (see Chapter 16) and were thus hot enough to release large amounts of gases such as carbon dioxide, nitrogen, and water vapor. Also, some theorists believe the last stages of the solar nebula saw the planets accreting planetesimals rich in volatiles, including water. Thus the primeval atmosphere must have been rich in carbon dioxide and water vapor.

The first oceans were small, but as more gases escaped and Earth cooled, more and more water rained out of the atmosphere. As the oceans grew larger, carbon dioxide began to dissolve in the wa-ter. Carbon dioxide is highly soluable in water—the reason carbonated beverages are inexpensive—but the primitive oceans could not have absorbed all the atmospheric carbon dioxide had the gas not reacted with dissolved compounds to form silicon dioxide, limestone, and other mineral sediments. Thus the oceans transferred the carbon dioxide from the atmosphere to the seafloor and left the air rich in nitrogen.

This removal of carbon dioxide is critical to us because a carbon dioxide–rich atmosphere can trap heat by a process called the **greenhouse effect.** When sunlight shines through the glass roof of a greenhouse, it heats the benches and plants inside (Figure 17-7). The warmed interior radiates heat in the form of infrared radiation, but the infrared photons cannot get out through the glass. Heat is trapped within the greenhouse, and the temperature climbs until the glass itself grows warm enough to radiate heat away as fast as the sunlight enters.* In the case of a planet, carbon

*A greenhouse also grows warm because the walls prevent the warm air from mixing with the cooler air outside.

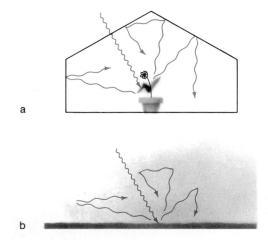

Figure 17-7 The greenhouse effect. (a) Short-wavelength light can enter a greenhouse and heat its contents, but the longer-wavelength infrared cannot get out, and the greenhouse grows warmer. (b) The same process can heat a planet if its atmosphere contains carbon dioxide, which is transparent to light but opaque to infrared radiation.

dioxide in the atmosphere admits sunlight and the surface grows warm. Carbon dioxide, however, is not transparent to infrared radiation, so heat is trapped and the planet's surface temperature rises.

Earth's oceans and its plant life have spent roughly 4 billion years removing carbon dioxide from our atmosphere and thus keeping Earth's climate from overheating. That carbon dioxide was buried in the form of carbonates such as limestone and as carbon-rich deposits such as coal, oil, and natural gas. In the last century or so, human civilization has become possessed with the need to generate energy by digging up carbon and burning it. This is steadily increasing the carbon dioxide concentration in our atmosphere and warming our climate. A temperature rise of only a few degrees could begin melting the polar caps and cause a rise in sea levels by as much as 50 m. Florida, for example, could vanish. When we discuss Venus later in this chapter, we will see what the greenhouse effect can do when a planet's atmosphere becomes rich in carbon dioxide.

Although Earth's oceans removed carbon dioxide from the atmosphere, any methane and ammonia in the atmosphere was removed by a different process. Ultraviolet radiation from the sun would have quickly destroyed these molecules. Earth's

lower atmosphere is now protected from ultraviolet radiation by a layer of ozone (O_3) 15–30 km (10–20 miles) above the surface. However, the secondary atmosphere did not at first contain free oxygen, so an ozone layer could not form, and the sun's ultraviolet radiation penetrated deep into the atmosphere. The energetic ultraviolet photons broke up water, methane, and ammonia molecules, and the hydrogen escaped to space. The carbon formed carbon dioxide and dissolved in the oceans, leaving nitrogen to accumulate in the atmosphere. By the time Earth was about 2.5 billion years old, the atmosphere had been cleared of methane and ammonia and was composed mostly of nitrogen.

This illustrates another effect human society is having on Earth's atmosphere. The ozone layer now protects us from harmful ultraviolet radiation, but certain compounds called chlorofluorocarbons (CFCs) are used in industrial processes and can destroy ozone when they leak into the atmosphere. Since the late 1970s, the ozone concentration has been falling, and this may eventually expose us to levels of ultraviolet radiation that could cause skin cancers and other problems. When we discuss Mars, we will see the effects of an atmosphere without ozone.

Earth's atmosphere has an ozone layer because it contains oxygen, and the origin of that oxygen is linked to the origin of life, the subject of Chapter 20. It is sufficient here, however, to note that life must have originated within a billion years of Earth's formation. This life did not significantly alter the atmosphere, however, until photosynthesis evolved about 3.3 billion years ago. Photosynthesis absorbs carbon dioxide from the air and utilizes it for plant growth, releasing oxygen back into the atmosphere. Because oxygen is a very reactive element, it combines easily with other compounds, and thus the oxygen abundance grew slowly at first. Apparently, the development of large, shallow seas along the continental margins half a billion years ago allowed ocean plants to manufacture oxygen faster than chemical reactions could consume it. Atmospheric oxygen then increased rapidly, and it is still increasing at the rate of about 1 percent every 36 million years.

While our society struggles to reduce the amount of carbon dioxide and CFCs released into our atmosphere, astronomers focus on other worlds, from which we can learn more about the evolution of planets and their atmospheres. Ironically, the next world we will discuss has no atmosphere at all.

17-2 THE MOON

Although the moon is technically only a satellite, we have good reason to choose it to begin our study of other planets. It is the only celestial body that humans have visited, and it is the only body other than Earth from which scientists have samples of known origin that can be analyzed and dated in terrestrial laboratories. Although the moon is too small to have any atmosphere, it is large enough to have a fascinating geological history.

Lunar Geology   The moon is small, only one quarter the diameter of Earth, and its size has dramatically affected the way the moon has evolved. (See Data File 3.) In general, the smaller a body is, the more rapidly it loses its internal heat. Earth's core is large, contains a great deal of heat, and is insulated by the large bulk of the earth. By contrast, the moon's core is small and has much less insulation around it. Thus the moon has lost much of its internal heat and is probably now solid rock from its surface well down toward its center. Seismographs left on the moon by the Apollo astronauts have detected moonquakes originating deep within the lunar interior, which suggest it may still have a small, partially molten core (Figure 17-8).

Because the moon's mantle is not plastic, its surface is not broken into plates. In fact, the lunar surface has never been affected by moving plates of crust. There are no lunar mountain ranges marking places where the crust has buckled because of colliding plates. The only mountains are those blasted out of the surface by the impact of large meteorites and small, cold volcanoes dating from the moon's youth.

The moon's small mass means it has a low escape velocity (see Chapter 11), and gas atoms at its surface can escape into space. Thus the moon was unable to retain an atmosphere for any length of time, and consequently erosion has never been important. Wind has not disturbed the dusty surface, and water has never formed ocean beds or river

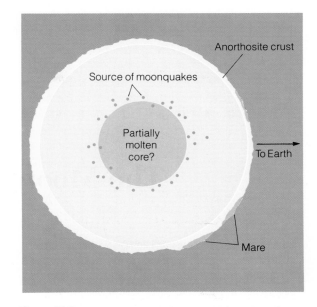

Figure 17-8   Cross section of the moon shows deep location of moonquakes near what may be a partially molten core. Note that the lunar crust is thinner on the side toward Earth.

channels. The only erosion comes from heating and cooling of the surface rocks and cratering. The constant bombardment by billions of microscopic meteorites has sand-blasted the surface into dust. Also, the impact of larger meteorites scatters pulverized rock called **ejecta** over large areas of the surface. This is especially obvious where this debris forms **rays,** white streamers radiating from craters such as Copernicus and Kepler (see Data File 3, figure a).

Because the moon's surface has never been modified by moving plates or strongly affected by erosion, it preserves a record of the first three stages of planetary formation unaltered by the passage of 3 billion years. The Apollo explorations recovered samples of **vesicular basalt,** a form of solidified lava containing holes that were once gas bubbles in the molten rock (see Data File 3, figure c). Like the gas bubbles in a newly opened bottle of soda, these bubbles must have formed when the pressure on the molten lava was released—when it flowed out onto the surface. Thus the lunar surface was once marked by volcanism, and lava flows were once active, but the interior cooled so rapidly the

The Moon

a. Photo composite of the moon (inverted as in a telescope) shows major maria and craters. (Lick Observatory photograph)

Average distance from Earth	384,400 km (center to center)
Eccentricity of orbit	0.055
Maximum distance from Earth	405,500 km
Minimum distance from Earth	363,300 km
Inclination of orbit to ecliptic	5°9′
Average orbital velocity	1.022km/sec
Orbital period (sidereal)	27.321661 days
Orbital period (synodic)	29.5305882 days
Inclination of equator to orbit	6°41′
Equatorial diameter	3476 km
Mass	7.35×10^{22} kg (0.0123 $M_\oplus$)
Average density	3.36 g/cm³ (3.35 g/cm³ uncompressed)
Surface gravity	0.167 Earth gravities
Escape velocity	2.38 km/sec (0.21 $V_\oplus$)
Surface temperature	−170° to 130°C (−274° to 266°F)
Absolute albedo	0.07

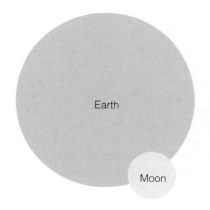

b. The moon's diameter is roughly 27 percent that of Earth.

c. A lunar sample of vesicular basalt—solidified lava containing bubbles. (NASA)

d. The lunar soil is rock finely powdered by 4.5 billion years of meteorite impacts. (NASA)

lava flows ended long ago. Since then the only influence on the surface has been cratering.

The lunar terrain divides into two strikingly different regions—the lowlands and the highlands (Figures 17-9 and 17-10). The lowlands are smooth, dark plains with generally circular outlines and few craters. These lowlands are called **maria** (singular **mare**), meaning "seas," because the first astronomers to examine the moon with telescopes thought they were oceans. In fact, they are low regions filled by successive flows of dark lava. The first Apollo missions landed on these plains and found the surface covered by a thick layer of pulverized lava. The samples returned to Earth indicate that the lava solidified about 3.2–3.3 billion years ago.

The lunar highlands are the lighter-colored, heavily cratered regions that lie about 3 km (2 miles) higher than the lowlands. Because of their height, they were not flooded by the lava that formed the maria, and thus they represent an earlier stage in the moon's history. Because of the interest in the older highlands, the last two Apollo missions risked landing among the jumbled craters. They found that the rock there was not the basalts of the maria but aluminum and calcium silicates called **anorthosite,** a rock type that is lighter in color and lower in density than basalt. As in the case of the maria, the surface in the highlands was covered by a layer of dust and bits of rock debris from the extensive cratering. Many lunar rocks are **breccias,** masses of broken fragments cemented together. Nowhere did the explorers find unbroken pieces of the original lunar crust.

Although we say the highlands are lighter-colored than the lowlands, we should note that the average **albedo** of the moon is only 0.06. That is, it reflects only 6 percent of the light that hits it. By contrast, Earth's albedo, thanks mostly to its bright clouds, is 0.39. The moon looks bright in the sky in contrast to the blackness of space, but moon rocks are nearly all dark gray-brown in color.

The Apollo missions returned 800 pounds of rock samples, of which only 15 percent have been analyzed so far, but that analysis permits us to summarize the history of the moon.

The History of the Moon The four-stage history of the moon is dominated by a single fact—the moon is small. It cooled rapidly and developed a thick, rigid crust.

Figure 17-9 (a) The lunar highlands are the oldest parts of the moon and are heavily battered by craters. (b) The maria are younger and contain few craters. They were formed by lava flows that occurred after the end of most cratering. (Lick Observatory photographs)

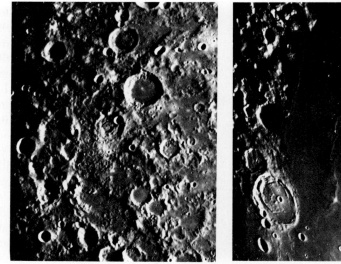

a

b

a

b

Figure 17-10 (a) Apollo 11 astronaut Edwin E. Aldrin, Jr., stands on the dusty surface of Mare Tranquillitatis in the lunar lowlands, site of the first lunar landing. The flat lava plain is almost featureless and the horizon is straight. (b) In the lunar highlands at Taurus–Littrow the surface is mountainous and irregular. Note the horizon. Apollo 17 scientist–astronaut Harrison Schmitt is shown passing huge boulders. (NASA)

During the first stage of planet formation, the moon formed and differentiated. The moon is poor in iron, but what it had sank to form a small core, and the lowest-density minerals floated upward on a sea of magma to form a low-density anorthosite crust. The surface solidified 4.1–4.6 billion years ago, as shown by the mineral content and radiometric ages of the oldest rocks from the highlands.

The second stage, cratering, began as soon as the crust solidified, and the record of the older

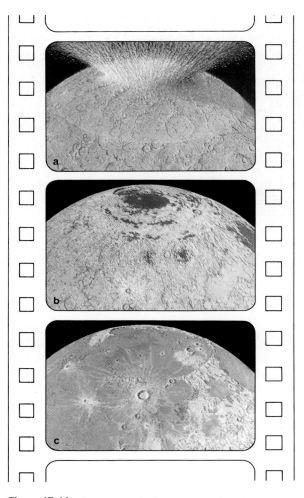

Figure 17-11 The evolution of the lunar surface is illustrated in this artist's conception of the formation of Mare Imbrium. A major impact (a) excavated a large basin, which was repeatedly flooded by lava (b). Since the end of the lava flows, minor cratering has created the surface we see (c). (Courtesy Don Davis)

blasted out by the impact of an object about the size of Rhode Island. This Imbrium event occurred about 4 billion years ago and hurled ejecta 1400 km (870 miles) away from the site in all directions, blanketing 16 percent of the lunar surface. Between 3.9 and 4.1 billion years ago, the cratering rate fell rapidly to its current rate, roughly 1000 times less than during the early heavy bombardment.

The tremendous impacts that formed the lunar basins cracked the anorthosite crust to great depths and led to the third stage. Though the moon cooled rapidly after its formation, some process, perhaps radioactive decay, heated the subsurface material, and part of it melted, producing lava that followed the cracks in the crust up from a depth of about 300 km into the basins. The basins were flooded by successive lava flows of dark basalts from 3.8 to 3.2 billion years ago (Figure 17-12).

When spacecraft first circled the moon, **NASA** scientists were surprised to discover that their orbits were perturbed by great masses located just below the lunar maria. These **mascons** (mass concentrations) are caused by the deep layers of dense lava filling the mare basins, and the fact that they have not sunk into the lower-density crust shows how rigid the lunar crust has become.

Studies of the lunar crust show that it is thinner on the side toward Earth. This may be due to Earth's tidal effects on the moon. The lava flooded the basins on the earthward side, where the crust was thinner, but it was unable to flood the lowlands on the far side. Mare Orientale is a gigantic impact basin at the extreme western edge of the moon, as seen from Earth. It contains only small areas filled with lava.

On Earth, the deeper basins were flooded first with lava, then with water, but liquid water seems never to have existed on the moon. Thus when the moon cooled further and volcanism died down, the airless surface stopped changing. With no water there has been almost no erosion. Cratering still occurs on the moon. Indeed, a few meteorites found on Earth have been identified as moon rocks ejected from the moon by impacts within the last few million years. Such large impacts are rare now, but the bombardment of micrometeorites goes on pulverizing the lunar surface. Except for these

highlands shows that the cratering was intense during the first 0.5 billion years. This **early heavy bombardment** was caused by the wealth of debris in the young solar system, and we will see its effects on other worlds as well. The moon's anorthosite crust was shattered to a depth of about 2 km (1.2 miles) by this rain of meteorites. The largest impacts formed gigantic crater basins hundreds of kilometers in diameter (Figure 17-11). The basin that became Mare Imbrium, for instance, was

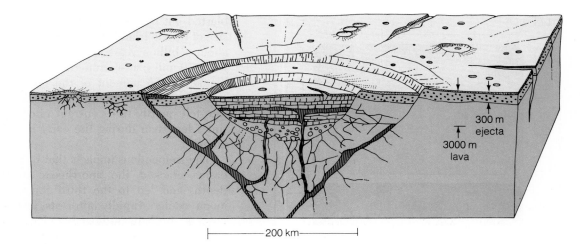

Figure 17-12 Major impacts on the moon broke the crust to great depth and produced large basins. Lava flowed up through the broken crust and flooded the basins to produce maria. (Adapted from William Hartmann, *Astronomy: The Cosmic Journey,* Belmont, Calif.: Wadsworth Publishing Company, 1991)

processes, the moon is geologically dead, frozen between stages 3 and 4.

The Origin of the Moon The four-stage history of the moon can be read from its surface, but until recently, astronomers could not agree on how the moon first formed. The three traditional hypotheses did not survive comparison with observation.

The **fission hypothesis** proposed that the moon broke from a rapidly spinning proto-Earth. If this fission occurred after the proto-Earth had differentiated, the moon would have formed from iron-poor crustal material. Thus the hypothesis explained the moon's low average density and lack of iron. The hypothesis did not survive because the moon rocks were found to differ chemically from those of Earth. Also, if the proto-Earth had spun fast enough to break up, the Earth–moon system should now contain a great deal of angular momentum—much more than is observed.

The **condensation hypothesis** suggested that Earth and the moon condensed simultaneously from the same cloud of material in the solar nebula. This hypothesis did not survive because Earth and the moon have different densities and compositions. The moon, for example, is very poor in volatiles—materials such as water, which are easily vaporized.

The **capture hypothesis** suggested that the moon formed elsewhere in the solar nebula and was later "captured" by Earth. This hypothesis has been unpopular because it requires a number of coincidences to permit Earth to capture the moon. Also, upon encountering Earth, the moon would be moving so rapidly that Earth's gravity would be unable to capture it without ripping it to fragments through tidal forces.

Until the mid-1980s, this left astronomers with no acceptable way to explain the origin of the moon, and they occasionally joked that the moon could not exist. But a new hypothesis has survived the first tests and may eventually win general acceptance.

The **large-impact hypothesis** supposes that the moon formed from debris ejected into a disk around Earth by the impact of a large body (Figure 17-13). The impacting body may have been as large as Mars. Instead of saying that Earth was hit by a large body, we may be more nearly correct in saying that Earth and the moon resulted from the collision and merger of two very large planetesimals. The resulting large body became Earth, and the ejected debris formed the moon.

This would explain a number of phenomena. The collision had to occur at a steep angle to eject enough matter to make the moon. That is, the ob-

Figure 17-13 A Mars-sized planetesimal strikes the young, molten Earth, blasting iron-poor mantle material into orbit. In this simulation, the iron core of the planetesimal falls back into Earth within 4 hours, and the moon begins to form within about 10 hours.

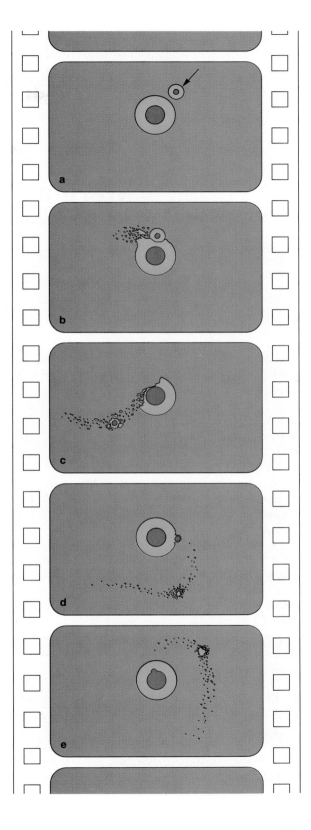

jects did not collide head-on. Such a glancing collision would have spun the resulting material rapidly and would explain the present angular momentum in the Earth–moon system. If the colliding planetesimals had already differentiated, the ejected material would be mostly iron-poor mantle and crust. Calculations show that the iron core of the impacting body would have fallen into Earth within 4 hours. This would explain why the moon is so poor in iron and why the abundances of other elements are so similar to rocks from Earth's mantle. Also, the material that eventually became the moon would have remained in a disk long enough for volatile elements, which the moon lacks, to be lost to space.

We have studied the moon in detail for two reasons. It makes a striking contrast to the more massive, water-rich Earth, but it also serves as a basis of comparison for our study of other planets. Mercury, for example, resembles the moon more than it resembles Earth.

17-3 MERCURY

Mercury is only 4878 km (3047 miles) in diameter, about 40 percent larger than the moon and 38 percent the diameter of Earth. (See Data File 4.) Like the moon, it has cooled too quickly to have developed plates in motion on a plastic mantle. Thus it too is a dead world.

Mariner 10 at Mercury Because Mercury's orbit keeps it near the sun, it is difficult to observe from Earth, and little was known about its surface until 1974–75, when Mariner 10 looped through the inner solar system and visited Venus once and Mercury three times. The photos of Mercury revealed a planet whose surface is heavily cratered, much like the moon's (Data File 4, figure c). Careful analysis of the photos shows that large areas of the surface have been flooded by lava and subsequently cratered.

Mercury

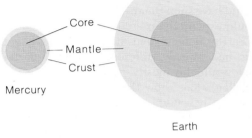

a. In proportion to its size, Mercury has a very large metallic core.

Average distance from the sun	0.387 AU (5.79 × 10⁷ km)
Eccentricity of orbit	0.2056
Maximum distance from the sun	0.467 AU (6.97 × 10⁷ km)
Minimum distance from the sun	0.306 AU (4.59 × 10⁷ km)
Inclination of orbit to ecliptic	7°00′16″
Average orbital velocity	47.9 km/sec
Orbital period	0.24085 y (87.969 days)
Period of rotation	58.646 days (direct)
Inclination of equator to oribit	0°
Equatorial diameter	4878 km (0.382 $D_\oplus$)
Mass	3.31 × 10²³ kg (0.0558 $M_\oplus$)
Average density	5.44 g/cm³ (5.4 g/cm³ uncompressed)
Surface gravity	0.38 Earth gravities
Escape velocity	4.3 km/sec (0.38 $V_\oplus$)
Surface temperature	−173° to 430°C (−279° to 806°F)
Average albedo	0.1
Oblateness	0

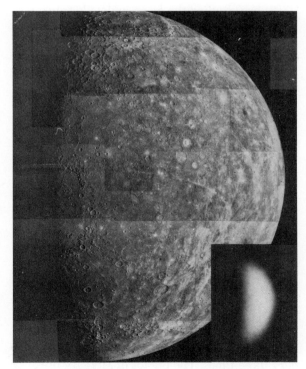

b. Photomosaic of Mercury made by the Pioneer 10 spacecraft. Caloris Basin is in shadow at left. Almost no surface detail is visible from Earth (inset). (NASA; inset courtesy Lowell Observatory)

The largest impact feature on Mercury is the Caloris Basin, a ringed area 1300 km (800 miles) in diameter (Figure 17-14), which looks much like Mare Orientale, the ringed basin on the moon. Both features consist of concentric rings of cliffs formed by impact.

Though Mercury looks like the moon, it does have one characteristic feature the moon lacks. Mariner 10 photos revealed great cliffs, called scarps, up to 3 km (1.9 miles) high and curving as far as 500 km (310 miles) across the surface (Data File 4, figure c). Because the cliffs are curved, planetary scientists refer to them as **lobate scarps.** The Mariner 10 encounters did not photograph the entire planet, but everywhere the probe looked it found lobate scarps cutting through the landscape. The scarps even cut through craters, indicating that they had formed after most of the early heavy bombardment. Apparently, Mercury's interior cooled and shrank by a few kilometers, forcing the crust to wrinkle into lobate scarps, much like the skin of a drying raisin.

In 1991, radar observations of Mercury revealed that the north pole of the planet, which was tipped slightly toward Earth at the time of the observations, is radar bright. The observers suggest this may indicate that Mercury has deposits of ice frozen into its soil near its poles. Although Mercury is close to the sun and exposed to large temperature extremes, the poles may remain cold enough for ice to survive in the soil. We will see evidence of similar ice deposits in the soil of Mars.

A History of Mercury   Mercury is small, and that fact has determined much of its history. It is too small to retain a true atmosphere, though it does "borrow" gases from the solar wind to form a tenuous atmosphere less dense than any vacuum on Earth. With no atmosphere for protection, the temperature in direct sunlight reaches 700 K (800°F) and falls in darkness to 100 K (−280°F).

The Mariner 10 data and the history of Earth's moon permit us to sketch the four-stage history of Mercury. In the first stage, Mercury differentiated. Its high density implies that it has a large metallic core, and Mariner 10 discovered a magnetic field with a strength of about 10^{-4} that of Earth—further evidence of a metallic core.

Cratering battered the crust, and lava flows welled up to fill the lowlands just as they did on the

c. Hero Rupes, a lobate scarp, cuts through craters in smooth plains. (NASA)

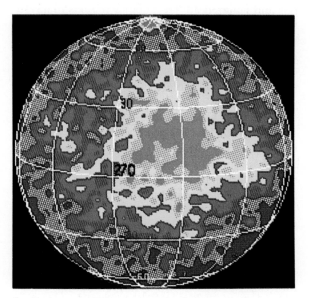

d.   Radar map of Mercury reveals a bright region at its north pole (top). Observers suspect polar ice. (Martin Slade, JPL)

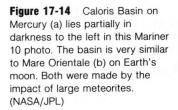

Figure 17-14 Caloris Basin on Mercury (a) lies partially in darkness to the left in this Mariner 10 photo. The basin is very similar to Mare Orientale (b) on Earth's moon. Both were made by the impact of large meteorites. (NASA/JPL)

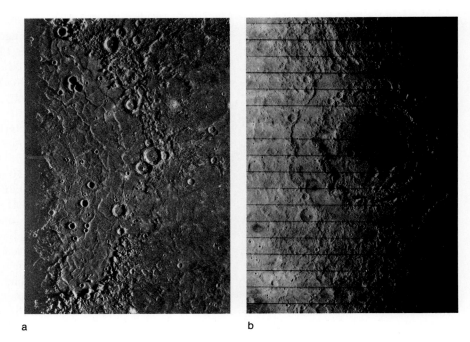

a

b

moon. As the small world lost internal heat, it contracted and its crust broke to form the lobate scarps. Lacking an atmosphere to provide erosion, Mercury has changed little since the last lava hardened.

17-4 VENUS

We might expect Venus to be much like Earth. It is 95 percent Earth's diameter, has a similar average density, and is located only 30 percent closer to the sun. Unfortunately, the surface of Venus is perpetually hidden below thick clouds, and only in the last decade have astronomers discovered that Venus is not Earth's twin. (See Data File 5.) In fact, Venus seems quite unearthly; its surface temperature is about 745 K (880°F) and its geology appears to be quite different from Earth's.

The Surface of Venus Although the clouds of Venus are opaque to visible light, they are transparent to radio waves, so it has been possible to map the surface using radar. The first radar maps were made in 1965 from Earth and had low resolution. In 1978, the Pioneer Venus spacecraft orbited Ve-

nus and mapped all but its polar regions to a resolution of 25 km. Later, the Soviet Venera 15 and 16 spacecraft mapped the north polar region at a resolution of 1–2 km. In 1990, the Magellan spacecraft began a 5-year mission to map the planet at a resolution of 100 meters.

Even the lower-resolution radar maps show that the surface of Venus is different from the surface of Earth (Figure 17-15a). Seafloors cover 65 percent of Earth, and continents, which rise about 4 km higher, cover 35 percent. Thus Earth's surface is divided between two major types of terrain—seafloors and continents. By contrast, Venus has only one major type of terrain. Roughly 70 percent of the planet is covered by gently **rolling plains.** Only 20 percent is lowlands and 10 percent is highlands.

The highlands resemble continents only in that they rise above the average surface level. One of these areas, Ishtar Terra,* is larger than the United States and contains some of the most dramatic terrain in the solar system (Figure 17-15b). At its eastern edge, a great mountain called Maxwell Montes thrusts up 12 km. (Mount Everest is only

*Ishtar is the Babylonian goddess of love. Except for Maxwell, Alpha Regio, and Beta Regio, place names on Venus are female.

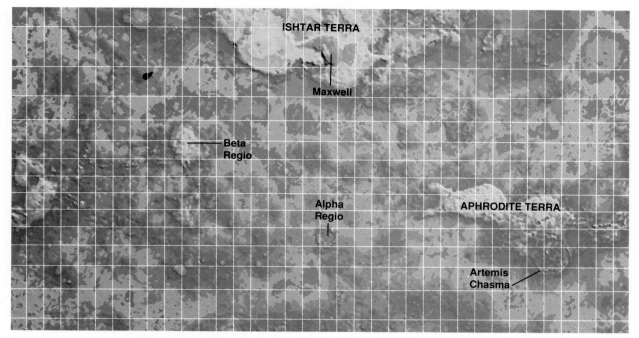

ISHTAR TERRA

Maxwell

Beta
Regio

Alpha
Regio

APHRODITE TERRA

Artemis
Chasma

a

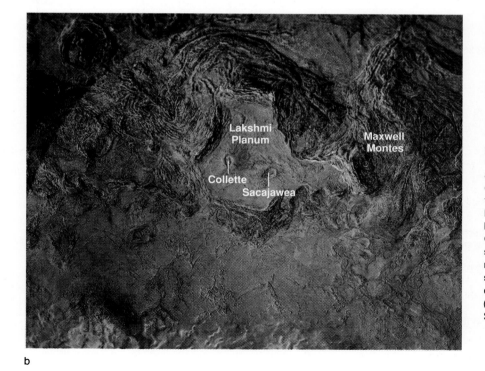

Lakshmi
Planum

Maxwell
Montes

Collette
Sacajawea

b

Figure 17-15 (a) This Pioneer Venus radar map covers all but the polar regions at a resolution of about 25 km. The lowest regions are violet and the rolling plains are blue. The highlands consist mainly of Ishtar Terra in the north and Aphrodite Terra at right. Beta Regio is a pair of volcanic peaks. Although large parts of this radar map are blue, recall that Venus has no oceans and is in fact a desert planet. (NASA) (b) Lakshmi Planum near the volcano Maxwell appears to be a volcanic plain produced by lava flooding from vents such as Colette. Folded mountain ranges suggest limited crustal movement. Here violet is the smoothest, green rougher, and orange the roughest terrain. (United States Geological Service)

Venus

Average distance from the sun	0.7233 AU (1.082 × 10^8 km)
Eccentricity of orbit	0.0068
Maximum distance from the sun	0.7282 AU (1.089 × 10^8 km)
Minimum distance from the sun	0.7184 AU (1.075 × 10^8 km)
Inclination of orbit to ecliptic	2°23'40″
Average orbital velocity	35.03 km/sec
Orbital period	0.61515 y (224.68 days)
Period of rotation	243.01 days (retrograde)
Inclination of equator to orbit	177°
Equatorial diameter	12,104 km (0.95 $D_\oplus$)
Mass	4.870 × 10^{24} kg (0.815 $M_\oplus$)
Average density	5.24 g/cm^3 (4.2 g/cm^3 uncompressed)
Surface gravity	0.903 Earth gravities
Escape velocity	10.3 km/sec (0.92 $V_\oplus$)
Surface temperature	472°C (882°F)
Albedo (cloud tops)	0.76

a. A photograph of Venus taken from Earth shows it in a typical quarter phase. Cloud features are faintly visible. (Lick Observatory photograph)

b. An ultraviolet photograph taken by the Pioneer Venus Orbiter shows weather patterns but no trace of the surface. (NASA)

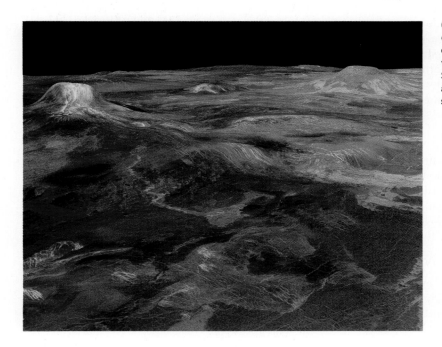

c. Below the clouds. A computer-generated image based on Magellan radar data shows volcanoes Gula Mons (left) and Sif Mons (right) and their associated lava flows. Vertical scale enlarged. (NASA)

d. This Magellan radar map shows impact craters on Venus. Volcanic peaks lie at lower right. (NASA)

8.8 km high.) Maxwell appears to be a very large volcano. Wrinkled ridges near Maxwell suggest the crust has been compressed there, but there is little evidence of major plate motion on Venus. Maxwell and Ishtar, like much of the highlands, are dominated by volcanism not plate tectonics.

Ishtar Terra is bounded by mountain ranges in the north and west, and the western portion also contains a great plateau called Lakshmi Planum. This plateau seems to have been formed by lava flows from the collapsed caldera Colette and Sacajawea (Figure 17-15b). The southern edge of Lakshmi Planum is marked by steep flanks rising 4 km above the rolling plains.

Volcanism is clearly important on Venus. Gula Mons and Sif Mons (Data File 5, figure c), for example, are volcanic peaks 3 km (9800 ft) and 2 km (6600 ft) high, respectively. Such peaks appear to occur over plumes of hot magma rising below the crust. Thus these volcanic cones are **shield volcanoes** much like those that formed the Hawaiian islands on Earth. Shield volcanoes are formed by fluid lava and have shallow slopes. (The vertical scale in figure c in Data File 5 is exaggerated.) Long lava flows are visible extending 500 km across the surface. Although no volcanic eruption has been detected in progress, it seems safe to assume that some of the volcanoes on Venus are currently active.

Smaller volcanic peaks are common in the Magellan radar maps. Also, the surface is broken by numerous faults and by sunken regions produced when subsurface magma solidified and contracted or drained away and allowed the terrain to collapse. All of this is clear evidence of subsurface circulation of molten magma.

Much of Venus is marked by circular features that are clearly not craters. The Venera spacecraft located circular features up to 1000 km in diameter called **coronae**. These slightly domed planes are bounded by concentric ridges and fractures, and appear to be bulges in the surface caused by rising plumes of molten rock below the crust (Figure 17-16). Small volcanic peaks occur along faults in the coronae.

The Magellan spacecraft located similar, smaller features called **arachnoids** because of their spider web appearance (Figure 17-17). Volcanic peaks and lava flows near the arachnoids suggest they too are caused by rising magma below the crust.

The number of impact craters on the plains of Venus amounts to about 15 percent of those on the lunar maria. While that is more than on Earth, it is much less than on the moon. This shows that the surface of Venus is not as rapidly renewed as is Earth's surface, where erosion and plate tectonics recycle the crust quickly. But the volcanism on Venus does recycle the crust over longer periods of time. Old crust and old craters are gradually buried under successive lava flows. No features on the surface seem older than 1 billion years.

Craters on Venus are slightly different from those on the moon. The thick atmosphere tends to break up in-falling meteorites, so there are no craters smaller than 2 km in diameter. Craters sometimes occur in groups, showing that a larger meteorite broke up in the atmosphere. Also, craters on Venus tend to have small ejecta sheets around them because the thick atmosphere prevents rock from being thrown very far from the impact (Data File 5, figure d). The high temperature means that rock melted by the impact does not cool rapidly, so some of the ejecta remains soft enough to ooze downhill, producing flow patterns around craters.

No one has ever stood on the surface of Venus, but a few robot spacecraft have landed there. The heat and atmospheric pressure destroyed the instruments after about an hour, but in that hour they transmitted data and pictures back to earth. Only four of the spacecraft carried cameras, but those photos reveal a rocky, desert world bathed in the orange glow of sunlight filtering through the thick atmosphere and clouds (Figure 17-18).

In addition to photographing the surface, seven Soviet landers have carried instruments to analyze the soil. Although the soil is difficult to match to known rock types on Earth, it is clear that all of the probes landed on basalt surfaces. Because basalts are produced by solidified magma, these results confirm that the surface of Venus has been dominated by volcanism and that most of the surface is covered by basaltic lava flows of various ages.

The geology of Venus has evidently been dominated by convection in the molten rock below the crust. Blobs of molten magma rise below the crust to produce coronae, arachnoids, and, where they burst through, volcanoes and lava flows. Thus plate tectonics does not seem important on Venus, and some Magellan scientists enjoy referring to "blob tectonics" instead.

Although Venus is Earth's twin in size, its sur-

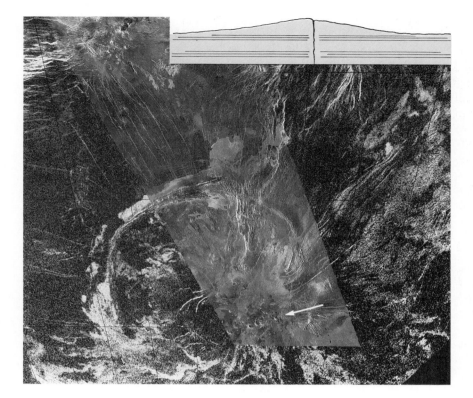

Figure 17-16 The Magellan spacecraft produced the central band in this radar map where features as small as 100 m are visible. The larger, lower-resolution map was made by Earth-based radar. The large, circular feature in the lower center of this image is Quetzal-petlatl Corona, approximately 700 km (420 miles) in diameter. Note the volcanic peaks (arrow). Volcanic peaks on Venus are shallow-sloped shield volcanoes (inset). (NASA)

face geology is dramatically different. When we turn our attention to its atmosphere, we find again that it is dramatically different from Earth's, but with one worrisome similarity.

The Atmosphere of Venus In composition, temperature, and density, the atmosphere of Venus is more reminiscent of Hades than of Heaven. About 96 percent of its atmosphere is carbon dioxide, and 3.5 percent is nitrogen. The remaining .5 percent is water, sulfuric acid (H_2SO_4), hydrochloric acid (HCl), and hydrofluoric acid (HF). In fact, the thick clouds that hide the surface are believed to be composed of sulfuric acid droplets and microscopic sulfur crystals. American and Russian spacecraft that have reached the surface report that the temperature is 745 K (880°F) and the atmospheric pressure is 90 times Earth's (Figure 17-19). Earth's atmosphere is 1000 times less dense than water, but on Venus the air is only 10 times less dense than water. The air is so dense that if we could survive its unpleasant composition, intense heat, and high pressure, we could strap wings to our arms and fly.

The present atmosphere of Venus is the result of the planet's proximity to the sun. It is 30 percent closer than Earth and receives twice as much solar energy. Thus, when it formed, Venus was warmer than Earth, and either this prevented liquid water from condensing and forming oceans, or it evaporated the oceans soon after they began to form. In either case, Venus lacked oceans to absorb carbon dioxide and convert it into mineral deposits. While Earth was removing carbon dioxide from its atmosphere via its oceans, Venus was adding more and more to its atmosphere via its volcanism. When its atmosphere grew rich in carbon dioxide, the fate of Venus was sealed.

We saw earlier in this chapter that carbon dioxide, because it is opaque to infrared radiation, can create a greenhouse effect in a planet's atmosphere. Venus, with its thick carbon dioxide atmosphere, was caught by a runaway greenhouse effect. The rising temperature baked more carbon dioxide out of the surface, and the atmosphere became even less transparent to infrared, which forced the temperature even higher. The surface is now so hot even chlorine, fluorine, and sulphur

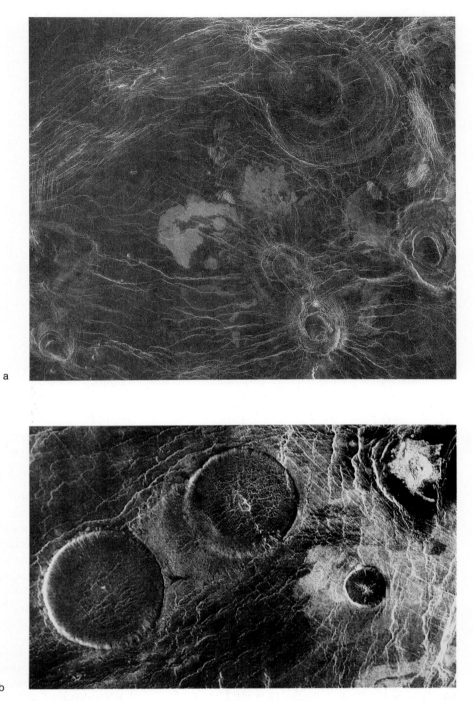

Figure 17-17 Evidence of subsurface magma on Venus: (a) Arachnoids are circular features of cracks and faults apparently caused by magma rising from below and bulging the crust upward. The largest Arachnoid here is 230 km (240 miles) in diameter. The central gray areas are lava flows, and numerous volcanic peaks dot the area. (b) Pancake domes of lava up to 65 km in diameter formed when viscous lava leaked through cracks on the surface. These features are roughly 1 km thick. (NASA)

Figure 17-18 The color of the surface of Venus appears orange in the original Venera 13 photograph (a), but analysis shows that the color is produced by the thick atmosphere. The corrected image (b) reveals slabs of gray rock and dark soil typical of iron oxides at high temperatures. (Courtesy of C. M. Pieters through the Brown/Vernadsky Institute to Institute Agreement and the U.S.S.R. Academy of Science)

a

b

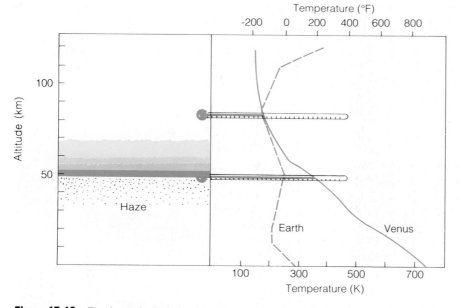

Figure 17-19 The four principal cloud layers in the atmosphere of Venus are shown at the left. Thermometers inserted into the atmosphere would register temperature as indicated in the graph at the right. The dashed line shows the temperature in Earth's atmosphere for comparison.

have baked out of the rock and formed hydrochloric, hydrofluoric, and sulfuric acid vapor.

Earth avoided this runaway greenhouse effect because it was farther from the sun and cooler. Thus it could form and preserve liquid water oceans to absorb the carbon dioxide, which left a nitrogen atmosphere that was relatively transparent in some parts of the infrared. If all of the carbon and oxygen in Earth's sediments were put back into the atmosphere as carbon dioxide, our air would be as dense as that of Venus.

And that is worrisome. Venus warns us that adding carbon dioxide to Earth's atmosphere can warm it significantly. We are in no danger of triggering a runaway greenhouse effect on Earth, but even a small temperature rise could make Earth into a very different world.

Venus is a fascinating twin of Earth, not because of similarities, but because of differences. Mars, significantly smaller than Earth, is in many ways a more hospitable planet.

17-5 MARS

Mercury and the moon are small. Venus and Earth are, for terrestrial planets, large. But Mars, with a diameter only 53 percent that of Earth's, is intermediate. (See Data File 6.) Its features reveal characteristics of both Earth and the moon.

The Geology of Mars In some ways Mars is much like Earth. A day on Mars is nearly the same length as on Earth—24 hours 37 minutes—and its year lasts 1.88 Earth years. Also, just as Earth's axis is tipped 23.5°, the axis of Mars is tipped 24°, so it has seasons. Telescopes on Earth can detect dust storms and, rarely, clouds in its carbon dioxide atmosphere (only 1 percent as dense as Earth's). Polar caps on Mars wax and wane with its passing seasons. But only vague surface features are visible from Earth (Data File 6, figure a), so it was not until the Mariner 4 spacecraft swept past Mars in

Figure 17-21 Olympus Mons, the tremendous shield volcano on Mars, is much larger than Mauna Loa, the largest volcano on Earth. Mauna Loa has sunk into the crust, producing a moat around its base. Olympus Mons has no moat, suggesting the Martian crust is stronger than Earth's. (NASA)

Figure 17-20 Impact craters in the southern hemisphere of Mars. The impact basin Argyre is 740 km in diameter (extending out of the photo to the left) and has a rim composed of jumbled mountains uplifted by the impact. The large crater at left is Galle, with a diameter of 210 km. Note the haze above the horizon. See also Figure 17-22. (NASA)

July 1965 that we learned much about the geology of Mars.

The Mariner 4 photos revealed a cratered surface swept by dusty winds. Later missions, beginning with Mariner 9 in 1971, revealed that the planet is a complex and geologically interesting place. The photographs show that the southern hemisphere is old and heavily cratered, giving it a lunar appearance (Figure 17-20). The northern hemisphere, however, has few craters, and those few are much sharper and less eroded, and thus younger. The region has been smoothed by repeated lava flows that have buried the original surface, and two areas of volcanic cones have been found. In addition, the northern hemisphere is the site of deformed crustal sections in the form of uplifted blocks and collapsed depressions that suggest geological activity.

Martian volcanoes are of the shield type, showing that the lava flowed easily. Nevertheless, the largest volcano, Olympus Mons, is a vast structure (Figure 17-21). Its base is 600 km (370 miles) in diameter and it towers 25 km (16 miles) above the surface. In contrast, the largest volcano on Earth is Mauna Loa in Hawaii. It rises only 10 km (6 miles) above its base on the Pacific Ocean floor, and its base is only 225 km (140 miles) in diameter. Mauna Loa is so heavy that it has sunk into Earth's crust, producing an undersea moat around its base, but Olympus Mons, 2.5 times higher, has no moat and is supported entirely by the Martian crust. Evidently the crust of Mars is much thicker than Earth's.

Olympus Mons is only one of a number of volcanoes that lie in the Tharsis region near the Martian equator (Figure 17-22). This region is about 10 km (6 miles) higher than the surrounding surface and appears to have been pushed up by subsurface activity. A similar uplifted volcanic plain, the Elysium region, appears to be older than the Tharsis

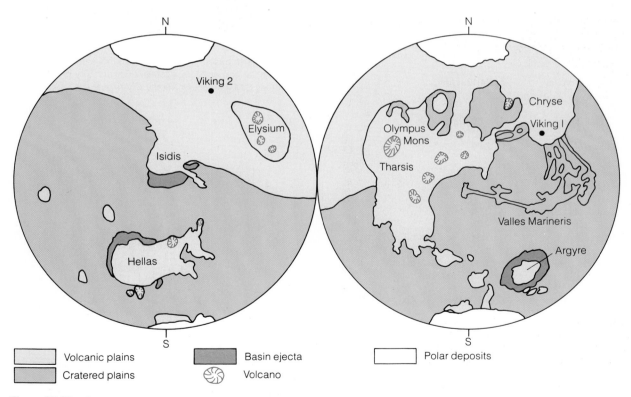

Figure 17-22 Geological map of Mars. Most of the southern hemisphere is old cratered terrain, while most of the northern hemisphere is younger volcanic plains. Note the volcanic areas, Tharsis and Elysium, and the large basins, Hellas and Argyre.

Mars

Average distance from the sun	1.5237 AU (2.279 × 10^8 km)
Eccentricity of orbit	0.0934
Maximum distance from the sun	1.6660 AU (2.492 × 10^8 km)
Minimum distance from the sun	1.3814 AU (2.066 × 10^8 km)
Inclination of orbit to ecliptic	1°51′09″
Average orbital velocity	24.13 km/sec
Orbital period	1.8808 y (686.95 days)
Period of rotation	24^{h}37^{m}22.6^s
Inclination of equator to orbit	23°59′
Equatorial diameter	6796 km (0.53 $D_\oplus$)
Mass	0.6424 × 10^{24} kg (0.1075 $M_\oplus$)
Average density	3.94 g/cm^3 (3.3 g/cm^3 uncompressed)
Surface gravity	0.379 Earth gravities
Escape velocity	5.0 km/sec (0.45 $V_\oplus$)
Surface temperature	−140° to 20°C (−220° to 68°F)
Albedo	0.16

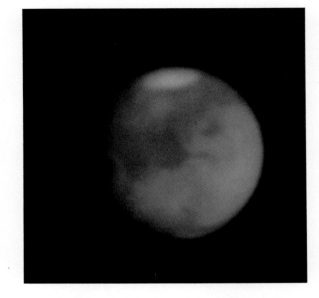

a. An Earth-based photo shows dark markings and the south polar cap on Mars. (Catalina Observatory photo courtesy S.M. Larson, University of Arizona)

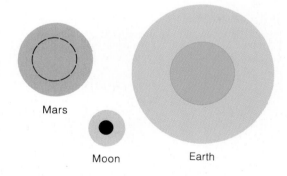

b. Mars is intermediate in size, and the nature of its core is unknown.

c. The Martian surface. Meteorite impacts have broken the surface rock to produce a rugged desert. The red color is caused by iron oxides. (NASA)

d. Photomosaic of Viking spacecraft photos of Mars reveals old, cratered terrain in the south (bottom) and younger, volcanically modified terrain in the north (top). (NASA/United States Geological Service)

region. It is more heavily cratered and eroded. The Elysium region was volcanically active about 1.5 billion years ago, but the Tharsis volcanoes may have been active as recently as 0.2 billion years ago.

When the crust of a planet is strained, it may break, producing faults. The Tharsis region is marked by a system of faults, and near it is a great valley, Valles Marineris, named after the Mariner spacecraft that first photographed it (Figure 17-23). The valley is apparently a block of crust that has dropped downward along parallel faults. Erosion and landslides have further modified the structure into a valley nearly 4000 km (2500 miles) long, stretching almost 19 percent of the way around the planet. On Earth, it would reach from New York to Los Angeles. At its widest it stretches 200 km (120 miles), and at its deepest it reaches down 6 km (4 miles), four times the depth of the Grand Canyon.

Where Valles Marineris begins, near the Tharsis region, it is marked by numerous faults, suggesting that the valley is related to the uplifted volcanic region (Figure 17-24). The number of craters in the valley indicates that it is 1–2 billion years old, placing its formation sometime before the end of volcanism in the Tharsis region. Some astronomers identify the Valles Marineris as the beginning of a crustal plate boundary. It looks like a rift valley, but it does not connect with other valleys to outline an entire plate. It is as if the crust began to break into plates, but cooled and thickened before the plates could separate and begin to move. Valles Marineris may be the frozen beginnings of this process.

While geologists were fascinated to see signs of an active surface on Mars, they were especially excited to see dry streambeds (Figure 17-25) much like the dry arroyos common in America's Southwest. The streambeds show many of the characteristic forms taken by riverbeds formed on Earth — sandbars, tributary streams, braided riverbeds cut by changing currents, and the meandering path characteristic of a river flowing through nearly flat terrain. Other features appear to be caused by sudden flooding, perhaps by the release of water frozen in the soil. Volcanism or large meteorite impacts could melt such water, and some craters show evidence of having formed in a water-laden surface (Figure 17-25).

Figure 17-23 The great canyon system Valles Marineris stretches across the lower third of this mosaic of 102 separate Viking 1 images. From end to end Valles Marineris is longer than the distance from New York to Los Angeles. The Tharsis volcanoes are visible at the lower left. (NASA/United States Geological Service, courtesy Alfred S. McEwen)

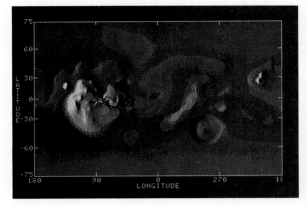

Mars

Mars to scale

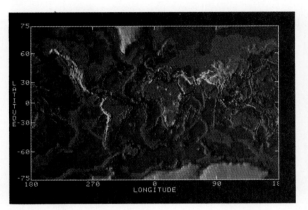

Earth

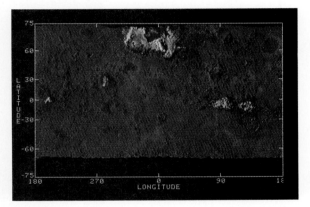

Venus

Figure 17-24 Earth, Venus, and Mars are shown in these maps, with the highest regions shaded beige and the lowest blue. Note the large volcanic highland, the Tharsis bulge, at left center on the Mars map and the Elysium volcanic uplift at extreme right. Two large impact basins on Mars, Argyre and Hellas, are visible at lower left and lower right. Were the Mars map reproduced to scale, it would be no larger than the map at upper right. (NASA)

The air pressure on Mars is now too low to keep water from boiling away, so liquid water could not have cut the flow channels recently. The number of craters superimposed on the channels shows that they are quite old, few younger than 3.8 billion years. Thus Mars appears to have once had a denser atmosphere with liquid water flowing over its surface, and it may still retain large amounts of water frozen in the soil. Estimates vary widely, but one study reports enough water frozen in the Martian crust to cover the planet to a depth of 400 m. (Earth's water could cover Earth to a depth of 3 km.)

The streambeds suggest that the Martian climate has changed, and careful photographic stud-

ies of the polar caps reveal more evidence of climatic variation (Figure 17-26). The polar caps consist of frozen carbon dioxide (dry ice) with frozen water beneath. When spring begins in a hemisphere, the corresponding polar cap begins to shrink as the carbon dioxide turns directly into gas. The water, however, never melts, but stays behind in a smaller, permanent ice cap. The region around this permanent cap is marked by layered terrain, evidently deposited by wind-borne dust that accumulates on the ice cap. When the cap shrinks, the material accumulates in layers. Each of the layers is 50–100 m (31–62 ft) thick. The variation in the layering suggests that some periodic change in climate, perhaps due to orbital changes, may affect

Figure 17-25 Evidence of water on Mars. Dry flow channels (bottom) meander into a lowland desert. Such flow features have all of the characteristics of dry river beds on Earth and may have been formed billions of years ago by moving water. Although liquid water can no longer survive on the Martian surface, large amounts of water may be trapped in the soil. The largest crater in this image is surrounded by a "splash" of ejected material, suggesting that the crust was rich in frozen water at the time of the impact. (NASA)

the frequency and intensity of planetwide dust storms, and thus alter the rate at which material is deposited. Recall that Earth's climate is believed to go through cycles caused by variations in its orbital motion. (See the Perspective in Chapter 3.)

The Four-Stage History of Mars The four-stage history of Mars is a case of arrested development. The planet began by differentiating into a crust and core. Mars has no appreciable magnetic field, even though it spins rapidly, so it cannot have a molten, conducting core. Since Mars is large enough to remain molten at its center, we must conclude that its core is very small or is not convective.

The crust of Mars is now quite thick, as shown by the mass of Olympus Mons, but it was thinner in the past. Cratering may have broken or at least weakened the crust, triggering lava flows that flooded some basins. Why most of the flows occurred in the northern hemisphere is unknown. Volcanism may have pushed up the Tharsis and Elysium regions, and broken the crust to form Valles Marineris, but moving crustal plates never developed because the planet cooled rapidly and the crust grew thick.

The large size of the volcanoes on Mars may be evidence that the plates do not move. On Earth, volcanoes like those that formed the Hawaiian Islands occur over rising currents of hot material beneath the crustal plate. Because the plate moves,

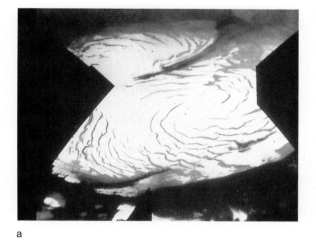

a

b

c

Figure 17-26 (a) A computer-enhanced photomosaic of the south polar cap of Mars shows the complexity of the frozen layers. (b) A summer photo taken near the north pole shows frozen water left behind when the dry ice vaporized. The varied extent of the layered terrain (arrows) is apparently caused by periodic changes in the Martian climate. (c) In places different sets of layers are superimposed. Layers are 10–50 m thick. (a. and b. NASA; c. adapted from a diagram by J. A. Cutts, K. R. Blasius, G. A. Briggs, M. H. Carr, R. Greeley, and H. Masursky)

the hot material breaks through the crust repeatedly and forms a chain of volcanoes instead of a single large feature. The Hawaiian Islands are merely the most recent of a series of volcanic islands called the Hawaiian–Emperor island chain (see Figure 17-4), which stretches nearly 3800 km (2300 miles) across the Pacific Ocean. If this idea is correct, then the lack of plate motion on Mars could have allowed Martian volcanoes to grow to gigantic proportions.

The fourth stage in the history of Mars has been one of slow decline. Volcanic activity probably reached a maximum about 1 billion years ago, and since then the planet has cooled and the crust has thickened. The plates appear never to have pushed against one another and built mountains, and now slow erosion is wearing away the volcanic and impact features.

The Atmosphere of Mars The present atmosphere of Mars is 95 percent carbon dioxide, 2–3 percent nitrogen, and about 2 percent argon. Its density is only about 1 percent of Earth's. This thin atmosphere may be the remains of a thicker blanket that allowed liquid water to carve the streambeds on the Martian surface.

The gases that now cover the surface of Mars were outgassed from its interior. Since Mars formed farther from the sun than Earth did, it may have incorporated more volatiles and then outgassed more. Most outgassing, however, occurred during the first billion years, when the planet's surface was active. Mars has cooled and now releases little gas.

How much atmosphere a planet has depends on how rapidly it releases internal gas and how rapidly it loses gas to space. Since Mars is no longer producing gas rapidly, its losses to space have thinned its atmosphere.

How rapidly a planet loses gas depends on its mass and temperature. The more massive the planet, the higher its escape velocity and the more difficult it is for gas atoms to leak into space. Mars has a mass less than 11 percent of Earth's, and its escape velocity is only 5 km/sec, less than half Earth's.

A planet's temperature is also important. If the gas is hot, its molecules have a higher average velocity and are more likely to exceed escape velocity. Thus a hot planet is less likely to retain an atmosphere than a cooler planet. However, the velocity of a molecule in a gas also depends on the mass of that molecule. On the average, a low-mass molecule travels faster than a high-mass molecule. Thus a planet loses its lightest gases more easily because they travel the fastest.

Applying these factors to individual planets, we find that all of the terrestrial planets are too small to retain hydrogen and helium. Those molecules are light and leak into space. Though Venus and Earth can hold onto their remaining gases, Mercury and the moon are too small and have lost all of their air. Mars is barely able to retain water vapor, methane, and ammonia, but it has a better gravitational grip on carbon dioxide.

Mars should have been able to retain some water and methane, but such molecules can be broken up by ultraviolet photons in sunlight. On Earth, the ozone layer protects the atmosphere from ultraviolet radiation, but Mars never developed an oxygen-rich atmosphere, and so it never had an ozone layer. Without that kind of protection, water and methane molecules absorbed ultraviolet photons and broke up. The hydrogen escaped to space, and the oxygen formed oxides in the soil or carbon dioxide in the air. Thus molecules that are too heavy to leak into space can be lost if they dissociate into lighter fragments.

Thus Mars, because of its small size, has lost much of its atmosphere, and additional supplies of water and carbon dioxide may be frozen in the soil and in the polar caps. Climatic cycles may periodically release such reservoirs and provide a thicker atmosphere. Studies show that Mars goes through cycles of change due to orbital and rotational changes just as Earth does. (See the Perspective in Chapter 3.)

The Moons of Mars Unlike Mercury or Venus, Mars has moons. Small and irregular in shape, Phobos (14 × 10 km) and Deimos (8 × 6 km) may be captured asteroids or they may have formed with Mars. In either case, the Mariner and Viking photographs illustrate some of the processes that might affect asteroids as they collide with one another in the asteroid belt.

The photographs reveal a unique set of narrow, parallel grooves on Phobos (Figure 17-27). Averaging 150 m (500 ft) wide and 25 m (80 ft) deep, the grooves run from Stickney, the largest crater, to an oddly featureless region on the opposite side of the satellite. One theory suggests that the grooves are deep fractures produced by the impact that formed the crater. High-resolution photographs show that the grooves are lines of pits, suggesting that the pulverized rock material on the surface has drained into the fractures, or that gas, liberated by the heat of the impact, escaped through the fractures, and blew the dusty soil away.

Deimos not only has no grooves, it also looks smoother because of a thicker layer of dust on its surface. This material partially fills craters and covers minor surface irregularities. It seems certain that Deimos experienced collisions in its past, so fractures may be hidden below the debris.

The debris on the surfaces of the moons raises an interesting question: How can the weak gravity of small bodies hold any fragments from meteorite impacts? The escape velocity on Phobos is only about 12 m/sec (40 ft/sec). An athletic astronaut who could jump 2 m (6 ft) high on Earth could jump 2.8 km (1.7 mi) on Phobos. Certainly most of the fragments from an impact should escape, but the slowest particles could fall back in the weak gravity and accumulate on the surface.

Since Deimos is smaller than Phobos, its escape velocity is smaller, so it seems surprising that it has more debris on its surface. This may be related to Phobos' orbit close to Mars. The Martian gravity is almost strong enough to pull loose material off of Phobos' surface, so the moon may be able to retain little of its cratering debris.

An ambitious plan to rendezvous with Phobos, zap it with a laser, and drop instrumented landers on its surface went awry when the Soviet spacecraft Phobos 1 and Phobos 2 both failed. Phobos 1 was fatally crippled by a flawed instruction sent to the spacecraft soon after launch. Phobos 2 reached Mars in early 1989 and made some observations before it failed due to a computer malfunction. Before the breakup of the Soviet Union, the Soviet space program planned further robotic missions to Mars, but the fate of those plans is now in question. In the United States, NASA is planning robotic mis-

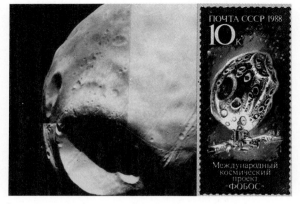

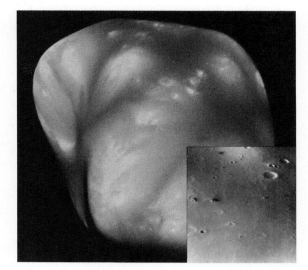

Figure 17-27 (a) Phobos, the larger satellite of Mars, is irregular in shape, heavily cratered, and marked by grooves. The inset shows the Soviet stamp honoring the ill-fated Phobos probes. (b) Deimos, smaller than Phobos, looks more uniform because of dust covering smaller features (inset). (NASA)

b

sions to Mars as forerunners of piloted exploration early in the next century.

With Mars we complete our survey of the terrestrial planets. We have seen the effects of cratering, plate tectonics, and volcanism. As we turn our attention to the Jovian planets in the next chapter, we will see entirely new effects.

SUMMARY

All terrestrial planets pass through a four-stage history: (1) differentiation, (2) cratering, (3) flooding of the crater basins by lava, water, or both, and (4) slow surface evolution. The importance of each of these processes in the evolution of a planet depends on the planet's mass and temperature.

Earth is the largest terrestrial planet in our solar system, and it has passed through all four stages. Studies show that Earth has differentiated into a metallic core and a silicate crust. Currents in the molten portion of the core produce Earth's magnetic field.

Because Earth is still partially molten, its surface is active. Plate tectonics refers to the motion of large crustal sections. As new crust appears in the midocean rifts, it pushes the massive plates apart and destroys old crust where plates slide over one another.

Earth's atmosphere has changed drastically since its formation. The first atmosphere was rich in carbon dioxide. As the carbon dioxide dissolved into the oceans and was incorporated in the bottom sediment, the atmosphere was left rich in nitrogen. Later, plant life produced oxygen.

When the moon formed, it differentiated into a dense core and a low-density crust of anorthosite. Cratering broke up this crust, and lava flows of dark basalt filled the lowlands, producing the maria. Cratering has pulverized the surface of the highlands and the lowlands, covering the lunar surface with ejecta.

Mercury, like the moon, has no atmosphere and is covered by craters, but unlike the moon, its surface is marked by lobate scarps. The long, curved cliffs suggest that, after formation, Mercury cooled, contracted, and its crust wrinkled.

The surface of Venus is hidden below a dense, cloudy atmosphere of carbon dioxide. The thick atmosphere creates a strong greenhouse effect, and the surface temperature is high enough to melt lead. Radar maps reveal that its geology is dominated by rising currents of molten rock below the surface. Volcanism has flooded much of the surface with lava flows. There is no evidence that plate tectonics is important on Venus.

Because Mars is small, it has cooled relatively rapidly and is no longer active. Nevertheless, its surface contains lava plains, volcanoes, and a long feature, possibly a rift valley. These suggest that the surface was active in the past.

Venus and Mars illustrate two principles of planetary atmospheres. A thick atmosphere that is opaque to infrared radiation can trap heat via the greenhouse effect and produce a high surface temperature. Earth avoided this by dissolving its CO_2 in the oceans, but Venus was too warm for liquid water to form. The second principle is that the density and composition of a planet's atmosphere depend on the mass and temperature of the planet. A warm, low-mass planet has difficulty retaining an atmosphere. Also, since light gases leak away most easily, a planet may be able to retain heavier gases like CO_2, while H_2 and He leak into space.

NEW TERMS

comparative planetology	anorthosite
mantle	breccia
plastic	albedo
dynamo effect	early heavy bombardment
plate tectonics	mascon
basalt	fission hypothesis
midocean rise	condensation hypothesis
midocean rift	capture hypothesis
rift valley	large-impact hypothesis
primeval atmosphere	lobate scarp
greenhouse effect	rolling plains
ejecta	shield volcano
ray	coronae
vesicular basalt	arachnoids
mare	

REVIEW QUESTIONS

1. Summarize the four stages in the development of a terrestrial planet.

2. How does plate tectonics create and destroy Earth's crust?

3. How did Earth's atmosphere evolve?

4. What are the stages in the development of the moon?

5. What kinds of erosion are now active on the moon?

6. What are lobate scarps?

7. How did Earth avoid the greenhouse effect that made Venus so hot?

8. What evidence do we have that Venus and Mars had active crusts?

9. Why is the atmosphere of Venus so rich in carbon dioxide?

10. What evidence do we have that the climate on Mars has changed?

DISCUSSION QUESTIONS

1. If we visited a planet in another solar system and discovered oxygen in its atmosphere, what might we guess about the planet's surface?

2. Why might we propose the following hypothesis: Any terrestrial planet as small or smaller than Mars will not have mountain ranges like the Allegheny Mountains in the eastern United States?

PROBLEMS

1. If the Atlantic seafloor is spreading at 30 mm/year and is now 6400 km wide, how long ago were the continents in contact?

2. Why do small planets cool faster than large planets? (HINT: Compare surface area to volume.)

3. The smallest detail visible through Earth-based telescopes is about 1 second of arc in diameter. What size is this on the moon? (HINT: See Box 3-2.)

4. Midocean rifts and the trenches where the sea-floor slips downward are 1 km or less wide. Could Earth-based telescopes resolve such features on the moon? Why are we sure such features are not present on the moon?

5. How long would it take radio signals to travel from Earth to Venus and back if Venus were at its farthest point from Earth? Why are such observations impractical?

6. Repeat Problem 5 for Mercury.

7. Imagine that we have sent a spacecraft to land on Mercury, and it has transmitted radio signals to us at a wavelength of 10 cm. If we see Mercury at its greatest angular distance west of the sun, to what wavelength must we tune our radio telescope to detect the signals? (HINTS: See Data File 4 to find Mercury's orbital velocity, and then see Box 6-4.)

8. The smallest feature visible through an Earth-based telescope has an angular diameter of about 1 second of arc. If a crater on Mars is just visible when Mars is nearest to Earth, what is its linear diameter? (HINT: See Box 3-2.)

9. What is the maximum angular diameter of Phobos as seen from Earth? What surface features should we be able to see using Earth-based tele-scopes? (HINT: See Box 3-2.)

10. Phobos orbits Mars at a distance of 9380 km from the center of the planet and has a period of 0.3189 days. Calculate the mass of Mars. (HINTS: Convert to AU and years, and then see Box 8-5.)

RECOMMENDED READING

ALLEN, DAVID A. "Laying Bare Venus' Dark Secrets." *Sky and Telescope* 74 (October 1987), p. 350.

BAKER, VICTOR R. *The Channels of Mars.* Austin: University of Texas Press, 1982.

BEATTY, J. KELLY. "The Amazing Olympus Mons." *Sky and Telescope* 64 (November 1982), p. 420.

———. "Report from a Torrid Planet." *Sky and Telescope* 63 (May 1982), p. 452.

———. "Venus: The Mystery Continues." *Sky and Telescope* 63 (February 1982), p. 134.

———. "Radar Views of Venus." *Sky and Telescope* 67 (February 1984), p. 110.

———. "A Radar Tour of Venus." *Sky and Telescope* 69 (June 1985), p. 507.

———. "The Making of a Better Moon." *Sky and Telescope* 72 (December 1986), p. 558.

———. "Venus in the Radar Spotlight." *Sky and Telescope* 82 (July 1991), p. 24.

BEATTY, J. KELLY, BRIAN O'LEARY, and ANDREW CHAIKIN. *The New Solar System*, 3rd ed. Cambridge, Mass.: Sky, 1990.

BEN-AVRAHAM, SVI. "The Movement of Continents." *American Scientist* 69 (May/June 1981), p. 291.

BURNHAM, ROBERT. "New Views of Mars and Phobos." *Astronomy* 17 (September 1989), p. 28.

CADOGAN, PETER H. *The Moon: Our Sister Planet.* New York: Cambridge University Press, 1981.

———. "The Moon's Origin." *Mercury* 12 (March/April 1983), p. 34.

CARR, MICHAEL H. "The Surface of Mars: A Post-Viking View." *Mercury* 12 (January/February 1983), p. 2.

CORDELL, BRUCE M. "Mercury: The World Closest to the Sun." *Mercury* 13 (September/October 1984), p. 136.

DIETZ, ROBERT. "In Defense of Drift." *The Sciences* 23 (November/December 1983), p. 22.

ESPOSITO, LARRY W. "Does Venus Have Active Volcanoes?" *Astronomy* 18 (July 1990), p. 42.

GOLDMAN, STUART W. "What's New on Mars?" *Sky and Telescope* 77 (May 1989), p. 471.

GRIEVE, RICHARD. "Impact Craters Shape Planetary Surfaces." *New Scientist* 100 (17 November 1983), p. 516.

HARTMANN, W. K. "Cratering in the Solar System." *Scientific American* 236 (January 1977), p. 84.

LANZEROTTI, LOUIS J., and STAMATIOS M. KRIMIGIS. "Comparative Magnetospheres." *Physics Today* 38 (November 1985), p. 24.

MARAN, STEPHEN P. "Quakes on the Moon." *Natural History* 91 (February 1982), p. 82.

MILLMAN, PETER M. "Names on Other Worlds." *Sky and Telescope* 67 (January 1984), p. 23.

PARKER, DONALD C., CHARLES F. CAPEN, and JEFF D. BEISH. "Exploring the Martian Arctic." *Sky and Telescope* 65 (March 1986), p. 218.

PHILLIPS, ROGER J., ROBERT E. GRIMM, and MICHAEL C. MALIN. "Hot-Spot Evolution and the Global Tectonics of Venus." *Science* 252 (3 May 1991), p. 651.

REGISTER, BRIDGET MINTZ. "The Fate of the Moon Rocks." *Astronomy* 13 (December 1985), p. 14.

ROBINSON, MARK S. "Surveying the Scars of Ancient Martian Floods." *Astronomy* 17 (October 1989), p. 38.

RUBIN, ALAN E. "Whence Came the Moon?" *Sky and Telescope* 68 (November 1984), p. 389.

RUNCORN, S. K. "The Moon's Ancient Magnetism." *Scientific American* 257 (December 1987), p. 60.

SAUNDERS, STEPHEN. "Venus: The Hellish Place Next Door." *Astronomy* 18 (March 1990), p. 18.

———. "The Surface of Venus." *Scientific American* 263 (December 1990), p. 60.

———. "The Exploration of Venus." *Mercury* 20 (September/October 1991), p. 130.

SCHUBERT, GERALD, and CURT COVEY. "The Atmosphere of Venus." *Scientific American* 245 (July 1981), p. 66.

SCHULTZ, PETER H. "Polar Wandering on Mars." *Scientific American* 253 (December 1985), p. 94.

SQUYRES, STEVEN W. "Searching for the Water on Mars." *Astronomy* 17 (August 1989), p. 20.

TAYLOR, STUART ROSS. "The Origin of the Moon." *American Scientist* 75 (September/October 1987), p. 469.

WASHBURN, MARK. "The Moon—A Second Time Around." *Sky and Telescope* 69 (March 1985), p. 209. See also *Sky and Telescope* 69 (May 1985), p. 395.

ZAKHAROV, ALEKSANDR V. "Close Encounters with Phobos." *Sky and Telescope* 76 (July 1988), p. 17.

FURTHER EXPLORATION

VOYAGER, the Interactive Desktop Planetarium, can be used for further exploration of the topics in this chapter. This chapter corresponds with the following projects in *Voyages Through Space and Time:* Project 16, Planetary Phases; Project 17, Mercury and Venus; Project 18, Retrograde Motion; Project 19, The View from Mars.

C H A P T E R

18

Worlds of the Outer Solar System

There wasn't a breath in that land of death . . .

Robert Service,
"The Cremation of Sam McGee"

The sulfuric acid fogs of Venus seem totally alien, but compared to the planets of the outer solar system, Venus is a tropical oasis. Of the five planets beyond the asteroid belt, four have no solid surface and one is so far from the sun that much of its atmosphere lies frozen on its surface.

The four Jovian planets—Jupiter, Saturn, Uranus, and Neptune—are strikingly different from the terrestrial planets. They are giant planets as much as 11 times Earth's diameter (Figure 18-1), and their thick atmospheres of hydrogen, helium, methane, and ammonia are filled with clouds of ammonia, and methane crystals. Explorers descending into the atmosphere of a Jovian planet would find the atmosphere and its clouds blending gradually with the planet's liquid hydrogen interior. Thus the Jovian planets have no solid surface and are not subject to the four-

stage development described in the previous chapter. In addition, the thick atmosphere of Jupiter is governed by weather patterns unlike those on Earth, and similar processes probably operate on the other Jovian planets.

Orbiting these giant worlds are swarms of smaller worlds—the satellites of the outer solar system. Data sent back to Earth by planetary probes show that these moons are complicated bodies with unearthly geologies.

Beyond the Jovian planets lies Pluto, a planetary enigma. It is not a Jovian planet, but it is so distant we know little about it. Some argue that it is not a true planet. Perhaps it is more nearly related to the moons of the outer solar system than to the planets.

18-1 JUPITER

Jupiter is a giant planet over 11 times the diameter of Earth and

Figure 18-1 The Jovian planets are giant worlds. Jupiter (lower left), the largest, is 11.18 times Earth's diameter, and Saturn is 9.42 times larger than Earth. In this composite photograph, the planets are reproduced to scale. Notice Earth at the center. (Stephen P. Meszaros, NASA)

over 317 times more massive. (See Data File 7.) It is truly the king of the planets, as it contains 70 percent of all planetary matter in our solar system. We will examine it in detail so we may later use it as a basis for comparison with the other Jovian planets.

The Interior Most of Jupiter's mass is hydrogen and helium. In the outer 1.3 percent of its radius, the material forms a gaseous atmosphere, but the pressure at deeper layers forces the hydrogen in the interior into a liquid state. But no one will ever sail over Jupiter's liquid hydrogen ocean. The upper layers of the liquid interior merge gradually with the densest layers of the gaseous atmosphere, and there is no liquid surface as in the case of Earth's oceans.

Theoretical models of Jupiter predict a core of heavy elements such as iron and silicates. Although this is commonly referred to as a "rocky core," the high temperature (30,000 K) and extreme pressure

(300 million times Earth's atmospheric pressure at sea level) make the material unlike any rock on Earth. Jupiter's rocky core occupies less than 1 percent of the planet's total volume. Thus Jupiter is truly a liquid planet.

Careful measurements of the amount of energy flowing out of Jupiter reveal that it emits about twice as much energy as it absorbs from the sun. This is evidently heat left over from the formation of the planet. In Chapter 16, we saw that Jupiter should have grown very hot when it formed, and some of this heat remains trapped in its interior.

Jupiter's high internal pressure converts the liquid hydrogen into a different form called liquid metallic hydrogen. It is still a liquid, but it is a very good conductor of electricity. This vast globe of conducting liquid, stirred by convection currents and spun by the planet's rapid rotation, generates a powerful magnetic field through the dynamo effect (see Chapter 17). The Pioneer and Voyager spacecraft detected a field over 10 times stronger than Earth's and extending through a vast region around the planet.

This magnetic field traps charged particles from the solar wind into large, doughnut-shaped radiation belts that surround the planet just as Earth is surrounded by its Van Allen radiation belts (Figure 18-2a). Because Jupiter's magnetic field is stronger, its trapped radiation is 1 million times more intense than Earth's. The Pioneer 10 spacecraft flew through these belts and received 427 times the radiation exposure considered lethal for a human.

These radiation belts are only one aspect of Jupiter's **magnetosphere**—the volume of space that is controlled by its magnetic field rather than by the solar wind. All planets with magnetic fields have magnetospheres, but because of the strength of Jupiter's field, its magnetosphere is 100 times larger than Earth's. If we could walk outside on a dark night and see Jupiter's magnetosphere, it would be 6 times larger than the full moon.

Earth's magnetosphere interacts with the solar wind and generates powerful electric currents and high-energy particles that flow downward into the magnetic poles, where they excite gas atoms in the upper atmosphere to emit photons. The glowing layers are familiar to residents of high latitudes as aurora, and are especially impressive when a flare on the sun (see Chapter 7) causes surges in the

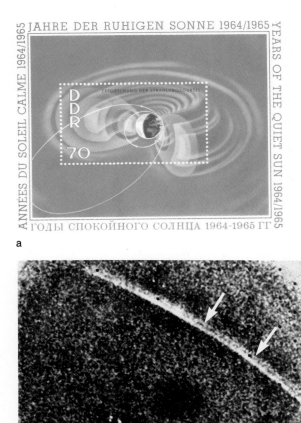

a

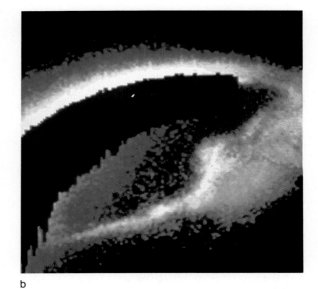

b

Figure 18-2 (a) Earth's magnetic field traps high-energy particles to form doughnut-shaped radiation belts as shown in the cross section on this stamp. Jupiter's radiation belts are 100 times larger and much more intense. A planet's magnetic field guides electrical currents down into its atmosphere and can produce aurora. (b) This true-color image of Earth's aurora was made from space. (L. A. Frank, University of Iowa/NASA) (c) Voyager 1 detected both aurora (arrows) and lightning (lower center) on Jupiter's night side. (NASA)

c

solar wind. When Voyager 1 passed Jupiter, it looked back at the planet's night side and saw great sheets of glowing aurora (Figure 18-2c).

Jupiter's magnetosphere is a source of radio noise. The radio radiation has a wavelength of about 10 cm (4 inches) and is produced by synchrotron radiation from high-energy electrons spiraling through the magnetic field. Radio waves with wavelengths of about 10 m (32 ft) appear to come from electrical currents in Jupiter's atmosphere. Spacecraft near the planet have detected lightning bolts a million times more powerful than a large bolt on Earth (Figure 18-2c).

The long-wave radio signals depend in a complex way on the position of Io, the innermost of Jupiter's four large moons. Io's orbit lies inside the inner radiation belt, and it apparently interacts with the magnetosphere to beam radio waves in specific directions. Thus only when Io is in certain positions in its orbit does one of the radio beams point toward Earth.

Atmosphere Although Jupiter's outer layers are mostly hydrogen and helium (Table 18-1), what we see are layers of clouds composed of ammonia and ammonia hydrosulfide crystals (Figure 18-3). Through gaps in these clouds, Earth-based telescopes and spacecraft have seen deeper, warmer

Jupiter

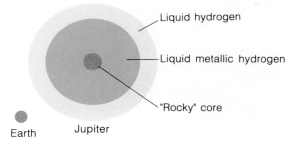

a. A theoretical model of the interior of Jupiter. Note the oblateness of the planet.

Average distance from the sun	5.2028 AU (7.783 × 10⁸ km)
Eccentricity of orbit	0.0484
Maximum distance from the sun	5.455 AU (8.160 × 10⁸ km)
Minimum distance from the sun	4.951 AU (7.406 × 10⁸ km)
Inclination of orbit to ecliptic	1°18′29″
Average orbital velocity	13.06 km/sec
Orbital period	11.867 y (4334.3 days)
Period of rotation	9ʰ50ᵐ30ˢ
Inclination of equator to orbit	3°5′
Equatorial diameter	142,800 km (11.18 $D_\oplus$)
Mass	1.899 × 10²⁷ kg (317.83 $M_\oplus$)
Average density	1.34 g/cm³
Gravity at base of clouds	2.54 Earth gravities
Escape velocity	61 km/sec (5.4 $V_\oplus$)
Temperature at cloud tops	−110°C (−116°F)
Albedo	0.51
Oblateness	0.0637

b. Ganymede, one of Jupiter's Galilean moons, is a mixture of rock and ice. (NASA)

c. Jupiter and two of its satellites, Io (arrow) and Europa (right). Note the Giant Red Spot at left. (NASA)

TABLE 18-1
Composition of Jupiter and Saturn (by Mass)

Molecule	Jupiter (%)	Saturn(%)
H_2	78	88
He	19	11
H_2O	0.0001	—
CH_4	0.2?	0.6
NH_3	0.5?	0.02

ally appear brown or red, though they may shade into blue-green, and the zones are yellow-white (Figure 18-4). The cause of these colors is poorly known, but they may arise from molecules produced by sunlight or lightning interacting with ammonia and other compounds in Jupiter's atmosphere.

Jupiter's belt-zone circulation is related to the high- and low-pressure areas we see on earthly weather maps. Zones are high-pressure regions of rising gas, and belts are low-pressure regions where gas sinks.

On Earth, the temperature difference between the equator and poles drives a wavelike wind pattern that organizes such high- and low-pressure regions into cyclonic circulations familiar from weather maps (Figure 18-5). On Jupiter, the equator and poles appear to be about the same temperature, perhaps because of heat rising from the interior. Thus there is no temperature difference to drive the wave circulation, and Jupiter's rapid rotation draws the high- and low-pressure regions into bands that circle the planet. On Earth, high- and low-pressure regions are bounded by winds induced by the wave circulation. On Jupiter, the

layers of water droplet clouds. Below that the atmosphere merges with the liquid hydrogen interior.

Jupiter's atmosphere is dominated by **belt-zone circulation,** which circles the planet parallel to its equator (see Data File 7, figure b, and Figure 18-1). Even a small telescope can reveal the dark belts and lighter zones on Jupiter. The belts gener-

Figure 18-3 The three principal cloud layers on Jupiter are shown at left. Thermometers inserted into the Jovian atmosphere would register temperature as in the graph at right.

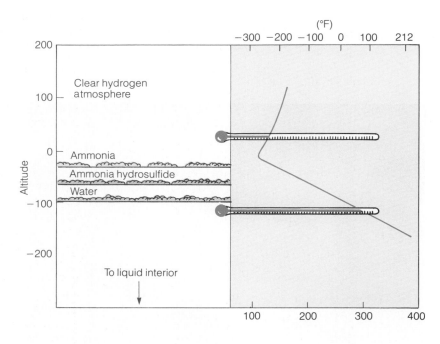

Figure 18-4 Turbulence in the belts and zones of Jupiter's atmosphere is highly complex. Cloud colors may arise from complex molecules formed from ammonia. Note the Giant Red Spot at upper right. Earth included for scale. (Stephen P. Meszaros, NASA/Goddard Space Flight Center)

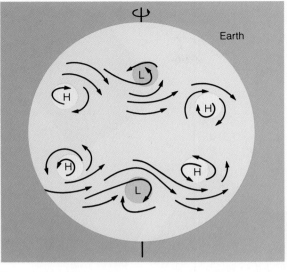

a

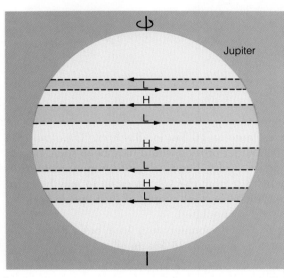

b

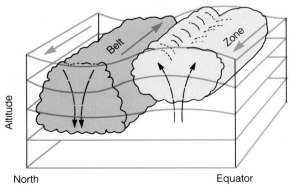

c

Figure 18-5 (a) In Earth's atmosphere, a wave circulation separates low- and high-pressure regions. (b) On Jupiter, low- and high-pressure regions take the form of dark belts of sinking gas and bright zones of rising gas. (c) Zones are bright because the rising gas forms clouds high in the atmosphere where sunlight is strong. On both worlds, winds circulate in the same way—counterclockwise around low-pressure regions in the northern hemisphere and clockwise in the southern hemisphere.

same circulation appears as high-speed winds that blow around the planet at the boundaries of the belts and zones. Above about 45° latitude, the belt-zone pattern becomes unstable and the circulation decays to turbulent whirlwinds.

Mixed among the belts and zones are light and dark spots, a few larger than Earth. Some appear and disappear in a few days, while others last for years. The largest is the Great Red Spot, a reddish oval that has been in one of the southern zones for at least 300 years (see Figure 18-4). Pioneer and Voyager photographs show that the Great Red Spot is a vast cyclonic storm where winds between adjacent belt and zone meet.

One of the interesting features of Jupiter does not lie within its cloudy atmosphere, but orbits above its equator. Jupiter, like the other three Jovian planets, has a ring.

Jupiter's Ring When Voyager 1 flew past Jupiter in March 1979, it was programmed to search for a possible ring as it swept through Jupiter's equatorial plane. The photograph it sent back to Earth showed that Jupiter does indeed have a ring with a thickness of less than 30 km and an outer edge at 1.81 planetary radii (Figure 18-6a). The ring is very tenuous, at least 100 times less opaque than Saturn's rings. The ring material is dark and reddish, so it can't be ices. It is most likely silicates.

The Voyager 2 photo in Figure 18-6b proves that many of the particles must be very small—no larger than 10 micrometers in diameter. The photo was taken looking back at Jupiter with the sun in the background. If the ring particles were large, baseball-sized, for instance, they would not scatter light forward toward the camera. Only very small particles, like those in cigarette smoke, scatter light forward. Known as **forward scattering**, this effect is clear evidence of small particles.

Computer-aided analysis of photos shows that

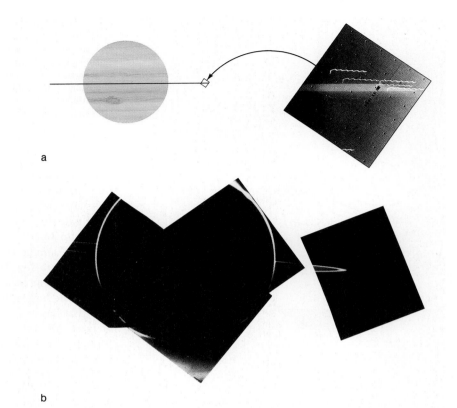

Figure 18-6 (a) The photograph taken by Voyager 1 on which Jupiter's ring was discovered. (b) A Voyager 2 photomosaic taken from within the planet's shadow. The planet is outlined by sunlight scattered from haze in the upper atmosphere. Forward scattering makes the ring appear very bright when illuminated from behind. (NASA)

a

b

the ring is most dense between 1.72 and 1.81 Jupiter radii and that a thin sheet of particles extends down to Jupiter's cloud tops. Also, a tenuous cloud of particles extends above and below the plane of the ring. These appear to be dust particles that have acquired electrostatic charges and are being pulled out of the ring by Jupiter's magnetic field.

The rings lie inside the **Roche limit,** the region near a planet where a moon would be pulled apart by the planet's tidal forces. Thus we might suspect that the rings are material that was unable to form a moon when Jupiter formed. However, it is clear that the ring particles cannot be very old because a number of processes remove such small specks from the ring. The pressure of sunlight and the magnetic forces in the magnetosphere alter the orbits of small particles, and if they don't fall into Jupiter, they will be blown out of the solar system. In addition, the deadly radiation grinds the dust specks down to nothing in a century or so. Thus the tiny particles we see in the ring must be continuously supplied from some source.

One set of sources are moonlets as small as 1 cm or as large as 10 km in diameter. They can't be much larger or they would have been seen. If such moonlets exist, bombardment by micrometeorites or by dust from the volcanoes on Io may be chipping off a constant supply of dust specks to keep the rings populated.

Two such moonlets may have been found. Adrastea is a moon 36 km in diameter that orbits Jupiter just at the edge of the ring. In fact, the small moon may prevent ring particles from drifting outward. Metis is another small satellite 40 km in diameter that lies within the ring at 1.79 Jupiter radii, just where the ring is brightest. Perhaps particles blasted from Metis keep Jupiter's ring bright at this radius.

Thus the ring of Jupiter is made up of billions of dust particles, each a separate satellite of the planet following its own orbit. In addition, Jupiter is orbited by a family of true moons large enough to have interesting geological histories.

Jupiter's Family of Worlds Jupiter's satellite system consists of at least 16 moons, some discovered by

Figure 18-7 The Galilean moons of Jupiter, clockwise from upper left: Io, Europa, Callisto, and Ganymede. Callisto and Ganymede are 50 percent water, and Europa contains an icy crust over a water mantle. Io is dry and sulfurous. Earth's moon is shown at center for scale. (NASA and Lick Observatory)

the Voyager 1 and Voyager 2 spacecraft that visited the planet in the late 1970s. As noted previously, many of these satellites are quite small, only a few dozen kilometers in diameter. In addition, the four outermost have retrograde (backward) orbits. That is, they revolve around Jupiter clockwise as seen from the north, backward compared to most motion in the solar system. These characteristics suggest that some of the smaller satellites are captured asteroids and did not form with Jupiter.

The four largest of Jupiter's moons are called the Galilean moons (Figure 18-7) after their discoverer Galileo (see Chapter 4). These moons, in order from outermost inward, are Callisto, Ganymede, Europa, and Io. They are a clearly related family that can give us an insight into how our solar system formed.

The outermost Galilean moon, Callisto, is 44

percent larger than Earth's moon but with a density of only 1.79 g/cm³. This low density means it must consist of a roughly 50-50 mixture of rock and ice, and the Voyager spacecraft photographs show a surface of dirty ice heavily pocked with impact craters (Figure 18-8). The newer craters are white because they are surrounded by cleaner ice ejected from below by the impact, but most of the craters are quite old. Callisto's crust is ancient and dates from the formation of the solar system.

The next Galilean moon inward, Ganymede, also is icy. Its density is 1.9 gm/cm³, which shows that it is roughly half ice. Ganymede is cratered, but it is marked by patterns of grooves that seem to be breaks in its icy crust (Figure 18-9; see also Data File 7, figure b). This **grooved terrain** does contain some craters but not as many as the ungrooved surface, and this shows that the grooves

Figure 18-8 Callisto, the most heavily cratered of the Galilean moons, has an ancient crust of dirty ices. Younger craters appear white because they expose cleaner ice from below the surface. (NASA)

formed after most of the cratering. If we visited the surface of Ganymede, we could walk across the grooves and not notice them; they are typically 3–4 km wide and only about 300 m deep, with very gentle slopes. They are, however, dramatic as seen from space and tell us something about the history of Ganymede.

Apparently, Ganymede cooled more slowly than Callisto and broke its crust repeatedly. Icy slush from below welled up into the breaks, forming the grooves and erasing any preexisting craters. Ganymede has been a cold, dead world for a few billion years, but it was once active, and faulting broke its crust.

Europa, the next Galilean satellite inward, also shows signs of activity. At 3.03 g/cm³, it is more dense than Callisto or Ganymede, which suggests a large proportion of rock, but it does contain a liquid water mantle and a thick, icy crust. Very few

craters are known on Europa, clear evidence of an active surface. Some process must be renewing the surface and erasing craters as fast as they form. The lines on the surface are typically 30 km wide and 1000 km long, and these suggest breaks in the crust where water from below has refrozen (Figure 18-10). A recent study of a Voyager 2 photo of Europa revealed faint traces of what could be a volcanic plume extending 100–150 km above the surface. Perhaps this is pressurized water spraying out through a crack in the crust, freezing in space and falling back. Such water volcanism could resurface the satellite continuously and erase craters rapidly.

Io, the innermost of the four moons, is also the most extreme. It has a density of 3.55 g/cm³, which shows that it is mostly rock. Its surface contains no impact features, but it is not icy. In fact, were we to visit Io's surface, we would find ourselves standing

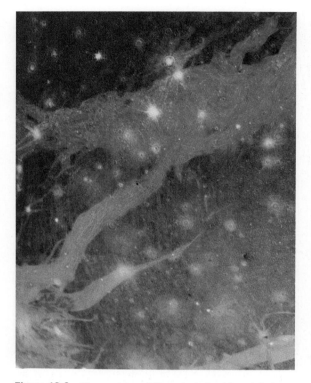

Figure 18-9 The surface of Ganymede is old and dark except where it is broken by the younger grooved terrain. Craters in the grooves show that they are 1–2 billion years old. (See Data File 7, figure b.) (NASA)

Figure 18-10 Europa contains an icy crust covering a mantle of liquid water. The lines on the surface appear to be cracks in the crust where water from the interior has refrozen. The almost total lack of craters suggests that some process erases craters as fast as they form. (NASA)

in a yellow sulfur desert constantly sprinkled by falling sulfur ash ejected from active volcanoes (Figure 18-11). The ash accumulates at a rate of about 1 mm per year, and any impact crater on Io would be buried by this ash in only a few million years. The Voyager spacecraft photographed nine active volcanoes circled by rings of ejected ash and many other inactive volcanic vents.

The heat to drive this volcanism on Io seems to come from **tidal heating.** Io is much too small to have retained heat from internal radioactivity. But Europa and Ganymede interact to force Io into an elliptical orbit, and as its distance from Jupiter changes, it is flexed by tidal forces that generate heat in its interior. This melts rock in the interior, which rises toward the surface and melts the sulfur deposits below the crust. Where this sulfur and sulfur dioxide gas burst through the crust, we see a volcano, and the various forms of liquid and solid

sulfur and the whitish frosts of sulfur dioxide give Io its colorful face. The suspected activity on Europa may also be driven by tidal heating.

The Galilean satellites appear to have been modified to some extent by tidal heating. The inner satellites are more active and contain less volatiles, especially water. Tidal heating could have helped drive off such volatiles.

But the differences among the moons may also be traced to their origin in a miniature solar nebula around Jupiter. When Jupiter formed, it must have been very hot, and its mass could have attracted a large nebula that would have flattened into a disk. According to the condensation sequence (see Chapter 16), the inner part of this disk, being warmer, would not have condensed much ice, but the outer, cooler regions could have formed large numbers of ice particles. Because of this, the moons that formed farther from Jupiter should

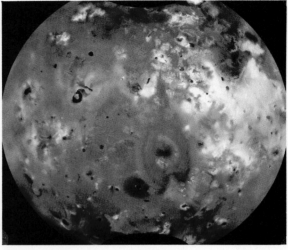

a

Figure 18-11 Io, Jupiter's innermost moon, is heated internally by tides, and volcanoes spew sulfur ash over its surface. (a) In this mosaic, the heart shape near center is the ash pattern around the volcano Pele. The black *D* above and left of center is a lake of molten sulfur 250 km long. (b) A computer-enhanced image of Io reveals the volcano Loki spewing ash above the surface. (NASA)

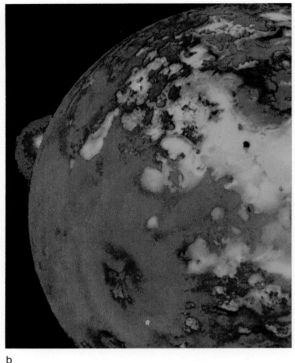

b

have been richer in ices. Thus the variation in density and ice content among the Galilean moons of Jupiter may be a miniature of the formation of the terrestrial and Jovian planets in the solar nebula.

18-2 SATURN

Saturn has played second fiddle to its own rings since Galileo first viewed it through a telescope in 1610. The rings are dramatic and easily seen through even a small telescope, and in 1979 the Voyager 1 and Voyager 2 spacecraft visited Saturn and revealed that the rings are much more complex than expected. The Voyager photos also showed that Saturn's moons are fascinating in their diversity. Yet even without its rings and moons, Saturn itself is a fascinating planet.

Saturn the Planet Slightly smaller and much colder, Saturn is a striking contrast to Jupiter. (See Data File 8.)

Seen from Earth, Saturn shows only faint evidence of belt-zone circulation, but the Voyager photos show that belts and zones are present and that the associated winds blow up to three times faster than on Jupiter. The belts and zones on Saturn are less visible because they occur deeper in the cold atmosphere below a layer of methane haze (Figure 18-12). Also, the low temperature may prevent the formation of some of the molecules responsible for the colors of Jupiter's clouds.

The observed density and size permit theorists to model Saturn's interior, and it is much like Jupiter's. Its lower mass produces lower internal pressures and there is thus less liquid metallic hydrogen. Perhaps because of this, Saturn's magnetic field is 20 times weaker than that around Jupiter. Like Jupiter, Saturn has a hot interior and radiates more energy than it receives from the sun.

All of the Jovian planets, being mostly liquid, are flattened by their rapid rotations. A planet's **oblateness** is the fraction by which its equatorial diameter exceeds its polar diameter. Saturn is the most oblate of the planets, which is evident in photographs (Data File 8, figure a).

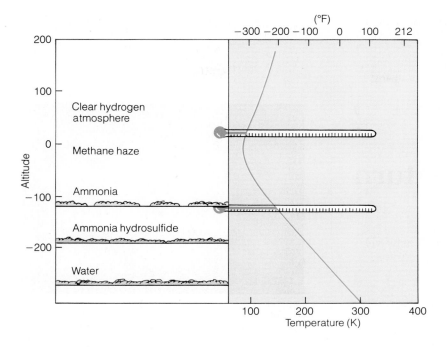

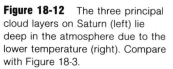

Figure 18-12 The three principal cloud layers on Saturn (left) lie deep in the atmosphere due to the lower temperature (right). Compare with Figure 18-3.

Figure 18-13 The transparency of Saturn's rings is evident where the rings cross in front of the planet (at left). Note the thinness of the C ring, the material in Cassini's division, and the F ring beyond the sharp edge of the A ring. (NASA/JPL)

Saturn is a bit bland as planets go, with its belts and zones buried deep in its hazy atmosphere, but what Saturn lacks in atmospheric detail, it makes up for in the splendor of its rings.

The Rings of Saturn Saturn's rings contain billions of icy particles. A few gaps, such as Cassini's division (Data File 8, figure c), are visible from Earth, but the true complexity of the rings was not evident until the Voyager 1 spacecraft flew past Saturn in November 1980. The Voyager 1 images showed that the rings were made up of almost 1000 ringlets, some as narrow as 2 km (Figure 18-13). There were even ringlets in Cassini's division.

In addition, the spacecraft discovered previously unseen rings. The E ring had been detected from Earth, but Voyager 1 confirmed its existence out to 8 Saturn radii and showed that it is most dense near the orbit of the satellite Enceladus.

Saturn

a. The belts and zones on Saturn are not as distinct as those of Jupiter because of overlying haze. (NASA)

Average distance from the sun	9.5388 AU (14.27 × 10⁸ km)
Eccentricity of orbit	0.0560
Maximum distance from the sun	10.07 AU (15.07 × 10⁸ km)
Minimum distance from the sun	9.005 AU (13.47 × 10⁸ km)
Inclination of orbit to ecliptic	2°29′17″
Average orbital velocity	9.64 km/sec
Orbital period	29.461 y (10,760 days)
Period of rotation	10ʰ13ᵐ59ˢ
Inclination of equator to orbit	26°24′
Equatorial diameter	120,660 km (9.42 $D_\oplus$)
Mass	5.69 × 10²⁶ kg (95.147 $M_\oplus$)
Average density	0.69 g/cm³
Gravity at base of clouds	1.16 Earth gravities
Escape velocity	35.6 km/sec (3.2 $V_\oplus$)
Temperature at cloud tops	−180°C (−292°F)
Albedo	0.61
Oblateness	0.102

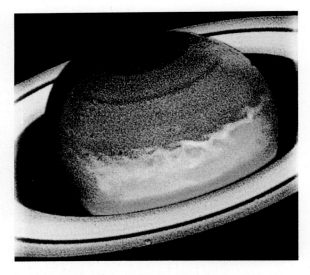

b. 1990 Hubble Space Telescope photograph of a vast storm on usually bland Saturn. (NASA)

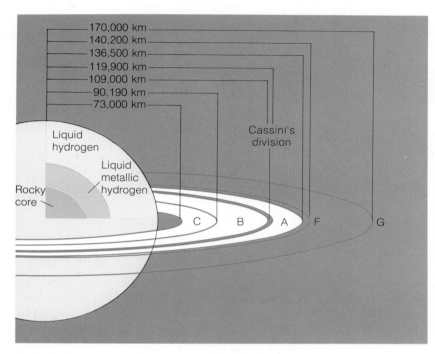

c. Saturn's interior, ring dimensions, and identifications.

d. Ringlets appear in the B ring like grooves in a record. (NASA)

e. A third the size of Earth's moon, Saturn's moon Dione is an icy, cratered world. (NASA)

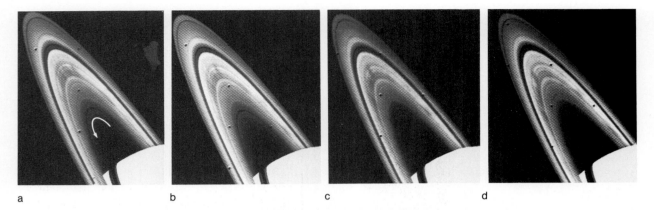

a   b   c   d

Figure 18-14 The radial spokes in the bright B ring are visible in this photo sequence as dark streaks oriented radially. The black dots are reference points in the camera. (NASA/JPL)

This moon may be ejecting ice particles somehow to maintain the E ring.

The spacecraft also photographed a threadlike ring lying outside the edge of the A ring (Figure 18-13). This F ring is forced into a narrow strand by two small moons of Saturn (Prometheus and Pandora) that orbit just inside and just outside. A ring particle wandering away from the F ring is shepherded back into the F ring by the gravitation of the moons. Thus they are known as **shepherd satellites** and may cause some of the irregularities seen along the F ring. A similar G ring is also presumed to be shepherded by small moons, and another moon shepherds the outer edge of the A ring and keeps the ring particles from spreading outward. Without such moons to confine the rings, they would gradually spread and disappear.

Major gaps in the rings are caused by resonances between ring particles and satellites. A particle in Cassini's division, for example, orbits Saturn twice in the time the moon Mimas takes to orbit once. Thus on every other orbit, the ring particle overtakes Mimas at the same place in the particle's orbit, and the gravitational tugs of Mimas gradually pull the particle into a slightly elliptical orbit. Elliptical orbits are dangerous for a ring particle because such orbits cross the paths of millions of other particles. Our unfortunate particle is almost certain to suffer a collision and have its orbit altered dramatically. Thus particles cannot remain in orbit at resonances with moons. Resonances can occur at various distances where the particles orbit two, three, or more times for each orbit of the moon.

This resonance theory could explain a few gaps but not the hundreds of ringlets. Theoretical calculations show that the ringlets can be produced by spiral density waves like those believed responsible for the spiral arms in galaxies (see Chapter 12). Another theory suggests that small moonlets lie inside the rings and sweep particles out of particular regions. One such moonlet has been found.

One surprise in the Voyager images was the presence of **spokes,** radially oriented features in the rings (Figure 18-14). The rings rotate differentially, with inner parts moving faster than the outer parts. This should erase radial features rapidly. But the spokes appear to be more stable. Apparently, Saturn's magnetic field can pull electrostatically charged dust up out of the ring plane, and the dust then rotates with the magnetic field, producing shadows on the rings that we see as spokes.

The origin of Saturn's rings is not well understood. They could be debris left over from the formation of the planet, but if Saturn formed hot, it should have driven such material away. The rings do lie inside the Roche limit, where a large satellite could not form. Some studies, however, suggest that icy particles are darkened by the impact of microscopic meteorites and could not remain so

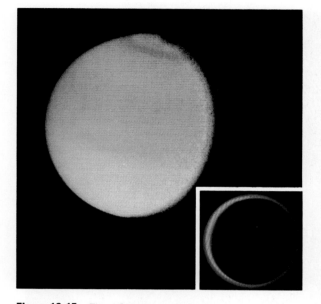

Figure 18-15 Titan, Saturn's largest moon, is 48 percent larger in diameter than Earth's moon and has a smoggy atmosphere of nitrogen and methane. Inset: The extent of Titan's atmosphere is evident in this photograph of the night side. (NASA)

bright for more than about 100 million years. This suggests that the material in Saturn's rings is debris from an icy moon of roughly 100 km diameter that was shattered by the impact of a comet. Saturn's moons certainly show signs of impacts.

The Moons of Saturn Saturn has at least 18 satellites, many of which are small bodies. All contain large amounts of ices mixed with rock.

The largest moon, Titan, is 5.6 percent larger than Mercury, and because it is so cold (−179°C at its surface) it has retained a nitrogen-rich atmosphere (Figure 18-15). Methane makes up no more than 10 percent of the atmosphere. Voyager photos of Titan show a world hidden below orange smog produced by sunlight interacting with the methane and nitrogen atmosphere. Theoretical models predict a surface pressure 1.3 times Earth's atmospheric pressure and a possible ocean of liquid methane and ethane up to 1 km deep.

Recent studies show that Titan, like Venus and Earth, is warmed by a greenhouse effect, in Titan's case caused by methane and hydrogen gas in its atmosphere. It also suffers from a small antigreenhouse effect, because a high organic haze absorbs much of the incoming sunlight but is nearly transparent to outgoing infrared. Earth's ozone layer has a similar but smaller effect. The greenhouse effect raises the temperature of Titan by 21 degrees while the antigreenhouse effect lowers the temperature 9 degrees. Without its haze, Titan would be much warmer at −170°C (−274°F) instead of its frigid −179°C (−290°F).

The remaining moons of Saturn are small and have no atmospheres. They are icy and heavily cratered, though there are signs of geological activity on a few.

Tethys, for example, is 1060 km (660 miles) in diameter and has an old, cratered crust (Figure 18-16a). However, a valley 3 km deep trails 1500 km across its surface—three quarters of the way around the satellite. Such cracks are seen on a few other satellites and appear to have formed long ago, when the interiors of the icy moons froze and expanded.

In contrast, Enceladus, though only 500 km (300 miles) in diameter, shows signs of recent activity (Figure 18-16c). Some parts of its surface contain 1000 times fewer craters than other regions, showing that these lightly cratered regions must be young. Its crust is also marked by parallel faults that suggest crustal activity. Enceladus orbits Saturn just where the E ring of ice crystals is most dense, and it seems possible that water volcanism on Enceladus periodically resurfaces regions of the crust and supplies fresh ice crystals to the E ring. The energy source is unknown, but tidal heating is likely.

Like almost all moons, Saturn's moons are bound to their parent planet and rotate to keep the same side facing the planet. This means one side of the moon always faces forward in its orbit, and in some cases this leading side is more heavily cratered than the trailing side. Iapetus shows an even more dramatic asymmetry (Figure 18-16b). Its cratered trailing side is about the color of dingy snow, but its leading side is as dark as fresh asphalt. One theory is that the dark material has flooded up from the interior through faults in the leading side.

Saturn, its rings, and its moons are a cold system rich in ices. But it is not the coldest Jovian planet. Uranus and Neptune lie even farther from the sun.

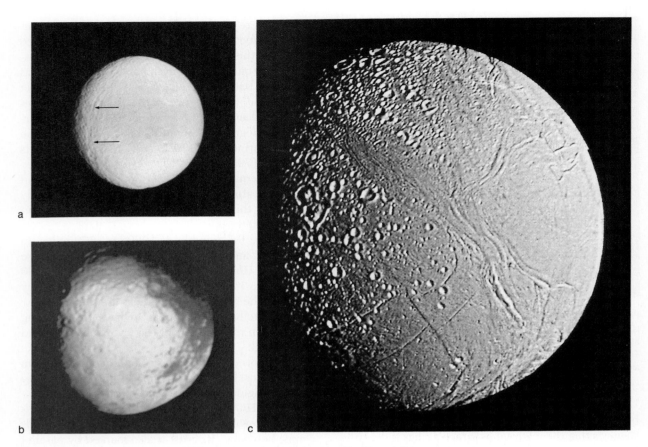

Figure 18-16 (a) Heavily cratered Tethys appears to be a dead world, although it is marked by a deep valley (arrows). (b) The leading side of Iapetus (facing upper right) is much darker than the trailing side. (c) Some portions of Enceladus have fewer craters. These areas apparently have been resurfaced recently, which suggests that Enceladus is an active moon. (NASA)

18-3 URANUS

Uranus is a Jovian planet, but it is dramatically different from Jupiter and Saturn. Its atmosphere, interior, rings, and satellites are profitable subjects for comparative planetology.

Uranus the Planet Uranus is small and cold. It is 36 percent the diameter of Jupiter, 4.6 percent as massive, and, being 3.7 times farther from the sun, is 111°C colder than Jupiter. (See Data File 9.) Size and temperature determine many of Uranus's characteristics.

Because Uranus is smaller than Jupiter, its internal pressure is lower, and it does not contain liquid metallic hydrogen. Nevertheless, it does have a magnetic field generated in a mantle of water containing ammonia and methane.

Uranus rotates on its side. Its axis of rotation is inclined almost 98°, and as Voyager 2 approached, the planet's south pole was pointed nearly at the sun (Data File 9, figure b). The reason for this peculiar rotation is not known, although most astronomers suppose that the axis was changed when Uranus collided with a very large planetesimal late in its formation.

The atmosphere of Uranus is a nearly featureless blue-green (Figure 18-17). The color arises from the absorption by methane in an atmosphere mostly composed of hydrogen. Only the most highly

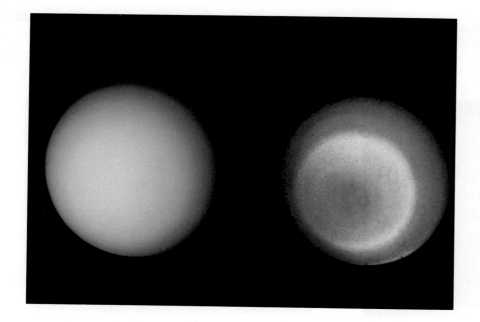

Figure 18-17 The planet Uranus as photographed by the Voyager 2 spacecraft shows almost no features in its atmosphere (left). Extreme image enhancement (right) reveals atmospheric banding concentric with the planet's axis of rotation. (NASA)

computer-enhanced images reveal evidence of belt-zone circulation. Evidently, the cloud layers lie very deep in the hazy atmosphere.

One reason Uranus lacks strong atmospheric circulation may be that, unlike Jupiter and Saturn, it has little heat flowing outward from its interior. Such heat flow can drive circulation that forms clouds and thus creates visible atmospheric features. Uranus has cooled and now radiates about the same amount of energy that it receives from the sun. Thus, little heat is flowing from its interior.

Although Uranus almost totally lacks visible belts and zones, it is attended by a fascinating family. Its rings are narrow and graceful, and its moons are heavily cratered.

The Rings of Uranus   The rings of Uranus were discovered in 1977 when astronomers watched the planet occult—cross in front of—a distant star. Repeatedly, the star dimmed momentarily as the planet approached and again after it passed. The astronomers quickly concluded that rings around Uranus had crossed in front of the star. Since then the rings have been studied from Earth when Uranus occulted other stars and by Voyager 2 when it visited Uranus in January 1986.

The rings of Uranus are difficult to detect. They are 100 to 1000 times more substantial than Jupiter's ring, but they total less mass than the thin material in Cassini's division, the largest gap in Saturn's rings. Also, they are composed of very dark material—as black as lumps of coal.

Nine rings have been detected from Earth, and Voyager 2 discovered two more (Figure 18-18a). These 11 rings are very narrow. The widest, the ϵ ring, varies from 12 to 60 km in width. The narrowness of the rings is a puzzle because planetary rings tend to spread out and form a smooth disk of particles. Astronomers speculated that small shepherd moons kept the rings confined in narrow bands, and Voyager 2 discovered two such moons 30 and 40 km in diameter herding the ϵ ring (Figure 18-18b). These moons may also explain why the ϵ ring is slightly elliptical. No other shepherd moons were found, but astronomers suspect they may be smaller than the 5- to 10-km resolution limit for the Voyager 2 images.

The black color of the rings may be caused by radiation. The rings are located inside radiation belts trapped in Uranus's magnetic field, and radiation could convert traces of frozen methane into a sticky, black organic polymer. Voyager 2 discovered 10 new moons inside the radiation belts, and they too are black.

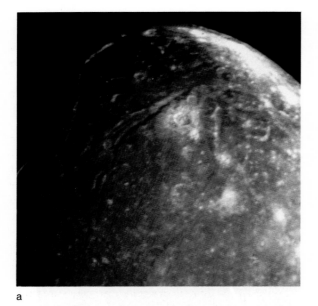

a

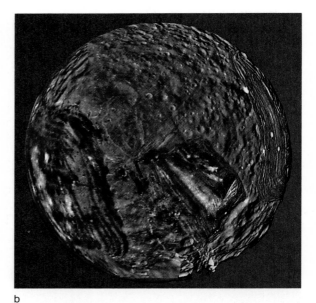

b

Figure 18-19 (a) Ariel, a satellite of Uranus, apparently had an active past, as shown by the broad, shallow valleys on its surface. (b) The face of Miranda is marred by ovoids, which may have formed as heavy blocks of rock sank through the icy mantle. Note the 5-km-high cliff at the lower edge of the moon. (NASA)

when the water-rich interior froze and expanded. Umbriel is 69 percent the diameter of Earth's moon and shows no faults or surface activity.

Ariel, the fourth moon inward, is slightly smaller than Umbriel, but its cratered crust is marked by broad, smooth-floored valleys that may have been cut by flowing ice (Figure 18-19a).

Miranda, the innermost moon, is only 480 km (300 miles) in diameter and marked by oval patterns of grooves called **ovoids** (Figure 18-19b). One theory is that Miranda was struck by a large body and broken into pieces that were pulled back together by their own gravity. If Miranda differentiated into a rocky core and an icy mantle before the impact, large blocks of rock could have been embedded in the icy crust of the reassembled satellite, forming ovoids as the heavy blocks sank and lighter ice rose. The crust has been seriously stressed, with faults and cliffs up to 5 km high.

Judging from featureless Uranus, bland Saturn, and strongly marked Jupiter, we might expect Neptune to be as lacking in detail as Uranus. But Neptune is in fact an active world.

18-4 NEPTUNE

Named after the Roman god of the sea, Neptune has a faint blue tint as seen through an Earth-based telescope. Hardly any other detail is visible. In August 1989, the Voyager 2 spacecraft passed only 4400 km above the cloud tops of Neptune and revealed that the planet, its rings, and its satellites are both beautiful and peculiar.

Planet Neptune Almost four times the diameter of Earth, Neptune was until recently thought of as a smaller, colder Uranus. The planets differ by only 4 percent in diameter, and both have rocky cores with deep mantles of liquid hydrogen. (See Data File 10.) They are thus both Jovian worlds. The striking differences between the two became apparent only when Voyager 2 flew past Neptune.

Neptune's atmosphere appears a deep blue-green (Figure 18-20) because, although it is mostly hydrogen and helium, it contains traces of methane—which absorbs yellow and red light. As light penetrates deep into the atmosphere and is scat-

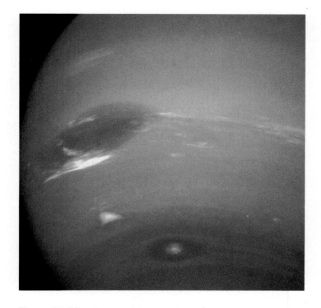

Figure 18-20 Neptune's blue-green color is caused by methane gas present in small amounts in its atmosphere of hydrogen and helium. Methane absorbs red and yellow photons leaving blue and green photons to give the planet its characteristic color. Note the Great Dark Spot (left) and the smaller spots named Scooter and D2 (bottom). (NASA)

tered back out, the blue and green photons are the ones most likely to survive.

The key to understanding Neptune may be that its atmosphere is dominated by the flow of heat out of the interior. The heat flow causes convection, which the rapid rotation of the planet converts into high-speed winds, high-level white clouds of methane ice crystals, and cyclonic circulations. Some of the cloud formations are carried 50 km above the lower cloud deck.

The largest circulation is the Great Dark Spot, named after Jupiter's Great Red Spot. It is roughly 10,000 km across and believed to be caused by rising currents of gas. Smaller spots such as Scooter (for its rapid rotation around the planet) and D2 are also visible in the planet's southern latitudes (Figure 18-20).

An off-center magnetic field inclined 58.6 degrees to the axis of rotation confirms that the planet has a conducting interior. But the high inclination, resembling that of the field around Uranus, is difficult to explain. One theory suggests that the fields of Uranus and Neptune may be generated in a conducting mantle rather than in the core of the planet.

While Voyager 2 results showed that the atmosphere of Neptune was unusually active, they also showed that the rings were active in that they have dense arcs, which are unique among rings in our solar system.

Neptune's Rings Earth-based astronomers long suspected that Neptune has rings. The Voyager photos revealed two narrow rings with radii of 63,000 km and 53,000 km and fainter, broader rings between. Small moons shepherd the narrow rings and keep the ring particles from spreading.

The narrow rings are brightest when illuminated from behind—forward scattering—so they must contain large amounts of dust. The broader rings, however, seem to contain less dust. As the ring particles are not very reflective, they are probably rocky rather than icy.

Within the outermost narrow ring are short arcs where the ring is denser. These ring arcs are difficult to explain because collisions among particles should smooth out the rings in relatively short times. During 1991, a planetary astronomer was able to build a mathematical model to explain the ring arcs. According to her model, the moon Galatea creates waves in the ring that confine the particles in certain small arcs. The model went on to predict the location of other, fainter arcs, and computer-enhanced images confirmed their existence (Figure 18-21). Thus the ring arcs are created by a gravitational interaction between the ring particles and one of Neptune's moons.

The Moons of Neptune Observations made from Earth found two moons, Triton in 1846 and Nereid in 1949. Both are in peculiar orbits. Nereid is a small moon in a large, elliptical orbit, taking 359.4 days to orbit Neptune once. Triton is a larger moon and orbits Neptune in only 5.877 days, but it orbits backward—clockwise as seen from the north. These odd orbits suggest that the system was disturbed long ago in an interaction with some other body.

Voyager photographs revealed six more moons orbiting among the rings. They are all small, dark, rocky bodies. No new moons were found beyond Triton. Some astronomers suggest that Triton in

Neptune

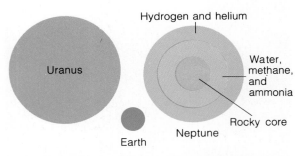

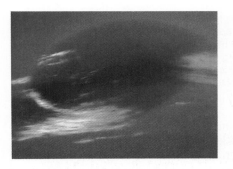

Average distance from the sun	30.0611 AU (44.971 × 10⁸ km)
Eccentricity of orbit	0.0100
Maximum distance from the sun	30.4 AU (45.4 × 10⁸ km)
Minimum distance from the sun	29.8 AU (44.52 × 10⁸ km)
Inclination of orbit to ecliptic	1°46′27″
Average orbital velocity	5.43 km/sec
Orbital period	164.793 y (60,189 days)
Period of rotation	16ʰ 3ᵐ
Inclination of equator to orbit	28°48′
Equatorial diameter	49,500 km (3.93 $D_\oplus$)
Mass	1.030 × 10²⁶ kg (17.23 $M_\oplus$)
Average density	1.66 g/cm³
Gravity	1.19 Earth gravities
Escape velocity	25 km/sec (2.2 $V_\oplus$)
Temperature at cloud tops	−216°C (−357°F)
Albedo	0.35
Oblateness	0.027

a. A model of the interior of Neptune, containing water and ammonia ices.

b. The Great Dark Spot on Neptune is a cyclonic disturbance about the size of Earth. (NASA)

c. The rings of Neptune are narrow and contain short segments called arcs. The overexposed planet is at lower right. (NASA)

Figure 18-21 This computer-enhanced and -colored image shows the arcs in Neptune's rings. A mathematical model shows that the arcs are created by the gravitational influence of Neptune's moon Galatea. Arrows point to fainter arcs whose existence was predicted by the model. (NASA image courtesy Carolyn C. Porco)

its retrograde orbit would have consumed such moons.

Although Triton is only 78 percent the radius of Earth's moon, it is so cold (37 K or $-393°F$) it can retain an atmosphere of nitrogen and methane about 10^5 times less dense than Earth's. Triton is composed of a rocky interior and an icy crust. The southern pole of Triton has been turned toward sunlight for the last 30 years, and deposits of nitrogen frost appear to be vaporizing and refreezing in the darkness of the northern pole (Figure 18-22a).

Many features on Triton suggest it has had an active past. It has few craters on its surface, but it does have long faults that appear to have formed when the icy crust broke. Some roughly round basins about 400 km in diameter appear to have been flooded time after time by liquids from the interior (Figure 18-22b). Most exciting of all are the dark smudges visible in the southern polar cap (Figure 18-22a). Analysis of the photos reveals that these are deposits produced when liquid nitrogen in the crust, warmed by the sun, erupts through vents and spews up to 8 km high into the atmosphere.

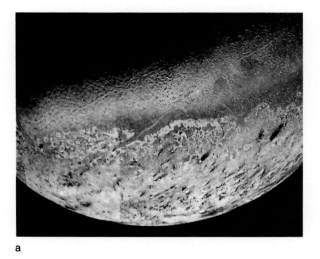

a

b

Figure 18-22 (a) Triton's southern polar cap is formed of nitrogen frost. Note the absence of craters and the dark smudges caused by nitrogen volcanism. (b) Such round basins on Triton appear to have been repeatedly flooded by liquid from the interior. (NASA)

Methane in the gases is converted into dark deposits that fall to the surface leaving the black smudges. Thus Triton joins Earth and Jupiter's moon Io as one of the three worlds known to have active volcanism.

By counting craters on Triton, astronomers have concluded that volcanic activity has been resurfacing the moon as recently as 1 million years ago. Triton may still be active. The energy source for this volcanism could come from radioactive decay. Its mass is two-thirds rock, and although such a small world would not be able to generate sufficient radioactive decay to keep molten rock flowing to its surface, frigid Triton may suffer from water–ammonia volcanism. A mixture of water and ammonia could melt at very low temperatures and resurface parts of the moon.

Low temperatures dominate the evolution of the planets and moons that exist in the outer solar system, and no more frigid world could be imagined than icy Pluto.

18-5 PLUTO

Pluto is not a Jovian planet because it is not a gas giant, but it is also not a terrestrial planet because it is not made mostly of rock and metal. What is Pluto? That question has bothered astronomers since the first moment Clyde Tombaugh of Lowell Observatory discovered it on photographic plates in 1930.

Pluto as a Planet Because Pluto is so distant, we know very little about it. Seen from Earth, Pluto's angular diameter is slightly larger than 0.1 second of arc, and even the best photographs do not show more than a tiny dot whose diameter is not apparent. Pluto is in fact only 65 percent the diameter of Earth's moon. (See Data File 11.) It is clearly not a Jovian world.

Most of the planetary orbits in our solar system are nearly circular, but Pluto's is quite elliptical. (See page 10.) Pluto's average distance from the sun is 39.52 AU, but it can come as close as 29.64 AU. In fact, from January 21, 1979, to March 14, 1999, Pluto will be closer to the sun than Neptune. The planets will never collide, however, because

Pluto's orbit is inclined 17° to the plane of the solar system. Orbiting so far from the sun, Pluto is cold enough to freeze most compounds that we think of as gases.

Spectroscopic observations of Pluto in the near infrared show that there is solid methane on its surface either as frost or ice. When Pluto occulted a star, astronomers found evidence of a thin atmosphere consisting of gaseous methane mixed with an unidentified gas that may be nitrogen or carbon monoxide. There is no direct evidence of water ice on Pluto, but most astronomers believe there must be water frozen into the interior. A further clue appeared when an astronomer found that Pluto has a moon.

Pluto's Moon In 1978, James W. Christy of the United States Naval Observatory, while examining a photographic plate, discovered a faint image beside Pluto's. The image turned out to be a moon orbiting the planet once every 6.387 days at an average distance of about 17,000 km (10,000 miles). The new moon was named Charon after the mythological ferryman who transports souls across the river Styx into the underworld.

The discovery of a new object in the solar system is always exciting, but the discovery of Charon was especially important because it made it possible to determine Pluto's mass. The same analysis that reveals the total mass of binary star systems reveals that the total mass of Pluto and its moon is about 7×10^{-9} solar masses, or about 0.002 Earth masses. By analyzing the brightness variations of the Pluto–Charon system as they would an eclipsing binary star (see Chapter 8), astronomers have been able to find the diameters of both bodies. Pluto is about 2300 km in diameter and Charon is about half that size.

Combining the mass of Pluto with its observed diameter reveals that its density is low, about 1.84 gm/cm^3. Thus Pluto must be an ice–rock mixture, which would be consistent with its formation in the outer solar nebula, where ices would have been stable. Although its composition seems appropriate for its location in the solar system, astronomers wonder how it could have formed.

The Origin of Pluto Pluto's eccentric, highly inclined orbit and its small size have led some astronomers to suggest that it is an escaped satellite of

Pluto

Average distance from the sun	39.44 AU (59.00 × 10⁸ km)
Eccentricity of orbit	0.2484
Maximum distance from the sun	49.24 AU (73.66 × 10⁸ km)
Minimum distance from the sun	29.64 AU (44.34 × 10⁸ km)
Inclination of orbit to ecliptic	17°9′3″
Average orbital velocity	4.73 km/sec
Orbital period	247.7 y (90,465 days)
Period of rotation	6ᵈ9ʰ21ᵐ
Inclination of equator to orbit	118° (estimated)
Equatorial diameter	2294 km (0.18 $D_\oplus$)
Mass	1.2 × 10²² kg (0.002 $M_\oplus$)
Average density	1.84 g/cm³
Surface gravity	0.06 Earth gravities
Escape velocity	1.2 km/sec (0.11 $V_\oplus$)
Surface temperature	−230°C (−382°F)
Albedo	0.4

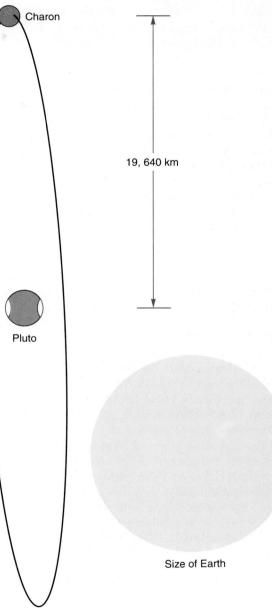

Charon

19, 640 km

Pluto

Size of Earth

Charon orbits Pluto in a highly inclined, circular orbit only a few times larger than the Earth.

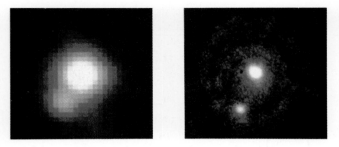

Figure 18-23 A high-quality, ground-based photo (left) shows Pluto and its moon, Charon. Hubble Space Telescope images (right) permit more accurate measurements of the position of the moon. (NASA/European Space Agency)

Neptune. That theory has had fewer supporters, however, since the discovery of Charon.

Charon is very large compared to the size of its host planet. Seen from Pluto, Charon appears 6.7° in diameter, almost seven times the angular diameter of Earth's moon. Charon and Pluto are similar in density and composition, and orbit so close to each other they are tidally locked, both rotating to keep the same side facing the other. This has led some astronomers to think of Pluto and Charon as a double planet (Figure 18-23).

The similarity between Pluto and Charon extends to other icy bodies in the outer solar system. Pluto and Triton, Neptune's largest moon, are nearly the same size and density, and perhaps have similar compositions. Recently, astronomers have begun discussing the possibility that Pluto, Charon, and Triton are merely the remains of a large number of small, icy bodies that once populated the outer solar system. Tentatively called Plutons or ice dwarfs, these bodies could explain a number of peculiarities.

The tilted axis of rotation of Uranus and Pluto and the disturbed orbits of Neptune's moons Triton and Nereid suggest collisions with other objects. Such collisions seem highly improbable. If, however, a thousand or so ice dwarfs existed in the outer solar system during or soon after planet formation, then such collisions would not be so unlikely. Most of those bodies would have been incorporated into planets or ejected into the far fringes of the solar system, but we might see the last remaining ice dwarfs as a few of the larger icy satellites such as Triton and as Pluto and its companion Charon.

This hypothesis is highly speculative and may not survive the search for small, icy bodies beyond Pluto's orbit. The hypothesis is important not because it might be correct, but because it points the way to further understanding of the origin of our solar system. The terrestrial planets, the Jovian planets, and Pluto are all remnants of the solar nebula, and they are the final product of interactions between vast numbers of small bodies made of rock and ice. Thus, to learn more about the planets, we must study the smaller bodies left behind when the solar nebula dissipated—the meteorites, asteroids, and comets.

SUMMARY

The Jovian planets are so massive they have been able to retain much of their initial hydrogen and helium. This excess of light elements gives the Jovian planets densities much lower than those of the terrestrial planets. In their interiors, the tremendous pressure forces the hydrogen into a liquid state, and in Jupiter and Saturn the pressure is so high the hydrogen becomes a liquid metal that is a good conductor of electricity. Electric currents within this material produce magnetic fields that trap particles from the solar wind to form radiation belts around the planets. Both Jupiter and Saturn are known to have radiation belts similar to Earth's Van Allen belts, and Jupiter's belts are very intense.

The atmosphere of Jupiter is marked by high- and low-pressure regions in the form of belts and zones that encircle the planet. Chemical reactions in the clouds are believed to produce compounds that color the belts. The Great Red Spot on Jupiter is believed to be a cyclonic disturbance much like an earthly hurricane.

Jupiter's ring is a tenuous sheet of dust particles orbiting in the planet's equatorial plane. Because certain processes rapidly remove dust from the ring, astronomers believe it must be continuously resupplied by new dust chipped from small satellites.

Jupiter's smaller moons may be captured asteroids, but its four largest, the Galilean moons, clearly formed with the planet. The outermost moon, Callisto, is an icy, dead world, and the next moon inward, Ganymede, shows grooves produced by ancient breaks in its icy crust. Europa is less rich in ice, but it does have an icy crust free of craters. Thus its surface must be active. Io is very dry and has active sulfur volcanos powered by tidal heating.

Saturn's cold atmosphere has belt-zone circulation, though it is partially hidden below methane haze. Saturn's rings are made of icy particles shepherded by small satellites and broken into ringlets by spiral density waves. The rings may be as old as the planet, but they may need an occasional supply of fresh ice from impacts on Saturn's moons. Most of Saturn's moons are small, icy, dead worlds, but at least one, Enceladus, shows signs of an active surface.

Uranus and Neptune are smaller than Jupiter and thus have lower internal pressures. They have no liquid metallic hydrogen, but they do have magnetic fields generated in liquid interiors.

Uranus rotates on its side, perhaps because it once collided with a large planetesimal late in its formation. Its atmosphere is nearly featureless, but highly computer-enhanced Voyager 2 images reveal traces of belt-zone circulation deep below the haze. Uranus's narrow rings of dark boulders appear to be confined by shepherd satellites. Although it has many small moons, only five are large enough to be easily visible from Earth. Voyager 2 images revealed traces of past impacts, and Miranda may once have been totally disrupted by such an impact.

Heat is still flowing outward from the interior of Neptune, and that drives strong weather patterns in its atmosphere. Like Jupiter, it has belt-zone circulation and a large cyclonic disturbance called the Great Dark Spot. Neptune's rings are thin hoops with denser arcs caused by a resonance with one of its smaller moons. Triton, its largest moon, is an icy world where methane–water volcanism may still be active.

Pluto and its moon Charon are small and made of a mixture of ice and rock. The origin of Pluto and Charon is unknown, but they are similar in many ways to Triton, and astronomers now speculate that such icy bodies may once have been common in the outer solar system.

NEW TERMS

magnetosphere	tidal heating
belt-zone circulation	oblateness
forward scattering	shepherd satellite
Roche limit	spokes
grooved terrain	ovoid

REVIEW QUESTIONS

1. Why is Jupiter so much richer in hydrogen and helium than Earth?

2. How does the dynamo effect account for the magnetic fields of Jupiter and Saturn?

3. How do Jupiter's belts and zones resemble earthly weather patterns?

4. Why are the belts and zones on Saturn less distinct than those on Jupiter?

5. What evidence do we have that small satellites can affect the structure of planetary rings?

6. How do the rings of Jupiter, Uranus, and Neptune resemble one another and differ from those of Saturn?

7. What evidence do we have that some satellites in the outer solar system are geologically active?

8. If we visited a Jovian planet that had no satellites at all, would you expect it to have a ring system? Why or why not?

9. What evidence do we have that major impacts influence the evolution of moons in the outer solar system?

10. What evidence do we have to suggest that Neptune's satellite system has been disturbed in some way?

9. How long did it take radio commands to travel from the Earth to Voyager 2 as it passed Neptune?

10. Use the orbital radius and orbital period of Charon to calculate the mass of the Pluto–Charon system. (HINTS: Express orbital radius in AU and period in years. Then see Box 8-5.)

DISCUSSION QUESTIONS

1. Describe the location of the equinoxes and solstices in the Uranian sky. What are seasons like there?

2. Why do no terrestrial planets have rings?

PROBLEMS

1. What is the maximum angular diameter of Jupiter as seen from Earth? Repeat this calculation for Saturn and Pluto. (HINTS: See Data Files 7, 8, and 11, and also Box 3-2.)

2. What is the angular diameter of Jupiter as seen from Callisto? (HINT: See Box 3-2.)

3. Measure the photograph in Figure 18-1 and calculate the oblateness of Saturn.

4. If we observe light reflected from Saturn's rings, we should see a red shift at one edge of the rings and a blue shift at the other edge. If we observe a spectral line and see a difference in wavelength of 0.056 nm, and the unshifted wavelength (observed in the laboratory) is 500 nm, what is the orbital velocity of particles at the outer edge of the rings? (HINT: See Box 6-4.)

5. One way to recognize a distant planet is by its motion along its orbit. If Uranus circles the sun in 84 years, how many seconds of arc will it move in 24 hours? (HINT: Ignore the motion of Earth.)

6. If the ϵ ring is 50 km wide and the orbital velocity of Uranus is 6.81 km/sec, how long a blink should we expect to see when the ring crosses in front of a star?

7. What is the angular diameter of Pluto as seen from the surface of Charon? (HINT: See Appendix D for the size of Charon's orbit.)

8. If Pluto has a surface temperature of 50 K, at what wavelength will it radiate the most energy? (HINT: See Box 6-2.)

RECOMMENDED READING

ARAKI, SUGURU. "Dynamics of Planetary Rings." *American Scientist* 79 (January-February 1991), p. 44.

BEATTY, J. KELLY. "Pluto and Charon: The Dance Begins." *Sky and Telescope* 69 (June 1985), p. 501.

———. "A Place Called Uranus." *Sky and Telescope* 71 (April 1986), p. 333.

———. "Uranus' Last Stand." *Sky and Telescope* 71 (January 1986), p. 11.

———. "Voyager 2's Triumph." *Sky and Telescope* 72 (October 1986), p. 336.

———. "Pluto and Charon: The Dance Goes On." *Sky and Telescope* 74 (September 1987), p. 248.

———. "Welcome to Neptune." *Sky and Telescope* 79 (October 1989), p. 358.

———. "Getting to Know Neptune." *Sky and Telescope* 80 (February 1990), p. 146.

BEATTY, J. KELLY, BRIAN O'LEARY, and ANDREW CHAIKIN. *The New Solar System*, 3rd ed. Cambridge, Mass.: Sky, 1990.

BEEBE, RETA F. "Queen of the Giant Storms." *Sky and Telescope* 80 (October 1990), p. 359.

BENNETT, GARY L. "Return to Jupiter." *Astronomy* 15 (January 1987), p. 6.

BENZEL, RICHARD. "Pluto." *Scientific American* 262 (June 1990), p. 50.

BERRY, RICHARD. "Voyager: Discovery at Uranus." *Astronomy* 14 (May 1986), p. 6.

———. "Triumph at Neptune." *Astronomy* 17 (November 1989), p. 20.

———. "Neptune Revealed." *Astronomy* 17 (December 1989), p. 22.

BROWN, ROBERT HAMILTON. "Exploring the Uranian Satellites." *The Planetary Report* 6 (November/December 1986), p. 4.

CHAIKIN, ANDREW. "Voyager Among the Ice Worlds." *Sky and Telescope* 71 (April 1986), p. 338.

CHAPMAN, CLARK R. "Encounter! *Voyager 2* Explores the Uranian System." *The Planetary Report* 6 (March/April 1986), p. 8.

CROSSWELL, KEN. "Pluto: Enigma on the Edge of the Solar System." *Astronomy* 14 (July 1986), p. 6.

Cuzzi, Jeffrey N., and Larry W. Esposito. "The Rings of Uranus." *Scientific American* 257 (July 1987), p. 52.

Dowling, Timothy. "Big Blue: The Twin Worlds of Uranus and Neptune." *Astronomy* 18 (October 1990), p. 42.

Elliot, J. *The Ring Tape: A Voice Recording.* Cambridge, Mass.: MIT Press, 1984.

Elliot, J., and R. Kerr. *Rings: Discoveries from Galileo to Voyager.* Cambridge, Mass.: MIT Press, 1984.

Henbest, Nigel. "Uranus After Voyager." *New Scientist* 106 (31 July 1986), p. 42.

Hunt, Garry. "Voyager 2 Investigates the Atmosphere of Uranus." *The Planetary Report* 6 (November/December 1986), p. 14.

Ingersoll, Andrew P. "Uranus." *Scientific American* 256 (January 1987), p. 38.

Johnson, Torrence V., Robert Hamilton Brown, and Laurence A. Soderblom. "The Moons of Uranus." *Scientific American* 256 (April 1987), p. 48.

Kinoshita, June. "Neptune." *Scientific American* 261 (November 1989), p. 82.

Kohlhause, Charles E. "Aiming at Neptune." *Astronomy* 15 (November 1987), p. 6.

Lanzerotti, Louis J., and Stamatios M. Krimigis. "Comparative Magnetospheres." *Physics Today* 38 (November 1985), p. 24.

Lunine, J. I. "Origin and Evolution of Outer Solar System Atmospheres." *Science* 245 (14 July 1989), p. 141.

Miner, Ellis O. "Voyager's Last Encounter." *Sky and Telescope* 79 (July 1989), p. 26.

———. "Voyager 2's Encounter with the Gas Giants." *Physics Today* 43 (July 1990), p. 40.

Morrison, Nancy. "A Refined View of Miranda." *Mercury* 18 (March/April 1989), p. 55.

Morrison, N., and S. Gregory. "The Exotic Atmosphere of Titan." *Mercury* 14 (September/October 1985), p. 154.

Ness, Norman F. "The Magnetosphere of Uranus." *The Planetary Report* 6 (November/December 1986), p. 8.

O'Meara, Stephen James. "Saturn's Great White Spot Spectacular." *Sky and Telescope* 81 (February 1991), p. 144.

Osterbrock, D. "The Nature of Saturn's Rings: James Keeler and the Doppler Principle." *Mercury* 14 (March/April 1985), p. 46.

Owen, Tobias. "Titan." *Scientific American* 246 (February 1986), p. 98.

Porco, Carolyn. "Voyager 2 and the Uranian Rings." *The Planetary Report* 6 (November/December 1986), p. 11.

Sanchez-Lavega, Augustin. "Saturn's Great White Spots." *Sky and Telescope* 78 (August 1989), p. 141.

Squyres, Steven W. "Ganymede and Callisto." *American Scientist* 71 (January/February 1983), p. 56.

Stevenson, David. "An Ocean in Uranus." *The Planetary Report* 6 (November/December 1986), p. 11.

Tombaugh, Clyde W. "The Discovery of Pluto: Some Generally Unknown Aspects of the Story." *Mercury* 15 (May/June 1986), p. 66; Part II, *Mercury* 15 (July/August 1986), p. 98.

———. "Plates, Pluto, and Planets X." *Sky and Telescope* 81 (April 1991), p. 360.

FURTHER EXPLORATION

VOYAGER, the Interactive Desktop Planetarium, can be used for further exploration of the topics in this chapter. This chapter corresponds with the following project in *Voyages Through Space and Time:* Project 21, Planetary Rings.

Meteorites, Asteroids, and Comets

When they shall cry "PEACE, PEACE" then cometh sudden destruction! COMET'S CHAOS?— What terrible events will the Comet bring?

From a religious pamphlet predicting the end of the world because of the appearance of Comet Kohoutek, 1973

In the first months of 1910, Comet Halley passed through the inner solar system and was spectacular in the sky. On the night of May 19, Earth actually passed through the tail of the comet, and millions of people panicked. The spectrographic discovery, only a few years earlier, of cyanide gas in the tails of comets led many to believe that all life on Earth would end. Householders in Chicago stuffed rags around doors and windows to keep out the gas, and many bought supplies of bottled oxygen. Con artists in Texas sold comet pills and inhalers to ward off the noxious fumes. An Oklahoma newspaper reported (in what was apparently a hoax) that a religious sect called the Select Followers tried to sacrifice a virgin to the comet and were prevented by the last-minute arrival of the sheriff.

Throughout history, comets have been seen as portents of doom. Even in 1973 and 1974,

Comet Kohoutek (Figure 19-1) was hailed by one mystic as "Star-Seed," an interstellar spaceship bringing light and love to Earth. One group tried to contact the comet by ESP, while others claimed the comet announced the second coming of Christ.

In 1986, Comet Halley returned and, being fainter and farther south than in 1910, did not stimulate much public concern. Astronomers, however, studied the comet closely and confirmed the theory that a comet is a lump of dirty ices vaporizing in the glare of the inner solar system.

To an astronomer, a comet is a messenger from the age of planet building. By studying comets we can learn about the conditions in the solar nebula from which the planets formed. In addition to the planets, the solar nebula left behind lumps of ice, which we see as comets, and lumps of rock, which we see as asteroids. Because we cannot easily visit comets and as-

Figure 19-1 Comet Kohoutek swept through the inner solar system in 1973, causing a number of predictions of the end of the world. Shown here as seen through the slit of a telescope dome (red at lower right), the comet was dramatically beautiful but did not become as bright as first predicted. (Courtesy Stephen M. Larson)

teroids, we begin our discussion with the fragments of those bodies that fall into our atmosphere —the meteorites.

19-1 METEORITES

We discussed meteorites in Chapter 16 when we discussed the age of the solar system. There we saw that the solar system is filled with small particles called meteoroids, which can fall into Earth's atmosphere where friction heats them to glowing and we see them as meteors. If a meteoroid is big enough and strong enough, it can survive its plunge through the atmosphere and reach Earth's surface.

Such objects, known as meteorites, contain clues to the origin of the planets.

Inside Meteorites Meteorites can be divided into two broad categories: *Iron* meteorites are solid chunks of iron and nickel. *Stony* meteorites are silicate masses that resemble Earth rocks.

When iron meteorites are sliced open, polished, and etched with nitric acid, they reveal regular bands called **Widmanstätten patterns** (Figure 19-2). The patterns arise from crystals of nickel–iron alloys that have grown very large, indicating that the meteorite cooled no faster than a few degrees per million years. Explaining how iron meteorites could have cooled so slowly will be a major step in analyzing their history.

Figure 19-2 An iron meteorite sliced open, polished, and etched with acid reveals a Widmanstätten pattern. Such patterns are evidence that the iron cooled from a molten state no faster than a few degrees per million years. (Courtesy Jordan Marché, North Museum, Franklin and Marshall College)

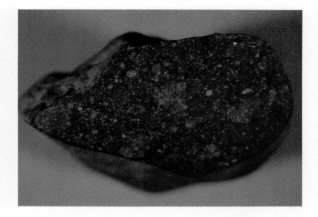

Figure 19-3 A stony meteorite sliced open and polished to show the small, spherical inclusions called chondrules. (Courtesy Jordan Marché, North Museum, Franklin and Marshall College)

In contrast to iron meteorites, most stony meteorites appear never to have been heated hot enough to melt. In fact, we can classify stony meteorites into three types, according to the degree to which they have been heated. **Chondrites** are stony meteorites that contain **chondrules,** rounded bits of glassy rock not much larger than a pea (Figure 19-3). The origin of chondrules is unknown, but they appear to be very old, and since melting would have destroyed them, their presence in chondrites indicates that these meteorites never melted.

Nevertheless, chondrites have been heated slightly. The rock in which the chondrules are imbedded is also old, but it contains no volatiles. The condensing solar nebula should have incorporated carbon compounds and water into the forming solids, but no such material is present in the chondrites. They have been heated enough to drive off these volatile compounds but not enough to melt the chondrules.

The **carbonaceous chondrites** contain both chondrules and volatile compounds. Had they been heated, even slightly, the volatiles would have evaporated. Thus the carbonaceous chondrites are the least altered remains of the solar nebula.

Stony meteorites of the third type are called **achondrites** because they contain no chondrules.

They also lack volatiles and appear to have been subjected to intense heat that melted chondrules and drove off volatiles, leaving behind rock with compositions similar to Earth's lavas.

The Origin of Meteors and Meteorites The mineral composition of the meteorites gives us a clue to the origins of these objects. Somehow they have survived almost unchanged since the age of planet building in the solar nebula.

Observations of meteors (as opposed to meteorites) give us further clues. Although we can see 10–15 meteors per hour on any dark night, we can sometimes observe **meteor showers** of up to 60 meteors per hour. (See Observational Activity: Observing Meteors at the end of this chapter.) These showers are known to occur when Earth passes near the orbit of a comet, and thus the meteors in meteor showers must be caused by dust and debris released by the icy head of the comet (Figure 19-4). The Infrared Astronomy Satellite telescope has detected the dusty orbits of a number of comets glowing in the far infrared because of the sun-warmed dust scattered along the orbits.

Studies of meteors show that most of the meteors we see on any given night (whether or not there is a shower) are produced by tiny bits of debris from comets. These are so small and so weak that they are vaporized completely in the atmosphere

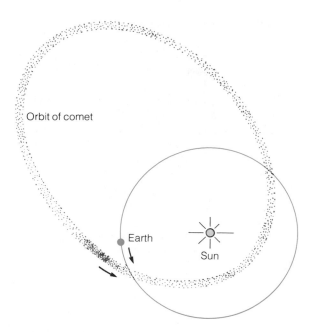

Orbit of comet

Earth

Sun

Figure 19-4 As a comet's ices evaporate, it releases rocky and metallic bits of material that spread along its orbit. If Earth passes through such material, it experiences a meteor shower.

and never reach the ground. Thus nearly all meteors come from comets.

The meteorites that reach Earth's surface are structurally much stronger than comet dust. They appear to have had a more complex history, and the different kinds of meteorites suggest that they are fragments of larger bodies that grew hot enough to melt and differentiate to form a molten iron core and a silicate crust (Figure 19-5).

If an object were 100 km in diameter, then the outer layers of rock would insulate the iron core. It would lose heat slowly, and the iron would cool so gradually that large crystals could grow. A sample of such material sliced, polished, and etched

Figure 19-5 Production of various kinds of meteoroids from asteroids. If it melted, an asteroid (a) could differentiate into an iron core surrounded by layers of different silicate compositions (b). Cratering (c), collisions (d), and fragmentation (e) could break these layers up and produce various kinds of meteorites. (Based on ''The Nature of Asteroids'' by C. R. Chapman. Copyright © 1975 by Scientific American, Inc. All rights reserved)

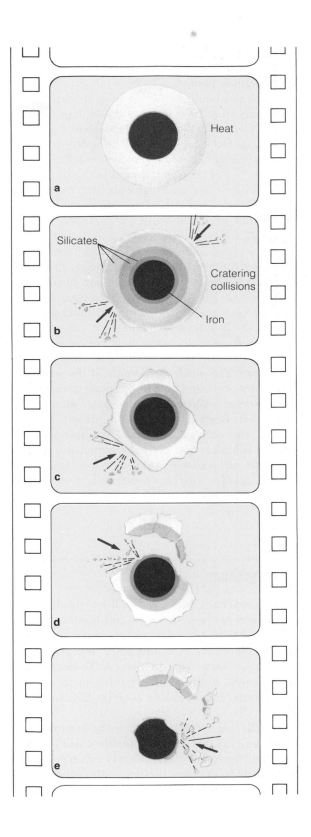

Heat

Silicates

Cratering collisions

Iron

a

b

c

d

e

would show Widmanstätten patterns. Thus the material in iron meteorites could have formed in the cores of such bodies.

Collisions could break up such an object, and fragments from the core would look like the iron meteorites. Fragments of the object's mantle, which was hot enough to melt chondrules, would make achondrites. Also, some achondrites appear to be pieces of lavas that flowed out onto the surfaces of larger bodies. Fragments from a crust that has never been hot enough to melt the chondrules might look very much like the meteorites called chondrites.

The carbonaceous chondrites may have formed in bodies that were smaller and cooler. Planetesimals farther from the sun would have been cooler and could have retained volatiles more easily. They could not have been large, or heating and differentiation would have altered them.

These theories trace the origin of meteorites to planetesimal-like parent bodies, but they produce a mystery. The small meteoroids in the solar system cannot be fragments of the planetesimals that formed the planets because small meteoroids would have been swept up by the planets in only a billion years or less. They could not have survived for 4.6 billion years. Thus the meteorites now in museums all over the world must have broken off planetesimals somewhere in our solar system within the last billion years. Where are these planetesimals?

19-2 ASTEROIDS

The Titius–Bode rule (see Box 16-1) predicts that a planet exists between Mars and Jupiter at a distance of 2.8 AU from the sun. Ancient astronomers saw no planet there, but in 1801, the first of the asteroids, Ceres, was discovered. Thousands are now known, and counting pebble-sized objects, the asteroid belt probably contains billions of objects.

In Chapter 16, we identified the asteroids as a characteristic that any acceptable theory of the origin of the solar system should explain. We are now ready to examine the properties of the asteroids and see how they could arise from the solar nebula.

Properties of Asteroids   Studies of the shapes of asteroids have been limited by their small size and distance. They cannot be resolved by conventional Earth-based telescopes, and thus we can see no surface details.

By measuring the amount of infrared radiation the asteroids emit, planetary astronomers have been able to calculate their sizes with some accuracy. Ceres, the largest, is about 30 percent the diameter of our moon, and Pallas, the next largest, is only 15 percent the diameter of the moon. Most asteroids are much smaller.

Because the brightness of the typical asteroid varies over periods of hours, astronomers conclude that most asteroids are not spherical. As their irregular shapes rotate, they reflect varying amounts of sunlight and their brightness varies. Presumably, most are small, cratered bodies with too little gravity to pull themselves into a spherical shape (Figure 19-6).

The irregular shape of asteroids was confirmed dramatically on October 29, 1990, when the Galileo spacecraft on its way to Jupiter passed only 16,000 km from the asteroid Gaspra. In a carefully choreographed series of moves, the spacecraft aimed its cameras and recorded the first photograph ever made of surface detail on an asteroid (Figure 19-6a). As expected, the asteroid proved to be irregularly shaped and cratered, but it was smoother than expected. In spite of its weak gravity, it seems to be covered by a layer of shattered rock soil about 1 m deep and about twice as reflective as the surface of the moon.

Infrared photometry of asteroids permits us to classify them (Figure 19-7). S-type asteroids, including Gaspra, are bright and have a reddish tint. They may be silicates mixed with metals, or they may resemble chondrites. C-type asteroids are very dark—about as bright as a lump of coal. They appear to be carbonaceous. M-type asteroids are bright, but they are not red. They appear to be mostly iron–nickel alloys. S types are common in the inner belt, and C types are very common in the outer belt. That distribution is a clue to the origin of the asteroids.

The Origin of the Asteroids   An old theory proposed that asteroids are the remains of a planet that exploded. Planet-shattering death rays may make for exciting science fiction movies, but in reality planets do not explode. The gravitational field of a planet holds the mass tightly, and disrupting the planet would take tremendous energy. Shattering Earth, for example, would take all of the energy generated by the sun over a period of two weeks. In addition, the total mass of the asteroids is only about one-fifth the mass of the moon, hardly enough to be the remains of a planet.

Most astronomers now believe that the asteroids are the remains of material that was unable to form a planet at 2.8 AU because of the disturbing gravitational influence of Jupiter, the next planet outward. If this is true, then the asteroids are the remains of planetesimals fragmented by collisions with one another. This would explain why the C-type asteroids, which appear to be carbonaceous, are more common in the outer asteroid belt. It is cooler there, and the condensation sequence (see Chapter 16) predicts that carbonaceous material would form there more easily than in the inner belt.

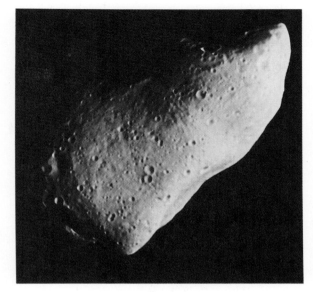

a

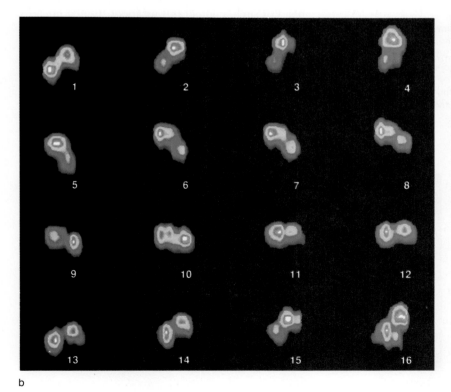

b

Figure 19-6 (a) Asteroid Gaspra as photographed from a distance of 16,000 km by the Galileo space craft. Gaspra is a cratered, irregular-shaped fragment about 19 × 12 × 11 km. In this color-enhanced image, color variations show differences in surface composition. (NASA/Galileo Imaging Team) (b) Asteroid 1989 PB was imaged 16 times by radar at 9-minute intervals. The images show that this asteroid is dumbbell-shaped and rotates in about 4 hours. (Courtesy Steven J. Ostro)

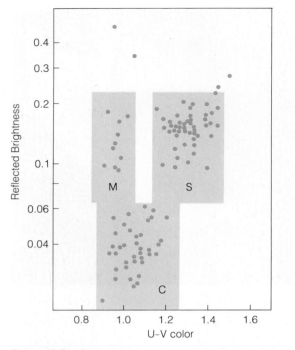

Figure 19-7 The three principal types of asteroids. In a diagram of reflected brightness versus ultraviolet minus visual color index, the most reflective asteroids lie near the top, and the reddest lie at the right. The S types, which are mixed silicates and metals or perhaps chondritic, are clearly redder than the M types, which seem to be metallic. The darkest asteroids, the C types, are believed to be similar to carbonaceous chondrites. (Diagram adapted from a figure by B. Zellner)

Now we can understand the compositions of the meteorites. They are fragments of planetesimals, the largest of which developed molten cores, differentiated, and cooled slowly. The asteroids are those planetesimals. But the solar nebula theory predicts that some planetesimals are not rock and metal but are made of ice. Can we find any of these icy planetesimals?

19-3 COMETS

Of all the fossils left behind by the solar nebula, comets are the most beautiful. Asteroids are dark, rocky worlds, and meteors are flitting specks of fire, but the comets move with the grace and beauty of a great ship at sea (Figure 19-8). Comet Halley, for instance, takes many weeks to move through the sky at each of its appearances in the inner solar system. Observations of the tail and head of such a comet tell us about the icy nucleus at its core and about the ancient solar nebula from which our planets formed.

Properties of Comets The head, or **coma,** of a comet is a vast cloud of gas and dust up to 100,000 km in diameter—seven times the diameter of Earth (Figure 19-9). The gas in the coma is made up of

Figure 19-8 In this visible-light photograph of Comet West, the tail is a transparent haze of gas and dust through which distant stars are visible. Note the streamers in the tail. (Celestron International)

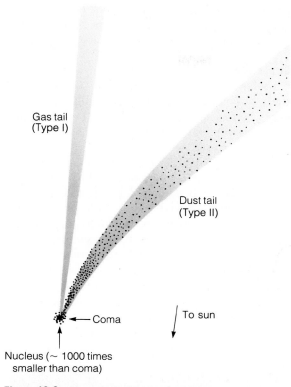

Figure 19-9 The principal parts of a comet.

Gas tail
(Type I)

Dust tail
(Type II)

Coma

To sun

Nucleus (~ 1000 times smaller than coma)

Figure 19-10 Comet Halley displayed both a gas tail and a dust tail. The gas tail is straight, blue, and brighter than the yellowish dust tail curving to the right. This color image was made by combining red, yellow, and blue plates taken at slightly different times. Thus stars are triple exposed in the three colors and shifted slightly due to the comet's motion. (Royal Observatory, Edinburgh)

H_2O, CO_2, CO, H, OH, O, S, C, and so on. Mixed with the gases of the coma are tiny specks of dust, most smaller than the particles in cigarette smoke. The dirty snowball theory, described in Chapter 16, proposes that all of the features of the coma are created by a central lump of dirty ices only a dozen kilometers in diameter. As the ices vaporize, dust is released and the molecules of the gas are broken into the atoms and ions we see in the coma.

The tail of a comet springs from the coma and typically extends 10^7–10^8 km through space, with the longest reaching 1 AU (1.5×10^8 km). Seen from Earth, the tail of a large comet can span 30° across the sky.

A comet can have two kinds of tails (Figure 19-10). A type I tail, also called a *gas tail*, is straight and wispy, and looks like a narrow trail of smoke. Its spectrum contains emission lines of ionized gases excited by the ultraviolet radiation from the sun and blown outward by the moving gas and the embedded magnetic field in the solar wind.

In the summer of 1985, NASA sent the International Cometary Explorer (ICE) plunging through the tail of Comet Giacobini–Zinner 8000 km behind the nucleus. Although it carried no cameras, the ICE spacecraft measured charged particles and magnetic fields in the solar wind and in the comet's tail, and the results confirmed that the gas tail is ionized gas trapped in a magnetic field. Evidently, the magnetic field in the solar wind becomes draped over the comet's nucleus like strands of seaweed draped over a fishhook. The ICE probe even detected the predicted change in the direction of the magnetic field as it crossed the central region of the tail. Thus we can be confident that the gas tail is controlled by magnetic forces related to the solar wind.

The type II tail, also known as a *dust tail,* is immune to magnetic fields. Its spectrum is identical to sunlight's, which means it is made up of solid bits of dust that reflect sunlight. The dust is evidently the dirt in the dirty snowball released as the ices vaporize. The Infrared Astronomy Satellite found that this dust fills the solar system, and it can be collected by spacecraft and high-flying aircraft.

The dust in the dust tail is not affected by magnetic fields but is pushed outward by the pressure of sunlight. It thus forms a long, curving tail with a much more uniform appearance than a gas tail. The dust tail curves because the comet moves along its orbit with a velocity that is comparable to the velocity of the dust. Like the stream of water from the moving nozzle of a lawn sprinkler, the dust tail appears curved. It lacks the wisps and twists of the gas tail because it is not affected by kinks and shifts in the magnetic field of the solar wind. Thus the gas tail and the dust tail both point roughly away from the sun, but for slightly different reasons.

Comet Geology Comet Halley returned to the inner solar system in the winter of 1985–86, and the study of comets changed dramatically. (See Data File 12.) Five spacecraft flew past the comet, and the measurements and photographs radioed back to Earth gave astronomers their first look at the nucleus of a comet. For the first time, astronomers could discuss the geology of cometary nuclei.

Two spacecraft launched by Japan carried no cameras, but the Soviet Vega 1 and Vega 2 and the European Giotto spacecraft photographed the nucleus and revealed that it was irregularly shaped ($16 \times 8 \times 7$ km) and so black that it reflects only 4 percent of the light that strikes it (Figure 19-11). A lump of coal reflects about 6 percent. Evidently, the icy core is coated with a layer of dark dust, and ices vaporize from only small active sites on the surface.

Figure 19-11 Sixty separate images from the Giotto spacecraft have been computer-enhanced and merged to form this image of the icy nucleus of Comet Halley. The surface is very dark and may resemble carbonaceous chondrites. Note the jets of gas and dust venting from the sunward side. (Photo courtesy Harold Reitsema, Ball Aerospace; copyright © 1986 Max Planck Institute)

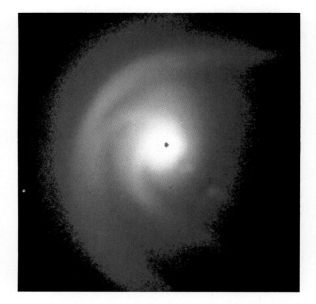

Figure 19-12 Jets of gas and dust, curved by the rotation of the nucleus, are sometimes visible in the coma of comets, as in this computer-enhanced image of Comet Halley from March 14, 1986. The jets issue from the nucleus and are bent back into the tail. The black dot at the center of the coma is caused by the computer enhancement and is not the nucleus. The diameter of the field of view here is about 7.3 Earth diameters. (Courtesy Steven Larson)

The density of the nucleus is only 0.1–0.25 gm/cm^3, which is much lower than the density of ice, about 1 gm/cm^3. Evidently, the nucleus is a fluffy mixture of frost and dust and not solid ices.

Astronomers suspect that the dark material on the surface of the nucleus resembles the meteorites called carbonaceous chondrites. These meteorites are dark in color because they are rich in carbon. Unfortunately, we cannot yet determine the chemical composition of the dark crust; however, future space missions to comets will be able to return samples to Earth.

Observations made over the past century have hinted at the peculiar geology of comet nuclei. The coma of a comet often contains what appear to be jets pointing roughly toward the sun and bending back to form the tail. As early as the 1835 appearance of Comet Halley, observers sketched such jets, and modern computer enhancements of comet photographs reveal jets and streamers extending back into the tails of the comets (Figure 19-12).

The spacecraft images of the nucleus of Comet Halley can be computer-enhanced to show jets of gas and dust venting from the sunlit side (Figure 19-13). The crust of the nucleus appears to insulate the ices below from the sun's heat, except in certain active regions where the ices vaporize. The crust may be thin or broken by faults in these areas. Jets issue from an active region so long as it remains in sunlight, but as the rotation of the nucleus carries an active region into darkness, the region stops venting gas and dust.

Studies of these jets show cyclical changes, with periods of 52.5 hours and 7.4 days. Apparently, the oddly shaped nucleus rotates with one period while precessing with the other like a badly thrown football.

The wasting of the icy surface apparently gives comet nuclei irregular shapes. Giotto images show traces of surface features probably caused by venting of subsurface ices and the collapse of the dusty crust (see Figure 19-11). A typical comet probably shrinks in radius by about 1 m on each orbit

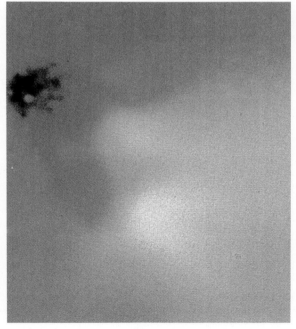

a

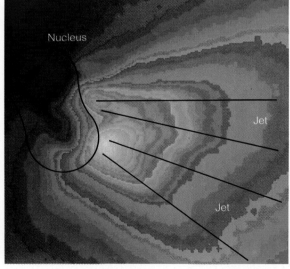

b

Figure 19-13 (a) The Giotto spacecraft recorded the image of Comet Halley's nucleus from a distance of only 600 km. The nucleus is visible as a dark body at upper left. (b) Computer-enhanced, the Giotto images reveal jets of gas and dust venting from the sunlit side of the nucleus. (© Max Planck Institute, courtesy H. U. Keller)

Comet Halley

Average distance from the sun	17.8 AU (2.66 × 10⁹ km)
Eccentricity of orbit	0.97
Maximum distance from the sun	35.1 AU (5.24 × 10⁹ km)
Minimum distance from the sun	0.53 AU (0.08 × 10⁹ km)
Inclination of orbit to ecliptic	162° (retrograde)
Orbital velocity maximum	56 km/sec
minimum	0.85 km/sec
Orbital period	76.1 y (27,800 days)
Period of rotation	52.5 hours (uncertain)
Size	16 km × 8 km × 7 km
Mass	5× 10¹³ to 2 × 10¹⁴ kg (estimated)
Average density	0.1 to 0.25 g/cm³ (estimated)
Surface gravity	0.0004 times Earth's
Escape velocity	6.3 m/sec (0.0006 $V_{\oplus}$)
Surface temperature	300 to 400 K (27° to 127°C or 80° to 160°F)

a. In reality, comets are pale white objects, but digital enhancement of this photo of Comet Halley produces a false-color map of intensity. Such images permit measurement of subtle brightness differences. (© 1986 Royal Observatory, Edinburgh)

b. Countries around the world marked the return of Comet Halley on their stamps.

c. Looking much as it did to the naked eye, Comet Halley passes in front of the Milky Way. The clouds of gas, dust, and stars in the Milky Way are a thousand or more light-years distant. The comet is only a few light-minutes from Earth. (National Optical Astronomy Observatories)

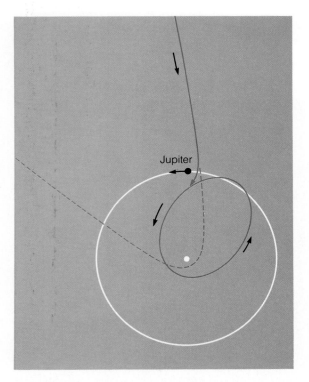

Figure 19-14 Massive Jupiter can alter the orbit of a comet (dashed) into a smaller orbit (ellipse) that keeps it in the inner solar system.

around the sun. Comets trapped in the inner solar system through encounters with Jupiter or other planets (Figure 19-14) rapidly lose their ices and cannot produce spectacular tails. Comet Halley will ultimately face such a fate. Its orbit has been altered by encounters with the planets, and it now stays within the solar system, never going farther than the orbit of Pluto. It will next appear in A.D. 2062.

The Origin of Comets The orbits of comets contain a clue to their origin. Of the approximately 600 comets with known orbits, about 100, including Comet Halley, stay in the solar system (Figure 19-15). Their periods are less than 200 years and their orbits lie within 30 degrees of the plane of the solar system, apparently because of encounters with the giant planets. Most comets, however, have very long, elliptical orbits that are randomly oriented in space.

The icy nuclei of the comets and their long,

randomly tipped orbits hint at a theory for their origin. According to the **Oort cloud** theory (named after the Dutch astronomer Jan Oort), the icy bodies of comets orbit the sun in a cloud extending from 10,000 AU to 100,000 AU (about 35 percent of the way to the nearest star) (Figure 19-16). There may be 200 billion objects in the cloud. At this distance the ices remain frozen, and the comets lack comas or tails. Their orbital velocities are only about 0.13 km/sec (about 300 mph), and the slight perturbations caused by the motions of nearby stars could eject a few of these icebergs into long elliptical orbits that carry them into the inner solar system. Thus the Oort cloud theory accounts for the random orientation of cometary orbits— they fall into the solar system from all directions.

The existence of the Oort cloud cannot be confirmed from Earth. The icy bodies, most of them less than 15 km in diameter, are totally undetectable beyond a dozen or so AU.

Many astronomers believe that comets now in the Oort cloud originated in the solar nebula among the Jovian planets, where it was cold enough for ices to solidify. The original planetesimals there must have been rich in ice. As the Jovian planets grew more massive, they could have swept up many of these planetesimals and ejected others from the solar system to form the Oort cloud.

This traditional theory was challenged in 1988 when a team of astronomers announced the results of computer simulations of cometary motion. Their computer used several months of computer time and simulated the motion of 5000 comets interacting with the four massive planets. The results indicated that the planets could not capture comets coming from the Oort cloud into short-period orbits in the plane of the solar system. Instead, the team suggested, the short-period comets come from icy debris in the plane of the solar system beyond the orbit of Neptune. If astronomers can detect these bodies with present telescopes or the Hubble Space Telescope, the new hypothesis will be confirmed, and our understanding of the origin of comets will be dramatically revised.

Whatever the source of the short-period comets, the evidence suggests that all comets are old, undisturbed objects. If they formed with the Jovian planets, they are as old as the solar system. They appear to be small enough not to have grown hot

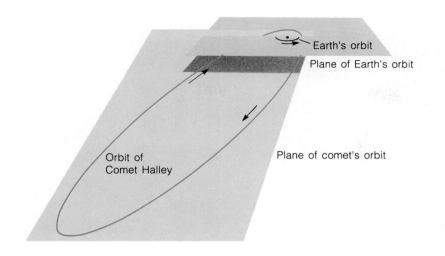

Figure 19-15 Comet Halley is a short-period comet whose orbit has been altered by close encounters with the giant planets. It never travels farther from the sun than the average distance to Pluto. Comets whose orbits are unaltered, the long-period comets, have much longer orbits randomly oriented to the plane of Earth's orbit.

from radioactive decay, and so should not have been seriously modified by differentiation and volcanism, as is the case with the planets. Thus the comets are undisturbed samples of the icy planetesimals from the ancient solar nebula.

It seems likely that other stars are surrounded by Oort clouds. Observations made by ground-based telescopes and the Infrared Astronomy Satellite are revealing that other young stars are surrounded by clouds or disks of cold matter. In some cases, this may be silicate dust, but in other cases, such as Vega, the material is cold enough to be frozen ices comparable to our own Oort cloud. In any case, such clouds of cold matter seem to be common, and thus planets should also be common.

Dinosaurs and Comets Occasionally, Earth must hit the head of a comet, and such a large impact could have devastating effects (Figure 19-17). Recent studies of sediments laid down 65 million years ago at the time of the extinction of the dinosaurs have found abundances of chemical elements and mineral forms that could have been produced by the impact of a very large meteorite or cometary nucleus. Such an impact would throw so much dust into the atmosphere that Earth would be plunged into a long winter during which many food chains could be disrupted. Thus the dinosaurs and over 75 percent of the species then on the Earth became extinct perhaps because of a collision with a meteorite or comet.

Very recent studies suggest that these mass ex-

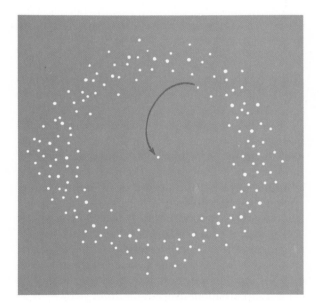

Figure 19-16 The Oort cloud of icy bodies is believed to extend out to about 100,000 AU from the sun. Collisions among the bodies or the effects of the motions of nearby stars could throw bodies into the inner solar system, where they become comets.

tinctions occur about every 26 million years, and some astronomers have suggested that they are caused by showers of comets. The Oort cloud contains a vast supply of comets, and a periodic disturbance could send a billion of them plunging through the inner solar system over a period as

Figure 19-17 The Barringer meteorite crater in Arizona was formed about 50,000 years ago by the impact of an iron meteorite no bigger than a large house. The impact released energy equivalent to a 3-megaton hydrogen bomb and dug a crater 1.2 km in diameter. Note the human figures near center for scale. The impact of the head of a comet 12 km in diameter could throw enough dust into the atmosphere to alter Earth's climate and trigger the extinction of many species. (Michael A. Seeds)

short as a million years. A few of these comets would hit Earth, alter the climate, and cause mass extinction of species.

Two theories have been suggested to explain the periodic triggering of this deadly comet shower. One team of astronomers has suggested that the sun has a faint, undiscovered companion in a long elliptical orbit with a period of 26 million years. When this star, dubbed Nemesis (after the Greek goddess of retribution and vengeance), comes close to the sun, it could disturb the comets in the Oort cloud. Another theory suggests that the shower of comets is triggered by the passage of the solar system up and down through the plane of the galaxy. This happens about every 30 million years—close enough given the accuracy with which the extinctions can be dated.

The theory that the extinction of the dinosaurs was caused by a major impact has been widely accepted. The theory that such extinctions occur every 26 million years is less secure, and the suggested triggers of comet showers are, at present, hardly more than exciting speculation. But they serve to illustrate how all life on Earth is linked to the stars.

Could life exist on planets circling other stars? That is the final question in our exploration of the universe. We will try to answer it in the next unit.

SUMMARY

The solar nebula theory proposes that the solar system formed from a disk of gas and dust around the forming sun, and that most of that nebula was blown away when the sun became luminous. The asteroids and comets appear to be rocky and icy debris left behind.

Meteorites can be classified according to their mineral content. Iron meteorites are mostly iron and nickel, and when sliced open, polished, and etched with acid, they show Widmanstätten patterns. These suggest the metal cooled from a molten state very slowly.

Stony meteorites can be classified as chondrites if they contain small, glassy particles called chondrules. If such a meteorite is rich in volatiles and carbon, it is called a carbonaceous chondrite. An achondrite is a stony meteorite that contains no chondrules. Achondrites appear to have been melted after they formed, but chondrites were never melted. Carbonaceous chondrites were never even warm enough to drive off volatiles.

This evidence suggests that the different kinds of meteorites were formed in bodies roughly 100 km in diameter. Such objects could have melted, differentiated, and cooled very slowly, and when they were later broken by collisions, their fragments would resemble the different kinds of meteorites. Also, we may suppose that the carbonaceous chondrites were formed farther from the sun, where it was cooler. This evidence suggests that meteorites are fragments of the asteroids.

Most of the meteors we see, however, are small, weak objects that never reach the ground. These bits of matter appear to be debris from comets, and we see meteor showers when Earth crosses the debris-laden orbit of a comet.

Most asteroids lie in the asteroid belt between Mars and Jupiter. Infrared studies show that asteroids in the outer belt are darker and redder than others. These may resemble the volatile-rich carbonaceous chondrites. Apparently, the asteroids are the remains of planetesimals that were unable to form a planet between Mars and Jupiter. The strong gravitational influence of Jupiter could have prevented the material from accumulating into a large body.

A comet is produced by a lump of dirty ices only a few dozen kilometers in diameter. In a long, elliptical orbit, the icy body stays frozen until the object draws close to the sun. Then the ices vaporize and release the imbedded dust and debris. The gas is caught in the solar wind and blown outward to form a type I, or gas, tail. The pressure of sunlight blows the dust away to form a type II, or dust, tail. The coma of a comet can be up to 100,000 km in diameter, and it contains jets issuing from the nucleus.

When spacecraft flew past Comet Halley in 1986, astronomers discovered that the nucleus was coated by a dark crust and that jets of vapor and dust were venting from active regions on the sunlit side. The low density of the nucleus showed that it was a fluffy mixture of ices and silicate dust.

Comets may have formed as icy planetesimals in the outer solar nebula and were later ejected by the Jovian planets to form the Oort cloud. The icy bodies in the Oort cloud remain frozen in orbits roughly 100,000 AU from the sun, but they can be disturbed to fall into the inner solar system, where we see them as comets.

NEW TERMS

Widmanstätten pattern

chondrite

chondrule

carbonaceous chondrite

achondrite

meteor shower

coma

Oort cloud

REVIEW QUESTIONS

1. What do Widmanstätten patterns tell us about the history of iron meteorites?

2. Why are there no chondrules in achondritic meteorites?

3. Why do astronomers refer to carbonaceous chondrites as "unmodified"?

4. How do observations of meteor showers reveal one of the sources of meteoroids?

5. How can most meteors be cometary if most, perhaps all, meteorites are asteroidal?

6. Why do we think that the asteroids were never part of a planet?

7. How is the composition of meteorites related to the formation and evolution of asteroids?

8. What is the difference between a type I tail and a type II tail?

9. Why do short-period comets tend to have orbits near the plane of the solar system?

10. If comets are icy planetesimals left over from the formation of the solar system, why haven't they all vaporized by now?

DISCUSSION QUESTIONS

1. Futurists suggest we may someday mine the asteroids for materials to build space colonies. What kinds of materials could we get from asteroids? (HINT: What are the S, M, and C asteroids made of?)

2. If cometary nuclei were heated by internal radioactive decay rather than by solar heat, how would comets differ from what we observe?

3. If infrared searches showed that no nearby stars had Oort clouds, would the solar nebula theory be strengthened or weakened?

PROBLEMS

1. Large meteorites are hardly slowed by Earth's atmosphere. Assuming the atmosphere is 100 km thick and that a large meteorite falls perpendicular to the surface, how long does it take to reach the ground? (**HINT:** About how fast do meteoroids travel?)

2. What is the orbital velocity of a meteoroid whose average distance from the sun is 2 AU? (**HINT:** Find the orbital period from Kepler's third law. See Table 4-1.)

3. If a single asteroid 1 km in diameter were fragmented into meteoroids 1 m in diameter, how many would it yield? (**HINT:** Volume of sphere = $\frac{4}{3}\pi r^3$.)

4. What is the orbital period of a typical asteroid? (**HINT:** Use Kepler's third law. See Table 4-1.)

5. If half a million asteroids each 1 km in diameter were assembled into one body, how large would it be? (**HINT:** Volume of a sphere = $\frac{4}{3}\pi r^3$.)

6. What is the maximum angular diameter of Ceres as seen from Earth? Could Earth-based telescopes detect surface features? Could the Hubble Space Telescope once it is repaired (see Chapter 5)? (**HINT:** See Box 3-2.)

7. If the velocity of the solar wind is about 400 km/sec and the visible tail of a comet is 10^8 km long, how long does it take an atom to travel from the nucleus to the end of the visible tail?

8. If you saw Comet Halley when it was 0.7 AU from Earth and it had a visible tail 5° long, how long was the tail in kilometers? Suppose that the tail was not perpendicular to your line of sight. Is your answer too large or too small? (**HINT:** See Box 3-2.)

9. What is the orbital period of a cometary nucleus in the Oort cloud? What is its orbital velocity? (**HINTS:** Use Kepler's third law. The circumference of a circular orbit = $2\pi r$.)

10. The mass of an average comet's nucleus is about 10^{12} kg. If the Oort cloud contains 200×10^9 cometary nuclei, what is the mass of the cloud in Earth masses? (**HINT:** Mass of Earth = 6×10^{24} kg.)

OBSERVATIONAL ACTIVITY: OBSERVING METEORS

Meteors, or shooting stars, are very common, and you may see one by accident any night. However, with a little planning, you can make systematic observations of meteors that will reveal something about their origins.

Observations Select a cloudless night when the moon will not be out and try to find a location away from bright lights. You can see meteors from the center of a city, but you will see more out in the country, away from city lights. Equip yourself with a reclining lawn chair, flashlight, clipboard, star charts, pencil, and refreshments.

Relax and watch the sky. You don't need a telescope or binoculars because you want to watch a large area of the sky continuously. When you see a meteor, note its position among the constellations and then sketch its path on your star charts.

In an hour or two, you may begin to see a pattern appearing on your star chart. Most of the meteors may seem to be radiating from a specific region of the sky. These meteors are part of a meteor shower, and the region from which they seem to come is the radiant of the shower. Meteors that do not seem to radiate from the radiant are not members of the shower.

Planning Ahead On any night you can see from 5 to 15 meteors an hour from a dark site, but these will not necessarily be part of a meteor shower. If you want to see more meteors and you want to be sure to find them coming from a radiant point, use the table of meteor showers in Appendix D to select a night when a shower is in progress.

Observing meteors is one of the most enjoyable activities in astronomy. Don't delay observing just because there is not a shower tonight. There are always meteors to be seen.

RECOMMENDED READING

BEATTY, J. KELLY. "An Inside Look at Halley's Comet." *Sky and Telescope* 71 (May 1986), p. 438.

———. "The High Road to Halley." *Sky and Telescope* 71 (March 1986), p. 244.

BERRY, RICHARD, and RICHARD TALCOT. "What Have We Learned from Comet Halley?" *Astronomy* 14 (September 1986), p. 6.

———. "Search for the Primitive." *Astronomy* 15 (June 1987), p. 6.

BINZEL, RICHARD P. "The Origins of the Asteroids." *Scientific American* 265 (October 1991), p. 88.

———. "Asteroids and Comets in Near-Earth Space." *The Planetary Report* 11 (November/December 1991), p. 8.

BRANDT, JOHN C., and MALCOME B. NEIDNER, JR. "The Structure of Comet Tails." *Scientific American* 254 (January 1986), p. 49.

CHAPMAN, CLARK, and DAVID MORRISON. "Cosmic Catastrophes." *Mercury* 18 (November/December 1989), p. 185.

DAVIES, JOHN. "Can Comets Become Asteroids?" *Astronomy* 13 (January 1985), p. 66.

———. "Is 3200 Phaethon a Dead Comet?" *Sky and Telescope* 70 (October 1985), p. 317.

DELSEMME, ARMAND H. "Whence Come Comets?" *Sky and Telescope* 77 (March 1989), p. 260.

DIETZ, ROBERT S. "The Demise of the Dinosaurs—A Mystery Solved?" *Astronomy* 19 (July 1991), p. 30.

DUNHAM, EDWARD. "Measuring the Diameter of Juno." *Sky and Telescope* 59 (April 1980), p. 276.

GEHRELS, TOM. "Asteroids and Comets." *Physics Today* 38 (February 1985), p. 11.

GRIEVE, RICHARD A. F. "Impact Cratering on the Earth." *Scientific American* 262 (April 1990), p. 66.

HARTMANN, WILLIAM L. "Vesta: A World of Its Own." *Astronomy* 11 (February 1983), p. 6.

———. "The Changing Face of Chiron." *Astronomy* 18 (August 1990), p. 44.

KERR, RICHARD. "Comet Source: Close to Neptune." *Science* 239 (18 March 1988), p. 1372.

KNACKE, ROGER. "Sampling the Stuff of a Comet." *Sky and Telescope* 73 (March 1987), p. 246.

LARSON, STEPHEN, and DAVID H. LEVY. "Observing Comet Halley's Near-Nucleus Features." *Astronomy* 15 (September 1987), p. 90.

MARAN, STEPHEN P. "The Comet That Wouldn't Die." *Natural History* 94 (January 1985), p. 84.

———. "Gaps in the Asteroid Belt." *Natural History* 95 (August 1986), p. 62.

———. "Off the Main Drag." *Natural History* 95 (June 1986), p. 70.

MENDIS, D. ASOKA. "The Science of Comets: A Post-Encounter Assessment." *The Planetary Report* 7 (March/April 1987), p. 5.

MORRISON, DAVID, and CLARK R. CHAPMAN. "Target Earth: It *Will* Happen." *Sky and Telescope* 79 (March 1990), p. 261.

ROMER, ALFRED. "Halley's Comet." *The Physics Teacher* 22 (November 1984), p. 488.

SAGAN, CARL, and ANN DRUYAN. *Comet*. New York: Random House, 1985.

SAGDEEV, ROALD Z., and ALBERT A. GALEEV. "Comet Halley and the Solar Wind." *Sky and Telescope* 73 (March 1987), p. 252.

TREFIL, JAMES. "Stop to Consider the Stones That Fall from the Sky." *Smithsonian* 20 (September 1989), p. 81.

WAGNER, JEFFREY K. "The Sources of Meteorites." *Astronomy* 12 (February 1984), p. 6.

WASSON, JOHN T. *Meteorites: Their Record of Early-Solar-System History*. New York: W. H. Freeman, 1985.

WHIPPLE, FRED L. "Flying Sandbanks Versus Dirty Snowballs: Discovery of the Nature of Comets." *Mercury* 15 (January/February 1986), p. 2.

———. "The Black Heart of Comet Halley." *Sky and Telescope* 73 (March 1987), p. 242.

YEOMANS, DONALD K. "Comets and the Perversity of Nature." *Sky and Telescope* 78 (September 1989), p. 253.

———. "Killer Rocks and the Celestial Police." *The Planetary Report* 11 (November/December 1991), p. 4.

Life

C H A P T E R

20

Life on Other Worlds

Did I solicit thee from darkness to promote me?

John Milton,
Paradise Lost

As living things, we have been promoted from darkness. We are made of heavy atoms that could not have formed at the beginning of the universe. Successive generations of stars fusing light elements into heavier elements have built the atoms so important to our existence. When a dark cloud of interstellar gas enriched in these heavy atoms fell together to form our sun, a small part of the cloud gave birth to the planet we inhabit.

Are there intelligent beings living on other planets? That is the last and perhaps the most challenging question in our study of astronomy. We will try to answer it in three steps, each dealing with a different aspect of life.

First, we must decide what we mean by *life*. Life is not so much a form of matter as it is a behavior. Living matter extracts energy from its environment to modify itself and its surroundings so as to preserve itself and to create off-

spring. Thus life is based on information, the recipe for the process of survival and reproduction.

Second, we must study the origin of life on Earth. If we can understand how life began on our planet, then we can try to estimate the likelihood that it has originated on other worlds as well.

Third, we must try to understand how the first primitive living organism on Earth could give rise to the tremendously diverse life forms that now inhabit our world. By understanding how living things evolve to fit their environment, we will see evidence that intelligence is a natural development in the evolution of life.

If life can originate on other worlds, and if intelligence is a natural result of the evolution of life forms, then we might expect that other intelligent races inhabit other worlds. Where might we look for such races? Our background in astronomy will help us select those stars most likely to be orbited by inhabited planets, and we will discuss the efforts that earthlings are now making to detect intelligent life on other worlds.

Alien life forms could be quite different from us, but if they are alive, then they must share with us certain characteristics. Thus the first step in our study is to try to identify the fundamental nature of life.

20-1 THE NATURE OF LIFE

What is life? Philosophers have struggled with that question for thousands of years, so it is unlikely that we will answer it here. But we must agree on a working model of life before we can speculate on its occurrence on other worlds. To that end, we will identify in living things two important aspects: a physical basis and a unit of controlling information.

The Physical Basis of Life On Earth, the physical basis of life is the carbon atom (Figure 20-1). Because of the way this atom bonds to other atoms, it can form long, complex, stable chains that are capable of extracting, storing, and utilizing energy. Other chemical bases of life may exist. Science fiction stories and movies abound with silicon creatures, living things whose body chemistry is based on silicon rather than carbon. However, silicon forms weaker bonds than carbon does, and it cannot form double bonds as easily. Consequently, it cannot form the long, complex, stable chains that carbon can. Silicon is 135 times more common on Earth than carbon is, yet there are no silicon creatures among us. All Earth life is carbon-based. Thus the likelihood that distant planets are inhabited by silicon people seems small, but we cannot rule out life based on noncarbon chemistry.

In fact, nonchemical life might be possible. All that nature requires is some mechanism capable of supporting the extraction and utilization of energy that we have identified as life. One could at least imagine life based on electromagnetic fields and

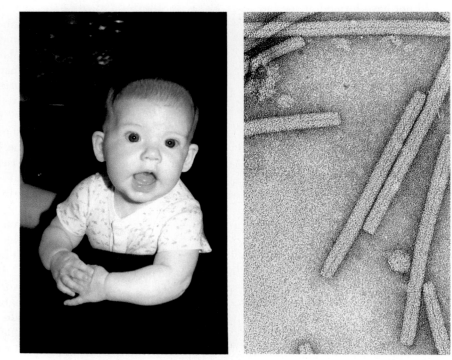

Figure 20-1 All living things on Earth are based on carbon chemistry. Even the long molecules that carry genetic information, DNA and RNA, have a framework defined by chains of carbon atoms. (a) Katie, a complex mammal contains about 30 astronomical units of DNA. (Michael Seeds) (b) Each rod of the tobacco mosaic virus contains a single spiral strand of RNA about 0.01 mm long. (L. D. Simon) All life on Earth stores its genetic information in such carbon-chain molecules.

a   b

ionized gas. No one has ever met such a creature, but science fiction writers conjure up all sorts.

Clearly, we could range far in space and time, theorizing about different bases for alien life, but to make progress we must discuss what we know best—carbon-based life on Earth. How can a lump of carbon-rich matter live? The answer lies in the information that guides its life processes.

Information Storage and Duplication The key to understanding life is information—the information the organism uses to control its utilization of energy. We must discover how life stores and uses that information and how the information changes and thus preserves the species.

The unit of life on Earth is the cell (Figure 20-2), the self-contained factory capable of absorbing nourishment from its surroundings, maintaining its own existence, and performing its task within the larger organism. The foundation of the cell's activity is a set of patterns that describe how it is to function. This information must be stored in the cell in some safe location, yet it must be passed on easily to new cells and be used readily to guide the cell's activity. To understand how matter can be alive, we must have an understanding of how the cell stores, reproduces, and uses this information.

The information is stored in long carbon-chain molecules called **DNA (deoxyribonucleic acid)**, most of which reside in the cell nucleus. The structure of DNA resembles a long, twisted ladder. The rails of the ladder are made of alternating phosphates and sugars; the rungs are made of pairs of molecules called bases (Figure 20-3). Only four kinds of bases are present in DNA, and the order in which they appear on the DNA ladder represents the information the cell needs to function. One human cell stores about 1.5 m of DNA, containing about 4.5 billion pairs of bases. Thus 4.5 billion pieces of information are available to run a human cell. That is enough to record all of the works of Shakespeare over 200 times. Because the human body contains about 60×10^{12} cells, the total **DNA** in a single human adult would stretch 9×10^{13} m, about 600 AU.

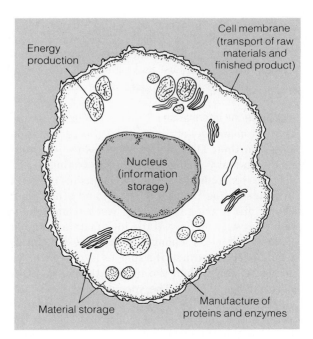

Figure 20-2 A living cell is a self-contained factory that absorbs raw materials from its surroundings and uses them to maintain itself and manufacture finished products for the use of the organism as a whole.

Energy production

Cell membrane (transport of raw materials and finished product)

Nucleus (information storage)

Material storage

Manufacture of proteins and enzymes

Figure 20-3 The DNA molecule consists of two rails of sugars and phosphates (dark) and rungs made of bases: adenine (A), cytosine (C), guanine (G), thymine (T).

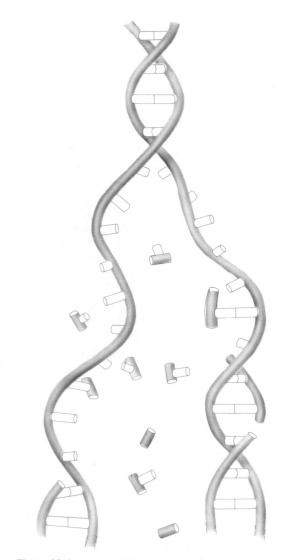

Figure 20-4 The DNA molecule can duplicate itself by splitting in half (top), assembling matching bases, sugars, and phosphates (center), and thus producing two DNA molecules (bottom). The actual duplication process is significantly more complex than in this schematic diagram.

Storing all this data in each cell does the organism no good unless the data can be reproduced and passed on to new cells. The **DNA** molecule is specially adapted for duplicating itself by splitting its ladder down the center of the rungs, producing two rails with protruding bases (Figure 20-4). These quickly bond with the proper bases, phosphates, and sugars to reconstruct the missing part of the molecule, and presto—the cell has two complete copies of the critical information. One set goes to each of the newly forming cells. Thus the **DNA** is the genetic information passed from parent to offspring.

Segments of the **DNA** molecules are patterns for the production of **proteins.** Many proteins are structural molecules—the cell might make protein to repair its cell wall, for example. **Enzymes** are special proteins that control other processes—growth, for example. Thus the **DNA** molecule contains the recipes to make all of the different molecules required in an organism.

Actually, the cell does not risk its precious **DNA** patterns by involving them directly in the manufacture of protein. The **DNA** stays safely in the cell nucleus, where it produces a copy of the patterns by assembling a long carbon-chain molecule called **RNA (ribonucleic acid).** This RNA carries the information out of the nucleus and then assembles the proteins from simple molecules called **amino acids,** the basic building blocks of protein. Thus the RNA acts as a messenger, carrying copies of the necessary plans from the central office to the construction site.

Though the information coded on the **DNA** must be preserved for the survival of the organism, it must be changeable or the species will become extinct. To see why, we must study evolution, the process that rewrites the data in the DNA.

Modifying the Information   If living things are to survive for many generations, then the information stored in their **DNA** must change as the environment gradually changes. A slight warming of the climate, for example, might kill a species of plant, in turn starving the rabbits, deer, and other plant eaters, and leaving the hawks, wolves, and mountain lions with no prey. If the information stored in **DNA** could never change, then these life forms would become extinct. If a species is to survive in a changing world, then the information in its **DNA** must change as well. The species must evolve.

Species evolve by **natural selection.** Each time an organism reproduces, its offspring receive the data stored in the **DNA,** but some variation is possible. For example, most of the rabbits in a litter may be normal, but it is possible for one to get a **DNA** recipe that gives it stronger teeth. If it has stronger teeth, it may be able to eat something other than the plant the others depend on, and if that plant is becoming scarce, the rabbit with stronger teeth has a survival advantage. It can eat other plants and so will be healthier than its litter mates and have more offspring. Some of these offspring may also have stronger teeth, as the altered DNA data are handed down to the new generation. Thus nature selects and preserves those attributes that contribute to the survival of the species. Those that are unfit die. Natural selection is merciless to the individual, but it gives the species the best possible chance to survive.

The only way nature can obtain new DNA patterns from which to select the best is to alter actual DNA molecules. This can happen through chance mismatching of base pairs—errors—in the reproduction of the DNA molecule. Another way this can occur is through damage to reproductive cells from exposure to radioactivity. Cosmic rays or natural radioactivity in the soil might perform this function. In any case, an offspring born with altered DNA is called a **mutant**. Most mutations are fatal, and the individual dies long before it can have offspring of its own. But in rare cases, a mutation may give a species a new survival advantage. Then natural selection makes it likely that the new **DNA** message will survive and be handed down, making the species more capable of surviving.

20-2 THE ORIGIN OF LIFE

If life on Earth is based on the storage of information in these long, complex, carbon-chain molecules, how could it have ever gotten started? Obviously, 4.5 billion chemical bases didn't just happen to drift together to form the DNA formula for a human being. The key is evolution. Once a life form begins to reproduce itself, natural selection preserves the most advantageous traits. Over long periods of time spanning thousands, perhaps millions, of generations, the life form becomes more fit to survive. This nearly always means the life form becomes more complex. Thus life could have begun as a very simple process that gradually became more sophisticated as it was modified by evolution.

We begin our search on Earth, where fossils and an intimate familiarity with carbon-based life give

Figure 20-5 Trilobites made their first appearance in the Cambrian oceans about 600 million years ago. This example, about the size of a human hand, lived 400 million years ago in an ocean floor that is now a limestone deposit in Pennsylvania. (Grundy Observatory photograph)

us a glimpse of the first living matter. Once we discover how earthly life could have begun, we can look for signs that life began on other planets in our solar system. Finally, we can speculate on the chances that other planets, orbiting other stars, have conditions that give rise to life.

The Origin of Life on Earth The oldest fossils hint that life began in the oceans. The oldest easily identified fossils appear in sedimentary rocks that formed between 0.6 and 0.5 billion years ago—the **Cambrian period.** Such Cambrian fossils were simple ocean creatures, the most complex of which were trilobites (Figure 20-5), but there are no Cambrian fossils of land plants or animals. Evidently, land surfaces were totally devoid of life until only 400 million years ago.

Precambrian deposits contain no obvious fossils, but microscopes reveal microfossils that were the ancestors of the Cambrian creatures. Fig tree chert* in South Africa is 3.0–3.3 billion years old, and chert from the Pilbara Block in northwestern Australia is 3.5 billion years old. Both contain

*Chert is a rock form that resembles flint.

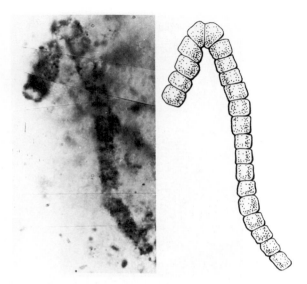

Figure 20-6 Among the oldest fossils known, this microscopic filament resembles modern bacterial forms (artist's reconstruction at right). This fossil was found in the 3.5-billion-year-old chert of the Pilbara Block in northwestern Australia. (Courtesy J. William Schopf)

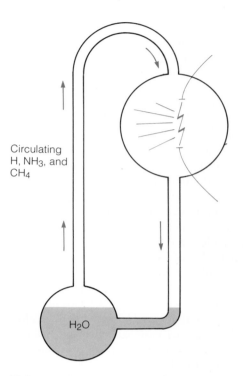

Circulating H, NH₃, and CH₄

H₂O

Figure 20-7 The Miller experiment circulated gases through water in the presence of an electric arc. This simulation of primitive conditions on Earth produced amino acids, the building blocks of protein.

structures that appear to be microfossils of bacteria or simple algae such as those that live in water (Figure 20-6). Apparently, life was already active in Earth's oceans a billion years after the planet formed.

The key to the origin of this life may lie in an experiment performed by Stanley Miller and Harold Urey in 1952. This **Miller experiment** sought to reproduce the conditions on Earth under which life began (Figure 20-7). In a closed glass container, the experimenters placed water (to represent the oceans), the gases hydrogen, ammonia, and methane (to represent the primitive atmosphere), and an electric arc (to represent lightning bolts). The apparatus was sterilized, sealed, and set in operation.

After a week, Miller and Urey stopped the experiment and analyzed the material in the flask. Among the many compounds the experiment produced, they found four amino acids that are common building blocks in protein, various fatty acids, and urea, a molecule common to many life processes. Evidently, the energy from the electric arc had molded the atmospheric gases into some of the basic components of living matter. Other energy

sources such as hot silica (to simulate hot lava spilling into the sea) and ultraviolet radiation (to simulate sunlight) give similar results.

Recent studies of the composition of meteorites and models of planet formation suggest that Earth's first atmosphere did not resemble the gases used in the Miller experiment. Earth's first atmosphere was probably composed of carbon dioxide, nitrogen, and water vapor. This finding, however, does not invalidate the Miller experiment. When such gases are processed in a Miller apparatus, a tarry gunk rich in organic molecules soon coats the inside of the chamber.

The Miller experiment did not create life, nor did it necessarily imitate the exact conditions on the young Earth. Rather, it is important because it shows that complex organic molecules form naturally in a wide variety of circumstances. The chemical deck is stacked to deal nature a hand of complex molecules. If we could travel back in time, we would probably find Earth's oceans filled with a

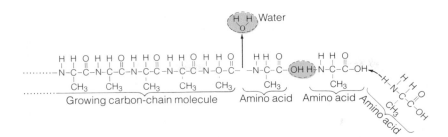

Figure 20-8 Amino acids can link together through the release of a water molecule to form long carbon-chain molecules. The amino acid in this hypothetical example is alanine, one of the simplest.

Growing carbon-chain molecule   Amino acid   Amino acid   Amino acid   Water

rich mixture of organic compounds in what some have called the **primordial soup.**

The next step on the journey toward life is for the compounds dissolved in the oceans to link up and form larger molecules. Amino acids, for example, can link together to form proteins. This linkage occurs when amino acids join together end to end and release a water molecule (Figure 20-8). For many years, experts have assumed that this process must have happened in sun-warmed tidal pools where evaporation concentrated the broth. But recent studies suggest that the young Earth was subject to extensive volcanism and large meteorite impacts that periodically modified the climate enough to destroy any life forms exposed on the surface. Thus the early growth of complex molecules likely took place among the hot springs along the midocean ridges. The heat from such springs could have powered the growth of long protein chains. Deep in the oceans, they would have been safe from climate changes.

Although these proteins might have contained hundreds of amino acids, they would not have been alive. Not yet. Such molecules would not have reproduced, but would have merely linked together and broken apart at random. Because some molecules are more stable than others, however, and because some molecules bond together more readily than others, this blind **chemical evolution** would have led to the concentration of the varied smaller molecules into the most stable larger forms. Eventually, somewhere in the oceans, a molecule took shape that could reproduce itself. At that point the chemical evolution of molecules became the biological evolution of living things.

An alternative theory proposes that primitive living things such as reproducing molecules did not originate on Earth but came here in meteorites or comets. Radio astronomers have found a wide vari-

Figure 20-9 A sample of the Murchison meteorite, a carbonaceous chondrite that fell in 1969 near Murchison, Australia. Analysis of the interior of the meteorite revealed evidence of amino acids. Whether the first building blocks of life originated in space is unknown, but the amino acids found in meteorites illustrate how commonly amino acids and other complex molecules occur through nonorganic means. (Courtesy Chip Clark, National Museum of Natural History)

ety of organic molecules in the interstellar medium, and some studies have found similar compounds inside meteorites (Figure 20-9). Such molecules form so readily that we would be surprised if they were not present in space. A few investigators, however, have speculated that living, reproducing molecules originated in space and came to Earth as a cosmic contamination. If this is true, every planet in the universe is contaminated with the seeds of life. However entertaining this theory may be, it is presently untestable, and an untestable theory is of little use in science. Experts studying the origin of life proceed on the assumption that life began as reproducing molecules in Earth's oceans.

Which came first, reproducing molecules or the cell? Because we think of the cell as the basic unit of life, this question seems to make no sense, but in fact the cell may have originated during chemical evolution. If a dry mixture of amino acids is heated, the acids form long, proteinlike molecules that, when poured into water, collect to form microscopic spheres that function in ways similar to cells (Figure 20-10). They have a thin membrane surface, they can absorb material from their surroundings, they grow in size, and they can divide and bud just as cells do. They contain no large molecule that copies itself, however. Thus the structure of the cell may have originated first and the reproducing molecules later.

An alternative theory proposes that the replicating molecule developed first. Such a molecule would have been exposed to damage if it had been bare, so the first to manufacture or attract a protective coating of protein would have had a significant survival advantage. If this was the case, the protective cell membrane was a later development of biological evolution.

The first living things must have been single-celled organisms much like modern bacteria and simple algae. Some of the oldest fossils known are **stromatolites,** structures produced by communities of blue-green algae or bacteria that grew in mats and, year by year, deposited layers of minerals that were later fossilized. One of the oldest such fossils known is 3.5 billion years old (Figure 20-11). If such algae were common when Earth was young, it may have been able to produce a small amount of oxygen in the early atmosphere. Recent studies suggest that only 0.1 percent would have been sufficient to provide an ozone screen that would protect organisms from the sun's ultraviolet radiation.

How evolution shaped these creatures to live in the ancient oceans, molded them into multicellular organisms, and caused them to develop sexual reproduction, photosynthesis, and respiration is a fascinating story, but we cannot explore it in detail here. (See Box 20-1.) We can see that life could have begun through simple chemical reactions building complex molecules, and that once some DNA-like molecule formed, it protected its own survival with selfish determination. Over billions of years, the genetic information stored in living things kept those qualities that favored survival

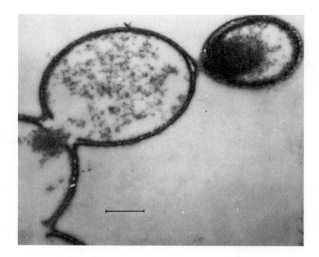

Figure 20-10 Proteinlike material added to water forms microspheres. Although these bodies do not contain DNA or related genetic information, they have many of the properties of cells, such as a double-layered boundary similar to a cell membrane. Thus the first cell structures may have originated through the self-ordering properties of the first complex molecules. The horizontal bar is 1 micrometer long. (Courtesy S. W. Fox, from S. W. Fox and K. Dose, *Molecular Evolution and the Origin of Life,* rev. ed., New York: Marcel Dekker, 1977)

and discarded the rest. As Samuel Butler said, "The chicken is the egg's way of making another egg." In that sense, all living matter on Earth is merely the physical expression of DNA's mindless determination to continue its existence.

Perhaps this seems harsh. Human experience goes far beyond mere reproduction. *Homo sapiens* has art, poetry, music, philosophy, religion, science. Perhaps all of the great accomplishments of our intelligence represent more than mere reproduction of DNA. Nevertheless, intelligence, the ability to analyze complex situations and respond with appropriate action, must have begun as a survival mechanism. For example, a fixed escape strategy stored in the DNA is a disadvantage for a creature that frequently moves from one environment to another. A rodent that always escapes from predators by automatically climbing the nearest tree would be in serious jeopardy if it met a hungry fox in a treeless clearing. Even a faint glimmer of

intelligence might allow the rodent to analyze the situation and, finding no trees, to choose running over climbing. Thus intelligence, of which *Homo sapiens* is so proud, may have developed in ancient creatures as a way of making them more versatile.

If life could originate on Earth and develop into intelligent creatures, perhaps the same thing could have happened on other planets. This raises three questions. First, could life originate if conditions were suitable? Second, if life begins on a planet, will it evolve toward intelligence? The answer to both questions seems to be yes. The process of chemical and biological evolution is directed toward survival, which should lead to versatility and intelligence. But what of the third question: Are suitable conditions so rare that life almost never gets started? The only way to answer that is to search for life on other planets. We begin, in the next section, with the other planets in our solar system.

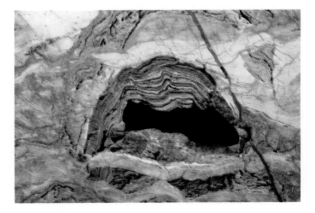

Figure 20-11   A 3.5-billion-year-old fossil stromatolite from western Australia is one of the oldest known fossils (above). Stromatolites were formed, layer by layer, by mats of blue-green algae or bacteria living in shallow water. Such algae may have been common in shallow seas when Earth was young (top). Stromatolites are still being formed today in similar environments. (Mural by Peter Sawyer; photos courtesy Chip Clark, National Museum of Natural History)

BOX 20-1
Geologic Time

Humanity is a new experiment on planet Earth. For most of its history, life on Earth was restricted to the sea. Living things began to populate the land slightly over 400 million years ago, and humans have existed for no more than 3 million years (Figure 20-12).

One way to represent the evolution of life is to compress the 4.6-billion-year history of Earth into the equivalent of a 1-year-long film. In such a film, Earth forms as the film begins on January 1, and through all of January and February, it cools and is cratered, and the first oceans form. But those oceans remain lifeless until sometime in March or early April, when the first living things develop. The 4-billion-year history of Precambrian evolution lasts until the film reaches mid-November, when primitive ocean life begins to evolve into complex organisms such as trilobites.

If we examine the land instead of the oceans, we find a lifeless waste. But once our film shows plant and animal life on the land, about November 28, evolution proceeds rapidly. Dinosaurs, for example, appear about December 12 and vanish by Christmas evening, as mammals and birds flourish.

Throughout the 1-year-run of our film, there are no humans, and even during the last days of the year as the mammals rise and dominate the landscape, there are no people. In the early evening of December 31, vaguely human forms move through the grasslands, and by late evening, they begin making stone tools. The Stone Age lasts until about 11:45 P.M., and the first signs of civilization, towns and cities, do not appear until 11:54 P.M. The Christian era begins only 14 seconds before the New Year, and the Declaration of Independence is signed with 1 second to spare.

Figure 20-12   If the entire history of Earth is represented in a single chart, the history of complex life must be magnified (left) to make details visible. If the story of Earth were a year-long film (right), life forms would not leave the oceans until late November.

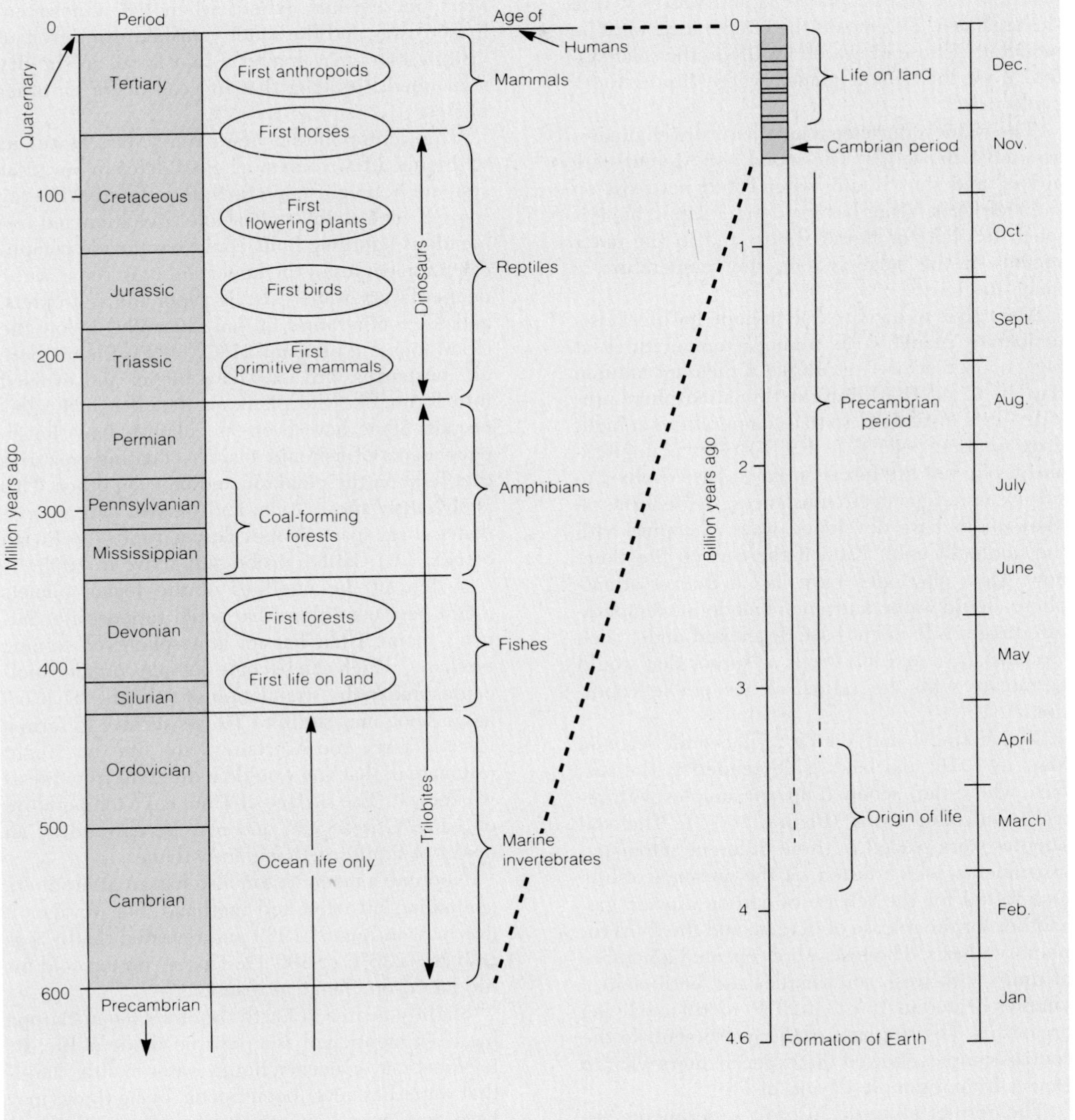

Period
Age of
Humans
Quaternary
Tertiary
First anthropoids
Mammals
First horses
Cretaceous
First flowering plants
Jurassic
First birds
Dinosaurs
Reptiles
Triassic
First primitive mammals
Permian
Pennsylvanian
Mississippian
Coal-forming forests
Amphibians
Devonian
First forests
First life on land
Silurian
Fishes
Ordovician
Trilobites
Ocean life only
Marine invertebrates
Cambrian
Precambrian
Million years ago
0
100
200
300
400
500
600
Billion years ago
0
1
2
3
4
4.6
Life on land
Cambrian period
Precambrian period
Origin of life
Formation of Earth
Dec.
Nov.
Oct.
Sept.
Aug.
July
June
May
April
March
Feb.
Jan

Life in Our Solar System   Though life based on something other than carbon chemistry may be possible, we must limit our discussion to life as we know it and the conditions it requires. The most important condition is the presence of liquid water, not only as part of chemical reactions but also as a medium to transport nutrients and wastes within the organism. This means the temperature must be moderate. Thus our search for life in the solar system must look for a planet where liquid water could exist.

The water requirement automatically eliminates a number of worlds. The moon and Mercury are airless, and thus liquid water could not exist on their surfaces. Venus has some water vapor, but it is much too hot for liquid water, and in the outer regions of the solar system, the temperature is much too low.

Mars gives us good reason to hope for life. It is a temperate planet with summer temperatures at high noon that are not unlike a pleasant autumn day (17°C, or 62°F). But the thin atmosphere provides little insulation, so that temperatures at night drop to about −88°C (−126°F). This is unpleasantly cold, but life forms on Earth have evolved to survive under harsh circumstances, so life forms on Mars might have developed ways of coping with the midnight cold. Even if there is no life there now, Mars may once have had a thicker atmosphere, liquid water, and a more moderate temperature range. Life might have developed under such circumstances and left seeds or spores that would germinate with the return of acceptable conditions.

The Viking 1 and Viking 2 spacecraft reached Mars in 1976, and landers descended to the surface, where they scooped up soil samples with remote-controlled arms (Figure 20-13). The soil samples were placed in three different automated instruments, which tested for the presence of life. One tested for the release of carbon dioxide gas, another for the release of oxygen, and the third for photosynthesis. The tests were repeated a number of times with fresh soil samples and with control samples heated to 160°C (320°F) to kill any living organisms. The response of the instruments to the control samples showed the experimenters what to expect from a sample devoid of life.

The results of those historic experiments are still debated, but clearly there was no obvious trace of life. If there is life on Mars now, its chemistry was not recognizable to the Viking experiments, or perhaps the organisms were not located at the sites where Viking landers collected samples.

Alternatively, life may have once existed on Mars but become extinct when the atmosphere leaked away and the water vanished. An astronaut visiting Mars may someday find fossils in the dry streambeds, proving that life can begin on other planets.

This disappointing news from Mars leaves us with only a few remaining candidates in our solar system. We might hope for life on Jupiter. Energy sources such as solar ultraviolet radiation and tremendous lightning bolts could have made complicated molecules in the atmosphere. In some of the deeper layers, water exists as vapor and as droplets, and the temperature 50 km (30 miles) below the cloud tops is a pleasant 27°C (80°F). Under these circumstances, life may have begun and evolved into forms like the plankton that live in Earth's oceans. Some have even speculated about Jovian equivalents of fish and sharks—floating creatures that feed on the plankton and on each other. This is obviously speculation, and we will have no evidence until space probes descend into the Jovian clouds. The Galileo probe will arrive in 1995.

A few of the satellites of the Jovian planets might have conditions that could support life. Saturn's moon Titan has an atmosphere containing methane, which can become complex organic molecules under the stimulation of sunlight. That organic smog may settle to Titan's surface to form a layer of tarry goo. Certainly, any life that might originate in that goo would be different from life as we know it. The surface of Titan has a temperature of −178°C (−288°F) and may be covered by an ocean of liquid methane and ethane.

Neptune's moon Triton also has an atmosphere containing nitrogen and methane, but Voyager 2 flew past in August 1989 and revealed that it is as cold as −236°C (−393°F). This seems too cold for life based on chemical reactions.

Slightly nearer to Earth, Jupiter's moon Europa has been mentioned as a possible abode of life. Its icy crust may conceal a liquid water mantle, and if that water has never been frozen, living things may have developed beneath the ice. Again, the only way to be sure is to drill through the ice to reach

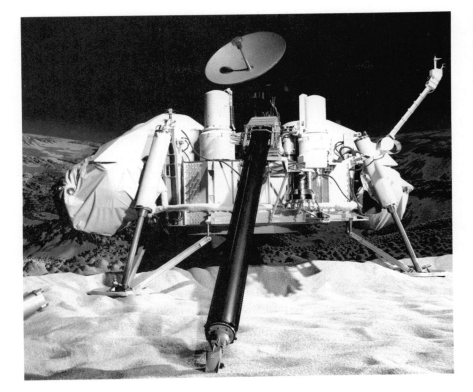

Figure 20-13 Viking 1 and Viking 2 landers reached the Martian surface in 1976. Among other experiments, they searched for signs of life in soil samples collected by a remote-controlled arm. (NASA)

any water that may lie underneath. The chance of life on Europa is probably slim, but it will be many years before that slim chance can be tested conclusively.

So far as we know now, the solar system is bare of life except for our planet. Consequently, our search for life in the universe takes us to other planetary systems.

Life in Other Planetary Systems Might life exist in other solar systems? To consider this question, let us try to decide how common planets are and what conditions a planet must fulfill for life to originate and evolve to intelligence. The first question is astronomical; the second is biological. Our ability to discuss the problem of life outside our solar system is severely limited by our lack of experience.

In Chapter 16, we concluded that planets form as a natural result of star formation. In addition, the process that gives rise to planets is probably related to the process that forms binary star systems. Had Jupiter been 100 times more massive, it would have been a star instead of a planet. Because about half of all stars are members of binary or multiple-star systems, it seems that this process is very common, implying that planetary systems are also common.

If a planet is to become a suitable home for life, it must have a stable orbit around its sun. This is simple in a solar system like our own, but in a binary system most planetary orbits are unstable. Most planets in such systems would not last long before they were swallowed up by one of the stars or ejected from the system.

Thus single stars are the most likely to have planets suitable for life. Because our galaxy contains about 10^{11} stars, half of which are single, there should be roughly 5×10^{10} planetary systems in which we might look for life.

A few million years of suitable conditions does not seem to be enough time to originate life. Our planet required at least 0.5–1 billion years to create the first cells and 4.6 billion years to create intelligence. Clearly, conditions on a planet must remain acceptable over a long time. This eliminates giant stars that change their luminosity rapidly as they evolve. It also eliminates massive stars that remain stable on the main sequence for only a

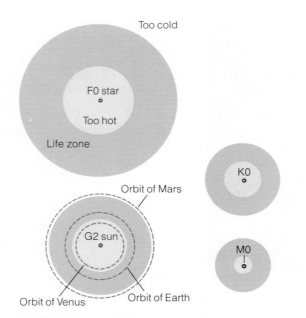

Figure 20-14 The life zone around a star is the region where a planet would have a moderate temperature. Too close to the star, and the planet will be too hot; too far from the star, and the planet will be too cool. The size of a life zone depends on the temperature of the star.

few million years. If life requires a few billion years to originate and evolve to intelligence, no star hotter than about F5 will do. This is not really a serious restriction, because upper-main-sequence stars are rare anyway.

In previous sections, we decided that life, at least as we know it, requires liquid water. That requirement defines a **life zone** (or ecosphere) around each star, a region within which a planet has temperatures that permit the existence of liquid water.

The size of the life zone depends on the temperature of the star (Figure 20-14). Hot stars have larger life zones because the planets must be more distant to remain cool. But the short main-sequence lives of these stars make them unacceptable. M stars have small life zones because they are extremely cool—only planets very near the star receive sufficient warmth. However, planets that are close to a star would probably become tidally coupled, keeping the same side toward the star. This might cause the water and atmosphere to freeze in the perpetual darkness of the planet's

night side and end all chance of life. Also, M stars are subject to sudden flares that might destroy life on a planet close to the star. Thus the life zone restricts our search for life to main-sequence G and K stars. Some of the cooler F stars and warmer M stars might also be good candidates.

Even a star on the main sequence is not perfectly stable. Main-sequence stars gradually grow more luminous as they convert their hydrogen to helium, and thus the life zone around a star gradually moves outward. A planet might form in the life zone and life might begin and evolve for billions of years only to be destroyed as the slowly increasing luminosity of its star moves the life zone outward and evaporates the planet's oceans and drives off its atmosphere. If a planet is to remain in the life zone for 4–5 billion years, it must form on the outer edge of the zone. This may be the most serious restriction we have yet discussed.

If all these requirements are met, will life begin? Early in this chapter we decided that life could begin through simple chemical reactions, so perhaps we should change our question and ask, What could prevent life from beginning? Given what we know about life, it should arise whenever conditions permit, and our galaxy should be filled with planets that are inhabited with living creatures. Then why haven't we heard from them?

20-3 COMMUNICATION WITH DISTANT CIVILIZATIONS

If other civilizations exist, perhaps we can communicate with them in some way. Sadly, travel between the stars is more difficult in real life than in science fiction—and may in fact be impossible. If we can't physically visit, perhaps we can communicate by radio. Again, nature places restrictions on such conversations, but the restrictions are not too severe. The real problem lies with the nature of civilizations.

Travel Between the Stars Practically speaking, roaming among the stars is tremendously difficult because of three limitations: distance, speed, and fuel. The distances between stars are almost beyond comprehension. It does little good to explain that if we use a Ping-Pong ball in New York City to represent the sun, the nearest star would be another Ping-Pong ball in Chicago. It is only slightly better to note that the fastest commercial jet would take about 4 million years to reach the nearest star.

The second limitation is a speed limit—we cannot travel faster than the speed of light. Though science fiction writers invent hyperspace drives so their heroes can zip from star to star, the speed of light is a natural and unavoidable limit that we cannot exceed. This, combined with the large distances between stars, makes interstellar travel very time consuming.

The third limitation is that we can't even approach the speed of light without using a fantastic amount of fuel. Even if we ignore the problem of escaping from Earth's gravity, we must still use energy stored in fuel to accelerate to high speed and to decelerate to a stop when we reach our destination. To return to Earth, assuming we wish to, we have to repeat the process. These changes in velocity require a tremendous amount of fuel. If we flew a spaceship as big as a large yacht to a star 5 light-years (1.5 pc) away and wanted to get there in only 10 years, we would use 40,000 times as much energy as the United States consumes in a year.

Travel for a few individuals might be possible if we accept very long travel times. That would require some form of suspended animation (currently unknown) or colony ships that carry a complete, though small, society in which people are born, live, and die generation after generation. Whether the occupants of such a ship would retain the social characteristics of humans over a long voyage is questionable.

These three limitations not only make it difficult for us to leave our solar system, they would also make it difficult for aliens to visit Earth. Reputable scientists have studied "unidentified flying objects" (UFOs) and related phenomena and have never found any evidence that Earth is being visited or has ever been visited by aliens from other worlds (see Appendix A). Thus humans are unlikely ever to meet an alien face to face. The only way we can communicate with other civilizations is via radio.

Radio Communication Nature places two restrictions on our ability to communicate with distant societies by radio. One has to do with simple physics, is well understood, and merely makes the communication difficult. The second has to do with the fate of technological civilizations, is still unresolved, and may severely limit the number of societies we can detect by radio.

Radio signals are electromagnetic waves that travel at the speed of light. Because even the nearest civilizations must be a few light-years away, this limits our ability to carry on a conversation with distant beings. If we ask a question of a creature 4 light-years away, we will have to wait 8 years for a reply. Clearly, the give-and-take of normal conversation will be impossible.

Instead, we could simply broadcast a radio beacon of friendship to announce our presence. Such a beacon would have to consist of a pattern of pulses obviously designed by intelligent beings, to distinguish it from natural radio signals emitted by nebulae, pulsars, and so on. For example, pulses counting off the first dozen prime numbers would do. In fact, we are already broadcasting a recognizable beacon. Short-wavelength radio signals, such as TV and FM, have been leaking into space for the last 50 years or so. Any civilization within 50 light-years might already have detected us.

If we intentionally broadcast such a signal, we could give listening aliens a good idea of what humanity is like by including coded data in the signal. For example, at the 1974 dedication of the new reflecting surface of the 1000-ft radio telescope at Arecibo, radio astronomers transmitted a series of pulses toward the million or so stars in the globular cluster M 13 (Figure 20-15). The number of data points in the message was 1679, a number selected because it can be factored only into 23 and 73. When the signal arrives at the globular cluster 26,000 years from now, any aliens who detect it will be able to arrange the data in only two ways—23 rows of 73 data points each, or 73 rows of 23 points each. The first way yields nonsense, but the second produces a picture that describes our solar system, the chemical basis of our life form, the general shape and size of the human body, and the number of humans on Earth. Whether there will still be humans on Earth in 52,000 years, when any reply to our message returns, cannot be predicted.

It took only minutes to transmit the Arecibo message. If more time were taken, a more detailed picture could be sent, and if we were sure our radio telescope was pointed at a listening civilization, we could send a long series of pictures. With

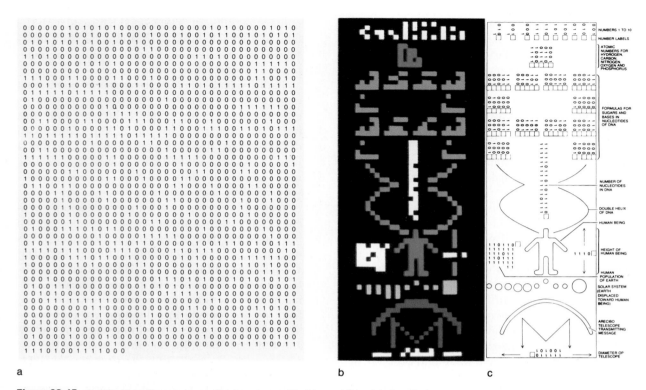

a	b	c

Figure 20-15 (a) The Arecibo message of pulses transmitted toward the globular cluster M 13 is shown as a series of 0s and 1s. (b and c) Arranged in 73 rows of 23 pulses each and represented as light and dark squares, the message would tell aliens about human life. (Color added for clarity.) (a. and c. from "The Search for Extraterrestrial Intelligence" by Carl Sagan and Frank Drake. Copyright © 1975 by Scientific American, Inc. All rights reserved. b. The Arecibo Message of November 1974 was prepared by the staff of the National Astronomy and Ionosphere Center, which is operated by Cornell University under contract with the National Science Foundation.)

pictures we could teach aliens our language and tell them all about our life, our difficulties, and our accomplishments.

If we can think of sending such signals, aliens can think of it too. If we point our radio telescopes in the right direction and listen at the right wavelength, we might hear other intelligent races calling out to one another. This raises two questions: Which stars are the best candidates, and what wavelengths are most likely? We have already answered the first question. Main-sequence G and K stars have the most favorable characteristics. But the second question is more complex.

Only certain wavelengths are useful for communication. We cannot use wavelengths longer than about 30 cm because the signal would be lost in the background radio noise from our galaxy.

Nor can we go to wavelengths much shorter than 1 cm because of absorption within our atmosphere. Thus only a certain range of wavelengths, a radio window, is open for communication (Figure 20-16).

This communications window is very wide, so a radio telescope would take a long time to tune over all the wavelengths searching for intelligent signals. Nature may have given us a way to narrow the search, however. Within the communications window lie the 21-cm line of neutral hydrogen and the 18-cm line of OH. The interval between these two lines has been dubbed the **water hole** because the combination of H and OH yields water (H_2O). Water is the fundamental solvent in our life form, so it might seem natural for similar water creatures to call out to each other at wavelengths in the water

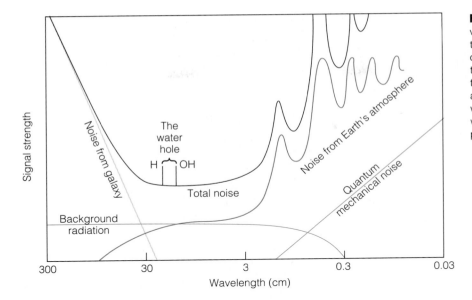

Figure 20-16 Radio noise from various sources makes it difficult to detect signals from distant civilizations at wavelengths longer than 30 cm or shorter than 3 cm. In this range, radio emission from H atoms and from OH mark a small wavelength range dubbed the water hole, which may be a likely place for communication.

hole. But even silicon creatures would be familiar with the 21-cm line of hydrogen. Thus they too might select wavelengths near the water hole.

This is not just idle speculation. A number of searches for extraterrestrial radio signals are now under way. Since 1985, **META**, Megachannel Extra-Terrestrial Assay, has been searching the entire sky using radio antennas at Harvard and near Buenos Aires, Argentina. Funded by the Planetary Society, the project uses a special receiver to search simultaneously 8.4 million adjacent radio frequencies in the water hole. A new project supported by the Planetary Society is called **BETA** because it is the Society's second search program. When **BETA** begins operation it will simultaneously survey 6 billion frequencies.

One ingenious search is called **SERENDIP**, Search for Extraterrestrial Radio Emission from Nearby Developed Intelligent Populations. This search uses a 65,000-channel receiver attached to the 92-m radio dish at Green Bank, West Virginia, but rather than dominating the telescope, SERENDIP rides piggyback. Wherever the radio astronomers point their telescope, SERENDIP samples the signal received looking for traces of intelligent signals.

NASA is currently developing a program dubbed **SETI**, Search for Extra-Terrestrial Intelligence. It will use a 10-million-channel receiver and the 34-m radio dishes that are part of the Deep Space Network to survey the entire sky in 5–7 years. In addition, it will use the 305-m Arecibo telescope to listen carefully to near-solar-type stars within 80 light-years.

The technology exists, but the most serious restriction on the search is an unanswered social question: How long can a civilization remain at a high enough technological level to engage in galactic communication? If other planets are like Earth, life takes 4.6 billion years to reach a technological level. If a society destroys itself within 100 years of the invention of radio (by nuclear war, nuclear pollution, chemical pollution, overpopulation, and so on), the chances of our communicating with it are very small. Most life forms in the galaxy would be on the long road up to civilization or on the path leading downward from the collapse of their technology. But if technological societies can solve their internal problems and remain stable for a million years, some would be at the proper stage to communicate with us. Estimates of the number of communicative civilizations in our galaxy range from 10 million to 1. (See Box 20-2.)

Are we the only thinking race? If we are, then we are the sole representatives of that state of matter called intelligence, and we bear the sole responsibility to understand and admire the universe. The mere detection of signals from another civilization, however, would demonstrate that we share the universe with others. Though we might

BOX 20-2

Communicative Civilizations in Our Galaxy

A simple formula first devised by radio astronomer Frank Drake can give us an estimate of the number of technological civilizations with which we might communicate. The formula for the number of communicative civilizations in a galaxy, N_c, is:

$$N_c = N^* \cdot f_P \cdot n_{LZ} \cdot f_L \cdot f_I \cdot F_S$$

N^* is the number of stars in a galaxy, and f_P represents the probability that a star has planets. If all single stars have planets, f_P is about 0.5. The factor n_{LZ} is the average number of planets in a solar system suitably placed in the life zone; f_L is the probability that life will originate if conditions are suitable; and f_I is the probability that the life form will evolve to intelligence. These factors can be roughly estimated, but the remaining factor is much more uncertain.

F_S is the fraction of a star's life during which the life form is communicative. Here we assume a star lives about 10 billion years. As explained in the text, if a society survives at a technological level for only 100 years, our chances of communicating with it are small. But a society that stabilizes and remains technologically advanced for a long time is much more likely to be in the communicative phase at the proper time to signal to us. If we assume technological societies destroy themselves in about 100 years, F_S is 100 divided by 10 billion or 10^{-8}. But if societies can remain technologically advanced for a million years, then F_S is 10^{-4}. The influence of this factor is shown in Table 20-1.

If the optimistic estimates are true, there may be a communicative civilization within a few dozen light-years of us, and we could locate it by searching through only a few thousand stars. On the other hand, if the pessimistic view is correct, we may be the only planet in our galaxy capable of communication. We may never know until we understand how technological societies function and how long they survive.

TABLE 20-1
The Number of Technological Civilizations per Galaxy

Variable		Estimates	
		Pessimistic	Optimistic
N^*	Number of stars per galaxy	2×10^{11}	2×10^{11}
f_P	Fraction of stars with planets	0.01	0.5
n_{LZ}	Number of planets per star that lie in life zone for longer than 4 billion years	0.01	1
f_L	Fraction of suitable planets on which life begins	0.01	1
f_I	Fraction of life forms that evolve to intelligence	0.01	1
F_S	Fraction of star's life during which a technological society survives	10^{-8}	10^{-4}
N_C	Number of communicative civilizations per galaxy	2×10^{-5}	10×10^6

never leave our solar system, such communication would end the self-centered isolation of humanity and stimulate a reevaluation of the meaning of our existence. We may never realize our full potential as humans until we communicate with nonhumans.

SUMMARY

To discuss life on other worlds, we must first understand something about life in general, life on Earth, and the origin of life. In general, we can identify three properties of living things: a process, a physical basis, and a controlling unit of information. The process must extract energy from the surroundings, maintain the organism, and modify the surroundings to promote the organism's survival. The physical basis is the arrangement of matter and energy that implements the life process. On Earth, all life is based on carbon chemistry. The controlling information is the data necessary to maintain the organism's function. Data for Earth life are stored in long carbon-chain molecules called DNA.

The DNA molecule stores information in the form of chemical bases linked together like the rungs of a ladder. When these patterns are copied by RNA molecules, they can direct the manufacture of proteins and enzymes. Thus the DNA information is the chemical formulae the cell needs to function. When a cell divides, the DNA molecule splits lengthwise and duplicates itself so that each of the new cells has a copy of the information. Errors in the duplication or damage to the DNA molecule can produce mutants, organisms that contain new DNA information and have new properties. Natural selection determines which of these new organisms are best suited to survive, and the species evolves to fit its environment.

The Miller experiment duplicated conditions in the Earth's primitive environment and suggests that energy sources such as lightning could have caused amino acids and other complex molecules to form. Chemical evolution would have connected these together in larger and more complex, but not yet living, molecules. When a molecule acquired the ability to produce copies of itself, natural selection perfected the organism through biological evolution. Though this may have happened in the first billion years, life did not become diverse and complex until the Cambrian period, about 0.6 billion years ago. Life emerged from the oceans about 0.4 billion years ago, and humanity developed only a few million years ago.

It seems unlikely that there is life on other planets in our solar system. Most of the planets are too hot or too cold. Mars may have had life long ago if its atmosphere was thicker and liquid water existed on its surface, but the Viking 1 and Viking 2 landers performed three kinds of experiments to look for life and found none. We can imagine how lightning bolts in Jupiter's atmosphere might have spawned life, but there are no data available on complex molecules in the Jovian atmosphere.

To find life, we must look beyond our solar system. Because we suspect that planets form from the leftover debris of star formation, we suspect that most stars have planets. The rise of intelligence may take billions of years, however, so short-lived massive stars and binary stars with unstable planetary orbits must be discarded. The best candidates are G and K main-sequence stars.

The distances between stars are too large to permit travel, but communication by radio could be possible. A certain wavelength range called a radio window is suitable, and a small range between the radio signals of H and OH, the so-called water hole, is especially likely.

We now have the technology to search for other intelligent life in the universe. Though there are some hazards such as culture shock, such communication would probably be beneficial. In any case, if life is common in the universe, and if technological societies do not destroy themselves too quickly, the discovery of radio signals from extraterrestrial life is inevitable.

Coming of Age in the Galaxy

If our galaxy is rich with life, if it is crowded with civilizations, we should be able to search out the nearest ones and begin a dialogue. Because we are just entering the communicative stage, we are beginners, and most of these civilizations would be far more advanced than we and should be able to help us in various ways. Before we announce our presence, however, we should remember that in virtually any city on Earth there are areas into which we would not venture alone after dark, and there may be similar rough neighborhoods in our galaxy.

Communication is fraught with both external hazards, such as invasion, and internal risks. Communication with an alien civilization may induce a difficult and irrevocable transformation for humanity, tearing us from our isolation and thrusting us into a larger society of civilized races. Like all comings of age, it could be both painful and beneficial.

Hazards The most obvious, though least likely, hazard of announcing our presence is invasion, that calamity so dear to science fiction fans. Transmission of signals marks Earth as habitable, identifies us as technological, and pinpoints our location. A fleet of armed interstellar starships would find Earth easy pickings.

Nature, however, seems to have given us natural safety factors that protect us from such invasion. Interstellar travel is at best difficult and tremendously expensive, and it may be impossible. In addition, Earth might not be attractive to aliens who have different body chemistries or who are far more advanced technologically.

Some people have suggested that we ensure our safety by not broadcasting signals but merely listening. If we could eavesdrop on the rest of the galaxy, we could test their good intentions. But we already leak TV and FM radio signals into space, and it is unlikely that we can stop the leak in the near future. We have no choice but to trust in the isolation of planetary systems.

The worst hazard we might face would be

culture shock, the bewildering impact of one society on another. The history of Earth is filled with examples of smaller, weaker societies that have crumbled when forced into contact with larger, more powerful cultures. Our only defense is the insulation of vast distance that makes conversation impossible and permits us time for thoughtful consideration of each new piece of information.

One aspect of culture shock would be the impact on our religious beliefs of the discovery of nonhumans. Many of Earth's religions teach that humanity is chosen, especially blessed, even made in the image of God. TV pictures of a race of intelligent squid with glowing antennae would challenge many of our most deeply held beliefs.

Another aspect of culture shock is the dependence we might develop if we began to receive all the answers to our problems via radio. George Wald, Nobel Prize–winning biologist, has said, "I can conceive of no nightmare as terrifying as establishing such communication with a so-called superior (or if you wish, advanced) technology in outer space." His point, that the human enterprise of trying to understand nature might wither, is a serious consideration. However, we should remember that interstellar distances are vast and conversations are impossible. The information we might receive would be minimal compared to the questions it would generate, questions we could not put to the aliens because we could not wait for the long-delayed answers. Communication with superior cultures might act as a stimulus to our human quest for understanding rather than as a suppressant. We should also note that refraining from communication for this reason in effect means preferring ignorance to knowledge. We would abandon our human enterprise in trying to preserve it.

The last and most dangerous hazard we might face is internal. Our planet is divided into nations that barely coexist. If one nation detected signals from a superior culture and began to learn advanced technology, it might hold a tremendous military advantage over its neighbors. If it kept its discovery secret, no other nation could learn it without precise knowledge of the proper star and radio frequency. Unfortunately, we have no defense against ourselves except international cooperation to ensure free access to such signals by all nations.

These hazards may seem so severe that we should close our ears and refuse to end our isolation on Earth. But the most chilling aspect of the venture is inevitability. If the galaxy is populated by many civilizations communicating with one another, it is only a matter of time until we discover them, on purpose or by accident, and then all these theoretical hazards will become real ones.

Benefits If the galaxy is richly populated with technological civilizations, we will discover them eventually, and we might as well look at the bright side. We could benefit tremendously. Such communication would open a galaxy of new ideas and might give us new insight into what it means to be human.

Probably the greatest benefit would be the end of the isolation of Earth. The process that Copernicus began when he said Earth was not at the center of the universe would be brought to completion by the realization that Earth is not the only home of life. If we could survive the culture shock, the discovery of other intelligent beings would be a growing experience for all humans.

One obvious benefit of interstellar communication would be survival information. If certain kinds of societies eventually destroy themselves, we are unlikely to discover one of them. The civilizations we might find would be the survivors, the ones that solved the critical problems and reached stability. We can hope their signals would contain survival information. They might warn us away from certain technological, economic, political, or social systems that lead inevitably to collapse. Our civilization faces so many critical situations it is hard to guess which might be fatal. Just knowing that some societies have solved their problems would be encouraging.

With the critical survival questions answered, a distant society might broadcast examples of their aesthetic expression. In addition to simple messages, it would be possible to transmit images or sound just as we transmit television. We might receive examples of painting and sculpture, music, literature, philosophy, and more. Are the laws of artistic composition the same for aliens? Is the theme of *Hamlet* universal?

Of course, the most obvious and perhaps most useful benefits of such communication would be technological. Such messages might contain directions for new methods of making steel, plastics, electronic circuits, and so on. After all, a distant civilization might be alien, but it would work with the same chemical elements and the same laws of physics, so its technological processes should be applicable to earthly problems. Medical technology might be another matter, however. Even if a distant race were carbon-based, it would probably have significantly different body chemistry, so its medical technology would not be immediately applicable to human ills.

Science, however, might benefit even more than technology. Communication with an advanced race might give us a much more sophisticated understanding of nature than we now have. Even if the civilization were no more advanced than us, communication would allow us to compare worlds. Thus we would no longer be confined to the study of a single planet and its life. We could compare Earth's geology, for example, with that of other planets in minute detail. We could create whole new areas of study such as comparative evolution, comparative ecology, comparative psychology, and so on.

If we assume that communication between civilizations is common, our galaxy may be filled with a network of signals carrying beneficial information. Some may have been transmitted long ago, and some recently. Some may be rebroadcasts received and retransmitted over and over as civilized races preserve our galactic heritage. To join that network would be a coming of age for humanity, an assumption of our heritage as intelligent beings.

Where Are They?   If the galaxy is filled with such a hum of communication, why haven't we heard it? To date, radio astronomers have tested more than 1000 stars for radio signals and have found none. But even if technological civilizations are common, we might have to test 10,000–100,000 stars before we find a civilization transmitting a signal. We have searched only 10^{-17} of all possible combinations of frequencies and directions. The search has just begun.

Nevertheless, some astronomers feel that life on other worlds is highly unlikely. The process that transforms chemical evolution into organic evolution is very complex, and it may require such special circumstances that biological evolution almost never gets started. If that is the case, we may be the only life in our galaxy.

Even if life begins easily, evolution may not lead to intelligence except in very special circumstances. In that case, there may be many planets where life exists, but we may be the only life that has the intelligence to wonder about the mystery of existence. Or life may inevitably progress to intelligence, but intelligence may be a self-destroying trait. Intelligent species may destroy themselves by altering their planet so drastically they cannot survive. This is the most frightening prospect of all. If it is correct, not only are we alone, but we also face an unavoidable catastrophe in the near future.

Before we abandon all hope, we should consider the comment of astronomer Martin Rees: "Absence of evidence is not evidence of absence." We have hardly begun to look for fellow beings, and until we search many stars and many different radio wavelengths, we can hardly conclude that we are alone.

NEW TERMS

DNA (deoxyribonucleic acid)

protein

enzyme

RNA (ribonucleic acid)

amino acid

natural selection

mutant

Cambrian period

Miller experiment

primordial soup

chemical evolution

stromatolite

life zone

water hole

culture shock

REVIEW QUESTIONS

1. If life is based on information, what is that information?

2. What would happen to a life form if the information handed down to offspring was always the same? How would that endanger the future of the life form?

3. How does the DNA molecule produce a copy of itself?

4. Why do we believe that life on Earth began in the sea?

5. What is the difference between chemical evolution and biological evolution?

6. How does intelligence make a creature more likely to survive?

7. What role did the control play in the Viking experiments? Why were controls necessary?

8. Why are upper-main-sequence stars unlikely sites for intelligent civilizations?

9. How does the stability of technological civilizations affect the probability that we can communicate with them?

10. What advantages would civilizations derive from communicating at frequencies in the water hole?

DISCUSSION QUESTIONS

1. In your opinion, where in our solar system is the most likely place to find life?

2. How would you change the Arecibo message if we lived on Mars instead of on Earth?

3. What argument can you make that we are alone in the galaxy?

PROBLEMS

1. A single human cell encloses about 1.5 m of DNA containing 4.5 billion base pairs. What is the spacing between these base pairs in nanometers? That is, how far apart are the rungs on the DNA ladder? (See Box 5-1.)

2. If we represent the history of Earth by a line 1 m long, how long a segment would represent the 400 million years since life moved onto the land? How long a segment would represent the 3-million-year history of human life?

3. If a human generation, the time from birth to childbearing, is 20 years, how many generations have passed in the last million years?

4. If a star must remain on the main sequence for at least 5 billion years for life to evolve to intelligence, how massive could a star be and still harbor intelligent life on one of its planets? (**HINT:** See Box 9-2.)

5. If there are about 1.4×10^{-4} stars like the sun per cubic light-year, how many lie within 100 light-years of Earth? (**HINT:** The volume of a sphere is $\frac{4}{3}\pi r^3$.)

6. Mathematician Karl Gauss suggested planting forests and fields in a gigantic geometric proof to signal to possible Martians that intelligent life exists on Earth. If Martians had telescopes that could resolve details no smaller than 1 second of arc, how large would the smallest element of Gauss's proof have to be? (**HINT:** See Box 3-2).

7. If we detected radio signals with an average wavelength of 20 cm and suspected that they came from a civilization on a distant planet, roughly how much of a change in wavelength should we expect to see because of the orbital motion of the distant planet? (**HINT:** See Box 6-4.)

8. Calculate the number of communicative civilizations per galaxy from your own estimates of the factors in Box 20-2.

RECOMMENDED READING

BAUGHER, JOSEPH F. *On Civilized Stars.* Englewood Cliffs, N.J.: Prentice-Hall, 1985.

BEATTY, J. KELLY. "The New, Improved SETI." *Sky and Telescope* 65 (May 1983), p. 411.

CARROLL, MICHAEL. "Digging Deeper for Life on Mars." *Astronomy* 16 (April 1988), p. 6.

COUPER, HEATHER. "In Search of Solar Systems." *New Scientist* (November 1986), p. 34.

CRICK, FRANCIS. *Life Itself.* New York: Simon & Schuster, 1981.

CROSWELL, KEN. "Does Alpha Centauri Have Intelligent Life?" *Astronomy* 19 (April 1991), p. 28.

DAWKINS, R. *The Selfish Gene.* New York: Oxford University Press, 1976.

DEARDROFF, JAMES W. "Possible Extraterrestrial Strategy for Earth." *Quarterly Journal of the Royal Astronomical Society of London* 27 (1986), p. 94.

DICK, STEVEN J. *Plurality of Worlds: The Origins of the Extraterrestrial Life Debate from Democritus to Kant.* Cambridge, England: Cambridge University Press, 1982.

FEINBERG, G., and R. SHAPIRO, eds. *Life Beyond Earth: The Intelligent Earthling's Guide to Life in the Universe.* New York: Morrow, 1980.

FINLEY, DAVID. "The Search for Extra Solar Planets." *Astronomy* 9 (December 1981), p. 90.

FINNEY, BEN R., and ERICH M. JONES, eds. *Interstellar Migration and the Human Experience.* Berkeley: University of California Press, 1985.

FOX, SIDNEY. "From Inanimate Matter to Living Systems." *The American Biology Teacher* 43 (March 1981), p. 127.

GILLETT, STEPHEN. "The Rise and Fall of the Early Reducing Atmosphere." *Astronomy* 13 (July 1985), p. 66.

GOLDSMITH, D., ed. *The Quest for Extra-Terrestrial Life: A Book of Readings.* Mill Valley, Calif.: University Science Books, 1980.

———. "SETI: The Search Heats Up." *Sky and Telescope* 75 (February 1988), p. 141.

GOLDSMITH, D., and T. OWEN. *The Search for Life in the Universe,* 2nd ed. Menlo Park, Calif.: Benjamin/Cummings, 1987.

GOODMAN, ALLAN E. "The Diplomatic Implications of Discovering Extraterrestrial Intelligence." *Mercury* 16 (March/April 1987), p. 56.

HORGAN, JOHN. "In the Beginning . . ." *Scientific American* 264 (February 1991), p. 116.

MCDONOUGH, THOMAS R. *The Search for Extraterrestrial Intelligence.* New York: Wiley, 1987.

MOOD, STEPHANIE. "Life on Europa?" *Astronomy* 11 (December 1983), p. 16.

OLSON, EDWARD C. "Intelligent Life in Space." *Astronomy* 13 (July 1985), p. 6.

PAPAGIANNIS, M. D. "The Search for Extra-Terrestrial Civilizations—A New Approach." *Mercury* 11 (January/February 1982), p. 12.

———. "Bioastronomy: The Search for Extra-Terrestrial Life." *Sky and Telescope* 67 (June 1984), p. 508.

———, ed. *The Search for Extraterrestrial Life: Recent Developments.* Boston: D. Reidel, 1985.

PARKER, B. "Are We the Only Intelligent Life in Our Galaxy?" *Astronomy* 7 (January 1979), p. 6. Reprinted in *Astronomy: Selected Readings,* M. A. Seeds, ed. Menlo Park, Calif.: Benjamin/Cummings, 1980, p. 155.

REGIS, E., ed. *Extraterrestrials: Science and Alien Intelligence.* Cambridge, England: Cambridge University Press, 1985.

SAGAN, C., and T. PAGE, eds. *UFOs: A Scientific Debate.* New York: Norton, 1972.

SCHECHTER, MURRAY. "Planets in Binary Star Systems." *Sky and Telescope* 68 (November 1984), p. 394.

SCHORN, R. A. "Extraterrestrial Beings Don't Exist." *Sky and Telescope* 62 (September 1981), p. 207.

STRAND, LINDA JOAN. "The Search for Life on Mars: Shots in the Dark." *Astronomy* 11 (December 1983), p. 66.

———. "The Star Tar in the Jupiter Jars." *Astronomy* 12 (June 1984), p. 66.

SWIFT, DAVID W. *SETI Pioneers.* Tucson: University of Arizona Press, 1990.

TIPPLER, FRANK. "Extraterrestrial Beings Do Not Exist." *Physics Today* 34 (April 1982), p. 9. But see also *Physics Today* 35 (March 1982), p. 26.

———. "The Most Advanced Civilization in the Galaxy Is Ours." *Mercury* 11 (January/February 1982), p. 5.

WALDROP, M. MITCHELL. "Goodbye to the Warm Little Pond." *Science* 250 (23 November 1990), p. 1078.

WOESE, CARL R. "Archaeobacteria." *Scientific American* 244 (June 1981), p. 98.

CHAPTER

21

Afterword

Supernatural is a null word.

Robert A. Heinlein,
The Notebooks of Lazarus Long

Our journey is over, but before we part company, there is one last thing to discuss—the place of humanity in the universe. Astronomy gives us some comprehension of the workings of stars, galaxies, and planets, but its greatest value lies in what it teaches us about ourselves. Now that we have surveyed astronomical knowledge, we can better understand our own position in nature.

To some, the word *nature* conjures up visions of furry rabbits hopping about in a forest glade dotted with pastel wildflowers. To others, nature is the blue-green ocean depths filled with creatures swirling in a mad struggle for survival. Still others think of nature as windswept mountaintops of gray stone and glittering ice. As diverse as these images are, they are all earthbound. Having studied astronomy, we can view nature as a beautiful mechanism composed of matter and energy interacting according to simple rules to form galaxies, stars, planets, mountaintops, ocean depths, and forest glades.

Perhaps the most important astronomical lesson is that we are a small but important part of the universe. Most of the universe is lifeless. The vast reaches between the galaxies appear to be empty of all but the thinnest gas, and the stars, which contain most of the mass, are much too hot to preserve the chemical bonds that seem necessary to allow life to survive and develop. Only on the surfaces of a few planets, where temperatures are moderate, could atoms link together to form living matter.

If life is special, then intelligence is precious. The universe must contain many planets devoid of life, planets where the wind has blown unfelt for billions of years. There may also be planets where life has developed but has not become complex, planets on which the wind stirs wide plains of grass and rustles dark forests. On some planets, insects, fish, birds, and animals may watch the passing days unaware of their own existence. It is intelligence, human or alien, that gives meaning to the landscape.

Science is the process by which intelligence tries to understand the universe. Science is not the invention of new devices or processes. It does not create home computers, cure the mumps, or manufacture plastic spoons—that is engineering and technology, the adaptation of scientific understanding for practical purposes. Science is understanding nature, and astronomy is understanding nature on the grandest scale. Astronomy is the science by which the universe, through its intelligent lumps of matter, tries to understand its own existence.

As the primary intelligent species on this planet, we are the custodians of a priceless gift—a planet filled with living things. This is especially true if life is rare in the universe. In fact, if Earth is the only inhabited planet, our responsibility is overwhelming. In any case, we are the only creatures who can take action to preserve the existence of life on Earth, and ironically, it is our own actions that are the most serious hazards.

The future of humanity is not secure. We are trapped on a tiny planet with limited resources and a population growing faster than our ability to produce food. In our efforts to survive, we have already driven some creatures to extinction and now threaten others. If our civilization collapses because of starvation, or if our race destroys itself somehow, the only bright spot is that the rest of the creatures on Earth will be better off for our absence.

But even if we control our population and conserve and recycle our resources, life on Earth is doomed. In 5 billion years, the sun will leave the main sequence and swell into a red giant, incinerating Earth. Earth will be lifeless long before that, however. Within the next few billion years, the growing luminosity of the sun will first alter Earth's climate and then boil away its atmosphere and oceans. Our Earth is, like everything else in the universe, only temporary.

To survive, humanity must leave Earth and search for other planets. Colonizing the moon and other planets of our solar system will not save us, since they face the same fate as Earth when the sun dies. But travel to other stars is tremendously difficult and may be impossible with the limited resources we have in our small solar system. We and all the living things that depend on us for survival may be trapped.

This is a depressing prospect, but a few factors are comforting. First, everything in the universe is temporary. Stars die; galaxies die; perhaps the entire universe will fall back in a "big crunch" and die. That our distant future is limited only assures us that we are a part of a much larger whole. Second, we have a few billion years to prepare, and a billion years is a very long time. Only a few million years ago, our ancestors were learning to walk erect and communicate with one another. A billion years ago our ancestors were microscopic organisms living in the primeval oceans. To suppose that a billion years hence we humans will still be human, or that we will still be the dominant species on Earth, or that we will still be the sole intelligence on Earth, is the ultimate conceit.

Our responsibility is not to save our race for all eternity, but to behave as dependable custodians of our planet, preserving it, admiring it, and trying to understand it. That will call for drastic changes in our behavior toward other living things and a revolution in our attitude toward our planet's resources. Whether we can change our ways is debatable —humanity is far from perfect in its understanding, abilities, or intentions. We must not imagine, however, that we and our civilization are less than precious. We have the gift of intelligence, and that is the finest thing this planet has ever produced.

Astrology, UFOs, and Pseudoscience

By taking a single course in astronomy, you have become, compared with the average person, an astronomer. With this knowledge comes the responsibility that all astronomers feel—the responsibility to help others understand the science of astronomy and evaluate such false sciences as astrology and the study of UFOs.

In this appendix, we will discuss such false sciences. We will examine why they are not scientific and why people believe in them anyway. Most of all, we will try to understand how we as astronomers can help people distinguish good science from bad.

A-1 ASTROLOGY

Astrology is an ancient superstition holding that a person's personality and life are influenced by the position of the sun, moon, planets, and stars at the moment of birth. In addition, astrologers claim that the daily changes in the location of the heavenly bodies can influence events in our lives. All of this is summarized in a horoscope, a diagram of the zodiac, showing the position of the heavenly bodies at a given time.

This belief originated over 2000 years ago in the sky-worshiping religions of Babylonia and was later modified by the Egyptians and Greeks. All these cultures believed that their gods were manifested in the moving heavenly bodies, so it was reasonable for them to imagine these bodies could affect lives on Earth. Modern people no longer worship the ancient sky gods, but many people today still profess a belief in the powers of those gods as represented in astrology.*

Although the ancient religions have been abandoned for 2000 years, astrology has survived nearly unchanged since the second century A.D. in spite of dozens of tests showing that it does not work. Perhaps it has survived because it appeals to our deepest fears and needs.

Does Astrology Work? We must consider two questions. First, is there any physical reason we should expect astrology to work? Second, is there any evidence that it actually does work?

According to the basic claim of astrology, a baby's character, personality, and future life are affected in some way by the location of the sun,

---

As recently as January 1984, the Vatican daily newspaper urged Roman Catholics not to believe in astrology lest they "subject the faith to a risk of defilement."

moon, and planets along the zodiac at the moment of birth. In fact, there seems to be no way for these objects to act on the body of the baby. The force of gravity is the most likely agent, but the gravitational field of the doctor is much larger than that of the planets. Thus it would seem more important where the doctor stands than where Mars is along the zodiac.

If gravity is not the agent of influence, we have only a few other effects to consider. Light and other electromagnetic radiation (see Chapter 5) from the sun, moon, and planets do not generally enter hospital delivery rooms, and the electrostatic force, the strong nuclear force, and the weak nuclear force are even less effective than gravity. No other forces, rays, or influences in nature are known, so there is no physical agency to explain the claims of astrology, and we have no physical reason to expect astrology to work.

In fact, astrology has been tested many times and no legitimate test has proven any astrological influence. A few positive tests have been reported by the grocery store tabloids, but when these tests were repeated by trained statisticians, astrology failed. Extensive tests have searched in vain for relationships between birth sign and such personal characteristics as blood type, bust size, handedness, profession, extroversion/introversion, divorce rate, and others. Astrology would be very useful if it worked, but no relationship has ever been found.

Should We Respect Astrology? As an astronomer, you must decide how to view astrology: as an alternative scientific view, as a religion, or as a harmless superstition.

Some astrologers claim that astrology is a science. They assert that it is based on scientific methods and they quote elaborate theories about unknown rays from the planets. If this were true, astrology would deserve our respect as a science, but there is no scientific evidence that astrology is valid. In fact, all scientific evidence argues against astrology.

In addition, astrology is not based on scientific methods. A scientist, above all else, must be willing to revise hypotheses in the face of contradictory evidence. Scientific theories are always open to revision, but the principles of astrology have not been revised for almost 2000 years. For instance,

precession has altered the location of the vernal equinox among the constellations so that the astrological signs of the zodiac are now one entire constellation out of place. Also, astrologers routinely ignore contradictory evidence from tests of astrology.

Even if we interpret *scientific* to mean *systematic*, astrology is not scientific. The astrological predictions you see in newspapers and magazines cannot have been "cast" according to systematic rules. To cast your horoscope, an astrologer would have to know the date, time, latitude, and longitude of your birth. Even if a newspaper horoscope did happen to be correct for you, it could be correct for no other person in the world.

We have only to examine the astrology in newspaper columns to see that it cannot have been derived from a systematic set of rules. The following predictions for Libra were extracted from newspapers on January 11, 1984. Both the zodiacal sign and the date were selected at random.

> Gain indicated through written word. Locate legal documents, special manuscripts and be aware of rights, permissions.—*Sydney Omarr*

> As best you can sort out the pieces.—*Clay R. Pollan*

> Exuberant moods sweep you up in a whirl of excitement. Make sure all financial negotiations are on the up and up. The evening promises much love and affection.—*Joyce Jillson*

> Renew worthwhile agreements with partners. A worldly matter could be confusing early. Be dynamic.—*Carroll Righter*

> You're able to lift the spirits of a loved one. Your belief in your own potential rubs off on others. Enter financial negotiations.—*Frances Drake*

> Stick close to home today. Those who can use public transportation should do so. Encourage teenagers to earn their own pocket money.—*Jeane Dixon*

Like all astrological predictions, these are vague but generally optimistic, give us good advice, and refer to common features in our daily lives such as friends, loved ones, financial matters, jobs, and so on. Even so, the disagreement among these predictions shows that astrology cannot claim to be systematic. Even the most popular astrologers cannot agree as to whether we should be exuberant and dynamic or stay home and study legal documents.

Astrology cannot claim to be a science. It is not open to revision as dictated by experiment, nor is it systematic and consistent in its methods. If someone tells us astrology is a science, we have every right to withhold our respect.

But perhaps we should respect astrology as a religion. Certainly the most devout believers in astrology have a faith in its principles that approaches religious intensity, and astrology did originate in ancient sky worship. Yet astrology is a false religion because it does not incorporate any standards of moral behavior. Indeed, true believers in astrology might steal and murder and then blame their crimes on the chance alignments of the stars at their birth. A true religious faith in astrology dilutes the responsibility we feel for our own actions.

If we astronomers cannot respect astrology as a science or as a religion, then perhaps we should treat it as an ancient and harmless superstition. After all, we knock on wood to attract the attention of the wood spirits that they might grant our wishes, and astrology is only another form of appealing to gods no longer worshiped. Yet astrology is not a harmless superstition. It asks us to believe that our personalities are fixed at the moment of birth. If that is true, then our fate is sealed and we who are shy can never make many friends. We who are weak can never grow strong. We who are sad can never be happy.

Worse yet, astrology as a superstition tells us that the daily events of our lives are beyond our control. If our love life falters one day, it is hopeless to try to save it. If our studies go well, it is not to our credit but is merely our stars at work. We are held hostage by our stars. In its most intense form, a superstitious belief in astrology is depersonalizing, and those who tell us they believe in astrology as a superstition deserve our sympathy rather than our respect.

An Attitude Toward Astrology Although each astronomer must develop an independent personal attitude toward astrology, it does seem clear that we cannot respect it as a science, as a religion, or as a superstition. Perhaps we can view it as a deep and ancient expression of our human need to believe that our lives have meaning beyond our daily existence.

However inevitable astrology is, we should not be content with its survival. It is an ancient belief that does not work. It is not science, not religion, but a potentially harmful superstition. Our universe harbors greater elegance and greater power than those forces astrology purports to describe.

A-2 UFO'S AND VISITORS FROM THE STARS

If your friends discover that you know astronomy, they may ask you about unidentified flying objects (UFOs) and visitors from outer space. Perhaps because astronomers watch the sky, your friends may expect you to be an expert on this subject, so we must try to decide what to say about UFOs.

The UFO Principle Those who believe in UFOs believe in something we might call the UFO principle—UFOs are spacecraft piloted by creatures from another world. We must ask ourselves if this is a reasonable belief and if any evidence exists to support it.

As astronomers, we know that life could be very common in the universe. Planets appear to be by-products of star formation (see Chapter 16), and it does seem possible that life could originate on any world where conditions are right (see Chapter 20). Once active, life modifies itself to fit its environment, so, although we don't know how common suitable planets are, we must admit that our galaxy could be well populated. Could such worlds, however, send spacecraft to visit ours?

No other planet in our solar system is suitable for intelligent life (see Chapter 20), so any visitors to Earth must come from other planetary systems orbiting other stars. This argues against the UFO principle because of the difficulty in traveling between stars. Other creatures might be different from us, but they must still obey the same laws of physics, and that means that traveling from their star system to ours would be almost impossible. The stars are very, very far apart.

Thus the astronomy we know tells us that the UFO principle is unreasonable, not because we don't believe in aliens on other worlds, but because we suspect they could never visit us. But we cannot discard a hypothesis because we find it unreasonable. We must look at the evidence of UFO sightings.

Sorting Sightings We can divide UFO sightings into three categories: hoaxes, natural phenomena, and the rest. To understand these sightings, we must think about the people who report them.

A few UFO sightings are hoaxes. These usually are the most dramatic, get the most press coverage, and often lead to books, lectures, and magazine articles. We must not forget that many people earn a good living from UFOs, and that a successful hoax is worth a lot of money to many people.

Most UFO sightings are made by honest people who mistake natural phenomena for something peculiar. The brightest stars and planets, when seen near the horizon, twinkle so violently they can appear to be flashing colored lights and rotating. Observatories receive many such reports when Venus is brilliant in the evening or morning sky. Other objects mistaken for UFOs include airplanes, weather balloons, clouds, and birds.

Once we discount hoaxes and mistakes, we have only a few sightings left. These are the ones that cannot be explained so easily but do appear to be honest reports. Some of these are misinterpretations of original descriptions. For example, Uncle George tells us he saw a bright light above his barn (Venus?), and we tell a reporter that our uncle saw a bright light hovering above his barn, and the reporter prints a story about a brilliant object hovering above a barn. Similarly, we all tend to exaggerate a story to make it just a bit better than the actual event. In the telling of the story, fishes get bigger and UFOs get more elaborate.

Some of the remaining sightings can be attributed to a well-known psychological principle—we see what we expect to see. Thus people who read about UFOs regularly are more likely to see one. Also, if we view, say, a meteor and in the first split second misidentify it as a spaceship, our brain quickly fits additional details into the spaceship pattern and we think we see a glowing spaceship with illuminated ports and a flaming exhaust. It is quite possible for an honest person to describe such a natural event in such a way that it cannot be identified with any natural phenomena.

Thus we should expect any large set of observations to contain some residual of unexplainable reports. But that does not prove that none of those reports is valid. To test such a conclusion, we would have to study all reports in great detail, and a number of teams of scientists have done exactly that. None of these studies has found even one sighting that is conclusive.

A few defenders of the UFO principle claim that the government, the military, and the scientific community are covering up the truth in a grand conspiracy. This hypothesis is very difficult to accept. Recent history tells us that governments are not very good at covering things up, even when only a few people know the truth, and a grand conspiracy to hide the truth about UFOs would involve the government, the military, the press, the airline industry, and the scientific community not only in the United States but all around the world. A visit from aliens from another world would be the news story of all time, the scientific discovery to assure a scientist of everlasting fame. It is very difficult to believe that such a conspiracy of silence could survive.

In fact, a visit from people beyond Earth would be a wonderful experience for our planet and might usher in a new age of understanding among the peoples of our world. Such aliens might exist, and it is possible though highly unlikely that they could travel to Earth. But at present there is not a single observation to support the idea that UFOs are spacecraft from other worlds.

A-3 SCIENCE AND PSEUDOSCIENCE

UFOs and astrology are only two of the subjects that we can describe as pseudosciences. A *pseudoscience* is something that pretends to be a science but does not obey the rules of good conduct common to all sciences. Thus such subjects are false sciences.

True science is a method of studying nature. It is a set of rules that prevents scientists from lying to one another or to themselves. Hypotheses must be open to testing and must be revised in the face of contradictory evidence. All evidence must be considered and all alternative hypotheses must be explored. The rules of good science are nothing more than the rules of good thinking—that is, the rules of intellectual honesty.

You have probably read of some subjects that are pseudosciences. A few years ago, popular paperbacks claimed that all the planets would line up and Earth would be shocked by earthquakes and storms. Nothing happened. A few years before that, popular books claimed that ancient astronauts (aliens from another world) had visited Earth and taught primitive humans how to be civilized. A few decades ago, a popular book argued that many of the events in the Bible could be explained if the planet Venus had been expelled from Jupiter and had collided with Earth. You can still buy paperbacks claiming that Earth is hollow and inhabited by a superrace. Of course, every generation has its pseudoscientists who claim to cure our worst diseases with quack treatments. Quite a number of people make their living promoting pseudoscientific ideas.

When you meet such a fantastic hypothesis, ask yourself if it is pseudoscience. Does the author discuss all of the relevant evidence or only that which supports the hypothesis? Does the author consider all alternative hypotheses? That is, is the author being intellectually honest? Such questions can usually separate pseudoscience from real science.

Ultimately, you and I must reach our own conclusions, but we can deal most effectively with astrology, UFOs, and other pseudosciences if we remember that people often believe in things for reasons that go beyond the rules of good logic. If we are to be astronomers and help others understand their place in the universe, we must help them use the same rules of intellectual honesty that we demand of ourselves.

RECOMMENDED READING

BOK, BART, and L. JEROME. *Objections to Astrology.* Buffalo, N.Y.: Prometheus Books, 1975.

CULVER, ROGER B., and PHILIP A. IANNA. *The Gemini Syndrome.* Tucson, Ariz.: Pachart, 1979.

GOLDBERG, S. "Is Astrology Science?" *The Humanist* 39 (March 1974), p. 9.

JEROME, L. *Astrology Disproved.* Buffalo, N.Y.: Prometheus Books, 1977.

KELLY, IVAN. "The Scientific Case Against Astrology." *Mercury* 9 (November/December 1980), p. 135.

SAGAN, CARL, and THORNTON PAGE, eds. *UFOs: A Scientific Debate.* New York: Norton, 1974.

B

Astronomical Coordinates and Time

The rotation of Earth on its axis is of great importance in astronomy. It defines a system of latitude and longitude used to locate places on Earth, and that system, projected into the sky, provides an astronomical coordinate system. In addition, the rotation of Earth is the basis of our timekeeping.

B-1 ASTRONOMICAL COORDINATES

Earth is girded by a mesh of imaginary lines that define the latitude and longitude of every spot on Earth's surface. An observer's latitude is measured north or south from Earth's equator. People living on Earth's equator—in Quito, Ecuador, for instance—have latitude 0°, while an encampment of explorers at Earth's North Pole are at latitude 90°. The people of New York City live at latitude 40°45′, and those in Los Angeles at 34°05′. Because latitude is measured north from the equator, people living south of the equator have negative latitudes. Thus we can draw lines around Earth parallel to the equator and refer to them as circles of latitude (Figure B-1a).

Longitude is measured east or west of Greenwich, England. Thus we can say that New York is 74° west of Greenwich. A north–south line through New York City extending from the North Pole to the South Pole is a line of longitude, and anyone on that line has the same longitude as New York City. Consequently, we can cover Earth with north–south lines of longitude (Figure B-1b).

It is common to express a longitude as a time instead of an angle. This arises because Earth rotates through 360° in 24 hours. Thus 15° is equivalent to 1 hour. Instead of saying New York City is 74° west of Greenwich, we could say it is 4^h56^m west.

A similar system of lines on the sky defines the celestial coordinate system, but instead of latitude we refer to a star's **declination,** and instead of its longitude we refer to its **right ascension.** These are often abbreviated **Dec.** and **R.A.** In some books and tables, you may find declination abbreviated δ and right ascension abbreviated α.

Declination is measured north or south from the celestial equator. A star on the celestial equator has declination 0°, and a hypothetical star at the north celestial pole has declination 90°. Polaris, the North Star, is not precisely at the pole, so its declination, 89°11′, is slightly less than 90°. Stars located south of the celestial equator have negative declinations, such as Rigel at −8°13′. Thus lines drawn around the celestial sphere parallel to the

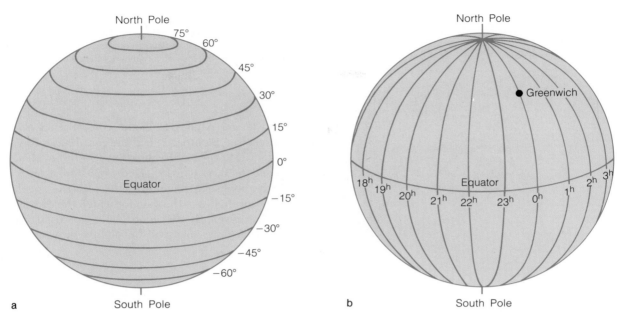

Figure B-1 (a) Lines on Earth parallel to Earth's equator are lines of constant latitude. (b) North–south lines extending from Earth's North Pole to its South Pole are lines of constant longitude. Greenwich, England, is the reference point from which longitude is measured.

celestial equator are lines of constant declination (Figure B-2a).

Before we can measure right ascension, we must have some reference mark on the sky from which to measure. Longitude is measured east or west from Greenwich, but there are no cities on the sky, and the stars are all moving slowly as they orbit the center of our galaxy. Faced with this dilemma, astronomers choose as their reference point the vernal equinox, the point on the celestial equator where the sun crosses into the northern sky in the spring. As we will see later, even the vernal equinox is moving slowly.

Unlike longitude, right ascension is always measured eastward from the reference mark and is nearly always expressed in units of time. We can say the star Sirius has a right ascension of 6^h44^m, meaning it lies 6 hours and 44 minutes east of the vernal equinox. Thus we divide the celestial equator into 24 equal divisions called hours of right ascension, and we draw 24 equally spaced north–south lines from the north celestial pole to the south celestial pole to serve as lines of constant right ascension (Figure B-2b).

Using this mesh of declination and right ascension lines, we can define the declination and right ascension of any object. For instance, the nearest star to the sun, α Centauri, is located at R.A. 14^h38^m, Dec. $-60°46'$. We could also give the right ascension and declination of the sun, moon, and planets, but since these objects move relatively rapidly, their coordinates change from day to day. Thus we would have to list them in an almanac, giving the location of the object on specific dates.

This coordinate system is based on the rotation of Earth, but Earth's axis of rotation is not fixed in space. It precesses in a conical motion, taking 26,000 years for one circuit (see Figure 3-3). Because this motion changes the orientation of Earth's axis, it also changes the direction toward the celestial poles and the celestial equator. Thus the mesh of imaginary lines that define the celestial coordinate system slips slowly across the sky. As a result, the vernal equinox, the intersection of the celestial equator and the ecliptic, moves slowly westward along the celestial equator.

This is a slow process, but it significantly affects the coordinates of a celestial body. For example, in

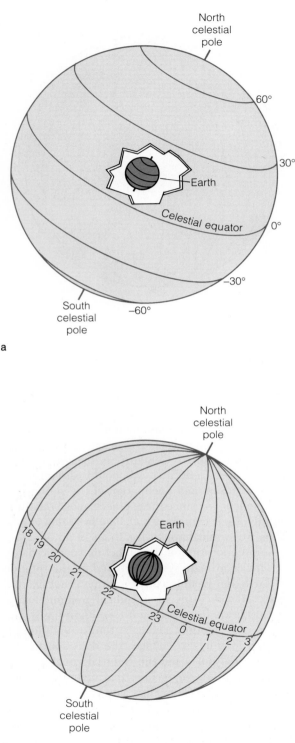

Figure B-2 The celestial coordinate system is the projection on the sky of latitude and longitude. (a) On the celestial sphere, lines of constant declination are parallel to the celestial equator. (b) Lines of constant right ascension run north–south from celestial pole to celestial pole.

a

b

1970 the star Alpheratz in the constellation Andromeda (see Figure 2-2) had the coordinates R.A. 0^{h}6^{m}48^s, Dec. +28°55′, but by 1980 they had changed to R.A. 0^{h}7^{m}18^s, Dec. +28°58′. Consequently, precise astronomical tables list the epoch of the coordinates—the date for which the coordinates are given. If the coordinates are used for some other date, they will be slightly in error.

B-2 ASTRONOMICAL TIME SYSTEMS

Though an object's coordinates specify the direction toward that object, the rotation of Earth makes the entire sky seem to rotate around Earth. Consequently, we cannot use an object's coordinates to point a telescope at it until we know which direction the earth is facing in its daily rotation.

Since the sky is divided into 24 equal intervals of right ascension, it is simple to set up a 24 hour clock to read 0:00 hours when the vernal equinox is on our local celestial meridian—the north–south line that goes through the observer's zenith. As time goes by, the rotation of Earth will cause the sky to move westward, and the right ascension of objects on the local celestial meridian will gradually increase. If the clock is properly adjusted, it will keep pace with this change, and we can tell by glancing at its dial the right ascension of objects currently on the local celestial meridian. This kind of clock would keep **sidereal time**—time based on the motion of the stars across the sky.

However, we would not want to strap a sidereal clock to our wrist to govern our daily activities. To see why, imagine that we set our clocks at noon on the day of the vernal equinox. Both the sun and the vernal equinox would be on the local celestial meridian, so we would set a normal clock to read 12:00 noon, and we would set our sidereal clock to read 0:00 hours. The next day, when our sidereal clock read 0:00 hours again, the vernal equinox would be back on the local celestial meridian.

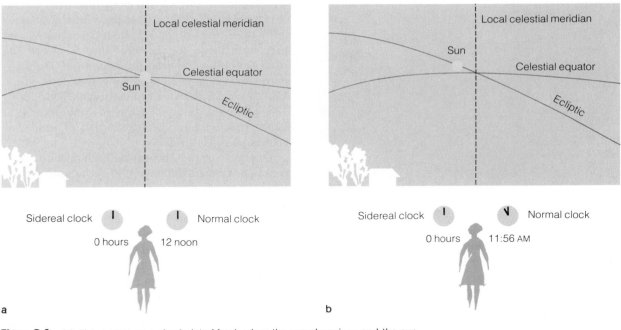

Figure B-3 (a) About noon on a day in late March when the vernal equinox and the sun are on the local celestial meridian, we set a normal clock to 12:00 noon and a sidereal clock to 0:00 hours. (b) One day later, Earth has rotated once on its axis, the vernal equinox is back on the local celestial meridian, and our sidereal clock reads 0:00 hours. The sun, however, has moved eastward along the ecliptic, and our normal clock reads 11:56 A.M.

During the interval of one day, however, Earth would have moved along its orbit, and consequently the sun would have moved eastward about 1° along the ecliptic. It would not be at the vernal equinox, but about 1° east. Since Earth requires 4 minutes to rotate 1°, our normal clock would read 11:56 A.M. (Figure B-3). Thus a sidereal clock that keeps track of the stars must run 4 minutes a day faster than a normal clock that keeps track of the sun. If we tried to run our lives according to a sidereal clock, it would gain 4 minutes a day, and in 6 months we would be eating supper at sunrise and breakfast at sunset.

B-3 ZONE TIME

It would, in fact, be impossible to regulate a wristwatch to keep precise time according to the sun. Because Earth's orbit is slightly elliptical, it moves more rapidly when it is closer to the sun and more slowly when it is farther away. This makes the sun appear to move more rapidly along the ecliptic in January than in July. This and other factors make the sun's motion around the sky slightly nonuniform. A wristwatch, running at a steady rate, could not keep track of the exact position of the sun.

This is one reason a sundial does not usually agree with our wristwatch. The sundial reads the true apparent solar time. Our wristwatch gives the time, assuming the sun moves at a constant rate around the sky. The two can differ by as much as 15 minutes.

Another reason sundials don't agree with our wristwatches is related to time zones. If one family lived 50 miles west of another family, their local celestial meridians would differ. The sun would first cross the local celestial meridian of the eastern family, and their sundial would read noon. However, the western family would see the sun slightly east of their meridian, and their sundial would say

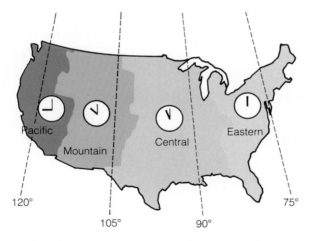

Figure B-4 For convenience, we divide Earth into time zones, and all clocks within a zone keep the same time as a clock on the zone's central meridian. Adjacent zones differ by 1 hour.

1883), and every clock in a given time zone was set to the same time. Philadelphia and Pittsburgh have local times that differ by about 20 minutes, but they both lie in the Eastern Time Zone, so the clocks in both cities read the same time. The standard time adopted in each time zone is the time kept by clocks on its central meridian (Figure B-4). Thus, when it is 12:00 noon in New York, it is 11:00 A.M. in Kansas, 10:00 A.M. in Colorado, and 9:00 A.M. in California.

One kind of time is especially important in astronomy. **Universal Time (UT)** is the same as the local mean time on the prime meridian passing through Greenwich, England—that is, it is the same as **Greenwich Mean Time (GMT).** Most astronomers use UT in their work no matter what time zone they live in, and most astronomical almanacs list the times of celestial events in UT. Universal Time is expressed on a 24-hour clock with 0:00 hours occurring at 7:00 A.M. Eastern Standard Time.

Whatever time zone we live in, when we compare our wristwatches with a sundial, we compare zone time to local time. The sundial is slow by 4 minutes for every degree of longitude it lies west of the central meridian of the time zone. This combined with the uneven motion of Earth along its orbit can add up to an appreciable difference between zone time and a sundial.

it was a few minutes before noon. Thus every person with a different longitude has a different time called local time.

This made no difference in the distant past, but with the advent of rapid communication and transportation it became very inconvenient for the clocks in one town to be a few minutes different from the clocks in the next town to the west. To solve this problem, time zones were established (in

NEW TERMS

declination (Dec.)
right ascension (R.A.)
sidereal time

Universal Time (UT)
Greenwich Mean Time
   (GMT)

FURTHER EXPLORATION

VOYAGER, the Interactive Desktop Planetarium, can be used for further exploration of the topics in Appendix B. This appendix corresponds with the following projects or chapters in *Voyages Through Space and Time:* Project 2, The Horizon System; Project 4, The Equatorial System; Project 9, Types of Time; Project 10, The Equation of Time; Project 11, Sidereal Time; Chapter 2, Locating Objects on the Earth and in the Sky; Chapter 3, Telling Time by the Stars.

APPENDIX
C

Systems of Units

SCIENTIFIC NOTATION

It is often convenient to write very large numbers using **scientific notation,** that is, with powers of 10. For example, the nearest star is about 43,000,000,000,000 km from the sun. It is easier to write this number as 4.3×10^{13} km.

Very small numbers can also be written with powers of 10. For example, the wavelength of visible light is about 0.0000005 m. In powers of 10 this becomes 5×10^{-7} m.

The powers of 10 used in scientific notation are shown below. The exponent tells us how to move the decimal point. If the exponent is positive, we move the decimal point to the right. If the exponent is negative, we move the decimal point to the left. Thus 2×10^3 equals 2000.0 and 2×10^{-3} equals 0.002.

	·	10^0	$= 1$
	·	10^{-1}	$= 0.1$
	·	10^{-2}	$= 0.01$
$10^5 =$	100,000	10^{-3}	$= 0.001$
$10^4 =$	10,000	10^{-4}	$= 0.0001$
$10^3 =$	1,000	·	
$10^2 =$	100	·	
$10^1 =$	10	·	

Temperature Scales

	Kelvin (K)	Centigrade (°C)	Fahrenheit (°F)
Absolute zero	0 K	−273°C	−459°F
Freezing point of water	273 K	0°C	32°F
Boiling point of water	373 K	100°C	212°F

Conversions $K = °C + 273$ $°C = \frac{5}{9}(°F - 32)$ $°F = \frac{9}{5}°C + 32$

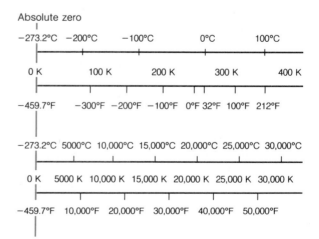

Units Used in Astronomy

1 Ångstrom (Å)	$= 10^{-8}$ cm
	$= 10^{-10}$ m
1 astronomical unit (AU)	$= 1.495979 \times 10^{11}$ m
	$= 92.95582 \times 10^{6}$ miles
1 light-year (ly)	$= 6.3240 \times 10^{4}$ AU
	$= 9.46053 \times 10^{15}$ m
	$= 5.9 \times 10^{12}$ miles
1 parsec (pc)	$= 206265$ AU
	$= 3.085678 \times 10^{16}$ m
	$= 3.261633$ ly
1 kiloparsec (kpc)	$= 1000$ pc
1 megaparsec (Mpc)	$= 1,000,000$ pc

Constants

astronomical unit (AU)	$= 1.495979 \times 10^{11}$ m
parsec (pc)	$= 206265$ AU
	$= 3.085678 \times 10^{16}$ m
	$= 3.261633$ ly
light-year (ly)	$= 9.46053 \times 10^{15}$ m
velocity of light (c)	$= 2.997925 \times 10^{8}$ m/sec
gravitational constant (G)	$= 6.67 \times 10^{-11}$ N·m^2/kg^2
mass of Earth ($M_\oplus$)	$= 5.976 \times 10^{24}$ kg
equatorial radius of Earth	$= 6378.164$ km
mass of sun ($M_\odot$)	$= 1.989 \times 10^{30}$ kg
radius of sun ($R_\odot$)	$= 6.9599 \times 10^{8}$ m
solar luminosity ($L_\odot$)	$= 3.826 \times 10^{26}$ J/sec
mass of moon	$= 7.350 \times 10^{22}$ kg
radius of moon	$= 1738$ km
mass of H atom	$= 1.67352 \times 10^{-27}$ kg

Metric Units and Conversion Factors

Length		U.S. Equivalent
1 meter	$= 100$ cm	$= 39.36$ inches
	$= 1000$ mm	$= 1.0933$ yards
1 kilometer (km)	$= 1000$ m	$= 0.6214$ miles
1 centimeter (cm)	$= 10$ mm	$= 0.39$ inches
	$= 1/100$ m	
1 millimeter (mm)	$= 10^{-3}$ m	$= 0.039$ inches
	$= 1/1000$ m	
1 micrometer (μm) [1 micron (μ)]	$= 10^{-6}$ m	$= 0.00039$ inches
1 nanometer (nm)	$= 10^{-9}$ m	

Mass		
1 gram (g)	$= 1/1000$ kg	$= 0.035$ ounces $= 1/453.9$ pounds
1 kilogram (kg)	$= 1000$ g	$= 2.205$ pounds

Astronomical Data

The Constellations

		Approximate Position	
		R. A.* (h)	Dec.* (°)
Andromeda (And)	The Princess	1	+40
Antlia (Ant)	The Air Pump	10	−35
Apus (Aps)	The Bird of Paradise	16	−75
Aquarius (Aqr)	The Water Bearer	23	−15
Aquila (Aql)	The Eagle	20	+5
Ara (Ara)	The Atlar	17	−55
Aries (Ari)	The Ram	3	+20
Auriga (Aur)	The Charioteer	6	+40
Boötes (Boo)	The Bear Driver	15	+30
Caelum (Cae)	The Sculptor's Chisel	5	−40
Camelopardus (Cam)	The Giraffe	6	+70
Cancer (Cnc)	The Crab	9	+20
Canes Venatici (CVn)	The Hunting Dogs	13	+40
Canis Major (CMa)	The Greater Dog	7	−20
Canis Minor (CMi)	The Smaller Dog	8	+5
Capricornus (Cap)	The Sea Goat	21	−20

*R. A. and Dec. refer to right ascension and declination (see Appendix B).

		Approximate Position	
		R. A.* (h)	Dec.* (°)
Carina (Car)	The Keel	9	−60
Cassiopeia (Cas)	The Seated Queen	1	+60
Centaurus (Cen)	The Centaur	13	−50
Cepheus (Cep)	The King	22	+70
Cetus (Cet)	The Whale	2	−10
Chamaeleon (Cha)	The Chameleon	11	−80
Circinus (Cir)	The Compasses	15	−60
Columba (Col)	The Dove	6	−35
Coma Berenices (Com)	Berenice's Hair	13	+20
Corona Australis (CrA)	The Southern Crown	19	−40
Corona Borealis (CrB)	The Northern Crown	16	+30
Corvus (Crv)	The Crow	12	−20
Crater (Crt)	The Cup	11	−15
Crux (Cru)	The Southern Cross	12	−60
Cygnus (Cyg)	The Swan	21	+40
Delphinus (Del)	The Dolphin	21	+10
Dorado (Dor)	The Swordfish	5	−65
Draco (Dra)	The Dragon	17	+65
Equuleus (Equ)	The Foal	21	+10
Eridanus (Eri)	The River	3	−20
Fornax (For)	The Laboratory Furnace	3	−30
Gemini (Gem)	The Twins	7	+20
Grus (Gru)	The Crane	22	−45
Hercules (Her)	Hercules	17	+30
Horologium (Hor)	The Clock	3	−60
Hydra (Hya)	The Water Serpent	10	−20
Hydrus (Hyi)	The Water Snake	2	−75
Indus (Ind)	The American Indian	21	−55
Lacerta (Lac)	The Lizard	22	+45
Leo (Leo)	The Lion	11	+15
Leon MInor (LMi)	The Lion Cub	10	+35
Lepus (Lep)	The Hare	6	−20
Libra (Lib)	The Scales	15	−15
Lupus (Lup)	The Wolf	15	−45
Lynx (Lyn)	The Lynx	8	+45
Lyra (Lyr)	The Lyre	19	+40
Mensa (Men)	The Table Mountain	5	−80

		Approximate Position	
		R. A.* (h)	Dec.* (°)
Microscopium (Mic)	The Microscope	21	−35
Monoceros (Mon)	The Unicorn	7	−5
Musca (Mus)	The Fly	12	−70
Norma (Nor)	The Carpenter's Square	16	−50
Octans (Oct)	The Octant	22	−85
Ophiuchus (Oph)	The Serpent Holder	17	0
Orion (Ori)	The Great Hunter	5	0
Pavo (Pav)	The Peacock	20	−65
Pegasus (Peg)	The Winged Horse	22	+20
Perseus (Per)	The Hero	3	+45
Phoenix (Phe)	The Phoenix	1	−50
Pictor (Pic)	The Painter's Easel	6	−55
Pisces (Psc)	The Fishes	1	+15
Piscis Austrinus (PsA)	The Southern Fish	22	−30
Puppis (Pup)	The Stern	8	−40
Pyxis (Pyx)	The Compass Box	9	−30
Reticulum (Ret)	The Net	4	−60
Sagitta (Sge)	The Arrow	20	+10
Sagittarius (Sgr)	The Archer	19	−25
Scorpius (Sco)	The Scorpion	17	−40
Sculptor (Scl)	The Sculptor's Workshop	0	−30
Scutum (Sct)	The Shield	19	−10
Serpens (Ser)	The Serpent	17	0
Sextans (Sex)	The Sextant	10	0
Taurus (Tau)	The Bull	4	+15
Telescopium (Tel)	The Telescope	19	−50
Triangulum (Tri)	The Triangle	2	+30
Triangulum Australe (TrA)	The Southern Triangle	16	−65
Tucana (Tuc)	The Toucan	0	−65
Ursa Major (UMa)	The Greater Bear	11	+50
Ursa Minor (UMi)	The Smaller Bear	15	+70
Vela (Vel)	The Sail	9	−50
Virgo (Vir)	The Maiden	13	0
Volans (Vol)	The Flying Fish	8	−70
Vulpecula (Vul)	The Fox	20	+25

The Brightest Stars

Star	Name	Apparent Visual Magnitude (m_v)	Spectral Type	Absolute Visual Magnitude (M_v)	Color Index (B-V)	Distance (ly)
α CMa A	Sirius	−1.47	A1	1.4	0.00	8.7
α Car	Canopus	−0.72	F0	−3.1	0.15	98
α Cen	Rigil Kentaurus	−0.01	G2	4.4	0.71	4.3
α Boo	Arcturus	−0.06	K2	−0.3	1.23	36
α Lyr	Vega	0.04	A0	0.5	0.00	26.5
α Aur	Capella	0.05	G8	−0.6	0.80	45
β Ori A	Rigel	0.14	B8	−7.1	−0.03	900
α Cmi A	Procyon	0.37	F5	2.7	0.42	11.3
α Ori	Betelgeuse	0.41	M2	−5.6	0.03	520
α Eri	Achernar	0.51	B3	−2.3	−0.16	118
β Cen AB	Hadar	0.63	B1	−5.2	−0.23	490
α Aql	Altair	0.77	A7	2.2	0.22	16.5
α Tau A	Aldebaran	0.86	K5	−0.7	1.54	68
α Cru	Acrux	0.90	B2	−3.5	−0.25	260
α Vir	Spica	0.91	B1	−3.3	−0.23	220
α SCO A	Antares	0.92	M1	−5.1	1.83	520
α PsA	Fomalhaut	1.15	A3	2.0	0.09	22.6
β Gem	Pollux	1.16	K0	1.0	1.00	35
α Cyg	Deneb	1.26	A2	−7.1	0.09	1600
β Cru	Beta Crucis	1.28	B0.5	−4.6	−0.23	490

The Nearest Stars

Name	Absolute Magnitude (M_V)	Distance (ly)	Spectral Type	Color Index (B-V)	Apparent Visual Magnitude (m_V)
Sun	4.83		G2	0.65	−26.8
α Cen A	4.38	4.3	G2	0.68	0.1
B	5.76	4.3	K5	0.88	1.5
Barnard's Star	13.21	5.9	M5	1.74	9.5
Wolf 359	16.80	7.6	M6	2.01	13.5
Lalande 21185	10.42	8.1	M2	1.51	7.5
Sirius A	1.41	8.6	A1	0.00	−1.5
B	11.54	8.6	white dwarf	—	7.2
Luyten 726-8A	15.27	8.9	M5	—	12.5
B (UV Cet)	15.8	8.9	M6	—	13.0
Ross 154	13.3	9.4	M5	—	10.6
Ross 248	14.8	10.3	M6	1.92	12.2
ε Eri	6.13	10.7	K2	0.88	3.7
Luyten 789-6	14.6	10.8	M7	1.96	12.2
Ross 128	13.5	10.8	M5	1.76	11.1
61 CYG A	7.58	11.2	K5	1.17	5.2
B	8.39	11.2	K7	1.37	6.0
ε Ind	7.0	11.2	K5	1.05	4.7
Procyon A	2.64	11.4	F5	0.42	0.3
B	13.1	11.4	white dwarf	—	10.8
Σ 2398 A	11.15	11.5	M4	1.54	8.9
B	11.94	11.5	M5	1.59	9.7
Groombridge 34 A	10.32	11.6	M1	1.46	8.1
B	13.29	11.6	M6	—	11.0
Lacaille 9352	9.59	11.7	M2	1.51	7.4
τ Ceti	5.72	11.9	G8	0.72	3.5
BD +5° 1668	11.98	12.2	M5	1.56	9.8
L 725-32	15.27	12.4	M5	1.59	11.5
Lacaille 8760	8.75	12.5	M0	1.38	6.7
Kapteyn's Star	10.85	12.7	M0	1.56	8.8
Kruger 60 A	11.87	12.8	M3	1.62	9.7
B	13.3	12.8	M4	1.80	11.2

Properties of Main-Sequence Stars

Spectral Type	Absolute Visual Magnitude (M_v)	Luminosity*	Temperature (K)	λ max (nm)	Mass*	Radius*	Average Density (gm/cm³)
O5	−5.8	501,000	40,000	72.4	40	17.8	0.01
B0	−4.1	20,000	28,000	100	18	7.4	0.1
B5	−1.1	790	15,000	190	6.4	3.8	0.2
A0	+0.7	79	9900	290	3.2	2.5	0.3
A5	+2.0	20	8500	340	2.1	1.7	0.6
F0	+2.6	6.3	7400	390	1.7	1.4	1.0
F5	+3.4	2.5	6600	440	1.3	1.2	1.1
G0	+4.4	1.3	6000	480	1.1	1.0	1.4
G5	+5.1	0.8	5500	520	0.9	0.9	1.6
K0	+5.9	0.4	4900	590	0.8	0.8	1.8
K5	+7.3	0.2	4100	700	0.7	0.7	2.4
M0	+9.0	0.1	3500	830	0.5	0.6	2.5
M5	+11.8	0.01	2800	1000	0.2	0.3	10.0
M8	+16	0.001	2400	1200	0.1	0.1	63

*Luminosity, mass, and radius are given in terms of the sun's luminosity, mass, and radius.

Planets: Physical Properties (Earth = ⊕)

Planet	Equatorial Radius (km)	Equatorial Radius (⊕ = 1)	Mass (⊕ = 1)	Average Density (g/cm³)	Surface Gravity (⊕ = 1)	Escape Velocity (km/sec)	Sidereal Period of Rotation	Inclination of Equator to Orbit
Mercury	2439	0.38	0.0554	5.44	0.378	4.2	59^d	<28°
Venus	6052	0.95	0.815	5.24	0.894	10.3	244.3^d	177°
Earth	6378	1.00	1.00	5.497	1.00	11.2	23^{h}56^{m}04.1^s	23°27′
Mars	3398	0.53	0.1075	3.9	0.379	5.0	24^{h}37^{m}22.6^s	23°59′
Jupiter	71,400	11.18	317.83	1.34	2.54	61	9^{h}50^{m}30^s	3°5′
Saturn	60,330	9.42	95.147	0.7	1.07	37	10^{h}14^m	26°44′
Uranus	25,559	4.01	14.54	1.19	0.919	22	17^{h}14^m	97°55′
Neptune	24,750	3.93	17.23	1.66	1.19	25	16^{h}3^m	28°48′
Pluto	1151	0.18	0.0022	1.84	0.06	1.2?	6^{d}9^{h}21^m	122°

Planets: Orbital Properties

Planet	Semimajor Axis (a) (AU)	(10⁶ km)	Orbital Period (P) (y)	(days)	Average Orbital Velocity (km/sec)	Orbital Eccentricity	Inclination to Ecliptic
Mercury	0.3871	57.9	0.24084	87.96	47.89	0.2056	7°0′26″
Venus	0.7233	108.2	0.61515	224.68	35.03	0.0068	3°23′40″
Earth	1	149.6	1	365.26	29.79	0.0167	0°0′14″
Mars	1.5237	227.9	1.8808	686.95	24.13	0.0934	1°51′09″
Jupiter	5.2028	778.3	11.867	4334.3	13.06	0.0484	1°18′29″
Saturn	9.5388	1427.0	29.461	10,760	9.64	0.0560	2°29′17″
Uranus	19.1914	2871.0	84.014	30,685	6.81	0.0461	0°46′23″
Neptune	30.0611	4497.1	164.793	60,189	5.43	0.0100	1°46′27″
Pluto	39.44	5900	247.7	90,465	4.74	0.2484	17°9′3″

Satellites of the Solar System

Planet	Satellite	Radius (km)	Distance from Planet (10³ km)	Orbital Period (days)	Orbital Eccentricity	Orbital Inclination
Earth	Moon	1738	384.4	27.322	0.055	5°8′43″
Mars	Phobos	14 × 10	9.38	0.3189	0.018	1°.0
	Deimos	8 × 6	23.5	1.262	0.002	2°.8
Jupiter J16	Metis	20	126	0.29	0.0	0°.0
J15	Adrastea	18	128	0.294	0.0	0°.0
J5	Amalthea	135 × 78	182	0.4982	0.003	0°.45
J14	Thebe	38	223	0.674	0.0	1°.3
J1	Io	1820	422	1.769	0.000	0°.3
J2	Europa	1565	671	3.551	0.000	0°.46
J3	Ganymede	2640	1071	7.155	0.002	0°.18
J4	Callisto	2420	1884	16.689	0.008	0°.25
J13	Leda	~4	11,110	240	0.146	26°.7
J6	Himalia	~85	11,470	250.6	0.158	27°.6
J10	Lysithea	~10	11,710	260	0.12	29°
J7	Elara	~30	11,740	260.1	0.207	24°.8
J12	Ananke	8?	20,700	617	0.169	147°
J11	Carme	12?	22,350	692	0.207	163°
J8	Pasiphae	14?	23,300	735	0.40	147°
J9	Sinope	10?	23,700	758	0.275	156°

Planet	Satellite	Radius (km)	Distance from Planet (10^3 km)	Orbital Period (days)	Orbital Eccentricity	Orbital Inclination
Saturn	Pan	10	133.570	0.574	0.000	0°
S15	Atlas	25 × 10	137.7	0.601	0.002	0°.3
S14	Prometheus	70 × 40	139.4	0.613	0.003	0°.0
S13	Pandora	55 × 35	141.7	0.629	0.004	0°.05
S11	Epimetheus	70 × 50	151.42	0.694	0.009	0°.34
S10	Janus	110 × 80	151.47	0.695	0.007	0°.14
S1	Mimas	196	185.54	0.942	0.020	1°.5
S2	Enceladus	250	238.04	1.370	0.004	0°.0
S3	Tethys	530	294.67	1.888	0.000	1°.1
S16	Calypso	17 × 11	294.67	1.888	0.0	~1°?
S17	Telesto	17 × 13	294.67	1.888	0.0	~1°?
S4	Dione	560	377	2.737	0.002	0°.0
S12	Helene	18 × 15	377	2.74	0.005	0°.15
S5	Rhea	765	527	4.518	0.001	0°.4
S6	Titan	2575	1222	15.94	0.029	0°.3
S7	Hyperion	205 × 110	1484	21.28	0.104	~0°.5
S8	Iapetus	730	3562	79.33	0.028	14°.72
S9	Phoebe	110	12,930	550.4	0.163	150°
Uranus	Cordelia	25	49.8	0.3333	~0	~0°
	Ophelia	25	53.8	0.375	~0	~0°
	Bianca	25	59.1	0.433	~0	~0°
	Cressida	30	61.8	0.462	~0	~0°
	Desdemona	30	62.7	0.475	~0	~0°
	Juliet	40	64.4	0.492	~0	~0°
	Portia	40	66.1	0.512	~0	~0°
	Rosalind	30	69.9	0.558	~0	~0°
	Belinda	30	75.2	0.621	~0	~0°
	Puck	85 ± 5	85.9	0.762	~0	~0°
U5	Miranda	242 ± 5	129.9	1.414	0.017	3°.4
U1	Ariel	580 ± 5	190.9	2.520	0.003	0°
U2	Umbriel	595 ± 10	266.0	4.144	0.003	0°
U3	Titania	805 ± 5	436.3	8.706	0.002	0°
U4	Oberon	775 ± 10	583.4	13.463	0.001	0°
Neptune	Naiad	30	48.2	0.296	~0	~0°
	Thalassa	40	50.0	0.312	~0	~0°
	Despina	90	52.5	0.333	~0	~0°
	Galatea	75	62.0	0.396	~0	~0°
	Larissa	95	73.6	0.554	~0	~0°
	Proteus	205	117.6	1.121	~0	~0°
N1	Triton	1352	354.59	5.875	0.00	160°
N2	Nereid	170	5588.6	360.125	0.76	27.7°
Pluto	Charon	596	19.7	6.38718	~0	122°

Meteor Showers

Shower	Dates	Hourly Rate	Radiant R. A.	Dec.	Associated Comet
Quadrantids	Jan. 2–4	30	15^h24^m	50°	
Lyrids	April 20–22	8	18^h4^m	33°	1861 I
η Aquarids	May 2–7	10	22^h24^m	0°	Halley?
δ Aquarids	July 26–31	15	22^h36^m	−10°	
Perseids	Aug. 10–14	40	3^h4^m	58°	1982 III
Orionids	Oct. 18–23	15	6^h20^m	15°	Halley?
Taurids	Nov. 1–7	8	3^h40^m	17°	Encke
Leonids	Nov. 14–19	6	10^h12^m	22°	1866 I Temp
Geminids	Dec. 10–13	50	7^h28^m	32°	

The Messier Objects

M	NGC*	Right Ascension (1950) (h)	(m)	Declination (1950) (°)	(')	Apparent Visual Magnitude (m_v)	Description
1	1952	5	31.5	+21	59	8.4	Crab Nebula in Taurus; remains of supernova
2	7089	21	30.9	−1	02	6.4	globular cluster in Aquarius
3	5272	13	39.8	+28	38	6.3	globular cluster in Canes Venatici
4	6121	16	20.6	−26	24	6.5	globular cluster in Scorpius
5	5904	15	16.0	+2	16	6.1	globular cluster in Serpens
6	6405	17	36.8	−32	10	5.3	open cluster in Scorpius
7	6475	17	50.7	−34	48	4.1	open cluster in Scorpius
8	6523	18	00.1	−24	23	6.0	Lagoon Nebula in Sagittarius
9	6333	17	16.3	−18	28	7.3	globular cluster in Ophiuchus
10	6254	16	54.5	−4	02	6.7	globular cluster in Ophiuchus
11	6705	18	48.4	−6	20	6.3	open cluster in Scutum
12	6218	16	44.7	−1	52	6.6	globular cluster in Ophiuchus
13	6205	16	39.9	+36	33	5.9	globular cluster in Hercules
14	6402	17	35.0	−3	13	7.7	globular cluster in Ophiuchus
15	7078	21	27.5	+11	57	6.4	globular cluster in Pegasus
16	6611	18	16.1	−13	48	6.4	open cluster with nebulosity in Serpens
17	6618	18	17.9	−16	12	7.0	Swan or Omega Nebula in Sagittarius

* New General Catalog number.

M	NGC	Right Ascension (1950)		Declination (1950)		Apparent Visual Magnitude	Description
		(h)	(m)	(°)	(')	(m_v)	
18	6613	18	17.0	−17	09	7.5	open cluster in Sagittarius
19	6273	16	59.5	−26	11	6.6	globular cluster in Ophiuchus
20	6514	17	59.4	−23	02	9.0	Trifid Nebula in Sagittarius
21	6531	18	01.6	−22	30	6.5	open cluster in Sagittarius
22	6656	18	33.4	−23	57	5.6	globular cluster in Sagittarius
23	6494	17	54.0	−19	00	6.9	open cluster in Sagittarius
24	6603	18	15.5	−18	27	11.4	open cluster in Sagittarius
25	IC4725†	18	28.7	−19	17	6.5	open cluster in Sagittarius
26	6694	18	42.5	−9	27	9.3	open cluster in Scutum
27	6853	19	57.5	+22	35	7.6	Dumb-bell Planetary Nebula in Vulpecula
28	6626	18	21.4	−24	53	7.6	globular cluster in Sagittarius
29	6913	20	22.2	+38	21	7.1	open cluster in Cygnus
30	7099	21	37.5	−23	24	8.4	globular cluster in Capricornus
31	224	0	40.0	+41	00	4.8	Andromeda galaxy
32	221	0	40.0	+40	36	8.7	elliptical galaxy; companion to M 31
33	598	1	31.0	+30	24	6.7	spiral galaxy in Triangulum
34	1039	2	38.8	+42	35	5.5	open cluster in Perseus
35	2168	6	05.7	+24	21	5.3	open cluster in Gemini
36	1960	5	33.0	+34	04	6.3	open cluster in Auriga
37	2099	5	49.1	+32	33	6.2	open cluster in Auriga
38	1912	5	25.3	+35	47	7.4	open cluster in Auriga
39	7092	21	30.4	+48	13	5.2	open cluster in Cygnus
40	—	12	20.0	+58	20	—	close double star in Ursa Major
41	2287	6	44.9	−20	41	4.6	loose open cluster in Canis Major
42	1976	5	32.9	−5	25	4.0	Orion Nebula
43	1982	5	33.1	−5	19	9.0	northeast portion of Orion Nebula
44	2632	8	37.0	+20	10	3.7	Praesepe; open cluster in Cancer
45	—	3	44.5	+23	57	1.6	The Pleiades; open cluster in Taurus
46	2437	7	39.5	−14	42	6.0	open cluster in Puppis
47	2422	7	34.3	−14	22	5.2	loose group of stars in Puppis
48	2458	8	11.0	−5	38	5.5	open cluster in Hydra
49	4472	12	27.3	+8	16	8.5	elliptical galaxy in Virgo
50	2323	7	00.6	−8	16	6.3	loose open cluster in Monoceros
51	5194	13	27.8	+47	27	8.4	Whirlpool spiral galaxy in Canes Venatici
52	7654	23	22.0	+61	20	7.3	loose open cluster in Cassiopeia
53	5024	13	10.5	+18	26	7.8	globular cluster in Coma Berenices
54	6715	18	51.9	−30	32	7.3	globular cluster in Sagittarius

† Index Catalog (IC) number.

M	NGC	Right Ascension (1950)		Declination (1950)		Apparent Visual Magnitude	Description
		(h)	(m)	(°)	(′)	(m_v)	
55	6809	19	36.8	−31	03	7.6	globular cluster in Sagittarius
56	6779	19	14.6	+30	05	8.2	globular cluster in Lyra
57	6720	18	51.7	+32	58	9.0	Ring Nebula; planetary nebula in Lyra
58	4579	12	35.2	+12	05	8.2	barred spiral galaxy in Virgo
59	4621	12	39.5	+11	56	9.3	elliptical spiral galaxy in Virgo
60	4649	12	41.1	+11	50	9.0	elliptical galaxy in Virgo
61	4303	12	19.3	+4	45	9.6	spiral galaxy in Virgo
62	6266	16	58.0	−30	02	6.6	globular cluster in Ophiuchus
63	5055	13	13.5	+42	17	10.1	spiral galaxy in Canes Venatici
64	4826	12	54.2	+21	57	6.6	spiral galaxy in Coma Berenices
65	3623	11	16.3	+13	22	9.4	spiral galaxy in Leo
66	3627	11	17.6	+13	16	9.0	spiral galaxy in Leo; companion to M 65
67	2682	8	48.4	+12	00	6.1	open cluster in Cancer
68	4590	12	36.8	−26	29	8.2	globular cluster in Hydra
69	6637	18	28.1	−32	24	8.9	globular cluster in Sagittarius
70	6681	18	40.0	−32	20	9.6	globular cluster in Sagittarius
71	6838	19	51.5	+18	39	9.0	globular cluster in Sagitta
72	6981	20	50.7	−12	45	9.8	globular cluster in Aquarius
73	6994	20	56.2	−12	50	9.0	open cluster in Aquarius
74	628	1	34.0	+15	32	10.2	spiral galaxy in Pisces
75	6864	20	03.1	−22	04	8.0	globular cluster in Sagittarius
76	650	1	38.8	+51	19	11.4	planetary nebula in Perseus
77	1068	2	40.1	−0	12	8.9	spiral galaxy in Cetus
78	2068	5	44.2	+0	02	8.3	small reflection nebula in Orion
79	1904	5	22.1	−24	34	7.5	globular cluster in Lepus
80	6093	16	14.0	−22	52	7.5	globular cluster in Scorpius
81	3031	9	51.7	+69	18	7.9	spiral galaxy in Ursa Major
82	3034	9	51.9	+69	56	8.4	irregular galaxy in Ursa Major
83	5236	13	34.2	−29	37	10.1	sprial galaxy in Hydra
84	4374	12	22.6	+13	10	9.4	S0 type galaxy in Virgo
85	4382	12	22.8	+18	28	9.3	S0 type galaxy in Coma Berenices
86	4406	12	23.6	+13	13	9.2	elliptical galaxy in Virgo
87	4486	12	28.2	+12	40	8.7	elliptical galaxy in Virgo
88	4501	12	29.4	+14	42	10.2	spiral galaxy in Coma Berenices
89	4552	12	33.1	+12	50	9.5	elliptical galaxy in Virgo
90	4569	12	34.3	+13	26	9.6	spiral galaxy in Virgo
91*	4571(?)	—	—	—	—	—	

* Item of doubtful identification.

M	NGC	Right Ascension (1950)		Declination (1950)		Apparent Visual Magnitude	Description
		(h)	(m)	(°)	(')	(m_V)	
92	6341	17	15.6	+43	12	6.4	globular cluster in Hercules
93	2447	7	42.4	−23	45	6.0	open cluster in Puppis
94	4736	12	48.6	+41	24	8.3	spiral galaxy in Canes Venatici
95	3351	10	41.3	+11	58	9.8	barred galaxy in Leo
96	3368	10	44.1	+12	05	9.3	spiral galaxy in Leo
97	3587	11	12.0	+55	17	12.0	Owl Nebula; planetary nebula in Ursa Major
98	4192	12	11.2	+15	11	10.2	spiral galaxy in Coma Berenices
99	4254	12	16.3	+14	42	9.9	spiral galaxy in Coma Berenices
100	4321	12	20.4	+16	06	10.6	spiral galaxy in Coma Berenices
101	5457	14	01.4	+54	36	9.6	spiral galaxy in Ursa Major
102*	5866(?)	—	—	—	—	—	
103	581	1	29.9	+60	26	7.4	open cluster in Cassiopeia
104	4594	12	37.4	−11	21	8.3	spiral galaxy in Virgo
105	3379	10	45.2	+13	01	9.7	elliptical galaxy in Leo
106	4258	12	16.5	+47	35	8.4	spiral galaxy in Canes Venatici
107	6171	16	29.7	−12	57	9.2	globular cluster in Ophiuchus

* Item of doubtful identification.

The Greek Alphabet

A, α alpha	H, η eta	N, ν nu	T, τ tau
B, β beta	Θ, θ theta	Ξ, ξ xi	Υ, υ upsilon
Γ, γ gamma	I, ι iota	O, o omicron	Φ, ϕ phi
Δ, δ delta	K, κ kappa	Π, π pi	X, χ chi
E, ϵ epsilon	Λ, λ lambda	P, ρ rho	Ψ, ψ psi
Z, ζ zeta	M, μ mu	Σ, σ sigma	Ω, ω omega

Glossary

ā pay   ē be   ī pie   ō so   ŭ cut
ă hat   ĕ pet   ĭ pit   ŏ pot
ä father   î pier   ô paw
$\overline{oo}$ food

Numbers refer to the page where the term is first discussed in the text.

absolute visual magnitude (M_v) (155) Intrinsic brightness of a star. The apparent visual magnitude the star would have if it were 10 pc away.

absorption line (118) A dark line in a spectrum. Produced by the absence of photons absorbed by atoms or molecules.

absorption spectrum (dark-line spectrum) (118) A spectrum that contains absorption lines.

accretion (359) The sticking together of solid particles to produce a larger particle.

accretion disk (230) The whirling disk of gas that forms around a compact object such as a white dwarf, neutron star, or black hole as matter is drawn in.

achondrite (ā·kŏn′drīt) (440) Stony meteorite containing no chondrules or volatiles.

achromatic lens (84) A telescope lens composed of two lenses ground from different kinds of glass and designed to bring two selected colors to the same focus and correct for chromatic aberration.

active-core radio galaxy A galaxy that emits radio energy from its core rather than from lobes. See also **double-lobed radio galaxy.**

active galactic nuclei (AGN) (304) The centers of active galaxies that are emitting large amounts of excess energy. See also **active galaxy.**

active galaxy (304) A galaxy whose center emits large amounts of excess energy, often in the form of radio emission. Active galaxies are suspected of having massive black holes in their centers into which matter is flowing.

active optics (88) Thin telescope mirrors that are controlled by computers to maintain proper shape as the telescope moves.

AGN See **active galactic nuclei.**

albedo (379) The ratio of the light reflected from an object divided by the light that hits the object. Albedo equals 0 for perfectly black and 1 for perfectly white.

alt-azimuth mounting (90) A telescope mounting that allows the telescope to move in altitude (perpendicular to the horizon) and in azimuth (parallel to the horizon). See also **equatorial mounting.**

amino acid (ŭ·mē′nō) (462) Carbon-chain molecule that is the building block of protein.

Angstrom (Å) (ăng′strŭm) (83) A unit of distance. 1 Å $= 10^{-10}$ m. Commonly used to measure the wavelength of light.

angular momentum (230) A measure of the tendency of a rotating body to continue rotating. Mathematically, the product of mass, velocity, and radius.

annular eclipse (47) A solar eclipse in which the solar photosphere appears around the edge of the moon in a bright ring, or annulus. The corona, chromosphere, and prominences cannot be seen.

anorthosite (ăn·ôr′thŭ·sīt) (379) Rock of aluminum and calcium silicates found in the lunar highlands.

aphelion (ŭ·fē′le·ŭn) (36) The orbital point of greatest distance from the sun.

apogee (47) The point farthest from Earth in the orbit of a body circling Earth.

apparent visual magnitude (m_v) (23) The brightness of a star as seen by human eyes on Earth.

arachnoid (390) On Venus, geological patterns of circular and radial faults resembling spider webs. Believed to be caused by the intrusion of magma into the crust.

association (184) Group of widely scattered stars (10 to 1000) moving together through space. Not gravitationally bound into clusters.

asterism (20) A named grouping of stars that is not one of the recognized constellations, such as the Big Dipper or the Pleiades.

asteroid (355) Small, rocky world. Most asteroids lie between Mars and Jupiter in the asteroid belt.

astronomical unit (AU) (9) Average distance from Earth to the sun; 1.5×10^8 km, or 93×10^6 mi.

atmospheric window (83) Wavelength region in which our atmosphere is transparent—at visual, infrared, and radio wavelengths.

aurora (ô·rôr′ŭ) (145) The glowing light display that results when a planet's magnetic field guides charged particles toward the north and south magnetic poles, where they strike the upper atmosphere and excite atoms to emit photons.

autumnal equinox (33) The point on the celestial sphere where the sun crosses the celestial equator going southward. Also, the time when the sun reaches this point and autumn begins in the Northern Hemisphere—about September 22.

Babcock model (141) A model of the sun's magnetic cycle in which the differential rotation of the sun winds up and tangles the solar magnetic field in a 22-year cycle. This is thought to be responsible for the 11-year sunspot cycle.

Balmer series (121) Spectral lines in the visible and near-ultraviolet spectrum of hydrogen produced by transitions whose lowest energy level is the second.

Barnard's star (260) A star located near the sun in space. Consequently, it has a high proper motion. Some believe it has planets in orbit around it.

barred spiral galaxy (284) A spiral galaxy with an elongated nucleus resembling a bar from which the arms originate.

basalt (373) Dark igneous rock characteristic of solidified lava.

belt-zone circulation (411) The atmospheric circulation typical of Jovian planets. Dark belts and bright zones encircle the planet parallel to its equator.

big bang theory (326) The theory that the universe began in a high-density, high-temperature state from which the expanding universe of galaxies formed.

binary stars (163) Pairs of stars that orbit around their common center of mass.

binding energy (113) The energy needed to pull an electron away from its atom.

bipolar flow (186) Jets of gas flowing away from a central object in opposite directions. Usually applied to protostars.

birth line (185) In the H–R diagram, the line above the main sequence where protostars first become visible.

black-body radiation (116) Radiation emitted by a hypothetical perfect radiator. The spectrum is continuous, and the wavelength of maximum emission depends only on the body's temperature.

black dwarf (218) The end state of a white dwarf that has cooled to low temperature.

black hole (244) A mass that has collapsed to such a small volume that its gravity prevents the escape of all radiation. Also, the volume of space from which radiation may not escape.

blaser (308) See **BL Lac objects.**

BL Lac objects (308) Objects that resemble quasars. Thought to be highly luminous cores of distant active galaxies.

Bok globule (186) Small, dark cloud only about 1 ly in diameter that contains 10–1000 solar masses of gas and dust. Believed to be related to star formation.

breccia (brĕch′ē·ŭ) (379) Rock composed of fragments of earlier rocks bonded together.

bright-line spectrum See **emission spectrum.**

brown dwarf (199) A star whose mass is too low to ignite nuclear fusion. Heated by contraction.

calibration (261) The establishment of the relationship between a parameter that is easily determined and a parameter that is more difficult to determine. For example, the periods of Cepheid variables have been calibrated to reveal absolute magnitudes, which can then be used to find distance. Thus astronomers say Cepheids have been calibrated as distance indicators.

Cambrian period (kăm′brē·ŭn) (463) A geological period 0.6–0.5 billion years ago during which life on Earth became diverse and complex. Cambrian rocks contain the oldest easily identifiable fossils.

capture hypothesis (382) The theory that Earth's moon formed elsewhere in the solar nebula and was later captured by Earth.

carbonaceous chondrite (kär·bŭ·nā′shŭs kŏn′drīt) (440) Stony meteorite that contains both chondrules and volatiles. These chondrites may be the least-altered remains of the solar nebula still present in the solar system.

carbon detonation (219) The explosive ignition of carbon burning in some giant stars. A possible cause of some supernova explosions.

carbon–nitrogen–oxygen (CNO) cycle (191) A series of nuclear reactions that use carbon as a catalyst to combine four hydrogen atoms to make one helium atom plus energy. Effective in stars more massive than the sun.

Cassegrain telescope (kăs′ŭ·grān) (85) A reflecting telescope in which the secondary mirror reflects light back down the tube through a hole in the center of the objective mirror.

CCD See **charge-coupled device.**

celestial equator (26) The imaginary line around the sky directly above Earth's equator.

celestial pole (north or south) (26) One of the two points on the celestial sphere directly above Earth's poles.

celestial sphere (24) An imaginary sphere of very large radius surrounding Earth and to which the planets, stars, sun, and moon seem to be attached.

center of mass (164) The balance point of a body or system of masses. The point about which a body or system of masses rotates in the absence of external forces.

Cepheid (sĕ′fē·ĭd) (258) Variable star with a period of 1–60 days. Period of variation related to luminosity.

Chandrasekhar limit (shăn′drä·sä′kär) (218) The maximum mass of a white dwarf, about 1.4 solar masses. A white dwarf of greater mass cannot support itself and will collapse.

charge-coupled device (CCD) (94) An electronic device consisting of a large array of light-sensitive elements used to record very faint images.

chemical evolution (465) The chemical process that led to the growth of complex molecules on primitive Earth. This did not involve the reproduction of molecules.

chondrite (kŏn′drīt) (440) A stony meteorite that contains chondrules.

chondrule (kŏn′drool) (440) Round, glassy body found in some stony meteorites. Believed to have solidifed very quickly from molten drops of silicate material.

chromatic aberration (krō·măt′ĭk) (83) A distortion found in refracting telescopes because lenses focus different colors at slightly different distances. Images are consequently surrounded by color fringes.

chromosphere (krō′mŭ·sfīr) (43) Bright gases just above the photosphere of the sun. Responsible for the emission lines in the flash spectrum.

circumpolar constellation (north or south) (27–28) A constellation so close to one of the celestial poles that it never sets or never rises as seen from a particular latitude.

closed universe (325) A model universe in which the average density is great enough to stop the expansion and make the universe contract.

cluster method (292) The method of determining the masses of galaxies based on the motions of galaxies in a cluster.

CNO cycle See **carbon–nitrogen–oxygen cycle.**

color index (B-V) A numerical measure of the color of a star in the form of the B magnitude minus the V magnitude.

coma (444) The glowing head of a comet.

comet (355) One of the small, icy bodies that orbit the sun and produce tails of gas and dust when they near the sun.

comparative planetology (367) The study of planets in relation to one another.

comparison spectrum (95) A spectrum of known

spectral lines used to identify unknown wavelengths in an object's spectrum.

condensation (358) The growth of a particle by addition of material from surrounding gas, atom by atom.

condensation hypothesis (382) The theory that Earth and the moon condensed from the same cloud of material in roughly their present orbital relationship.

condensation sequence (358) The sequence in which different materials condense from the solar nebula as we move outward from the sun.

constellation (19) One of the stellar patterns identified by name, usually of mythological gods, people, animals, or objects. Also, the region of the sky containing that star pattern.

continuity of energy law (195) One of the basic laws of stellar structure. The amount of energy flowing out of the top of a shell must equal the amount coming in at the bottom plus whatever energy is generated within the shell.

continuity of mass law (195) One of the basic laws of stellar structure. The total mass of the star must equal the sum of the masses of the shells, and the mass must be distributed smoothly through the star.

continuous spectrum (116) A spectrum in which there are no absorption or emission lines.

Copernican principle (322) The belief that Earth is not in a special place in the universe.

corona (43, 390) On the sun, the faint outer atmosphere composed of low-density, high-temperature gas. On Venus, large, round geological faults in the crust caused by the intrusion of magma below the crust.

coronal hole (146) An area of the solar surface that is dark at X-ray wavelengths. Thought to be associated with divergent magnetic fields and the source of the solar wind.

cosmological principle (322) The assumption that any observer in any galaxy sees the same general features of the universe.

cosmology (319) The study of the nature, origin, and evolution of the universe.

coudé focus (koo·dāy) (85) The focal arrangement of a reflecting telescope in which mirrors direct the light to a fixed focus beyond the bounds of the telescope's movement, typically in a separate room. Usually used for spectroscopy.

Coulomb barrier (koo·lôn) (190) The electrostatic force of repulsion between bodies of like charge. Commonly applied to atomic nuclei.

Coulomb force (113) The electrostatic force of repulsion or attraction between charged bodies.

critical density (331) The average density of the universe needed to make its curvature flat.

culture shock (479) The bewildering impact of an advanced society on a less sophisticated society.

dark-line spectrum See **absorption spectrum.**

dark matter (292) The undetected matter that some astronomers believe makes up the missing mass of the universe.

declination (490) A coordinate used on the celestial sphere just as latitude is used on Earth. An object's declination is measured from the celestial equator— positive to the north and negative to the south.

deferent (děf'ŭr·ŭnt) (60) In the Ptolemaic theory, the large circle around Earth along which the center of the epicycle was thought to move.

degenerate matter (212) Extremely high-density matter in which pressure no longer depends on temperature due to quantum mechanical effects.

density wave theory (270) Theory proposed to account for spiral arms as compressions of the interstellar medium in the disk of the galaxy.

diamond ring effect (44) During a total solar eclipse, the momentary appearance of a spot of photosphere at the edge of the moon, producing a brilliant glare set in the silvery ring of the corona.

differential rotation (141, 263) The rotation of a body in which different parts of the body have different periods of rotation. This is true of the sun, the Jovian planets, and the disk of the galaxy.

differentiation (360) The separation of planetary material according to density.

diffraction fringe (91) Blurred fringe surrounding any image, caused by the wave properties of light. Because of this, no image detail can be seen smaller than the fringe.

dirty snowball theory (355) The theory that the nuclei of comets are kilometer-sized chunks of ice with imbedded silicates in long, elliptical orbits around the sun.

disk component (261) All material confined to the plane of the galaxy.

distance indicator (286) Object whose luminosity or diameter is known. Used to find the distance to a star cluster or galaxy.

distance modulus (157) The difference between the apparent and absolute magnitude of a star. A measure of how far away the star is.

DNA (deoxyribonucleic acid) (461) The long carbon-chain molecule that records information to govern the biological activity of the organism. DNA carries the genetic data passed to offspring.

Doppler effect (127) The change in the wavelength of radiation due to relative radial motion of source and observer.

double-exhaust model (304) The theory that double radio lobes are produced by pairs of jets emitted in opposite directions from the centers of active galaxies.

double galaxy method A method of finding the masses of galaxies from orbiting pairs of galaxies.

double-lobed radio galaxy (304) A galaxy that emits radio energy from two regions (lobes) located on opposite sides of the galaxy.

dynamo effect (369) The theory that Earth's magnetic field is generated in the conducting material of its molten core.

early heavy bombardment (381) The intense cratering during the first 0.5 billion years in the history of the solar system.

eclipsing binary (168) A binary star system in which the stars eclipse each other.

ecliptic (32) The apparent path of the sun around the sky.

Einstein ring (315) A gravitational lens effect in which a distant object is imaged as a complete ring.

ejecta (377) Pulverized rock scattered by meteorite impacts on a planetary surface.

electromagnetic radiation (81) Changing electric and magnetic fields that travel through space and transfer energy from one place to another, such as light or radio waves.

electron (113) Low-mass atomic particle carrying negative charge.

ellipse (68) A closed curve around two points called the foci such that the total distance from one focus to the curve and back to the other focus remains constant.

elliptical galaxy (282) A galaxy that is round or elliptical in outline and contains little gas and dust, no disk or spiral arms, and few hot, bright stars.

emission line (120) A bright line in a spectrum caused by the emission of photons from atoms.

emission nebula (181) A cloud of glowing gas excited by ultraviolet radiation from hot stars.

emission spectrum (bright-line spectrum) (120) A spectrum containing emission lines.

energy level (115) One of a number of states an electron may occupy in an atom, depending on its binding energy.

energy transport (197) Flow of energy from hot regions to cooler regions by one of three methods: conduction, convection, or radiation.

enzyme (462) Special protein that controls processes in an organism.

epicycle (ĕp′ŭ·sĭ·kŭl) (60) The small circle followed by a planet in the Ptolemaic theory. The center of the epicycle follows a larger circle (the deferent) around Earth.

equant (ē′kwŭnt) (60) In the Ptolemaic theory, the point off center in the deferent from which the center of the epicycle appears to move uniformly.

equatorial mounting (90) A telescope mounting that allows motion parallel to and perpendicular to the celestial equator.

escape velocity (244) The initial velocity an object needs to escape from the surface of a celestial body.

evening star (37) Any planet visible in the sky just after sunset.

event horizon (245) The boundary of the region of a black hole from which no radiation may escape. No event that occurs within the event horizon is visible to a distant observer.

evolutionary track (184) The path a star follows in the H–R diagram as it gradually changes its surface temperature and luminosity.

excited atom (116) An atom in which an electron has moved from a lower to a higher energy level.

eyepiece (83) A short-focal-length lens used to enlarge the image in a telescope. The lens nearest the eye.

false color image (96) A representation of graphical data with added or enhanced color to reveal detail.

filtergram (136) A photograph (usually of the sun) taken in the light of a specific region of the spectrum —for example, an H-alpha filtergram.

fission hypothesis (382) The theory that the moon and Earth formed when a rapidly rotating protoplanet split into two pieces.

flare (143) A violent eruption on the sun's surface.

flatness problem (333) In cosmology, the peculiar circumstance that the early universe must have contained almost exactly the right amount of matter to make space-time flat.

flat universe (325) A model of the universe in which space-time is not curved.

flocculent galaxy (272) A galaxy whose spiral arms have a woolly or fluffy appearance.

flux (155) A measure of the flow of energy out of a surface. Usually applied to light.

focal length (83) The focal length of a lens is the distance from the lens to the point where it focuses parallel rays of light.

forward scattering (413) The optical property of finely divided particles to preferentially direct light in the original direction of the light's travel.

galactic cannibalism (296) The theory that large galaxies absorb smaller galaxies.

galactic corona (264) The extended, spherical distribution of low-luminosity matter believed to surround the Milky Way and other galaxies.

galaxy seeds (338) The first irregularities in the universe, which stimulated the formation of the galaxies.

Galilean satellites (găl·ŭ·lē′ŭn) (352) The four largest satellites of Jupiter, named after their discoverer, Galileo.

geocentric universe (59) A model universe with Earth at the center, such as the Ptolemaic universe.

giant star (160) Large, cool, highly luminous star in the upper right of the H–R diagram. Typically 10–100 times the diameter of the sun.

glacial period An interval when ice sheets cover large areas of the land.

glitch (237) A sudden change in the period of a pulsar.

globular star cluster (258) A star cluster containing 50,000–1 million stars in a sphere about 75 ly in diameter. Generally old, metal-poor, and found in the spherical component of the galaxy.

grand unified theories (GUTs) (333) Theories that attempt to unify (describe in a similar way) the electromagnetic, weak, and strong forces of nature.

granulation (136) The fine structure of bright grains covering the sun's surface.

grating (94) A piece of material in which numerous microscopic parallel lines are scribed. Light encountering a grating is dispersed to form a spectrum.

gravitational lens (314) The focusing of light from a distant galaxy or quasar by an intervening galaxy to produce multiple images of the distant body.

gravitational red shift (247) The lengthening of the wavelength of a photon due to its escape from a gravitational field.

greenhouse effect (375) The process by which a carbon dioxide atmosphere traps heat and raises the temperature of a planetary surface.

Greenwich Mean Time (GMT) (494) The mean or average solar time at the longitude of Greenwich, England.

grooved terrain (415) Regions of the surface of Ganymede consisting of parallel grooves. Believed to have formed by repeated fracture and refreezing of the icy crust.

ground state (116) The lowest permitted electron energy level in an atom.

GUTs See **grand unified theories.**

halo (261) The spherical region of a spiral galaxy containing a thin scattering of stars, star clusters, and small amounts of gas.

head–tail radio galaxy (307) A radio galaxy with a contour consisting of a head and a tail. Believed to be caused by the motion of an active galaxy through the intergalactic medium.

heat of formation (360) In planetology, the heat released by in-falling of matter during the formation of a planetary body.

heliocentric universe (62) A model of the universe with the sun at the center, such as the Copernican universe.

helioseismology (142) The study of the interior of the sun by the analysis of its modes of vibration.

helium flash (211) The explosive ignition of helium burning that takes place in some giant stars.

Herbig–Haro objects (185) Small nebulae that vary irregularly in brightness. Believed to be associated with star formation.

Hertzsprung–Russell (H–R) diagram (hĕrt′sprŭng·rŭs·ŭl) (159) A plot of the intrinsic brightness versus the surface temperature of stars. It separates the effects of temperature and surface area on stellar luminosity. Commonly plotted as absolute magnitude versus spectral type, but also as luminosity versus surface temperature or color.

heterogeneous accretion (361) The formation of a planet by the accumulation of planetesimals of different composition—for example, first iron particles, then silicates. See also **homogeneous accretion.**

homogeneity (321) The assumption that, on the large scale, matter is uniformly spread through the universe.

homogeneous accretion (360) The formation of a planet by the accumulation of planetesimals of the same composition. See also **heterogeneous accretion.**

horizon problem (333) In cosmology, the circumstance that the primordial background radiation seems much more isotropic than can be explained by the standard big bang theory.

hot spot (304) In geology, a place on Earth's crust where volcanism is caused by a rising convection cell in the mantle below. In radio astronomy, a bright spot in a radio lobe.

H–R diagram See **Hertzsprung–Russell diagram.**

H II region (287) A region of ionized hydrogen around a hot star.

Hubble constant (H) (334) A measure of the rate of expansion of the universe. The average value of velocity of recession divided by distance. Presently believed to be between 50 and 100 km/sec/Mpc.

Hubble law (289) The linear relation between the distances to galaxies and their velocity of recession.

Hubble time (335) The age of the universe, equivalent to 1 divided by the Hubble constant. The Hubble time is the age of the universe if it has expanded since the big bang at a constant rate.

hydrostatic equilibrium (196) The balance between the weight of the material pressing downward on a layer in a star and the pressure in that layer.

hypothesis (61) A conjecture, subject to further tests, that accounts for a set of facts.

inflationary universe (333) A version of the big bang theory that includes a rapid expansion when the universe was very young. Derived from grand unified theories.

infrared cirrus (181) Wispy network of cold dust clouds discovered by the Infrared Astronomy Satellite.

instability strip (257) The region of the H–R diagram in which stars are unstable to pulsation. A star passing through this strip becomes a variable star.

interstellar medium (180) The gas and dust distributed between the stars.

interstellar reddening (180) The process in which dust scatters blue light out of starlight and makes the stars look redder.

inverse square law (76) A rule that the strength of an effect (such as gravity) decreases in proportion as the distance squared increases.

ion (113) An atom that has lost or gained one or more electrons.

ionization (113) The process in which atoms lose or gain electrons.

irregular galaxy (284) A galaxy with a chaotic appearance, large clouds of gas and dust, and both population I and II stars, but without spiral arms.

isotopes (113) Atoms that have the same number of protons but a different number of neutrons.

isotropy (ī·sŏt′rŭ·pē) (321) The assumption that in its general properties the universe looks the same in every direction.

joule (J) (jōōl) (118) A unit of energy equivalent to a force of 1 newton acting over a distance of 1 m. One joule per second equals 1 watt of power.

Jovian planet (351) Jupiterlike planet with a large diameter and low density.

kiloparsec (kpc) (261) A unit of distance equal to 1000 pc or 3260 ly.

Kirchhoff's laws (120) A set of laws that describe the absorption and emission of light by matter.

Lagrangian point (226) Points of stability in the orbital plane of a planet or moon. One is located 60° ahead and one 60° behind the orbiting bodies.

large-impact hypothesis (382) The theory that the moon formed from debris ejected during a collision between Earth and a large planetesimal.

life zone (472) A region around a star within which a planet can have temperatures that permit the existence of liquid water.

light curve (169) A graph of brightness versus time commonly used in analyzing variable stars and eclipsing binaries.

light-gathering power (90) The ability of a telescope to collect light. Proportional to the area of the telescope's objective lens or mirror.

lighthouse theory (237) The description of a pulsar as a neutron star producing pulses of radiation by sweeping radio beams around the sky as it spins.

light pollution (91) The illumination of the night sky by waste light from cities and outdoor lighting, which prevents the observation of faint objects.

light-year (ly) (12) A unit of distance. The distance light travels in one year.

liquid metallic hydrogen (352) A form of liquid hydrogen that is a good electrical conductor found in the interiors of Jupiter and Saturn.

lobate scarp (lō′bāt skärp) (385) A curved cliff such as those found on Mercury.

local hypothesis (313) The proposal that quasars are not great distances away but are nearby, that is, local.

look-back time (288) The amount by which we look into the past when we look at a distant galaxy. A time equal to the distance to the galaxy in light-years.

luminosity (L) (156) The total amount of energy a star radiates in 1 second.

luminosity class (161) A category of stars of similar luminosity. Determined by the widths of lines in their spectra.

lunar eclipse (41) The darkening of the moon when it moves through Earth's shadow.

Lyman series (lī′měn) (121) Spectral lines in the ultraviolet spectrum of hydrogen produced by transitions whose lowest energy level is the ground state.

magnifying power (91) The ability of a telescope to make an image larger.

Magellanic clouds (măj·ŭ·lăn′ĭk) (255) Small, irregular galaxies that are companions to the Milky Way. Visible in the southern sky.

magnetosphere (măg·nē′tō·sfîr) (408) The volume of space around a planet within which the motion of charged particles is dominated by the planetary magnetic field rather than the solar wind.

magnitude scale (22) The astronomical brightness scale. The larger the number, the fainter the star.

main sequence (159) The region of the H–R diagram running from upper left to lower right, which includes roughly 90 percent of all stars.

mantle (369) The layer of dense rock and metal oxides that lies between the molten core and Earth's surface. Also, similar layers in other planets.

mare (sea) (mä′rā) (379) One of the lunar lowlands filled by successive flows of dark lava.

mascon (381) Concentrations of mass found immediately below the maria on the moon.

mass (76) A measure of the amount of matter making up an object.

mass–luminosity relation (169) The more massive a star is, the more luminous it is.

Maunder butterfly diagram (môn′dŭr) (140) A graph showing the latitude of sunspots versus time. First plotted by W. W. Maunder in 1904.

Maunder minimum (môn′dŭr) (140) A period of less numerous sunspots and other solar activity between 1645 and 1715.

megaparsec (Mpc) (286) A unit of distance equal to 1,000,000 pc.

metals (265) In astronomical usage, all atoms heavier than helium.

meteor (356) A small bit of matter heated by friction to incandescent vapor as it falls into Earth's atmosphere.

meteorite (356) A meteor that survives its passage through the atmosphere and strikes the ground.

meteoroid (356) A meteor in space before it enters Earth's atmosphere.

meteor shower (440) A multitude of meteors that appear to come from the same region of the sky. Believed to be caused by comet debris.

midocean rift (373) Chasms that split the midocean rises where crustal plates move apart.

midocean rise (373) One of the undersea mountain ranges that push up from the seafloor in the center of the oceans.

Milankovitch hypothesis (53) Suggestion that Earth's climate is determined by slow periodic changes in the shape of its orbit, the angle of its axis, and precession.

Miller experiment (464) An experiment that reproduced the conditions under which life began on Earth and manufactured amino acids and other organic compounds.

millisecond pulsar (241) A pulsar with a pulse period of only a few milliseconds.

minute of arc (26) An angular measure. Each degree is divided into 60 minutes of arc.

missing mass Unobserved mass in clusters of galaxies believed to provide sufficient gravity to bind the cluster together.

model (57) A tentative description of a phenomenon for use as an aid to understanding.

molecular cloud (182) A dense interstellar gas cloud in which atoms are able to link together to form molecules such as H_2 and CO.

molecule (113) Two or more atoms bonded together.

morning star (37) Any planet visible in the sky just before sunrise.

mutant (463) Offspring born with altered DNA.

natural law (63) A theory that is almost universally accepted as true.

natural selection (462) The process by which the best traits are passed on, allowing the most able to survive.

neap tide (40) Ocean tide of low amplitude occurring at first- and third-quarter moon.

nebula (181) A glowing cloud of gas or a cloud of dust reflecting the light of nearby stars.

Newtonian focus (85) The focal arrangement of a reflecting telescope in which a diagonal mirror reflects light out the side of the telescope tube for easier access.

neutrino (192) A neutral, massless atomic particle that travels at the speed of light.

neutron (113) An atomic particle with no charge and about the same mass as a proton.

neutron star (235) A small, highly dense star composed almost entirely of tightly packed neutrons. Radius about 10 km.

node (50) The points where an object's orbit passes through the plane of Earth's orbit.

north celestial pole (26) The point on the celestial sphere directly above Earth's North Pole.

nova (230) From the Latin, meaning "new," a sudden brightening of a star making it appear as a new star in the sky. Believed to be associated with eruptions on white dwarfs in binary systems.

nuclear bulge (262) The spherical cloud of stars that lies at the center of spiral galaxies.

nuclear fission (189) Reactions that break the nuclei of atoms into fragments.

nuclear fusion (189) Reactions that join the nuclei of atoms to form more massive nuclei.

nucleus (of an atom) (113) The central core of an atom containing protons and neutrons. Carries a net positive charge.

objective lens (83) In a refracting telescope, the long-focal-length lens that forms an image of the object viewed. The lens closest to the object.

objective mirror (84) In a reflecting telescope, the principal mirror (reflecting surface) that forms an image of the object viewed.

oblateness (418) The flattening of a spherical body. Usually caused by rotation.

Olbers' paradox (ôl′bŭrs) (320) The conflict between observation and theory about why the night sky should or should not be dark.

Oort cloud (ōrt) (450) The hypothetical source of comets. A swarm of icy bodies believed to lie in a spherical shell 50,000 AU from the sun.

opacity (197) The resistance of a gas to the passage of radiation.

open star cluster (258) A cluster of 10–10,000 stars with an open, transparent appearance. The stars are not tightly grouped. Usually relatively young and located in the disk of the galaxy.

open universe (325) A model of the universe in which the average density is less than the critical density needed to halt the expansion.

oscillating universe theory (332) The theory that the universe begins with a big bang, expands, is slowed by its own gravity, and then falls back to create another big bang.

outgassing (361) The release of gases from a planet's interior.

ovoid (428) The oval features found on Miranda, a satellite of Uranus.

parallax (*p*) (58) The apparent change in position of an object due to a change in the location of the observer. Astronomical parallax is measured in seconds of arc.

parsec (pc) (pär′sĕk) (155) The distance to a hypothetical star whose parallax is 1 second of arc. 1 pc = 206,265 AU = 3.26 ly.

Paschen series (pä′shŭn) (121) Spectral lines in the infrared spectrum of hydrogen produced by transitions whose lowest energy level is the third.

penumbra (pĭ·nŭm′brŭ) (41) The portion of a shadow that is only partially shaded.

perigee (47) The point closest to Earth in the orbit of a body circling Earth.

perihelion (pĕ·ŭ·hē′lē·ŭn) (36) The orbital point of closest approach to the sun.

period–luminosity relation (258) The relation between period of pulsation and intrinsic brightness among Cepheid variable stars.

period of a variable star (257) The total time a variable star takes to complete one cycle from bright to faint and back to bright.

permitted orbit (114) One of the energy levels in an atom that an electron may occupy.

photometer (95) An instrument used to measure the intensity and color of starlight.

photon (81) A quantum of electromagnetic energy. Carries an amount of energy that depends inversely on its wavelength.

photosphere (43, 134) The bright visible surface of the sun.

planetary nebula (215) An expanding shell of gas ejected from a star during the latter stages of its evolution.

planetesimal (plăn·ŭ·tĕs′ŭ·mŭl) (358) One of the small bodies that formed from the solar nebula and eventually grew into protoplanets.

plastic (369) A material with the properties of a solid but capable of flowing under pressure.

plate tectonics (369) The constant destruction and

renewal of Earth's surface by the motion of sections of crust.

poor galaxy cluster (294) An irregularly shaped cluster that contains fewer than 1000 galaxies, many spiral, and no giant ellipticals.

population I (265) Stars rich in atoms heavier than helium. Nearly always relatively young stars found in the disk of the galaxy.

population II (265) Stars poor in atoms heavier than helium. Nearly always relatively old stars found in the halo, globular clusters, or the nuclear bulge.

precession (34) The slow change in the direction of Earth's axis of rotation. One cycle takes nearly 26,000 years.

prime focus (84) The point at which the objective mirror forms an image in a reflecting telescope.

primeval atmosphere (375) Earth's first air.

primordial background radiation (327) Radiation from the hot clouds of the big bang explosion. Because of its large red shift, it appears to come from a body whose temperature is only 2.7 K.

primordial soup (465) The rich solution of organic molecules in Earth's first oceans.

prominence (43) Eruption on the solar surface. Visible during total solar eclipses.

proper motion (259) The rate at which a star moves across the sky. Measured in seconds of arc per year.

protein (462) Complex molecule composed of amino acid units.

proton (113) A positively charged atomic particle contained in the nucleus of atoms. The nucleus of a hydrogen atom.

proton–proton chain (190) A series of three nuclear reactions that builds a helium atom by adding together protons. The main energy source in the sun.

protoplanet (360) Massive object resulting from the coalescence of planetesimals in the solar nebula and destined to become a planet.

protostar (184) A collapsing cloud of gas and dust destined to become a star.

pulsar (237) A source of short, precisely timed radio bursts. Believed to be spinning neutron stars.

quantum mechanics (114) The study of the behavior of atoms and atomic particles.

quasar (quasi-stellar object, or QSO) (kwā′zär) (310) Small, powerful sources of energy believed to be the active cores of very distant galaxies.

radial velocity (V_r) (128) That component of an object's velocity directed away from or toward Earth.

radiation pressure (362) The force exerted on the surface of a body by its absorption of light. Small particles floating in the solar system can be blown outward by the pressure of the sunlight.

radio galaxy (304) A galaxy that is a strong source of radio signals.

radio interferometer (98) Two or more radio telescopes that combine their signals to achieve the resolving power of a larger telescope.

rays (377) Ejecta from meteorite impacts forming white streamers radiating from some lunar craters.

recombination (330) The stage within 1 million years of the big bang, when the gas became transparent to radiation.

reflecting telescope (84) A telescope that uses a concave mirror to focus light into an image.

reflection nebula (182) A nebula produced by starlight reflecting off of dust particles in the interstellar medium.

refracting telescope (83) A telescope that forms images by bending (refracting) light with a lens.

relativistic red shift (312) The red shift due to the Doppler effect for objects traveling at speeds near the speed of light.

resolving power (90) The ability of a telescope to reveal fine detail. Depends on the diameter of the telescope objective.

retrograde motion (59) The apparent backward (westward) motion of planets as seen against the background of stars.

revolution (31) Orbital motion about a point located outside the orbiting body. See also **rotation.**

rich galaxy cluster (293) A cluster containing over 1000 galaxies, mostly elliptical, scattered over a volume about 3 Mpc in diameter.

rift valley (374) A long, straight, deep valley produced by the separation of crustal plates.

right ascension (R.A.) (490) A coordinate used on the celestial sphere just as longitude is used on Earth. An object's right ascension is measured eastward from the vernal equinox.

ring galaxy (295) A galaxy that resembles a ring around a bright nucleus. Believed to be the result of a head-on collision of two galaxies.

RNA (ribonucleic acid) (462) Long carbon-chain molecules that use the information stored in DNA to manufacture complex molecules necessary to the organism.

Roche limit (rōsh) (413) The minimum distance between a planet and a satellite that holds itself together by its own gravity. If a satellite's orbit brings it within its planet's Roche limit, tidal forces will pull the satellite apart.

Roche lobe (rōsh) (226) The volume of space a star controls gravitationally within a binary system.

Roche surface (226) The outer boundary of the volume of space that a star's gravity can control within a binary system.

rolling plains (386) On Venus, the relatively smooth lowlands that comprise the largest proportion of the planetary surface.

rotation (31) Motion around an axis passing through the rotating body. See also **revolution.**

rotation curve (263) A graph of orbital velocity versus radius in the disk of a galaxy.

rotation curve method (290) A method of determining a galaxy's mass by observing the orbital velocity and orbital radius of stars in the galaxy.

RR Lyrae variable star (är·är·lī'rē) (257) Variable star with periods of from 12 to 24 hours. Common in some globular clusters.

Sagittarius A (274) The powerful radio source located at the core of the Milky Way galaxy.

Saros cycle (sě'rōs) (50) An 18-year, 11-day period after which the pattern of lunar and solar eclipses repeats.

Schmidt camera (85) A photographic telescope that takes wide-angle photographs.

Schmidt–Cassegrain telescope (86) A Cassegrain telescope that uses a thin corrector plate at the entrance to the tube. A popular design for small telescopes.

Schwarzschild radius (*R*ₛ) (schwôrts'shēld) (246) The radius of the event horizon around a black hole.

scientific notation (8) The system of recording very large or very small numbers by using powers of 10.

second of arc (26) An angular measure. Each minute of arc is divided into 60 seconds of arc.

secondary mirror (85) In a reflecting telescope, the mirror that reflects the light to a point of easy observation.

seeing (91) Atmospheric conditions on a given night. When the atmosphere is unsteady, producing blurred images, the seeing is said to be poor.

self-sustaining star formation (272) The process by which the birth of stars compresses the surrounding gas clouds and triggers the formation of more stars. Proposed to explain spiral arms.

Seyfert galaxy (sē'fûrt) (307) An otherwise normal spiral galaxy with an unusually bright, small core that fluctuates in brightness. Believed to indicate the core is erupting.

shepherd satellite (422) A satellite that, by its gravitational field, confines particles to a planetary ring.

shield volcano (390) Wide, low-profile volcanic cone produced by highly liquid lava.

shock wave (183) A sudden change in pressure that travels as an intense sound wave.

sidereal drive (sī·dîr'ē·ŭl) (90) The motor and gears on a telescope that turn it westward to keep it pointed at a star.

sidereal period (37) The time a celestial body takes to turn once on its axis or revolve once around its orbit relative to the stars.

sidereal time (492) Time based on the rotation of Earth with respect to the stars. The sidereal time at any moment equals the right ascension of objects on the upper half of the local celestial meridian.

singularity (244) The object of zero radius into which the matter in a black hole is believed to fall.

solar eclipse (41) The event that occurs when the moon passes directly between Earth and the sun, blocking our view of the sun.

solar granulation The patchwork pattern of bright areas with dark borders observed on the sun. The tops of rising currents of hot gas in the convective zone.

solar nebula theory (347) The theory that the planets formed from the same cloud of gas and dust that formed the sun.

solar system (7) The sun and its planets, asteroids, comets, and so on.

solar wind (138) Rapidly moving atoms and ions that escape from the solar corona and blow outward through the solar system.

south celestial pole (26) The point on the celestial sphere directly above Earth's South Pole.

spectral class or type (126) A star's position in the temperature classification system O, B, A, F, G, K, M. Based on the appearance of the star's spectrum.

spectral sequence (126) The arrangement of spectral classes (O, B, A, F, G, K, M) ranging from hot to cool.

spectrograph (94) A device that separates light by wavelengths to produce a spectrum.

spectroscopic binary (166) A star system in which the stars are too close together to be visible separately. We see a single point of light, and only by taking a spectrum can we determine that there are two stars.

spectroscopic parallax (162) The method of determining a star's distance by comparing its apparent magnitude with its absolute magnitude as estimated from its spectrum.

spherical component (261) The part of the galaxy including all matter in a spherical distribution around the center (the halo and nuclear bulge).

spicule (spĭk′yōōl) (136) A small, flamelike projection in the chromosphere of the sun.

spiral arm (14, 261) Long spiral pattern of bright stars, star clusters, gas, and dust. Spiral arms extend from the center to the edge of the disk of spiral galaxies.

spiral galaxy (283) A galaxy with an obvious disk component containing gas; dust; hot, bright stars; and spiral arms.

spiral tracer (268) Object used to map the spiral arms—for example, O and B associations, open clusters, clouds of ionized hydrogen, and some types of variable stars.

spokes (422) Radial features in the rings of Saturn.

spring tide (40) Ocean tide of high amplitude that occurs at full and new moon.

steady state theory (328) The theory (now generally abandoned) that the universe does not evolve.

stellar density function (172) A description of the abundance of stars of different types in space.

stellar model (197) A table of numbers representing the conditions in various layers within a star.

stellar parallax (*p*) (153) A measure of stellar distance. See also **parallax**.

stromatolite (466) A layered fossil formation caused by ancient mats of algae or bacteria, which build up mineral deposits season after season.

strong force (189) One of the four forces of nature. The strong force binds protons and neutrons together in atomic nuclei.

summer solstice (33) The point on the celestial sphere where the sun is at its most northerly point. Also, the time when the sun passes this point, about June 22, and summer begins in the Northern Hemisphere.

sunspot (139) Relatively dark spot on the sun that contains intense magnetic fields.

supercluster (299) A cluster of galaxy clusters.

superconductor (239) A material that can conduct electricity with essentially zero resistance.

supergiant star (160) Exceptionally luminous star whose diameter is 10–1000 times that of the sun.

supergranule (137) Very large convective features in the sun's surface.

supernova remnant (224) The expanding shell of gas marking the site of a supernova explosion.

synchrotron radiation (sĭn′krŭ·trŏn) (224) Radiation emitted when high-speed electrons move through a magnetic field.

synodic period (sĭ·nŏd′ĭk) (38) The time a solar system body takes to orbit the sun once and return to the same orbital relationship with Earth. That is, orbital period referenced to Earth.

terrestrial planet (349) An Earthlike planet—small, dense, rocky.

theory (58) A system of assumptions and principles applicable to a wide range of phenomena that have been repeatedly verified.

tidal heating (417) The heating of a planet or satellite because of friction caused by tides.

time dilation (417) The slowing of moving clocks or clocks in strong gravitational fields.

Titius–Bode rule (349) A simple series of steps that produces numbers approximately matching the sizes of the planetary orbits.

transition (120) The movement of an electron from one atomic energy level to another.

triple-alpha process (192) The nuclear fusion process that combines three helium nuclei (alpha particles) to make one carbon nucleus.

T Tauri stars (tôrē) (185) Young stars surrounded by gas and dust. Believed to be contracting toward the main sequence.

tuning-fork diagram (282) A system of classification for elliptical, spiral, and irregular galaxies.

turn-off point (228) The point in an H–R diagram at which a cluster's stars turn off of the main sequence and move toward the red-giant region, revealing the approximate age of the cluster.

umbra (ŭm′brŭ) (41) The region of a shadow that is totally shaded.

uncompressed density (358) The density a planet would have if its gravity did not compress it.

uniform circular motion (59) The classical belief that the perfect heavens could only move by the combination of uniform motion along circular orbits.

Universal Time (UT) (494) The local mean time at the longitude of Greenwich, England. The same as Greenwich Mean Time.

universality (321) The assumption that the physical laws observed on Earth apply everywhere in the universe.

variable star (257) A star whose brightness changes periodically.

velocity dispersion method (292) A method of finding a galaxy's mass by observing the range of velocities within the galaxy.

vernal equinox (33) The place on the celestial sphere where the sun crosses the celestial equator moving northward. Also, the time of year when the sun crosses this point, about March 21, and spring begins in the northern hemisphere.

vesicular basalt (vŭ·sĭk′yŭ·lŭr) (377) A porous rock formed by solidified lava with trapped bubbles.

visual binary (164) A binary star system in which the two stars are separately visible in the telescope.

water hole (474) The interval of the radio spectrum between the 21-cm hydrogen radiation and the 18-cm OH radiation. Likely wavelengths to use in the search for extraterrestrial life.

wavelength (81) The distance between successive peaks or troughs of a wave. Usually represented by λ.

wavelength of maximum (λ_{max}) (117) The wavelength at which a perfect radiator emits the maximum amount of energy. Depends only on the object's temperature.

weak force (189) One of the four forces of nature. The weak force is responsible for some forms of radioactive decay.

white dwarf star (161) Dying star that has collapsed to the size of Earth and is slowly cooling off. At the lower left of the H–R diagram.

Widmanstätten pattern (wĭd′mŭn·stä·tŭn) (439) Bands in iron meteorites due to large crystals of nickel–iron alloys.

winter solstice (33) The point on the celestial sphere where the sun is farthest south. Also the time of year when the sun passes this point, about December 22, and winter begins in the Northern Hemisphere.

Zeeman effect (139) The splitting of spectral lines into multiple components when the atoms are in a magnetic field.

zero-age main sequence (ZAMS) (200) The locus in the H–R diagram where stars first reach stability as hydrogen-burning stars.

zone of avoidance (281) The region of the sky around the Milky Way where almost no galaxies are visible because our view is blocked by dust in our galaxy.

Answers to Even-Numbered Problems

Chapter 1
(page 18)

2. 3475 km
4. 1.3 seconds
6. 28 million
8. 20

Chapter 2
(page 29)

2. 2755
4. 14 mag or 400,000
6. none, all constellations north of the celestial equator

Chapter 3
(page 55)

2. 190,000
4. (a) full; (b) 1st quarter; (c) gibbous waxing; (d) crescent waxing
8. July 22, 1990, from Finland and northern Siberia

Chapter 4
(pages 78–79)

2. No. The observed ratio is 6.1 and the diagram implies a ratio of 1.5.
4. 32 years
6. 39.4 AU

Chapter 5
(pages 107–108)

2. 450 nm, 4.5×10^{-7} m
4. 500 times more
6. 11.3 m in diameter
8. 0.5 cm
10. 100 times less

Chapter 6
(pages 131–132)

2. 9700 nm, infrared
4. 250 nm
6. (a) 20,000 K; (b) 7500 K; (c) 3000 K; (d) 4500 K
8. 91 km/sec, receding

Chapter 7
(page 148)

2. 4.2 times
4. 4.3 days

Chapter 8
(page 176)

2. 28 m
4.

m	M_v	d	p
7	7	10	0.1
11	1	1000	0.001
1	−2	40	0.025
4	2	25	0.040

6. about B7
8. 160 pc
10. a, c, c, c, d

Chapter 9
(page 207)

2. 0.1 pc
4. 100 K
6. 9×10^{16}J, 28.5 years
10. about 10 million years

Chapter 10
(page 233)

2. 1.7 ly
4. 32,000 years
6. 1.2 million
8. about 475 km/sec
10. about 310 million years

Chapter 11
(pages 251–252)

2. 0.00096 AU if the total mass is 2 $M_\odot$
4. 3nm
6. 8 minutes

Chapter 12
(page 279)

2. less than 11 percent
4. 6300 pc
6. 25 pc
8. 20 times
10. 1500 K

Chapter 13
(page 302)

2. 2.6 Mpc
4. 28.6 Mpc
6. 165 million years

Chapter 14
(pages 317–318)

2. 7.6 million years
4. 170 Mpc
6. −28.5
8. 91,000 km/sec
10. 0.16 (The change in wavelength is 77.8 nm.)

Chapter 15
(pages 340–341)

2. 512 km/sec/Mpc
4. 3000 nm
6. 16 billion years, 11 billion years

Chapter 16
(page 365)

2. 22.6 mag (It will look $206,265^2$ times fainter, which is 26.6 magnitudes.)
4. 20 km
6. large amounts of methane and water ices
8. about 1400

Chapter 17
(pages 404–405)

4. No. Their angular diameter would be only 0.5 second of arc. They are not visible from orbit around the moon.
6. 23 minutes
8. 277 km (Minimum distance is 0.38 AU. See Data File 6.)
10. 6.5×10^{23} kg (3.25×10^{-7} $M_\odot$.)

Chapter 18
(page 436)

2. 4.4 degrees
4. 16.8 km/sec
6. 7 seconds
8. 60,000 nm (60 microns)
10. 9×10^{21} kg

Chapter 19
(page 454)

2. 21 km/sec
4. 4.9 years
6. 0.76 second of arc
8. 9 million km, too small
10. 0.03 $M_\oplus$

Chapter 20
(pages 481–482)

2. 9.6 cm, 0.65 mm
4. 1.3 $M_\odot$.
6. 380 km

Index

Ozone, atmospheric
  of Earth, 376
  of Mars, 402

Pallas (asteroid), 442
Pangaea, *374*
Parallax, **58**, *59*
  spectroscopic, **162**, 163
  stellar, **153**, 154
Parsec (pc), **155**
Paschen series, **121**
Pauli Exclusion Principle, 212
Pegasus (constellation), 22
  Great Square of, 20, *21*
Penumbra (Earth's shadow), **41**,
  42
Penzias, Arno, 326, *327*
Perigee of moon, **47**
Perihelion, **36**
Period–luminosity relationship, **258**
Period of variable star, **257**, *258*
PG 1613 (quasar), *316*
Phase transition, 338–39
Phobos (satellite), 402, *403*
Photography as astronomical tool,
  94
Photometer, **95**–96
Photon, **81**
  absorption of, by excited atom,
    115, *116*, 118, *119*
  big bang theory and, 328–29
  red vs. blue, 180, *181*
  stretching of, in space–time,
    323, *324*
  wavelength and energy of, 81
Photosphere, solar, **43**, *46*, **134**,
  136
  granulation on, *134*, **136**
Pioneer spacecrafts, 386, *387*, 408
Pisces–Cetus supercluster, **336**
Planet(s). *See also names of*
  individual planets
  comparative study of, 367
  formation of (*see* Planetary
    formation)
  formation of asteroids from
    unformed, 442–44
  Jovian, **351**–55, 362, 407, *408*
  orbits (*see* Orbits of planets)
  origin of, 346–48
  physical properties of, A18
  term, *60*

terrestrial, *349*, 351
Planetary formation, 186, 357–63
  chemical composition of solar
    nebula, 357–58
  condensation of solids from gas,
    358, 359 (table)
  end of solar nebula and,
    362–63
  formation of planetesimals,
    358–59, *360*
  four stages of, *360*, 380–81
  growth of protoplanets, 360–62
  solar nebula theory of, **347**, 348
  solar system characteristics and,
    357
Planetary motion, 36–37, 57
  concept of, rejected by early
    astronomers, 58, 59
  Kepler's laws of, 68, 69, 70
  Ptolemaic model of, 59, *60*, 61
  retrograde, 59, *60*, 63, *64*
Planetary nebulae, **215**, *216*, *217*
  as distance indicators, 287
Planetary systems, potential for
  life on other, 471–72
Planetesimals, formation of,
  358–60
Plastic materials, **369**
Plate tectonics, **369**, 372, *373*,
  374, *375*
Plato, 59
Pleiades (star cluster), *182*, *228*,
  258
Pluto (planet), 36, 351, 407,
  432–34
  data file, 433
  mass of, 432
  orbit of, 10, 69, 348, *349*, 432,
    433
  origin of, 432, 434
  satellite of, 432, *434*
Poe, Edgar Allan, 320
Polaris (star), 27
Poor galaxy cluster, **294**–95
Population I and Population II of
  Milky Way stars, **265**, *266*
Positron, 190, *191*
Precession of Earth's axis, **34**
  climate and, 53–54
Pressure-temperature thermostat,
  194–95
Prime focus of telescope, 84–85,
  86

Primeval atmosphere, Earth's, **375**
Primordial background radiation,
  326, **327**, *328*, 333, 337
  irregularities in, *338*, 339
Primordial soup, **465**
Principia (Newton), 76
Prism, 94, *95*
Prominences, solar, **43**, *46*, 135,
  142–44, *145*
Proper motions of stars, **259**, *260*
Proteins, **462**
Proton of atom, **113**
Proton-proton chain, **190**, *191*
Protoplanets, **360**–62
Protostar, **184**
  contraction of, *184*, *185*
  formation of, in Orion Great
    Nebula, 205
Proxima Centauri (star), distance
  to, 12
Prutenic Tables, 62, 65
Pseudoscience, A4–A5
PSR 1913+16 (pulsar), 241, *242*
PSR 1937+21 (pulsar), 240, 241
Ptolemy (Claudius Ptolemaeus),
  59–61, 65
Pulsar, **237**
  discovery of, 236, *237*
  Doppler effect of, *242*
  evolution of, *240*, 241
  glitches of, *237*, 238–39
  millisecond, 241, 244
  model of, 237, *238*, *239*
  optical, 241 (table)
Pythagoras, 59

QSO 0351+026, *313*
QSO 1059+730, *314*
Quantum mechanics, **114**
  big bang theory of universe and,
    333–34
  degenerate matter and laws of,
    212
Quasars, **310**–16
  discovery of, 310–13
  in distant galaxies, 313–15,
    337, *338*
  model of, 315–16

Radial velocity V_r, **128**
Radiation, 81–83. *See also* Light
  black-body, 116, *117*–18
  energy transport by, 197

Star Charts

To use the star charts in this book, select the appropriate chart for the date and time. Hold it overhead and turn it until the direction at the bottom of the chart is the same as the direction you face.

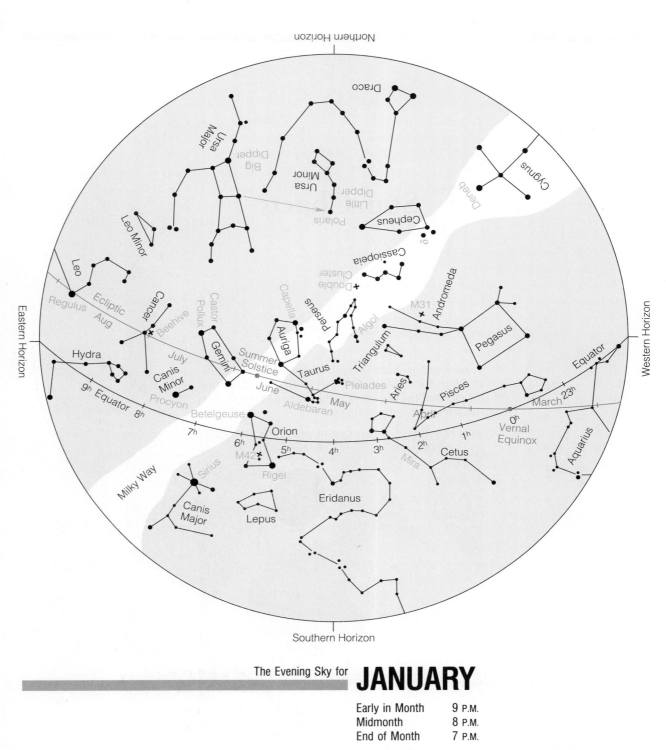

The Evening Sky for **JANUARY**

Early in Month	9 P.M.
Midmonth	8 P.M.
End of Month	7 P.M.

Months along the ecliptic show the location of the sun during the year.

Numbers along the celestial equator show right ascension.

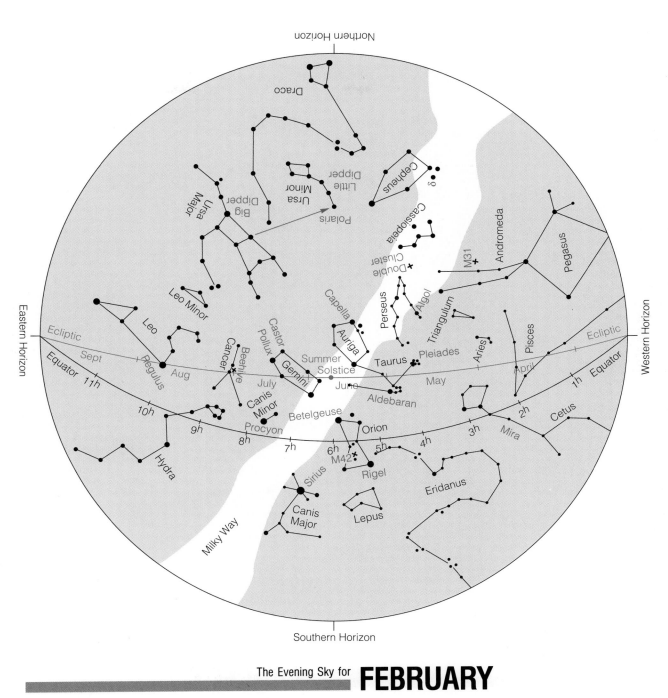

The Evening Sky for **FEBRUARY**

Early in Month	9 P.M.
Midmonth	8 P.M.
End of Month	7 P.M.

Months along the ecliptic show the location of the sun during the year.

Numbers along the celestial equator show right ascension.

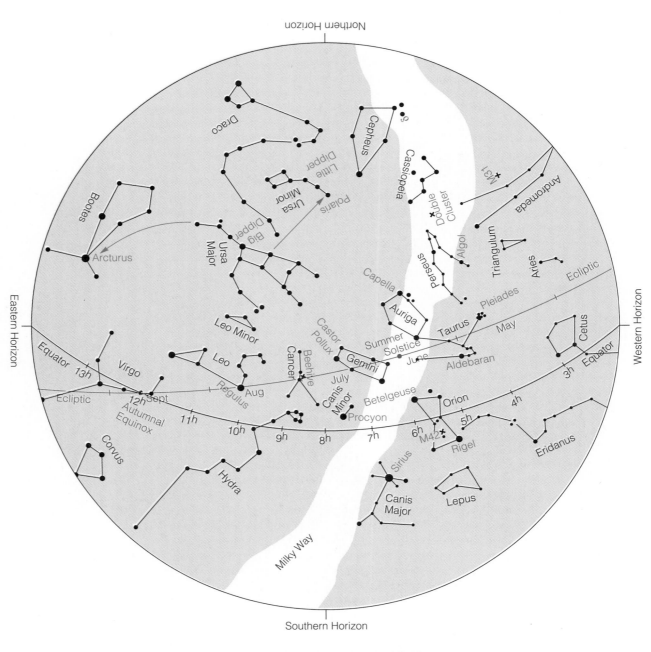

The Evening Sky for **MARCH**

Early in Month	9 P.M.
Midmonth	8 P.M.
End of Month	7 P.M.

Months along the ecliptic show the location of the sun during the year.

Numbers along the celestial equator show right ascension.

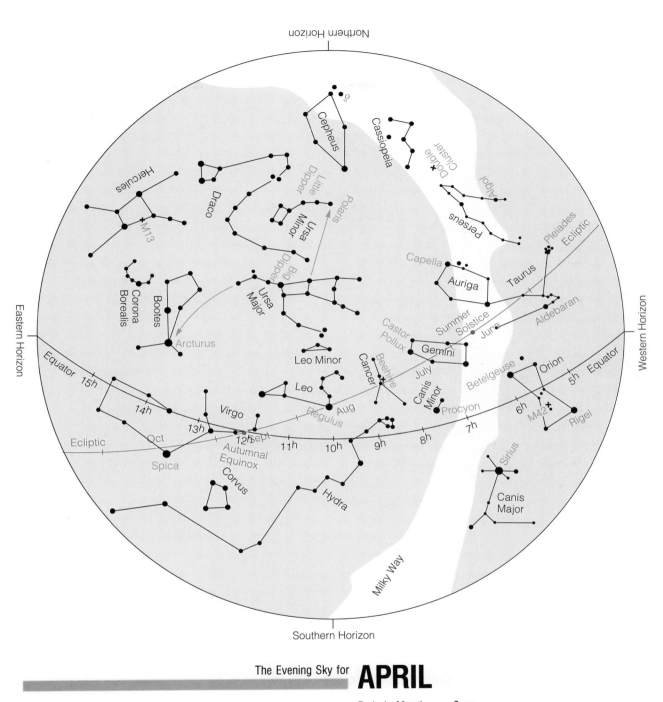

The Evening Sky for **APRIL**

Early in Month	9 P.M.
Midmonth	8 P.M.
End of Month	7 P.M.

Months along the ecliptic show the location of the sun during the year.

Numbers along the celestial equator show right ascension.

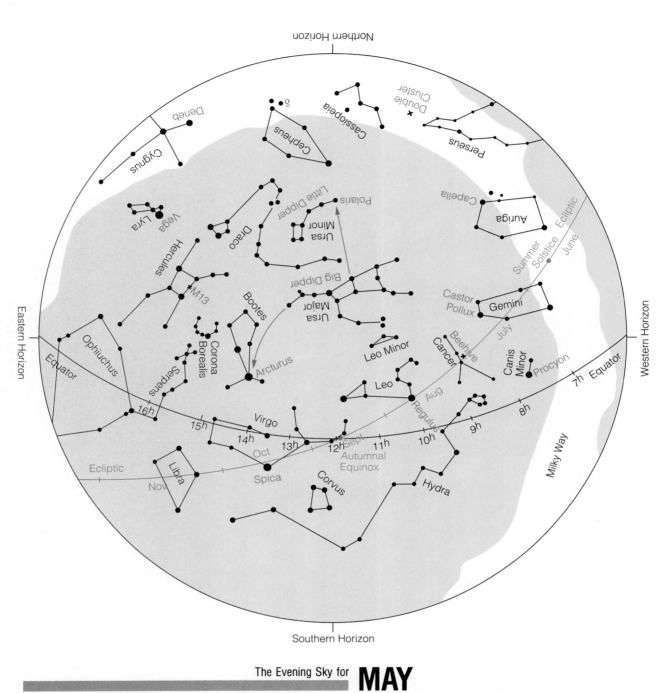

The Evening Sky for **MAY**

Early in Month	9 P.M.
Midmonth	8 P.M.
End of Month	7 P.M.

Months along the ecliptic show the location of the sun during the year.

Numbers along the celestial equator show right ascension.

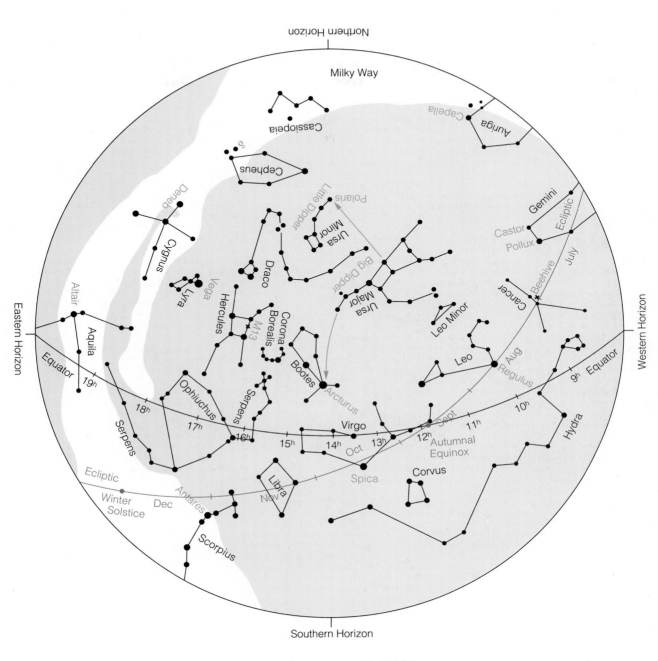

Northern Horizon

Milky Way

Cassiopeia

Capella

Auriga

Cepheus

Denck

Polaris

Little Dipper

Gemini

Ecliptic

Castor

Pollux

Cygnus

Ursa Minor

Big Dipper

July

Vega

Draco

Beehive

Lyra

Corona Borealis

Ursa Major

Cancer

Hercules

M13

Leo Minor

Altair

Aquila

Bootes

Aug

Regulus

Equator

19ʰ

Ophiuchus

Arcturus

Leo

Ecliptic

Eastern Horizon

18ʰ

Serpens

9ʰ

Equator

Western Horizon

17ʰ

16ʰ

Virgo

Hydra

Serpens

15ʰ

14ʰ

13ʰ

12ʰ

11ʰ

10ʰ

Oct

Sept

Autumnal
Equinox

Ecliptic

Libra

Spica

Corvus

Winter
Solstice

Dec

Antares

Nov

Scorpius

Southern Horizon

The Evening Sky for **JUNE**

Early in Month	9 P.M.
Midmonth	8 P.M.
End of Month	7 P.M.

Months along the ecliptic show the location of the sun during the year.

Numbers along the celestial equator show right ascension.

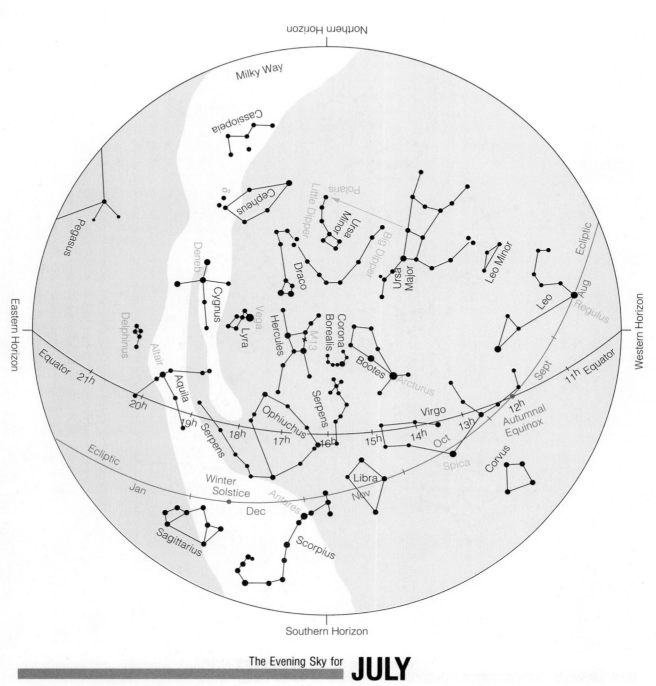

The Evening Sky for **JULY**

Early in Month	9 P.M.
Midmonth	8 P.M.
End of Month	7 P.M.

Months along the ecliptic show the location of the sun during the year.

Numbers along the celestial equator show right ascension.

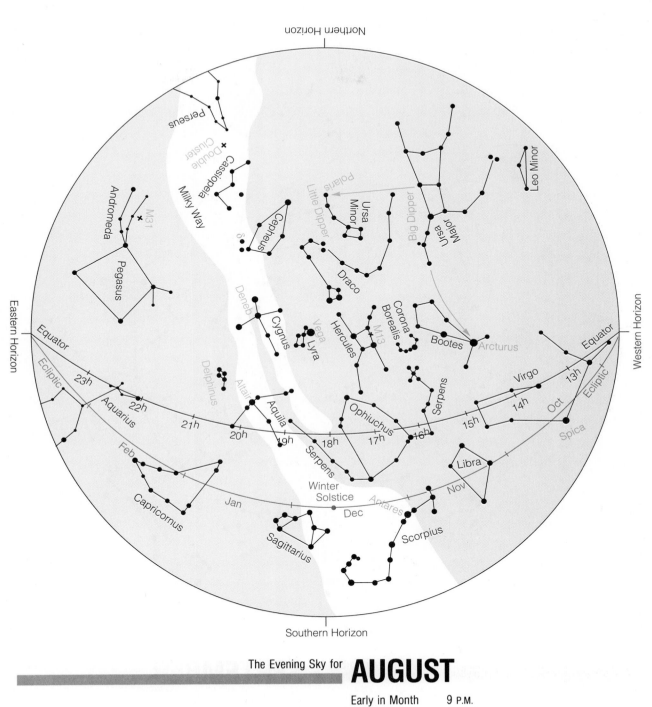

The Evening Sky for **AUGUST**

Early in Month   9 P.M.
Midmonth   8 P.M.
End of Month   7 P.M.

Months along the ecliptic show the location of the sun during the year.

Numbers along the celestial equator show right ascension.

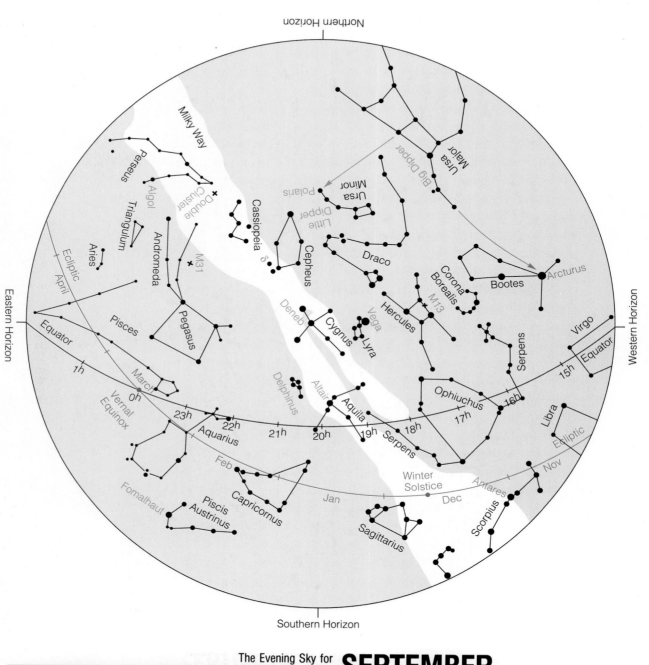

The Evening Sky for **SEPTEMBER**

Early in Month	9 P.M.
Midmonth	8 P.M.
End of Month	7 P.M.

Months along the ecliptic show the location of the sun during the year.

Numbers along the celestial equator show right ascension.

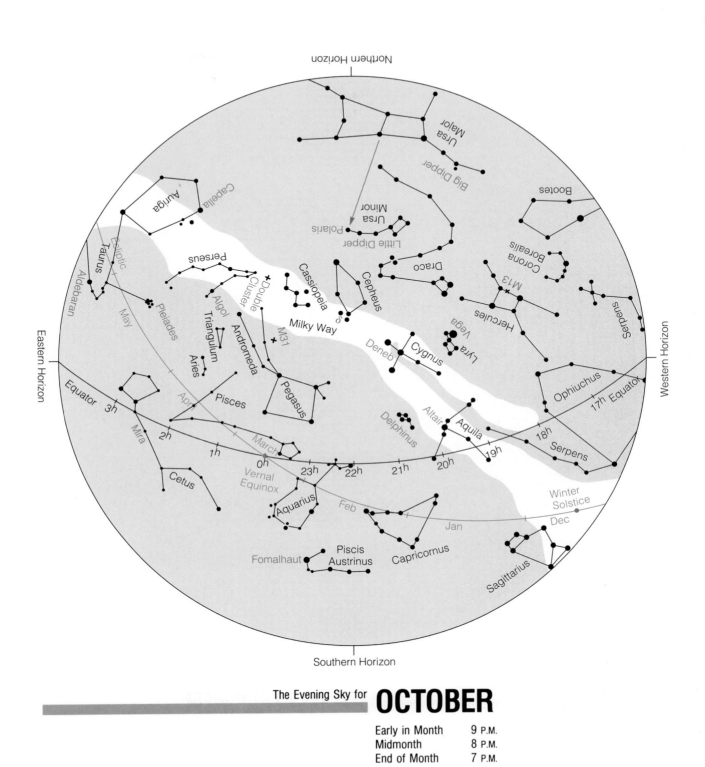

The Evening Sky for **OCTOBER**

Early in Month	9 P.M.
Midmonth	8 P.M.
End of Month	7 P.M.

Months along the ecliptic show the location of the sun during the year.

Numbers along the celestial equator show right ascension.

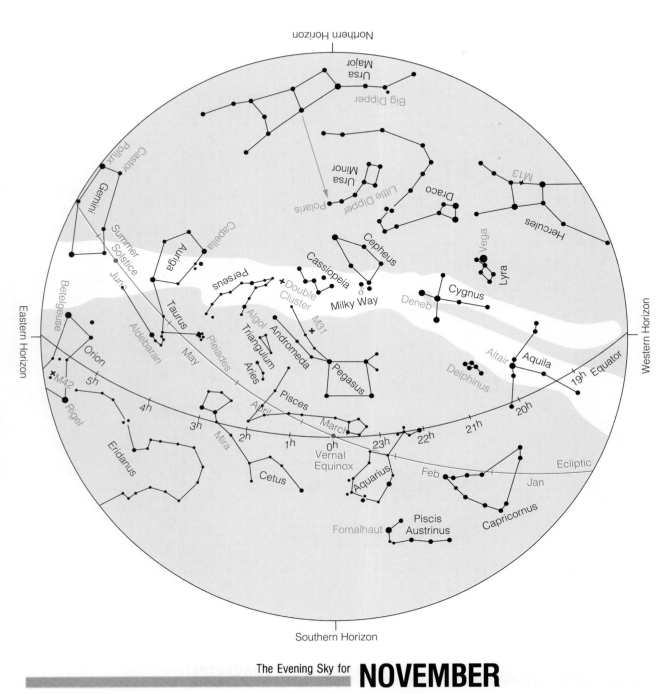

The Evening Sky for **NOVEMBER**

Early in Month	9 P.M.
Midmonth	8 P.M.
End of Month	7 P.M.

Months along the ecliptic show the location of the sun during the year.

Numbers along the celestial equator show right ascension.

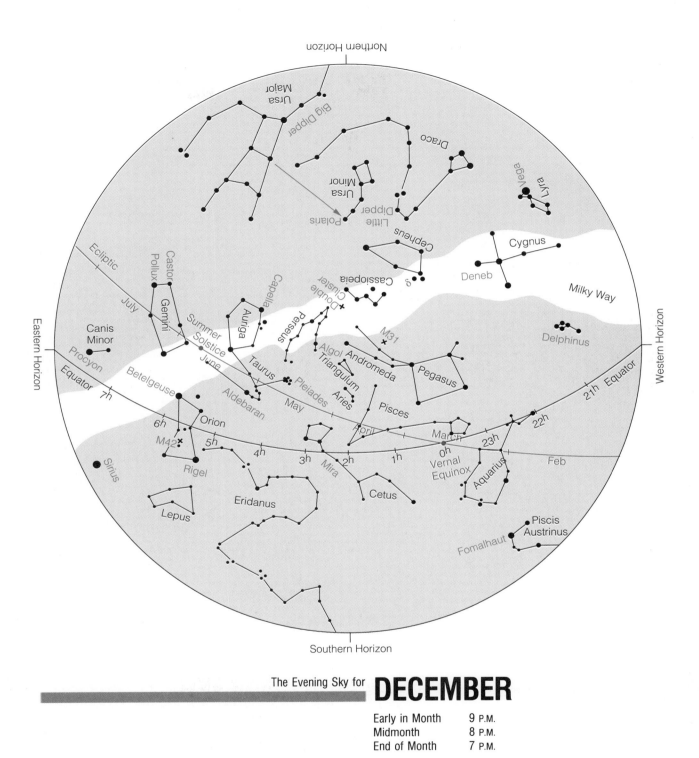

The Evening Sky for **DECEMBER**

Early in Month	9 P.M.
Midmonth	8 P.M.
End of Month	7 P.M.

Months along the ecliptic show the location of the sun during the year.

Numbers along the celestial equator show right ascension.